Introduction to Supergravity and Its Applications

This graduate textbook covers the basic formalism of supergravity, as well as its modern applications, suitable for a focused first course. Assuming a working knowledge of quantum field theory, Part I gives basic formalism, including on- and off-shell supergravity, the covariant formulation, superspace and coset formulations, coupling to matter, higher dimensions, and extended supersymmetry. A wide range of modern applications are introduced in Part II, including string theoretical (T- and U-duality, anti-de Sitter/conformal field theory (Ads/CFT), susy and sugra on the worldsheet, and superembeddings), gravitational (p-brane solutions and their susy, attractor mechanism, and Witten's positive energy theorem), and phenomenological (inflation in supergravity, supergravity no-go theorems, string theory constructions at low energies, and minimal supergravity and its susy breaking). The broader emphasis on applications than competing texts gives Ph.D. students the tools they need to do research that uses supergravity and benefits researchers already working in areas related to supergravity.

Horaţiu Năstase is a researcher at the Institute for Theoretical Physics, São Paulo State University. He completed his Ph.D. at S.U.N.Y. at Stony Brook with Peter van Nieuwenhuizen, a codiscoverer of supergravity. While at the Institute for Advanced Study in Princeton for a postdoc, in a 2002 paper with David Berenstein and Juan Maldacena, he started the pp-wave correspondence – a subarea of the AdS/CFT correspondence. He has written more than 100 scientific articles and 6 other books: *Introduction to the AdS/CFT Correspondence* (2015), *String Theory Methods for Condensed Matter Physics* (2017), *Classical Field Theory* (2019), *Introduction to Quantum Field Theory* (2019), *Cosmology and String Theory* (2019), and *Quantum Mechanics: A Graduate Course* (2022).

Introduction to Supergravity and Its Applications

Horaţiu Năstase
São Paulo State University

Shaftesbury Road, Cambridge CB2 8EA, United Kingdom

One Liberty Plaza, 20th Floor, New York, NY 10006, USA

477 Williamstown Road, Port Melbourne, VIC 3207, Australia

314–321, 3rd Floor, Plot 3, Splendor Forum, Jasola District Centre, New Delhi – 110025, India

103 Penang Road, #05–06/07, Visioncrest Commercial, Singapore 238467

Cambridge University Press is part of Cambridge University Press & Assessment,
a department of the University of Cambridge.

We share the University's mission to contribute to society through the pursuit of
education, learning and research at the highest international levels of excellence.

www.cambridge.org
Information on this title: www.cambridge.org/9781009445597

DOI: 10.1017/9781009445573

First published 2024

A catalogue record for this publication is available from the British Library

A Cataloging-in-Publication data record for this book is available from the Library of Congress

ISBN 978-1-009-44559-7 Hardback

To the memory of my mother,
who inspired me to become a physicist

Contents

Preface

Supergravity is by now a very well established field, and yet there are not many books that exclusively deal with supergravity. Usually, it is either in the larger context of string theory or within general supersymmetry (like the, by now dated, books Wess and Bagger's *Supersymmetry and Supergravity*, and Peter West's *Introduction to Supersymmetry and Supergravity*), with little applications and treating supergravity only sparingly. Peter van Nieuwenhuizen's "Supergravity" published in *Physics Reports* (1981) is still a very good reference but quite dated and without any applications. The 2012 book *Supergravity* by Freedman and van Proeyen is quite comprehensive, with an introduction to a variety of issues, but I found that their approach was not quite what I wanted; for instance, they avoided superspace and coset theory (except in mentioning a little about these in the appendices of the book) and did not consider superspace constraints and torsions and curvatures. Also, the number of applications they used was less than I would like.

So I wrote this book as a means to fill the need for a solid treatment of supergravity, with most of its applications, and dealing also with the formalism of superspace and coset theory, not just the component one. The book is intended as a one-semester graduate course on supergravity, with each chapter corresponding to a two-hour lecture, as it was presented in a course at my institute, the IFT-UNESP in Brazil, in 2022. The only parts that were added afterward are two long sections denoted with an asterisk in Chapter 16. The course goes through all the basic formalism, and its modern applications, such as providing the students with the tools they need to do research on supergravity. It can be followed by a Ph.D. student early in his or her studies, since it just assumes a working knowledge of quantum field theory, but very little (or even no) knowledge of general relativity and supersymmetry.

Acknowledgments

Above all, I would like to thank Peter van Nieuwenhuizen (codiscoverer of supergravity) from whom I learned most of the topics dealt with in this book, while I was his graduate student at Stony Brook University.

I want to thank all the people that helped me go on the path toward theoretical physics, starting with my mother Ligia, the first example I had of a physicist, from which I learned to know and love physics. My high school physics teacher, Iosif Sever Georgescu, helped start me in a career in physics, showing me that is something I can do very well. My student exchange advisor at the Niels Bohr Institute, Poul Olesen, first showed me about string theory and supergravity, and told me that is something I should look into. Of course, my PhD advisor, Peter van Nieuwenhuizen, taught me most of the things about supergravity and string theory I describe in this book, besides shaping me as a scientist in every way that matters.

There are many newer topics in this book about which I learned by working with my collaborators, students, and postdocs, so I want to thank all of them; especially Juan Maldacena, for making me understand the many ways supergravity is relevant to other fields, through holography.

Thanks to my editor at Cambridge University Press, Vince Higgs, who helped me get this book published, as well as my previous (now retired) editor Simon Capelin, for his encouragement in starting me in the path to publishing books. To all the staff at Cambridge University Press, thanks for making sure this book, as well as my previous ones, is as good as it can be.

Introduction

As explained in the Preface, this book is intended for graduate students, with a good working knowledge of quantum field theory, like from a good two-semester course, but which can have very little, or even no, knowledge of general relativity, since these are (quickly) described in the first five chapters. The other 10 chapters of Part I describe the formalism of supergravity, and the 16 chapters of Part II describe applications.

For the formalism part, I describe the most important issues, the on- and off-shell supergravity, the covariant formulation, superspace and coupling to matter, higher dimensions, and extended supersymmetry methods. Besides the standard component formalism, a large part of the presentation is devoted to superspace and coset theory, as well as superspace constraints in terms of torsions and curvatures.

For the applications part, I describe more standard ones such as T-dualities (though generalized to solution-generating techniques such as non-Abelian T-duality, TsT, and $O(d, d)$ transformations), extremal p-brane solutions and their susy algebras, the attractor mechanism, AdS/CFT and gravity duals, inflation with supergravity, compactification of low-energy string theory, and toward embedding the Standard Model in supergravity. In addition, I describe less standard ones such as U-duality, susy and integrable deformations, Penrose limits, supergravity on the string worldsheet, superembeddings, supergravity no-go theorems, and Witten's positive energy theorem.

My goal being to equip the graduate student with the tools and knowledge in the broad field of supergravity, I present a broad range of methods and applications, but I do not make a comprehensive analysis of each of them, rather I focus on the essentials.

After each chapter, I summarize a set of "Important concepts to remember," and present four exercises whose solution is meant to clarify the concepts in the chapter.

PART I

FORMALISM

Introduction to general relativity 1: Kinematics and Einstein equations

In this chapter, I will give a lightning review of the basics of general relativity, from how it is built, to its kinematics, and finally to its dynamics, given by the Einstein equation.

1.1 Intrinsically curved spacetime and the geometry of general relativity

I will start with the need for and meaning of intrinsically curved spacetime, which will lead us to the geometry of general relativity.

But since general relativity is a generalization of special relativity, I will review its basic ideas in order to be able to generalize it.

1.1.1 Special relativity

Special relativity was developed as a result of the experimental observation that the speed of light in a vacuum is equal to a constant in all inertial reference frames, where the constant can be put to 1, so that $c = 1$. This then becomes a postulate of special relativity.

As a result, we find that the line element, or the infinitesimal distance between two points, must be taken in *spacetime*, not just in space, in order to be invariant under transformations of coordinates between any inertial reference frames. This invariant distance is then

$$ds^2 = -dt^2 + d\vec{x}^2 = \eta_{\mu\nu}dx^\mu dx^\nu, \tag{1.1}$$

where $\eta_{\mu\nu} = \mathrm{diag}(-1, 1, ..., 1)$ is the Minkowski metric. This now takes the role of the invariant length $d\vec{x}^2$ in Newtonian physics, which is invariant under rotations of space at a given time.

The symmetry group that leaves ds^2 invariant is $SO(1, 3)$, or in a general dimension $SO(1, d - 1)$, called the Lorentz group. It is a generalization of the group $SO(d - 1)$ of spatial rotations that leaves $d\vec{x}^2$ invariant. The corresponding Lorentz transformations are linear transformations of the coordinates that generalize rotations, $x'^i = \Lambda^i{}_j x^j$, where $\Lambda \in SO(d - 1)$, which leaves invariant $d\vec{x}^2$. Now, instead, we have

$$x'^\mu = \Lambda^\mu{}_\nu x^\nu; \quad \Lambda^\mu{}_\nu \in SO(1, 3), \tag{1.2}$$

which leaves invariant ds^2.

In conclusion, special relativity is defined by the following statement: Physics is Lorentz invariant or covariant (under $SO(1, d - 1)$ transformations). It replaces the statement of Newtonian or Galilean physics that physics is invariant under the Galilean group, of spatial rotations, with no action on time.

1.1.2 General relativity

Now to define general relativity, we need to consider the most general line element

$$ds^2 = g_{\mu\nu}(x)dx^\mu dx^\nu, \tag{1.3}$$

where $g_{\mu\nu}(x)$ is a symmetric matrix of functions called "the metric." By extension, sometimes one calls the corresponding ds^2 the metric. Moreover, consider here that x^μ make up an arbitrary parametrization of spacetime, that is, are arbitrary coordinates on a manifold.

Example 1 S^2 **in angular coordinates.** To understand the notation, consider the usual case of a two-dimensional sphere, described in terms of angles. Then the line element is

$$ds^2 = d\theta^2 + \sin^2\theta d\phi^2, \tag{1.4}$$

so $x^\mu = (\theta, \phi)$. Then it follows that $g_{\theta\theta} = 1, g_{\phi\phi} = \sin\theta$, and $g_{\theta\phi} = 0$.

Example 2 S^2 **as an embedding in three-dimensional Euclidean space.** We can describe the sphere also by embedding it in three Euclidean dimensions, meaning as we usually understand it, as an object in three-dimensional space, with the metric

$$ds^2 = dx_1^2 + dx_2^2 + dx_3^2 \tag{1.5}$$

defined by the constraint

$$x_1^2 + x_2^2 + x_3^2 = R^2. \tag{1.6}$$

Differentiating the constraint, we obtain

$$2(x_1 dx_1 + x_2 dx_2 + x_3 dx_3) = 0$$
$$\Rightarrow dx_3 = -\frac{x_1 dx_1 + x_2 dx_2}{x_3} = -\frac{x_1}{\sqrt{R^2 - x_1^2 - x_2^2}}dx_1 - \frac{x_2}{\sqrt{R^2 - x_1^2 - x_2^2}}dx_2, \tag{1.7}$$

and by substituting it back into the Euclidean metric, we obtain the *induced metric on the S^2*,

$$ds^2_{\text{induced}} = dx_1^2\left(1 + \frac{x_1^2}{R^2 - x_1^2 - x_2^2}\right) + dx_2^2\left(1 + \frac{x_2^2}{R^2 - x_1^2 - x_2^2}\right) + 2dx_1 dx_2\frac{x_1 x_2}{R^2 - x_1^2 - x_2^2}$$
$$= g_{\mu\nu}(x^\rho)dx^\mu dx^\nu. \tag{1.8}$$

This was an example of a curved d-dimensional space obtained by embedding it into a flat (Euclidean or Minkowski) $(d + 1)$-dimensional space. We can ask: Is this always possible? The answer is no. To see this, first note that $g_{\mu\nu}$ is a symmetric matrix, with $d(d+1)/2$ arbitrary components. Then, the general coordinate transformations $x'^\mu = x'^\mu(x^\rho)$ correspond to d arbitrary functions, which can be used to put d components to zero, thus remaining $d(d-1)/2$ independent components of $g_{\mu\nu}$. On the other hand, if we were to embed the manifold M^d into $(d+1)$-dimensional Euclidean space E^{d+1}, there would be a unique coordinate x^{d+1} written as a function of the others, $x^{d+1} = x^{d+1}(x^\rho)$, as in the example of the sphere. We see that $d(d-1)/2 = 1$ is true only in the particular case of $d = 2$.

We note here that general coordinate transformations $x'^\mu = x'^\mu(x^\rho)$ act on *the fields* $g_{\mu\nu}(x)$, that is, the functions of spacetime, allowing us to fix their d components, so we have a redundancy similar to the one in gauge transformations in field theory; thus, we can say that general coordinate invariance is a kind of gauge invariance. We will see that we can turn this observation into a useful tool later on.

If we cannot always embed the manifold M^d into $(d+1)$-dimensional space, can we do it by adding more extra dimensions? At first sight, we would say yes, perhaps by adding not 1, but $d(d-1)/2$ dimensions in general. But actually, the situation is worse than that: We also need to make, *case by case*, a discrete choice of the *signature* of the space into which we are embedding a manifold.

Even in the simplest case of two-dimensional surfaces, we need to make this choice: Do we embed two-dimensional surfaces into a 3-dimensional Euclidean space like in the case of the sphere, with signature $(+,+,+)$, or into a three-dimensional Minkowski space, with signature $(-,+,+)$? Note that, since the multiplication of the metric by a sign changes only the convention, these are the only possibilities in three dimensions (the $(-,-,-)$ and $(-,-,+)$ ones are related by multiplication by a sign).

The example of embedding Lobachevsky space into Minkowski space is a famous one, defined by the constraint

$$x^2 + y^2 - z^2 = -R^2. \tag{1.9}$$

Lobachevsky space cannot be embedded into Euclidean space but only into Minkowski space with the metric

$$ds^2 = dx^2 + dy^2 - dz^2, \tag{1.10}$$

with the minus sign in the same place as in the constraint. We might think that this is because the signature on the two-dimensional Lobachevsky space is Minkowski, $(-,+)$ (equivalent to $(+,-)$), but that is wrong also: The signature on the space is two-dimensional Euclidean, so $(+,+)$ or equivalently, $(-,-)$. That is, $\det g_{\mu\nu} > 0$ and not < 0. Indeed, by differentiating the constraint, like for the sphere, we obtain

$$dz = \frac{xdx + ydy}{z} = \frac{xdx + ydy}{\sqrt{R^2 + x^2 + y^2}}, \tag{1.11}$$

and by replacing in the Minkowski metric, we obtain the induced metric on the Lobachevsky space,

$$ds^2_{\text{induced}} = dx^2 + dy^2 + \frac{(xdx + ydx)^2}{R^2 + x^2 + y^2} \equiv g_{\mu\nu}dx^\mu dx^\nu, \tag{1.12}$$

which is positive definite, so $\det g_{\mu\nu} > 0$.

Finally, this means that even two-dimensional surfaces of Euclidean signature can be embedded in three dimensions, but either in Euclidean or in Minkowski ones, depending on the surface. In higher dimensions, the number of choices for the signature becomes even larger, so defining spaces by embedding is possible, but very complicated and not very useful.

Instead, we must consider spaces as intrinsically curved, without embedding, and that in turn leads to non-Euclidean, Riemannian, geometry. This observation was believed to be first made by Gauss, who tried to measure if our space is actually curved (but failed, of course; on scales of even kilometers, space is flat to a very high accuracy).

In curved spaces, to define geometry, we must first define the analog of "straight lines" of Euclidean geometry, which are the geodesics, also defined as lines of the shortest distance $\int_a^b ds$ between two points a and b. In non-Euclidean geometry, a triangle made by two geodesics has the sum of its inner angles, $\alpha + \beta + \gamma \neq \pi$. In Euclidean geometry, of course, the sum is *equal to π* by a theorem.

On spaces like S^2 of "positive curvature," $R > 0$, we have $\alpha + \beta + \gamma > \pi$, as we can easily see in the following example: Consider a triangle made by two meridian lines starting from the North Pole and ending on an Equator line. The meridian lines with the Equator line make $\pi/2$ each, so $\alpha + \beta + \gamma > \pi$.

But that is not the only possibility. On a space like Lobachevsky space, we can check that $\alpha + \beta + \gamma < \pi$, and we call this a space of "negative curvature," $R < 0$. We will see in Section 1.3 what $R < 0$ and $R > 0$ means.

In conclusion, we see that for general relativity, we will need intrinsically curved space-times, with non-Euclidean geometry, with a general metric $g_{\mu\nu}(x)$, and acted upon by general coordinate transformations that act as gauge transformations.

1.2   Einstein's theory of general relativity

Einstein thought of defining general relativity in order to modify Newton's gravity at high gravitational acceleration $\vec{g}$ and high velocity $\vec{v}$ in order to make it compatible with special relativity. The need for that arose also because of experimental results: The deflection of light by the Sun using only special relativity is a factor of 1/2 off the actual result.

The construction of general relativity was based on two physical assumptions:

(1) **Gravity is geometry**

That is, matter follows geodesics (paths of shortest distance) in curved spacetimes, and to us, it appears as the effect of gravity.

Pictorially, consider a planar rubber sheet and put a heavy ball at a point on it: It will curve the sheet locally. Then, when throwing a light ball on the sheet, the local disturbance deflects it (think of a golfer doing a putt and the golf ball just missing the hole). Of course, this is just a pictorial way of describing the phenomenon; otherwise, it is a cheat: The sheet curves because of the terrestrial gravity it feels, and the curvature is only of space, not of spacetime. But this is a nice way of viewing what happens.

(2) **Matter sources gravity**

This means matter generates the gravitational field that is equated with the curvature of the geometry of spacetime from the first assumption.

These two physical assumptions were then translated into two physical principles with a mathematical formulation, defining the *kinematics* of general relativity, plus one equation for the dynamics, that is, Einstein's equation.

(A) Physics is invariant (or, more generally, covariant) under general coordinate transformations, which generalizes the Lorentz invariance or covariance in the case of special relativity.

For a general coordinate transformation $x'^{\mu} = x'^{\mu}(x^{\nu})$, we obtain

$$ds^2 = g_{\rho\sigma}(x)dx^{\rho}dx^{\sigma} = g'_{\mu\nu}dx'^{\mu}dx'^{\nu}, \tag{1.13}$$

giving the transformation rules for the field $g_{\mu\nu}$ (thought of as a field in spacetime),

$$g'_{\mu\nu}(x') = g_{\rho\sigma}(x)\frac{\partial x^{\rho}}{\partial x'^{\mu}}\frac{\partial x^{\sigma}}{\partial x'^{\nu}}. \tag{1.14}$$

This transformation is like a gauge invariance, and physics must be invariant or covariant with respect to it.

(B) The equivalence principle.

In Newtonian theory, there are *a priori* two masses: one is the inertial mass m_i, appearing in Newton's law of force, that is, $\vec{F} = m_i\vec{a}$, and the other is the gravitational mass m_g, appearing in Newton's gravitational law, that is, $\vec{F}_G = m_g\vec{g}$.

The equality of the two masses is the mathematical form of the equivalence principle, that is,

$$m_i = m_g. \tag{1.15}$$

In more physical terms, we say that "there is no difference between gravity and local acceleration." We can also explain this using Einstein's *gedanken (thought) experiment.* Consider a person inside a freely falling elevator with no windows. Then, by performing local experiments inside the elevator, the person cannot distinguish between being weightless and being inside a freely falling elevator. Of course, the *locality* condition is important, because if one is allowed to probe large regions of space, then he or she will note that there are tidal forces – gravity acting at different points in different directions (all pointing toward the center of the Earth). Also, locality in time is important; otherwise, eventually the elevator will hit the hard surface of the Earth, ending the experiment.

On the basis of the above principles, we now turn to constructing the kinematics of general relativity.

First, consider an infinitesimal general coordinate transformation, $x'^{\mu} = x^{\mu} - \xi^{\mu}$, with ξ^{μ} small, and we want to describe it as a gauge transformation. Then,

$$\begin{aligned} g'_{\mu\nu}(x^{\lambda} - \xi^{\lambda}) &= (\delta_{\mu}^{\rho} + \partial_{\mu}\xi^{\rho})(\delta_{\nu}^{\sigma} + \partial_{\nu}\xi^{\sigma})g_{\rho\sigma}(x) \\ &= g'_{\mu\nu}(x^{\lambda}) - (\partial_{\lambda}g'_{\mu\nu}(x))\xi^{\lambda}, \end{aligned} \tag{1.16}$$

where in the first equality, we used the transformation law of $g_{\mu\nu}$, and in the second equality, we used the Taylor expansion.

Equating the two, we obtain

$$\delta g_{\mu\nu}(x) \equiv g'_{\mu\nu}(x) - g_{\mu\nu}(x) \simeq \xi^\lambda \partial_\lambda g'_{\mu\nu}(x) + (\partial_\mu \xi^\rho)g_{\rho\nu}(x) + (\partial_\nu \xi^\sigma)g_{\mu\sigma}(x)$$
$$\simeq \xi^\lambda \partial_\lambda g_{\mu\nu}(x) + (\partial_\mu \xi^\rho)g_{\rho\nu}(x) + (\partial_\nu \xi^\rho)g_{\mu\rho}(x). \qquad (1.17)$$

In formula (1.17), the first term was from the Taylor expansion, so it is just a translation, while the last two terms correspond to a generalized gauge transformation with parameter ξ^ρ instead of the usual α of gauge theory (with $\delta A_\mu = \partial_\mu \alpha$). Since there are two indices on $g_{\mu\nu}$, unlike the case of A_μ, there are two terms, one with ∂_μ and the other with ∂_ν, and the extra metric is needed in order to lower the index on ξ^ρ.

Note that in the global case (with ξ^ρ independent of position), there is only the translation term. Therefore, we can say that *general coordinate transformations are a local version of translations*, and moreover, *General relativity is a "gauge theory of translations."*

1.3 Kinematics

We now move on to defining kinematics per se. We first ask: What is a good variable that corresponds to A_μ in our gauge theory analogy? And correspondingly, what is the respective field strength $F_{\mu\nu}$?

Our first guess would be the metric $g_{\mu\nu}$ itself. We saw that it has $(d(d-1)/2)$-*independent* components (or degrees of freedom, off-shell). However, we know that locally (in a small enough neighborhood), every space looks flat (which in our case means locally Minkowski). In mathematical terms, locally we can always find coordinates such that

$$g_{\mu\nu}(x) = \eta_{\mu\nu} + \mathcal{O}(x^2). \qquad (1.18)$$

This means also that locally we can define Lorentz transformations, and so there is an $SO(1,3)$ (or $SO(1,d-1)$ in general dimension) invariance, called the *local (x-dependent) Lorentz invariance*.

In any case, this means that $g_{\mu\nu}$ is not a good measure of the curvature of space, but also not quite like the gauge field A_μ either, since A_μ can locally be put to 0, whereas $g_{\mu\nu}$ can only be put to $\eta_{\mu\nu}$.

To understand better what happens, defining general relativity tensors through a simple generalization of special relativity tensors, we have:

– Contravariant tensors A^μ, that are the objects that transform as dx^μ,

$$dx'^\mu = \frac{\partial x'^\mu}{\partial x^\nu}dx^\nu \Rightarrow A'^\mu = \frac{\partial x'^\mu}{\partial x^\nu}A^\nu. \qquad (1.19)$$

– Covariant tensors B_μ that are the objects that transform as ∂_μ,

$$\partial'_\mu = \frac{\partial x^\nu}{\partial x'^\mu}\partial_\nu \Rightarrow B'_\mu = \frac{\partial x^\nu}{\partial x'^\mu}B_\nu. \qquad (1.20)$$

– Mixed tensors that transform as products, for example,

$$T'^{\mu}{}_{\nu}(x') = \frac{\partial x'^{\mu}}{\partial x^{\rho}}\frac{\partial x^{\sigma}}{\partial x'^{\nu}}T^{\rho}{}_{\sigma}(x), \qquad (1.21)$$

and with an obvious generalization to $T^{\mu_1,\dots,\mu_n}_{\nu_1,\dots,\nu_m}$.

Given these definitions, we turn back to the question of what is a good analog of the gauge field A_{μ}? We can now rephrase this question. Since in gauge theory the covariant derivative $D_{\mu} = \partial_{\mu} - iA_{\mu}$ transforms covariantly, that is, like a covariant vector, we can ask the same question in general relativity as follows: How do we construct a gravitationally covariant derivative?

Since, as we saw, the local Lorentz group is $SO(1, d − 1)$, and this is in some sense the gauge group we are looking for, we note that for an $SO(p, q)$ group, the adjoint representation, for the gauge field, is written in terms of the fundamental indices a, b as (ab) (antisymmetric in them), so the gauge covariant derivative on a generic field in the fundamental representation, ϕ^a, is (lowering one index b on the gauge field to have a match with the general relativity construction)

$$D_{\mu}\phi^a = \partial_{\mu}\phi^a + (A^a{}_b)_{\mu}\phi^b. \qquad (1.22)$$

In our case, we define something similar to that, with the only difference being that we identify fundamental gauge and spacetime indices, and write for the gravitationally covariant derivative of a contravariant tensor (so that the index is up, just like a on ϕ^a)

$$D_{\mu}T^{\nu} = \partial_{\mu}T^{\nu} + (\Gamma^{\nu}{}_{\sigma})_{\mu}T^{\sigma}, \qquad (1.23)$$

where the object $\Gamma^{\nu}{}_{\sigma\mu}$ is called the "Christoffel symbol," and in Equation (1.23), we put brackets around Γ, just like for the gauge field, but we did not need to, since the gauge and spacetime indices are the same. This object is then the "gauge field of gravity" that we were looking for.

We can easily generalize its action on tensors, by taking into account the position of the indices (only the sign in front is not defined this way), so that

$$D_{\mu}T^{\rho}{}_{\nu} = \partial_{\mu}T^{\rho}{}_{\nu} + \Gamma^{\rho}{}_{\sigma\mu}T^{\sigma}{}_{\nu} - \Gamma^{\sigma}{}_{\mu\nu}T^{\rho}{}_{\sigma}. \qquad (1.24)$$

To calculate $\Gamma^{\mu}{}_{\nu\rho}$ in terms of the metric $g_{\mu\nu}$, we consider the following: If $\Gamma^{\mu}{}_{\nu\rho}$ is a gauge field, then it should be possible to put it locally to zero by a general coordinate transformation (a gauge transformation), when the space becomes locally flat. At the same time, we saw that $g_{\mu\nu}$ is locally $\eta_{\mu\nu}$. Then

$$D_{\mu}g_{\nu\rho} = \partial_{\mu}g_{\nu\rho} - \Gamma^{\sigma}{}_{\nu\rho}g_{\sigma\rho} - \Gamma^{\sigma}{}_{\rho\mu}g_{\sigma\nu} = 0 \qquad (1.25)$$

locally, but we saw that a tensor transforms by multiplication under general coordinate transformations, so it must be that the result is 0 globally as well (in any coordinate system).

This is an equation whose unique solution is

$$\Gamma^{\mu}{}_{\nu\rho} = \frac{1}{2}g^{\mu\sigma}\left(\partial_{\nu}g_{\sigma\rho} + \partial_{\rho}g_{\nu\sigma} - \partial_{\sigma}g_{\nu\rho}\right). \qquad (1.26)$$

The proof of this is left as an exercise. Note that here we define the inverse metric $g^{\mu\nu}$ as the matrix inverse of $g_{\nu\rho}$, so $g^{\mu\nu}g_{\nu\rho} = \delta^{\mu}_{\rho}$.

Further, we define the Riemann tensor as the analog of the field strength of the $SO(p,q)$ gauge field, $F_{\mu\nu}^{ab}$, namely, since

$$F_{\mu\nu}^{ab} = \partial_\mu A_\nu^{ab} - \partial_\nu A_\mu^{ab} + A_\mu^{ac} A_\nu^{cb} - A_\nu^{ac} A_\mu^{cb}, \tag{1.27}$$

it follows that we can define

$$(R^\mu{}_\nu)_{\rho\sigma}(\Gamma) = \partial_\rho (\Gamma^\mu{}_\nu)_\sigma - \partial_\sigma (\Gamma^\mu{}_\nu)_\rho + (\Gamma^\mu{}_\lambda)_\rho (\Gamma^\lambda{}_\nu)_\sigma - (\Gamma^\mu{}_\lambda)_\sigma (\Gamma^\lambda{}_\nu)_\rho. \tag{1.28}$$

Here we have put brackets around the "gauge indices" to make the analogy with the gauge case more obvious, but, as in the case of the Christoffel symbol, this is not necessary, since gauge and spacetime indices are the same now.

Unlike the gauge case, now we can define the contractions of the Riemann tensor as the Ricci tensor,

$$R_{\mu\nu} = R^\rho{}_{\mu\rho\nu}, \tag{1.29}$$

and as the Ricci scalar,

$$R = R_{\mu\nu} g^{\mu\nu}. \tag{1.30}$$

Finally, the Ricci scalar, by virtue of being a scalar, is invariant under general coordinate transformations, so it is a true invariant measure of the curvature of space at a point, the object we were looking for. In particular, when we said that the sphere was an object of positive curvature $R > 0$ and the Lobachevsky space of negative curvature $R < 0$, we were referring to the Ricci scalar.

The symmetry properties of the Riemann tensor are as follows. First, there are a number of properties that are obvious from the gauge field strength analogy:

1. Since for a gauge field we have the Bianchi identity $(D_{[\mu} F_{\nu\rho]})^a = 0$, we now also have the gravitational Bianchi identity

$$D_{[\lambda} (R^\mu{}_\nu)_{\rho\sigma]} = 0, \tag{1.31}$$

where antisymmetry only acts on $[\lambda\rho\sigma]$.

2, 3. From the antisymmetry of the spacetime indices of the field strength, and of the fundamental indices in the adjoint of $SO(p,q)$, we have (note that we have lowered the first index with a metric on the Riemann tensor for simplicity)

$$R_{\mu\nu\rho\sigma} = -R_{\nu\mu\rho\sigma} = -R_{\mu\nu\sigma\rho}. \tag{1.32}$$

4. Not a symmetry property but the action on a tensor is defined through two covariant derivatives. Since for a gauge field we have $[D_\mu, D_\nu] = F_{\mu\nu}$, which can act on tensors, we now have

$$[D_\mu, D_\nu] T_\rho = -(R^\sigma{}_\rho)_{\mu\nu} T_\sigma = R_{\rho\sigma\mu\nu} T^\sigma. \tag{1.33}$$

5, 6. But then, we have other properties that are not obtained this way, and we must check them from the definition of the Riemann tensor:

$$R_{\mu\nu\rho\sigma} = R_{\rho\sigma\mu\nu}, \quad R_{\mu[\nu\rho\sigma]} = 0. \tag{1.34}$$

1.4 Actions in general relativity

We now move on to writing actions for general relativity fields. To generalize a special relativity action to a general relativity action, we first change special relativity tensors into general relativity tensors, in particular derivatives ∂_μ into gravitationally covariant derivatives D_μ and the metric $\eta_{\mu\nu}$ into the general $g_{\mu\nu}$. The only thing left is to generalize the integration measure from $d^d x$. From the transformation rule for dx^μ, we find the one for $d^d x$,

$$dx^\mu = \frac{\partial x^\mu}{\partial x'^\nu} dx'^\nu \Rightarrow d^d x = \det\left(\frac{\partial x^\mu}{\partial x'^\nu}\right) d^d x'. \tag{1.35}$$

To compensate this, we note that we have the transformation of the metric

$$g'_{\mu\nu}(x') = \frac{\partial x^\rho}{\partial x'^\mu} \frac{\partial x^\sigma}{\partial x'^\nu} g_{\rho\sigma}(x) \Rightarrow \det g'_{\mu\nu} \equiv g' = \left[\det\left(\frac{\partial x^\mu}{\partial x'^\nu}\right)\right]^2 g, \tag{1.36}$$

which means that the invariant measure is (the minus is for the reality of the square root in the Minkowski signature case)

$$\sqrt{-g'}d^d x' = \sqrt{-g}d^d x. \tag{1.37}$$

1.5 The Einstein–Hilbert action

Now we are ready to write an action for the dynamics of general relativity, the Einstein–Hilbert action, as promised. This is an *a priori* independent postulate, which doesn't follow from the previous ones. The role of the action for the dynamics is to match experiment, and there is no fundamental principle behind it, unlike the case of the kinematics.

As is familiar from quantum field theory, to construct actions, we go in increasing order of mass dimension of possible terms in the Lagrangian.

The simplest possibility is to integrate a constant (one times a dimensionful constant) with the invariant measure, so a term of dimension 0,

$$S_0 = \Lambda \int d^d x \sqrt{-g}. \tag{1.38}$$

This doesn't give the correct dynamics. In fact, by varying it, we obtain $\delta g^{\mu\nu} g_{\mu\nu} = 0$, so the equation of motion is nonsensical, $g_{\mu\nu} = 0$. We will see in Section 2.3 that such a term can in fact be added, with a very small constant Λ in front (of dimension d), called a cosmological constant term, but it is not understood as part of the gravity action, but usually as part of the matter action.

The next term, at dimension 2, is the Ricci scalar (since $R \sim \partial\Gamma + \Gamma\Gamma$, and $\Gamma \sim g^{-1}\partial g$, and $g_{\mu\nu}$ is dimensionless, it follows that R is dimension 2, as it has two derivatives), integrated with the invariant measure, with a certain dimensionful constant in front. This is in fact the Einstein–Hilbert action, the correct action for gravity,[*]

$$S_{E-H} = \frac{1}{16\pi G_N} \int d^d x \sqrt{-g} R. \tag{1.39}$$

Note that the factor in front had to involve G_N, since we need to obtain Newtonian gravity in the weak field, small velocities limit, and the actual factor is taken such that we obtain exactly the Newtonian potential $U_N(r)$. This action matches experiments to a very high accuracy: Every experiment we did until now confirms it. Note that we can write the coefficient in terms of the d-dimensional Planck mass,

$$\frac{1}{16\pi G_N} = \frac{M_{Pl}^{d-2}}{2}. \tag{1.40}$$

However, note that in principle we can have corrections to this action, coming from terms of higher mass dimension, and such terms in fact do appear because of quantum corrections in string theory or supergravity, for instance. The next possible terms, at mass dimension 4, are (with coefficients that are implicitly coming from quantum corrections, due to the power of M_{Pl})

$$\sim \int d^d x \sqrt{-g} M_{Pl}^{d-4} R^2, \tag{1.41}$$

where R^2 can mean the Ricci scalar squared, but also $R_{\mu\nu} R^{\mu\nu}$ or $R_{\mu\nu\rho\sigma} R^{\mu\nu\rho\sigma}$.

We are now ready to write Einstein's equations in vacuum, the equations of motion of the Einstein–Hilbert action. Writing the determinant of $g_{\mu\nu}$ as an exponential, we can calculate its variation,

$$g = \det g_{\mu\nu} = e^{\text{Tr} \log g_{\mu\nu}} \Rightarrow \frac{\delta \sqrt{-g}}{\sqrt{-g}} = -\frac{1}{2} g_{\mu\nu} \delta g^{\mu\nu}, \tag{1.42}$$

where we have used $g_{\mu\nu} \delta g^{\mu\nu} = -g^{\mu\nu} \delta g_{\mu\nu}$. Since $R = g^{\mu\nu} R_{\mu\nu}$, we must only calculate $\delta R_{\mu\nu}$.

But, as left to prove in Exercise 4, we have

$$g^{\mu\nu} \delta R_{\mu\nu} = D_\mu (g^{\nu\rho} \delta \Gamma^\mu_{\nu\rho} - g^{\mu\nu} \delta \Gamma^\rho_{\nu\rho}) \equiv D_\mu U^\mu. \tag{1.43}$$

Then, since

$$D_\mu U^\mu = \partial_\mu U^\mu + \Gamma^\mu_{\sigma\mu} U^\sigma, \tag{1.44}$$

and

$$\Gamma^\mu_{\sigma\mu} = \frac{1}{2} g^{\mu\lambda} \partial_\sigma g_{\mu\lambda} = \frac{\partial_\sigma \sqrt{-g}}{\sqrt{-g}}, \tag{1.45}$$

we have

$$\sqrt{-g} D_\mu U^\mu = \partial_\mu (\sqrt{-g} U^\mu), \tag{1.46}$$

and the term becomes a total derivative, that is, a boundary term.

---

[*] Note on conventions: If we use the $+ - - -$ metric, we get a $-$ in front of the action, since $R = g^{\mu\nu} R_{\mu\nu}$ and $R_{\mu\nu}$ is invariant under constant rescalings of $g_{\mu\nu}$.

Finally then, the variation of the Einstein–Hilbert action is

$$\delta S_{\text{E-H}} = \frac{1}{16\pi G_N} \int d^d x \sqrt{-g}\, \delta g^{\mu\nu} \left(R_{\mu\nu} - \frac{1}{2} g_{\mu\nu} R \right), \tag{1.47}$$

so the Einstein equations in vacuum are

$$R_{\mu\nu} - \frac{1}{2} g_{\mu\nu} R = 0. \tag{1.48}$$

When adding matter, for the energy–momentum tensor in curved space, we have the Belinfante formula,

$$T_{\mu\nu} \equiv -\frac{2}{\sqrt{-g}} \frac{\delta S_{\text{matter}}}{\delta g^{\mu\nu}}. \tag{1.49}$$

It is worth noting that, even if we are in flat space, we can formally introduce a nontrivial metric and vary with respect to it as in (1.49), after which we put back $g_{\mu\nu} = \eta_{\mu\nu}$, in order to obtain the Belinfante energy–momentum tensor, uniquely defined (even if, otherwise, for instance for electromagnetism, there are ambiguities in the definition of $T_{\mu\nu}$).

Now reading this formula in reverse, we can find the variation of the matter action as

$$\delta S_{\text{matter}} = -\frac{1}{2} \int d^d x \sqrt{-g}\, \delta g^{\mu\nu} T_{\mu\nu}, \tag{1.50}$$

so that the total variation of the action is

$$\delta (S_{\text{gravity}} + S_{\text{matter}}) = \frac{1}{16\pi G_N} \int d^d x \sqrt{-g}\, \delta g^{\mu\nu} \left(R_{\mu\nu} - \frac{1}{2} g_{\mu\nu} R - 8\pi G_N T_{\mu\nu} \right) = 0, \tag{1.51}$$

giving the Einstein equations with matter,

$$R_{\mu\nu} - \frac{1}{2} g_{\mu\nu} R = 8\pi G_N T_{\mu\nu}. \tag{1.52}$$

To understand better $T_{\mu\nu}$, we give a couple of examples. The kinetic action for a scalar in Minkowski space is

$$S_{M,\phi} = -\frac{1}{2} \int d^d x (\partial_\mu \phi)(\partial_\nu \phi) \eta^{\mu\nu}. \tag{1.53}$$

In curved space, ∂_μ becomes D_μ; however, on a scalar, they are the same, $\partial_\mu \phi = D_\mu \phi$, so the scalar kinetic action in curved space is

$$S_\phi = -\frac{1}{2} \int d^d x \sqrt{-g} (\partial_\mu \phi)(\partial_\nu \phi) g^{\mu\nu}. \tag{1.54}$$

Then the resulting energy momentum tensor for the scalar kinetic term is

$$T_{\mu\nu}^\phi = \partial_\mu \phi \partial_\nu \phi - \frac{1}{2} g_{\mu\nu} (\partial_\rho \phi)^2. \tag{1.55}$$

For electromagnetism, the action in flat space is

$$S_{M,\text{e-m}} = -\frac{1}{4} \int d^d x F_{\mu\nu} F_{\rho\sigma} \eta^{\mu\rho} \eta^{\nu\sigma}, \tag{1.56}$$

which easily translates into curved space as

$$S_{\text{e-m}} = -\frac{1}{4} \int d^d x \sqrt{-g} F_{\mu\nu} F_{\rho\sigma} g^{\mu\rho} g^{\nu\sigma}, \tag{1.57}$$

leading to the (Belinfante) energy–momentum tensor

$$T_{\mu\nu}^{\text{e-m}} = F_{\mu\rho}F_{\nu}{}^{\rho} - \frac{1}{4}g_{\mu\nu}F_{\rho\sigma}F^{\rho\sigma}. \tag{1.58}$$

Important concepts to remember

- In general relativity, space is intrinsically curved.
- In general relativity, physics is invariant under general coordinate transformations.
- Gravity is the same as curvature of space, or gravity = local acceleration, or $m_i = m_g$.
- General relativity can be thought of as a gauge theory of local translations.
- General relativity tensors are a generalization of special relativity tensors.
- The Christoffel symbol acts like a gauge field of gravity, giving the covariant derivative.
- Its field strength is the Riemann tensor, whose scalar contraction, the Ricci scalar, is an invariant measure of curvature.
- One postulates the action for gravity as $(1/(16\pi G_N)) \int d^d x \sqrt{-g} R$, giving Einstein's equations.
- Adding a matter action, obtained from the flat space action by generalization, we obtain the Einstein's equations with matter.

References and further reading

For a very basic (but not too explicit) introduction to general relativity, you can try the general relativity chapter in Peebles [1]. A good and comprehensive treatment is done in [2], which has a very good index, and detailed information, but one needs to be selective in reading only the parts you are interested in. An advanced treatment, with an elegance and concision that a theoretical physicist should appreciate, is found in the general relativity section of Landau and Lifshitz [3], though it might not be the best introductory book. A more advanced and thorough book for the theoretical physicist is Wald [4].

Exercises

(1) Parallel the derivation in the text to find the metric on the two-dimensional sphere in its usual form,

$$ds^2 = R^2(d\theta^2 + \sin^2\theta d\phi^2), \tag{1.59}$$

from the three-dimensional Euclidean metric, using the embedding in terms of θ, ϕ of the three-dimensional Euclidean coordinates.

(2) Show that the metric $g_{\mu\nu}$ is covariantly constant ($D_\mu g_{\nu\rho} = 0$) by substituting the Christoffel symbols.

(3) Prove that we have the relation

$$(D_\mu D_\nu - D_\nu D_\mu)A_\rho = R^\sigma{}_{\rho\mu\nu}A_\sigma, \tag{1.60}$$

if A_σ is a covariant vector.

(4) The Christoffel symbol $\Gamma^{\mu}_{\nu\rho}$ is not a tensor, and can be put to zero at any point by a choice of coordinates (Riemann normal coordinates, for instance), but $\delta\Gamma^{\mu}_{\nu\rho}$ is a tensor. Show that the variation of the Ricci scalar can be written as

$$\delta R = \delta^{\rho}_{\mu} g^{\nu\sigma} (D_{\rho}\delta\Gamma^{\mu}_{\nu\sigma} - D_{\sigma}\delta\Gamma^{\mu}_{\nu\rho}) + R_{\nu\sigma}\delta g^{\nu\sigma}. \tag{1.61}$$

2 Introduction to general relativity 2: Vielbein and spin connection, anti-de Sitter space, black holes

In this chapter, we will continue the short description of general relativity, with a new formulation, in terms of objects called vielbeins and spin connections, and then describe the most important (for us) solutions of general relativity, anti-de Sitter (AdS) and space and black holes.

2.1 The vielbein–spin connection formulation of general relativity

In the standard formulation of general relativity, we introduced the fundamental object that was the metric $g_{\mu\nu}$, and from it we constructed the Christoffel symbol $\Gamma^{\mu}{}_{\nu\rho}(g)$, which played the role of gauge field of gravity, and the Riemann tensor $R^{\mu}{}_{\nu\rho\sigma}(\Gamma)$, which played the role of field strength of the gauge field (we needed to identify gauge and spacetime indices to do that).

But there is another formulation that we can use, in terms of vielbeins e^a_μ and spin connections ω^{ab}_μ, which is a first-order formulation, with an extra, auxiliary, field (the spin connection).

A first-order formulation is one with an auxiliary field, where usually the two-derivative action becomes a one-derivative action. For example, for electromagnetism, with Lagrangian

$$-\frac{1}{4}F^2_{\mu\nu}, \quad F_{\mu\nu} = \partial_\mu A_\nu - \partial_\nu A_\mu, \tag{2.1}$$

one writes a first-order form

$$\frac{1}{4}\tilde{F}^2_{\mu\nu} - \frac{\tilde{F}^{\mu\nu}F_{\mu\nu}}{2}, \tag{2.2}$$

and if we solve for the auxiliary field $\tilde{F}_{\mu\nu}$ we obtain $\tilde{F}_{\mu\nu} = F_{\mu\nu}$, and substituting we go back to the second-order form.

The "vielbein" $e^a_\mu(x)$ is so called from "viel," which is German for "many," and "bein," meaning "legs," for the many "legs" in $\mu = 1, .., d$. Originally, it was "vierbein" in four dimensions, for "vier" means four, then one considers "ein-bein, zwei-bein, drei-bein," and so on, for one, two, three, and so on, or generally "vielbein."

As we said in Chapter 1, any manifold on a small enough scale (so locally) is flat, so there is a local Lorentz invariance acting on it, or rather, there is a flat tangent space to the manifold that approximates it in a small neighborhood. Therefore, we can act with a Lorentz transformation with parameter $\Lambda^a{}_b(x)$, where a, b are fundamental indices of the Lorentz group $SO(1, d-1)$, for the tangent space $(a, b = 0, 1, ..., d-1)$.

We make manifest the action of the local Lorentz group by introducing the vielbein, with index a on which it acts, with invariance under

$$e_\mu^a(x) \to e_\mu'^a(x) = \Lambda^a{}_b(x)e_\mu^b(x). \tag{2.3}$$

In particular, the metric must be invariant under this transformation, so the unique formula we can write is

$$g_{\mu\nu}(x) = e_\mu^a(x)e_\nu^b(x)\eta_{ab}, \tag{2.4}$$

which shows that the vielbein is a kind of square root of the metric. In e_μ^a, a is a "flat" index, acted on by local Lorentz transformations, while μ is a spacetime index, or "curved" index, acted on by general coordinate transformations. Then we see that $e_\mu^a(x)$ is a covariant tensor of general relativity and a contravariant tensor of local Lorentz symmetry. Thus,

$$\begin{aligned} \delta_{\text{l.L.}} e_\mu^a(x) &= \delta\lambda^a{}_b(x)e_\mu^b(x) \\ \delta_{g.c.,\xi} e_\mu^a(x) &= (\xi^\rho \partial_\rho)e_\mu^a(x) + (\partial_\mu \xi^\rho)e_\rho^a, \end{aligned} \tag{2.5}$$

where we see that the first term in the general coordinate transformation is the usual translation term, and the second is the "gauge transformation" with parameter ξ^ρ, on the "gauge field" e_μ^a. The only difference is, e_μ^a has an index a in the *fundamental* representation of the local Lorentz group $SO(1, d-1)$, not the adjoint like for a gauge field. The adjoint representation of $SO(1, d-1)$ is written as (ab) (antisymmetric) in terms of the fundamental indices.

The general coordinate transformation on e_μ^a is consistent with the transformation of $g_{\mu\nu}$, meaning that substituting it in the definition $g_{\mu\nu} = e_\mu^a e_\nu^b \eta_{ab}$ we find the correct transformation of $g_{\mu\nu}$,

$$\delta_{g.c.,\xi} g_{\mu\nu}(x) = (\xi^\rho \partial_\rho)g_{\mu\nu}(x) + (\partial_\mu \xi^\rho)g_{\rho\nu}(x) + (\partial_\nu \xi^\rho)g_{\mu\rho}(x). \tag{2.6}$$

Finally, this means that the description in terms of the vielbein e_μ^a is completely equivalent with the description in terms of the metric $g_{\mu\nu}$. In fact, also the number of degrees of freedom off-shell matches. We saw that $g_{\mu\nu}$ had $d(d-1)/2$ off-shell degrees of freedom, obtained as follows: the symmetric matrix $g_{\mu\nu}$ has $d(d+1)/2$ components, but they are subject to general coordinate transformations with the parameter ξ^ρ, which means we can put to zero d components, so we have $d(d+1)/2 - d = d(d-1)/2$ degrees of freedom.

But also for the vielbein, we have the same number of degrees of freedom. Indeed, we have d^2 components for e_μ^a, subject to the same general coordinate transformations with parameter ξ^ρ, which can put to zero d components, leaving $d(d-1)$. But now we also have the local Lorentz transformations, with parameter $\Lambda^a{}_b(x)$, an antisymmetric matrix, so it can put to zero $d(d-1)/2$ components, leaving again $d(d-1)/2$.

In this equivalent formulation in terms of e_μ^a, the vielbein is sort of a gauge field, just that with the group index in the fundamental representation instead of the adjoint one. But there is a better formulation, closer to the gauge theory picture.

For that, we must go to a first-order formulation, and introduce an auxiliary field called the *spin connection* $\omega_\mu^{ab}(x)$. Connection is a math term for gauge field, so ω_μ^{ab} is a gauge field for the application on spinor fields. That is, the gravitational covariant derivative on a spinor is written as

$$D_\mu \psi = \partial_\mu \psi + \frac{1}{4}\Gamma_{ab}\omega_\mu^{ab}\psi, \tag{2.7}$$

where $\frac{1}{4}\Gamma_{ab}$ is the generator of $SO(1, d-1)$ (i.e., J_{ab}) in the spinor representation, so indeed this is the formula for the covariant derivative if ω_μ^{ab} is the gauge field.

Since the number of degrees of freedom in e_μ^a is just right, as we saw, it means that ω_μ^{ab} must be auxiliary, and therefore we have a second-order formulation, with a fixed functional form of ω_μ^{ab} in terms of the vielbein,

$$\omega_\mu^{ab} = \omega_\mu^{ab}(e). \tag{2.8}$$

Note that this is true only in the absence of dynamical fermions (with a kinetic term), otherwise we must have $\omega_\mu^{ab}(e, \psi)$, as we will see shortly. This will be the case in supergravity.

The functional form $\omega_\mu^{ab}(e)$ is fixed by the "vielbein postulate," or "no-torsion constraint," which states that the torsion, defined as below, vanishes,

$$T_{\mu\nu}^a \equiv 2D_{[\mu}e_{\nu]}^a = 2\partial_{[\mu}e_{\nu]}^a + 2\omega_{[\mu}^{ab}e_{\nu]}^b = 0. \tag{2.9}$$

We see that the torsion $T_{\mu\nu}^a$ is a sort of field strength of the sort of gauge field e_μ^a. The solution of the vielbein postulate is $\omega = \omega(e)$, namely

$$\omega_\mu^{ab}(e) = \frac{1}{2}e^{a\nu}\left(\partial_\mu e_\nu^b - \partial_\nu e_\mu^b\right) - \frac{1}{2}e^{b\nu}\left(\partial_\mu e_\nu^a - \partial_\nu e_\mu^a\right)$$
$$- \frac{1}{2}e^{a\rho}e^{b\sigma}\left(\partial_\rho e_{c\sigma} - \partial_\sigma e_{c\rho}\right)e_\mu^c. \tag{2.10}$$

The proof of it is left as an exercise.

Here we have defined the inverse vielbein, denoted by $e^{a\nu} = e^{-1,a\nu}$ as the matrix inverse of the vielbein, $e^{-1,\mu b}e_\mu^a = \eta^{ab}$, and note that in our convention of dropping the -1, the inverse vielbein is characterized by the position up of the curved index ν; the position of the flat index is irrelevant, since it can be raised or lowered with η_{ab}.

We could also impose, alternatively, the more general condition that the (not antisymmetrized) covariant derivative of the vielbein vanishes,

$$D_\mu e_\nu^a \equiv \partial_\mu e_\nu^a - \partial_\nu e_\mu^a + \omega_{\mu\nu}^{ab}e_\nu^b - \Gamma^\rho_{\mu\nu}e_\rho^c = 0. \tag{2.11}$$

Note that the term with the spin connection is there because e_μ^a is a tensor of $SO(1, d-1)$, so ω_μ^{ab} acts on the a index (by changing it), whereas the term with the Christoffel symbol is there because e_μ^a is also a general relativity tensor, so $\Gamma^\rho_{\mu\nu}$ acts on the μ index.

If we take the antisymmetric part in $\mu\nu$ of the above condition, we get back the vielbein postulate (since the Christoffel symbol is symmetric in $\mu\nu$), while if we take the symmetric part, after having substituted $\omega = \omega(e)$ from the antisymmetric part, we also obtain the solution for $\Gamma^\rho_{\mu\nu}(e)$.

From ω_μ^{ab} we can construct also its field strength, like for any $SO(p, q)$ group, as we have described in Chapter 1,

$$R_{\mu\nu}^{ab}(\omega) = \partial_\mu\omega_\nu^{ab} - \partial_\nu\omega_\mu^{ab} + \omega_\mu^{ac}\omega_\nu^{cb} - \omega_\nu^{ac}\omega_\mu^{cb}. \tag{2.12}$$

However, unlike the case of the Riemann tensor $R^\mu{}_{\nu\rho\sigma}(\Gamma)$, defined in terms of $\Gamma(g)$, space and gauge indices are separate, so the above is a *true field strength*. Then, as we can guess from the similarity between $R^{ab}_{\mu\nu}(\omega)$ and $R^\mu{}_{\nu\rho\sigma}(\Gamma)$, they are related by only "flattening" the indices (turning the curved indices into flat ones by multiplication with the vielbein), namely

$$R^{ab}_{\rho\sigma}(\omega(e)) = e^a_\mu e^{-1,\nu b} R^\mu{}_{\nu\rho\sigma}(\Gamma(g(e))). \tag{2.13}$$

Then we can also write a simple (and more transparent) similar formula for the Ricci scalar, appearing in the Einstein–Hilbert action,

$$R(\Gamma(g(e))) = R^{ab}_{\mu\nu}(\omega(e))e^{-1,\mu}_a e^{-1,\nu}_b. \tag{2.14}$$

To complete the rewriting of the Einstein–Hilbert action, we must rewrite the measure. Since $g_{\mu\nu} = e^a_\mu \eta_{ab} e^b_\nu$, or in matrix notation $g = e \cdot \eta \cdot e$, it means that $-\det g = (\det e)^2$ ($\det \eta = -1$), so finally the Einstein–Hilbert action in vielbein-spin connection formalism is

$$S_{\text{E–H}} = \frac{1}{16\pi G_N} \int d^d x (\det e) R^{ab}_{\mu\nu}(\omega(e)) e^{-1,\mu}_a e^{-1,\nu}_b. \tag{2.15}$$

This is a second-order formulation, with $\omega = \omega(e)$ being of a fixed form, so not truly auxiliary.

But we can also introduce a first-order formulation for gravity, with an *independent* ω^{ab}_μ, called the "Palatini formalism" (this is a slight abuse of notation, Palatini originally wrote a first-order formulation for an independent Christoffel Γ and the metric; but this abuse of notation is common in supergravity, and we call the first-order formulation with e^a_μ and ω^{ab}_μ the Palatini formalism).

We now note that if in the Einstein–Hilbert action above we make ω^{ab}_μ independent, its equation of motion, $\delta S_{\text{E–H}}/\delta\omega^{ab}_\mu = 0$ will give $T^a_{\mu\nu} = 0$, solved by $\omega = \omega(e)$, so we go back to the second-order form!

To prove this statement, we need first to rewrite the action. From the definition of the determinant,

$$\det e^a_\mu = \frac{1}{d!}\epsilon^{\mu_1...\mu_d}\epsilon_{a_1...a_d} e^{a_1}_{\mu_1}...e^{a_d}_{\mu_d}, \tag{2.16}$$

we can derive the formula (as one can easily check)

$$(\det e)e^{-1,[\mu_1}_{a_1}e^{-1,\mu_2]}_{a_2} = \frac{1}{2!(d-2)!}\epsilon^{\mu_1...\mu_d}\epsilon_{a_1...a_d} e^{a_3}_{\mu_3}...e^{a_d}_{\mu_d}. \tag{2.17}$$

In fact, the formula is more general, we can keep p inverse vielbeins on the left and $(d-p)$ vielbeins on the right, with coefficient $1/(p!(d-p)!)$, and it is still valid, but we only need the above. Substituting it in the action, we have

$$S_{\text{E–H}} = \frac{1}{16\pi G_N}\frac{1}{2!(d-2)!}\int d^d x \epsilon^{\mu_1...\mu_d}\epsilon_{a_1...a_d}R^{a_1 a_2}_{\mu_1\mu_2}(\omega)e^{a_3}_{\mu_3}...e^{a_d}_{\mu_d}. \tag{2.18}$$

This then is rewritten in form language, as a form integral, because of the antisymmetrization with the Levi-Civita tensor, as

$$S_{\text{E–H}} = \frac{1}{16\pi G_N}\int \epsilon_{a_1...a_d}R^{a_1 a_2}(\omega) \wedge e^{a_3} \wedge ... \wedge e^{a_d}. \tag{2.19}$$

Then the equation of motion of ω_μ^{ab} is easily found to be

$$\epsilon_{a_1...a_d}\epsilon^{\mu_1...\mu_d}(D_{\mu_2}e_{\mu_3}^{a_3})e_{\mu_4}^{a_4}...e_{\mu_d}^{a_d} = 0, \tag{2.20}$$

or, by "peeling off" the epsilon tensors and the vielbeins, via multiplying with epsilon tensors on the respective free indices, and after that with inverse vielbeins to get rid of the extra vielbeins, we obtain

$$T_{\mu\nu}^a = D_{[\mu}e_{\nu]}^a = 0, \tag{2.21}$$

as promised.

We note that we have used form language and Yang–Mills notation for gauge fields and field strengths, so we can ask if the above action for gravity means the theory is a Yang–Mills theory? The answer is that only in three dimensions this is true off-shell, and then we actually have a Chern–Simons (CS) theory, not a Yang–Mills one, since then the above action is of the type $\epsilon F \wedge A$. Outside three dimensions, the equality with a gauge theory is only valid partially on-shell, namely if we use the $\omega = \omega(e)$ equation of motion, and moreover we identify flat (a) and curved (μ) indices, so identify local Lorentz and general coordinate actions on fields.

In this latter case, the local Lorentz transformation of the spin connection,

$$(\omega'^a{}_b)_\mu = (\Lambda^{-1})^a{}_c(\omega^c{}_d)_\mu \Lambda^d{}_b + (\Lambda^{-1})^a{}_c \partial_\mu \Lambda^c{}_b, \tag{2.22}$$

is just the usual nonabelian gauge transformation,

$$A'_\mu = U^{-1}A_\mu U + U^{-1}\partial_\mu U. \tag{2.23}$$

This transformation leaves invariant the half-on-shell constraint $T_{\mu\nu}^a = 0$, so we still have $\omega = \omega(e)$, and then, since $R_{\mu\nu}^{ab}(\omega)$ is just the Yang-Mills curvature of ω (note that from now on I will speak interchangeably of "curvature" and "field strength"), so it transforms covariantly under the local Lorentz transformation, we have

$$(R'^a{}_b)_{\mu\nu} = (\Lambda^{-1})^a{}_c(R^c{}_d)_{\mu\nu}\Lambda^d{}_b. \tag{2.24}$$

Finally, we explain the advertised fact that, if there are dynamical fermions, the $\omega(e)$ turns into $\omega(e, \psi)$. Indeed, the kinetic term for the fermions in gravitational field will be with $\bar{\psi}\slashed{D}\psi$, so in the first-order formulation, when varying the action, $\delta S/\delta\omega_\mu^{ab}$, we get extra $\bar{\psi}\psi$ terms, so we have

$$\omega = \omega(e) + \bar{\psi}\psi \text{ terms} = \omega(e, \psi), \tag{2.25}$$

that is, we have nonzero fermionic torsion.

Note that it is *mistaken* to take a second-order formulation with only $\omega = \omega(e)$ if we have dynamical fermions; this leads to inconsistencies in the quantum theories (the symmetries are not preserved). We can understand that easily, since there is no good first-order formulation in this case.

Then we see that it is not useful to consider the second-order formulation; instead, we should start with the first-order formulation and *derive* the second-order formulation from it.

2.2 Anti-de Sitter space

Having defined general relativity, both in the standard formulation and in the formulation with vielbeins and spin connections, necessarily the one we will use in supergravity, since supergravity has gravity and fermions, we move on to the description of the most useful (for our purposes) solutions in gravity.

We start with the anti-de Sitter, or AdS, space. It is the unique space of Lorentzian signature with a constant and negative curvature, $R < 0$, so it is the Lorentzian signature equivalent of the Lobachevsky space (which also has constant curvature).

We have an AdS, because there is also a de Sitter space, or dS space, which is the unique Lorentzian signature space of constant positive curvature, $R > 0$, so the Lorentzian signature equivalent of the sphere (which also has constant curvature).

de Sitter space in d dimensions is defined by embedding it in $d + 1$ dimensions, now with a Minkowski metric and corresponding Minkowskian constraint, namely

$$ds^2 = -dX_0^2 + \sum_{i=1}^{d-1} dX_i^2 + dX_{d+1}^2,$$

$$R^2 = -X_0^2 + \sum_{i=1}^{d-1} X_i^2 + X_{d+1}^2. \tag{2.26}$$

Note that for the d-dimensional sphere we have the same formulas as for the above, only for the signature being Euclidean, so changing $-dX_0^2$ to $+dX_0^2$ in the embedding metric and $-X_0^2$ to $+X_0^2$ in the constraint.

From this construction it is obvious that the dS_d space is invariant under $SO(1, d)$, defined by linear rotations with $(1, d)$ signature of the embedding space,

$$X'^A = \Lambda^A{}_B X^B, \quad A, B = 0, 1, ..., d - 1; d + 1, \tag{2.27}$$

and we see that both the embedding metric and constraint are invariant under it. This is similar to the way the sphere S^d is invariant under $SO(d + 1)$.

Anti-de Sitter space in d dimensions, AdS_d, is also defined by embedding it in $d + 1$ dimensions, but with an extra minus in the signature of the embedding space, and $-R^2$ in the constraint,

$$ds^2 = -dX_0^2 + \sum_{i=1}^{d-1} dX_i^2 - dX_{d+1}^2,$$

$$-R^2 = -X_0^2 + \sum_{i=1}^{d-1} X_i^2 - X_{d+1}^2. \tag{2.28}$$

So this is a Lorentzian signature version of the Lobachevsky space in d dimensions, for which we would have $+dX_0^2$ instead of $-dX_0^2$ and $+X_0^2$ instead of $-X_0^2$ in the constraint.

It is also obviously invariant under $SO(2, d - 1)$, defined as rotations with $(2, d - 1)$ signature of the embedding space (2.27). Again both the embedding metric and the constraint are invariant under it.

For the analytical continuation to Euclidean signature, that is, the Wick rotation, of AdS_d and dS_d physics, useful in Quantum Field Theory, we see that Euclidean de Sitter is a sphere, $EdS_d = S^d$, and Euclidean AdS is Lobachevsky, $EAdS_d = Lob_d$.

In order to construct other coordinates for AdS space, we construct solutions of the embedding constraint. The Poincaré coordinates are obtained from the solution

$$X_0 = \frac{1}{2u}\left(1 + u^2(R^2 + \vec{x}^2 - t^2)\right)$$
$$X_i = Rux^i, \quad i = 1, ..., d-1$$
$$X_{d-1} = Rut,$$
$$X_{d+1} = \frac{1}{2u}\left(1 - u^2(R^2 - \vec{x}^2 - t^2)\right), \tag{2.29}$$

which leads to the metric

$$ds_d^2 = R^2\left[u^2\left(-dt^2 + \sum_{i=1}^{d-2} dx_i^2\right) + \frac{du^2}{u^2}\right]$$
$$= \frac{R^2}{x_0^2}\left(-dt^2 + \sum_{i=1}^{d-2} dx_i^2 + dx_0^2\right), \tag{2.30}$$

where in the second equality we have transformed $u = 1/x_0$. Note that the above metric is manifestly invariant under $ISO(1, d-2)$, the Poincaré group (Lorentz plus translations) in $d-1$ dimensions, and under $SO(1, 1)$ rotations.

Also, while the range of the coordinates is maximal, t and $x_i \in \mathbb{R}$ and $x_0 \in (0, +\infty)$, this doesn't cover the whole of AdS space, as defined by the full solution of the constraint equation. It only covers a patch known as the "Poincaré patch."

Another form of the Poincaré patch is obtained by the change of coordinates $x_0/R = e^{-y}$, leading to the metric

$$ds^2 = e^{+2y}\left(-dt^2 + \sum_{i=1}^{d-2} dx_i^2\right) + dy^2, \tag{2.31}$$

with a warp factor for the (t, x_i) coordinates depending only on y, or x_0.

To see that the Poincaré patch is not complete, we can send a light ray (the fastest object) to infinity, that is, consider a null trajectory $ds^2 = 0$ at fixed x_i, for which we obtain

$$t = \int dt = \int^\infty e^{-y}dy < \infty, \tag{2.32}$$

so we can say that "infinity is a finite distance in time away". Physically, it means that light can go further, and we can extend the space to the full, "global AdS."

The solution of the embedding constraint for the global (full) AdS space is

$$X_0 = R\cosh\rho\cos\tau$$
$$X_i = R\sinh\rho\,\Omega_i, \quad i = 1, ..., d-1$$
$$X_{d+1} = R\cosh\rho\sin\tau, \tag{2.33}$$

leading to the metric for global AdS space

$$ds_d^2 = R^2\left(-\cosh^2\rho\,d\tau^2 + d\rho^2 + \sinh^2\rho\,d\vec{\Omega}_{d-2}^2\right). \tag{2.34}$$

2.3 Cosmological constant

AdS space is a solution of the Einstein's equations with a constant energy–momentum tensor $T_{\mu\nu}$,

$$T_{\mu\nu} = -\Lambda g_{\mu\nu}, \tag{2.35}$$

which corresponds to a term in the action of the type

$$-\int d^dx\sqrt{-g}\,\Lambda, \tag{2.36}$$

as we have already noted. In the case of AdS we have $\Lambda < 0$ (negative cosmological constant). The equations of motion for this energy–momentum tensor are

$$\mathcal{R}_{\mu\nu} - \frac{1}{2}g_{\mu\nu}\mathcal{R} = 8\pi G_N(-\Lambda)g_{\mu\nu}, \tag{2.37}$$

where we have written $\mathcal{R}_{\mu\nu}$ on the Ricci tensor (and on the Ricci scalar) in order not to confuse with R, the parameter in the AdS metric. Then we obtain

$$\mathcal{R} = \frac{2d}{d-2}8\pi G_N\Lambda, \tag{2.38}$$

and we see that indeed, $\mathcal{R} < 0$ for $\Lambda < 0$. Using the Poincaré patch metric, it is easy to find that

$$\mathcal{R}_{\mu\nu} = -\frac{d-1}{R^2}g_{\mu\nu} \Rightarrow \mathcal{R} = -\frac{d(d-1)}{R^2}. \tag{2.39}$$

Then the cosmological constant of AdS space is

$$\Lambda_{\text{AdS}} = -\frac{(d-1)(d-2)}{16\pi G_N R^2} = -\frac{(d-1)(d-2)M_{\text{Pl,d}}^{d-2}}{2R^2}. \tag{2.40}$$

Now we can write the metric of global AdS space in a form that will be useful later, using the coordinate transformation

$$\sinh\rho = \frac{r}{R}, \quad \tau = \frac{\bar{t}}{R}, \tag{2.41}$$

leading to

$$ds^2 = -\left(1 + \frac{r^2}{R^2}\right)d\bar{t}^2 + \frac{dr^2}{1 + \frac{r^2}{R^2}} + r^2 d\vec{\Omega}_{d-2}^2$$

$$= -\left(1 - \frac{2\Lambda}{(d-1)(d-2)M_{\text{Pl,d}}^{d-2}}\right)d\bar{t}^2 + \frac{dr^2}{1 - \frac{2\Lambda}{(d-1)(d-2)M_{\text{Pl,d}}^{d-2}}} + r^2 d\vec{\Omega}_{d-2}^2. \tag{2.42}$$

Here I open a parenthesis about solutions of general relativity in general. Unlike field theory in special relativity, say classical electromagnetism, where one gives a set of sources (charges and currents for electromagnetism) and finds the corresponding fields from them, in general relativity one cannot give some sources and find the gravitational fields.

Instead, the solution corresponds *together* to the metric $g_{\mu\nu}$ (field) and energy–momentum tensor $T_{\mu\nu}$ (source). What one gives is a certain distribution of masses, charges, and some symmetries. But the energy–momentum tensor depends also on the metric, as seen in the example of the electromagnetic field ($T_{\mu\nu} = F_{\mu\rho}F_{\sigma\nu}g^{\rho\sigma} - \frac{1}{4}g_{\mu\nu}(F_{\rho\sigma}F_{\lambda\tau}g^{\rho\lambda}g^{\sigma\tau})$), so it is found from the solution.

This is perhaps even clearer in the above example. AdS space corresponds to a constant negative cosmological constant, which is what we give in order to find the solution. But then $T_{\mu\nu} = -\Lambda g_{\mu\nu}$ and $g_{\mu\nu}$ is unknown. So $g_{\mu\nu}$ and $T_{\mu\nu}$ are found together from the Einstein's equations.

2.4 Black holes

Another very important solution of general relativity is the Schwarzschild solution, found in 1916. It is a solution to Einstein's equations in the vacuum ($T_{\mu\nu} = 0$), so to

$$R_{\mu\nu} - \frac{1}{2}g_{\mu\nu}R = 0. \tag{2.43}$$

In fact, there is an important theorem known as the Birkhoff theorem, stating that the most general static solution of Einstein's equations in vacuum with spherical symmetry is the Schwarzschild solution,

$$ds_4^2 = -\left(1 - \frac{2MG_N}{r}\right)dt^2 + \frac{dr^2}{1 - \frac{2MG_N}{r}} + r^2 d\Omega_2^2, \tag{2.44}$$

in four dimensions, or in D dimensions,

$$ds_D^2 = -\left(1 - \frac{2C^{(D)}G_N^{(D)}M}{r^{D-3}}\right)dt^2 + \frac{dr^2}{1 - \frac{2C^{(D)}G_N^{(D)}M}{r^{D-3}}} + r^2 d\vec{\Omega}_{D-2}^2, \tag{2.45}$$

where

$$C^{(D)} = \frac{2\pi^{\frac{3-D}{2}}\Gamma\left(\frac{D-1}{2}\right)}{D-3}, \quad (M_{\text{Pl}}^{(D)})^{D-2} = \frac{1}{8\pi G_N^{(D)}}. \tag{2.46}$$

What this means is that one can always find a coordinate system (by a general coordinate transformation) such that the metric is of the above form. One might ask, but what is the mass parameter M in there? The point is that the metric is *outside* some spherical mass distribution, and the parameter M is the *total* mass inside it. Thus we have the same Gauss law we had in Newtonian gravity: The total mass inside the spherical distribution acts as if it concentrated in a point.

That is so, since we find the correct Newton potential in the Newtonian approximation. The Newtonian approximation of general relativity is for weak fields, $g_{\mu\nu} - \eta_{\mu\nu} = h_{\mu\nu} \ll 1$, and small velocities $v \ll 1$, and in this limit, we can find coordinates in which the metric is written in terms of the Newton potential U_N as

$$ds^2 \simeq -(1+2U_N)dt^2 + (1-2U_N)d\vec{x}^2 = -(1+2U_N)dt^2 + (1-2U_N)(dr^2 + r^2 d\Omega^2). \tag{2.47}$$

We see that in the weak field limit we indeed match the Schwarzschild solution against the above general form, and we find that

$$U_{N,4}(r) = -\frac{MG_N}{r}, \quad U_{N,D}(r) = -\frac{C^{(D)}MG_N^{(D)}}{r^{D-3}}, \tag{2.48}$$

which is the correct Newtonian potential of a spherical mass distribution of total mass M. So the Schwarzschild solution at least has the correct Newtonian limit.

We mentioned that the Schwarzschild solution is valid only outside the mass distribution, that is, in vacuum (inside, it will get modified: for instance, outside the Earth, we have – approximately – the Schwarzschild solution, but inside it, it is not valid). If, however, the solution is valid all the way down to $r = r_H \equiv 2MG_N$ (or $(2C^{(D)}MG_N^{(D)})^{\frac{1}{D-3}}$ in general), where g_{00} vanishes and g_{rr} blows up, we say we have a *(Schwarzschild) black hole*, and r_H is called its *event horizon*.

We have an *apparent* singularity of the metric at $r = r_H$, so is it really singular? And can we reach $r = 0$?

The solution is NOT singular at $r = r_H$. We can in fact calculate the Ricci scalar (the only invariant measure of curvature) there, and find that it is finite,

$$R \sim \frac{1}{r_H^2} = \frac{1}{(2MG_N)^2} = \text{finite}. \tag{2.49}$$

To be more precise, $R = 0$, since we are in vacuum, but the Riemann tensor is $R^\mu{}_{\nu\rho\sigma} \sim 1/r_H^2$.

This means that an observer falling through the black hole will see nothing singular at $r = r_H$.

But can something reach $r = 0$? Consider light propagation (which is the fastest signal), so $ds^2 = 0$, at fixed angle, $d\theta = d\phi = 0$. Then we find, near the event horizon,

$$dt = \frac{dr}{1 - \frac{2MG_N}{r}} \simeq 2MG_N \frac{dr}{r - 2MG_N} \Rightarrow t \simeq 2MG_N \ln(r - 2MG_N) \to \infty. \tag{2.50}$$

This means that light sent toward the black hole never reaches the event horizon *from the point of view of the time t*. But we see that at $r \to \infty$, the space is flat, and $g_{00} = -1$, which means that the time t is the time *measured at infinity*. Thus from the point of view of an observer at infinity, light never reaches the event horizon.

But a falling observer doesn't notice that, he falls through the event horizon like nothing is there. This is due to the time dilation experienced near the black hole: for an observer fixed at a point, $dr = d\Omega = 0$,

$$ds^2 = -\left(1 - \frac{2MG_N}{r}\right)dt^2 = -d\tau^2 \Rightarrow d\tau = \sqrt{-g_{00}}dt = dt\left(1 - \frac{2MG_N}{r}\right) \to 0, \tag{2.51}$$

so the observer near the event horizon is like "frozen in time" (a very small time experienced by it as a very large time is experienced at infinity).

If the falling observers see nothing special as they fall through the black hole, there must be some coordinates that show that. Indeed, there are, and they are called Kruskal coordinates: They go smoothly inside the horizon. To find them, we first define "tortoise coordinates,"

$$dr_* = \frac{dr}{1 - \frac{2MG_N}{r}},$$

(2.52)

then Eddington–Finkelstein coordinates,

$$u = t - r_*, \quad v = t + r_*,$$

(2.53)

after which the metric becomes

$$ds^2 = \left(1 - \frac{2MG_N}{r}\right)(-dudv) + r^2(u, v)d\Omega^2.$$

(2.54)

Then we finally define the Kruskal coordinates,

$$\bar{u} = -4MG_N \exp\left(-\frac{u}{4MG_N}\right),$$

$$\bar{v} = +4MG_N \exp\left(+\frac{v}{4MG_N}\right),$$

(2.55)

after which the metric becomes

$$ds^2 = -\frac{2MG_N}{r}e^{-\frac{r}{2MG_N}}d\bar{u}d\bar{v} + r^2 d\Omega^2.$$

(2.56)

As we see, there is nothing special here at $r = r_H$. Of course, here we must implicitly understand r as a function $r(\bar{u}, \bar{v})$.

Important concepts to remember

- Vielbeins are defined by $g_{\mu\nu}(x) = e_\mu^a(x)e_\nu^b(x)\eta_{ab}$, by introducing a Minkowski space in the neighborhood of a point x, giving local Lorentz invariance.
- The spin connection is the gauge field needed to define covariant derivatives acting on spinors. In the absence of dynamical fermions, it is determined as $\omega = \omega(e)$ by the vielbein postulate: The torsion $T_{\mu\nu}^a = D_{[\mu}e_{\nu]}^a$ is zero.
- The field strength of this gauge field, $R_{\mu\nu}^{ab}(\omega)$, is related to the Riemann tensor $R^\rho{}_{\sigma\mu\nu}$ by flattening two indices.
- In the first-order formulation (Palatini), the spin connection is independent and is determined from its equation of motion: It gives the same viebein postulate.
- In the case of dynamical fermions, $\omega = \omega(e, \phi)$ must be determined from the first-order formulation, from the equation of motion of the independent ω.
- de Sitter space is the Lorentzian signature version of the sphere; AdS space is the Lorentzian version of Lobachevski space, a space of negative curvature.
- Anti-de Sitter space in d dimensions has $SO(2, d - 1)$ invariance.
- The Poincaré coordinates only cover part of AdS space, despite having maximum possible range (over the whole real line).
- Anti-de Sitter space has a cosmological constant.
- The Schwarzschild solution is the most general solution with spherical symmetry and no sources. Its source is located behind the event horizon.
- If the solution is valid down to the horizon, it is called a black hole.
- Light takes an infinite time to reach the horizon, from the point of view of the far away observer, and one has an infinite time dilation at the horizon ("frozen in time").

- Classically, nothing escapes the horizon (quantum mechanically, there is Hawking radiation).
- The horizon is not singular, and one can analytically continue inside it via the Kruskal coordinates.

References and further reading

For the general relativity part, we have the same references as in the first chapter. The vielbein and spin connection formalism for general relativity is harder to find in standard general relativity books, but one can find some information for instance in the supergravity review [5]. For an introduction to black holes, the relevant chapters in [2] are probably the best. A very advanced treatment of the topological properties of black holes can be found in Hawking and Ellis [6].

Exercises

(1) Prove that the general coordinate transformation on $g_{\mu\nu}$,

$$g'_{\mu\nu}(x') = g_{\rho\sigma}(x)\frac{\partial x^\rho}{\partial x'^\mu}\frac{\partial x^\sigma}{\partial x'^\nu},\tag{2.57}$$

reduces for infinitesimal transformations to

$$\partial_\xi g_{\mu\nu}(x) = (\xi^\rho\partial_\rho)g_{\mu\nu} + (\partial_\mu\xi^\rho)g_{\rho\nu} + (\partial_\nu\xi^\rho)g_{\rho\mu}.\tag{2.58}$$

(2) Substitute the coordinate transformation

$$X_0 = R\cosh\rho\cos\tau; \quad X_i = R\sinh\rho\,\Omega_i; \quad X_{d+1} = R\cosh\rho\sin\tau,\tag{2.59}$$

to find the global metric of AdS space from the embedding $(2, d-1)$ signature flat space.

(3) Check that

$$\omega_\mu^{ab}(e) = \frac{1}{2}e^{av}(\partial_\mu e_\nu^b - \partial_\nu e_\mu^b) - \frac{1}{2}e^{bv}(\partial_\mu e_\nu^a - \partial_\nu e_\mu^a) - \frac{1}{2}e^{a\rho}e^{b\sigma}(\partial_\rho e_{c\sigma} - \partial_\sigma e_{c\rho})e_\mu^c\tag{2.60}$$

satisfies the no-torsion (vielbein) constraint, $T_{\mu\nu}^a = 2D_{[\mu}e_{\nu]}^a = 0$.

(4) Check that the transformation of coordinates $r/R = \sinh\rho, t = \bar{t}/R$ takes the AdS metric between the global coordinates

$$ds^2 = R^2(-dt^2\cosh^2\rho + d\rho^2 + \sinh^2\rho d\Omega^2)\tag{2.61}$$

and the coordinates (here $R = \sqrt{-3/\Lambda}$)

$$ds^2 = -\left(1 - \frac{\Lambda}{3}r^2\right)d\bar{t}^2 + \frac{dr^2}{1 - \frac{\Lambda}{3}r^2} + r^2d\Omega^2.\tag{2.62}$$

Introduction to supersymmetry 1: Wess–Zumino models, on-shell and off-shell supersymmetry

After the review of general relativity, in Chapters 4–6, I will review supersymmetry, which is the basis for supergravity. In this chapter, we will study the simplest model, the Wess–Zumino (WZ) model, in two and four dimensions, and its on-shell and off-shell supersymmetry.

First, what is supersymmetry and why people thought about it?

In the 1960s, people were wondering what kind of symmetries are possible. Certainly we have the spacetime symmetries, forming the Poincaré group $ISO(3, 1)$, with generators $J_{\mu\nu}$ (the Lorentz generators of $SO(3, 1)$) and P_μ (the translation generators). Its algebra is given by the Lorentz algebra,

$$[J_{\mu\nu}, J_{\rho\sigma}] = -(\eta_{\mu\rho}J_{\nu\sigma} + \eta_{\nu\sigma}J_{\mu\rho} - \eta_{\mu\sigma}J_{\nu\rho} - \eta_{\nu\rho}J_{\mu\sigma}), \tag{3.1}$$

together with the commutators

$$[P_\mu, J_{\nu\rho}] = \eta_{\mu\nu}P_\rho - \eta_{\mu\rho}P_\nu, \tag{3.2}$$

saying that P_μ is a Lorentz vector, and the commutation of translations, $[P_\mu, P_\nu] = 0$, saying basically that Minkowski space is flat.

There are also internal symmetries of particle physics, generically denoted by T_r, examples of which are the global $SU(2)$ of isospin, the local $U(1)$ of electromagnetism, and $SU(3)$ of color. They form some Lie algebra,

$$[T_r, T_s] = f_{rs}{}^t T_t. \tag{3.3}$$

The question that arose then was: can one combine spacetime and internal symmetries into a larger algebra, such that $[T_s, J_{\mu\nu}] \neq 0, [T_s, P_\mu] \neq 0$?

The answer turned out to be NO, in the form of the Coleman–Mandula theorem, which stated that if we combine them, then the resulting S-matrices for quantum fields are trivial, that is, vanish (have no interactions).

Note that we can embed $ISO(3, 1)$ into $SO(4, 2)$, the conformal group, which can be a symmetry of some quantum field theory, but $SO(4, 2)$ is still a spacetime symmetry, not an internal one.

3.1 Supersymmetry: a Bose–Fermi symmetry

But, as always, a theorem is only as strong as its assumptions, and the assumption that seemed obvious at the time was that internal symmetries correspond to Lie algebras, of the type (3.3). However, it turns out that that is not the only possibility: We can have a *graded Lie algebra*, and thus evade the theorem. Specifically, we will consider the simplest form of grading, a $\mathbb{Z}_2$ grading, that splits the generators in only two types: the "even" ones, the usual bosonic ones $(P_\mu, J_{\mu\nu}, T_r)$, and a new kind, the "odd" ones, denoted by Q_α^i, which obey anticommutation rules, so

$$\{Q_\alpha^i, Q_\beta^j\} = \text{others}, \tag{3.4}$$

and moreover, Q_α^i act as fermions, and we have the same usual commutation rules for fermions and bosons, namely

$$[\text{even, even}] = \text{even}; \quad \{\text{odd, odd}\} = \text{even}; \quad [\text{even, odd}] = \text{odd}. \tag{3.5}$$

Note that we can have higher gradings, like, for instance, a $\mathbb{Z}_4$ grading, that will be useful in the second part of the book.

From the graded Lie algebra definition, we obtain generalized Jacobi identities, just like the usual Jacobi identities are obtained from the double commutator, by cancelling individual terms. Now we have four types of generalized Jacobi identities,

$$[[B_1, B_2], B_3] + [[B_3, B_1], B_2] + [[B_2, B_3], B_1] = 0$$
$$[[B_1, B_2], F_3] + [[F_3, B_1], B_2] + [[B_2, F_3], B_1] = 0$$
$$\{[B_1, F_2], F_3\} + \{[F_3, B_2], F_2\} + \{\{F_2, F_3\}, B_1\} = 0$$
$$[\{F_1, F_2\}, F_3] + [\{F_3, F_1\}, F_2] + [\{F_2, F_3\}, F_1] = 0. \tag{3.6}$$

The generator Q_α^i must be in a representation of the Lorentz group, so its commutation relation with $J_{\mu\nu}$ must be of the type

$$[Q_\alpha^i, J_{\mu\nu}] = (...)Q_\beta^i. \tag{3.7}$$

But because of the anticommuting nature of Q_α^i, acting as a fermion, we choose it to be in the spinor representation, which means

$$[Q_\alpha^i, J_{\mu\nu}] = \frac{1}{2}(\gamma_{\mu\nu})_\alpha{}^\beta Q_\beta^i, \tag{3.8}$$

since $\frac{1}{2}\gamma_{\mu\nu}$ is the Lorentz generator in the spinor representation.

Then we see that α on Q_α^i is the spinor index, and i is a label. For $i = 1, ..., \mathcal{N}$, we have $\mathcal{N}$ supersymmetries. The parameter associated with Q_α^i will be ϵ_α^i, also a spinor.

Since a spinor times a boson gives a spinor, we see that Q_α, the generator of a symmetry called *supersymmetry*, takes us between a boson and a fermion and vice versa, so

$$\delta \text{ boson} = \text{fermion}; \quad \delta \text{ fermion} = \text{boson}. \tag{3.9}$$

Note that $\{Q_\alpha, Q_\beta\}$ is called the supersymmetry algebra, while the full graded Lie algebra is called the superalgebra.

3.2 Spinors in general dimensions

Before continuing, since we will deal with various dimensions, we must understand what are spinors in various dimensions.

Since the Lorentz group in d dimensions is $SO(1, d - 1)$, one will have a representation for it, denoted by χ_α, that is also a representation for the *Clifford algebra*,

$$\{\gamma_\mu, \gamma_\nu\} = 2g_{\mu\nu}, \tag{3.10}$$

where $g_{\mu\nu}$ is the Minkowski metric associated with $SO(1, d - 1)$. A representation of the Clifford algebra is defined in a similar way to how we define a representation of the Lie algebra: as a vector space on which the generators of the algebra, here the γ_μ matrices, act. Thus γ_μ maps a spinor into another,

$$(\gamma_\mu)^\alpha{}_\beta \chi^\beta = \tilde{\chi}^\alpha. \tag{3.11}$$

In d dimensions, the dimension of the representation of the Clifford algebra is $2^{[d/2]}$ complex one, called a Dirac spinor, but it is not an irreducible representation of the Lorentz group.

An irreducible representation (irrep) is obtained by imposing either the Weyl condition (or chirality condition) or the Majorana condition (or reality condition). In $d = 2$ and $d = 10$, we can actually impose both, and the irrep is a Majorana–Weyl spinor.

The Weyl condition is only defined for even dimensions, $d = 2n$. The Majorana condition also is not always valid, but when it is not, we can use instead the *modified Majorana condition*. So we can always define modified Majorana spinors. In fact, in supersymmetry and supergravity, we usually work with Majorana or modified Majorana spinors, rarely with Weyl ones (although the two are equivalent, in even dimensions).

In the simplest nontrivial case, of $d = 2$ Euclidean dimensions, we can use the Pauli matrices σ_i, $i = 1, 2, 3$, satisfying

$$\sigma_i \sigma_j = \delta_{ij} + i\epsilon_{ijk}\sigma_k, \tag{3.12}$$

so

$$\{\sigma_i, \sigma_j\} = 2\delta_{ij}, \tag{3.13}$$

as gamma matrices. For instance, we can choose $\gamma_1 = \sigma_1, \gamma_2 = \sigma_2$. Then, γ_{d+1} ("γ_3" in this case) as

$$\gamma_{d+1} = \gamma_3 = \sigma_3 = -i\sigma_1\sigma_2. \tag{3.14}$$

Then we also have

$$\{\gamma_i, \gamma_3\} = 0, \quad (\gamma_3)^2 = 1, \tag{3.15}$$

which means that also in $d = 3$, we can choose $\gamma_i = \sigma_i$.

In higher even dimensions, $d = 2n$, we can construct the gamma matrices as tensor products from the smaller even dimensions. For instance, if γ^a are gamma matrices in $d = 2n$, then in $d = 2n + 2$, we can consider the representation

$$\Gamma^a = \gamma^a \otimes \sigma_3, \quad a = 1, ..., 2n, \quad \Gamma^{2n+i} = \mathbb{1} \otimes \sigma^i, \quad i = 1, 2, \tag{3.16}$$

and as usual, we define

$$\Gamma_{d+1} = i\gamma_1...\gamma_{2n} \otimes \mathbb{1} = i\gamma_{2n+1} \otimes \mathbb{1}. \tag{3.17}$$

Weyl spinors are defined by using projectors onto Dirac fermions,

$$P_L = \frac{1 + \Gamma_{2n+1}}{2}, \quad P_R = \frac{1 - \Gamma_{2n+1}}{2}, \tag{3.18}$$

which are indeed projectors, since $P_L^2 = P_L, P_R^2 = P_R, P_L + P_R = \mathbb{1}$, and $P_L P_R = 0$. Then $\psi_L = P_L \psi$ and $\psi_R = P_R \psi$.

On the other hand, the Majorana condition is defined using a *charge conjugation matrix* C. We impose the condition that the usual Dirac conjugate $\bar{\chi}^D$ equals the Majorana conjugate $\bar{\chi}^C \equiv \chi^T C$, so

$$\bar{\chi}^D \equiv \chi^\dagger \tilde{\gamma}_t = \bar{\chi}^C \equiv \chi^T C, \tag{3.19}$$

where $\tilde{\gamma}_t$ refers to γ_0 (in the time direction) in the Minkowski case and to $\mathbb{1}$ in the Euclidean case. Note that

$$\bar{\chi}_\beta = \chi^\alpha C_{\alpha\beta}. \tag{3.20}$$

The charge conjugation matrix C transforms γ^μ into its transpose up to a sign and is either symmetric or antisymmetric, so

$$C\gamma^\mu C^{-1} = \sigma \gamma^{\mu T}, \quad C^T = aC, \quad \sigma = \pm, \quad a = \pm. \tag{3.21}$$

In even dimensions, there are two C matrices, C_+ and C_-, $\pm$ referring to σ, whereas in odd dimensions, there is only one.

In $d = 2$, we have

$$C_+^T = C_+, \quad C_-^T = -C_-, \tag{3.22}$$

so in both cases, $\sigma a = +1$, and we can use both. Indeed, this means that (as we will see shortly) $\bar{\epsilon}\chi = +\bar{\chi}\epsilon$, which is consistent (putting $\chi = \epsilon$, we don't get a contradiction). If $\sigma a = -1$, we get $\bar{\epsilon}\chi = -\bar{\chi}\epsilon$, which gives a contradiction.

In $d = 4$, we have

$$C_+^T = -C_+, \quad C_-^T = -C_-, \tag{3.23}$$

so only C_- can be used, as it has $\sigma a = +1$ (C_+ has $\sigma a = -1$).

In $d = 5, 6, 7$, the C's all have $\sigma a = -$, so cannot be used for a Majorana condition. But we can still define a *modified Majorana condition*, if we have an even number of spinors. The condition adds an extra antisymmetric matrix, Ω^{ij}, that acts on the spinor label, so

$$\bar{\chi}_C^i = \chi_j^T \Omega^{ji} C, \quad i, j = 1, 2, \quad \Omega = \begin{pmatrix} 0 & \mathbb{1} \\ -\mathbb{1} & 0 \end{pmatrix}. \tag{3.24}$$

We will use Majorana spinors in the following, which is easier and more convenient, in order to prove supersymmetry identities.

Then, Majorana indices are raised and lowered with C, as

$$\psi_\beta = \psi^\alpha C_{\alpha\beta}, \quad \psi^\beta = \psi_\alpha C^{-1\alpha\beta}. \tag{3.25}$$

Note that a convention is needed, since

$$\chi_\alpha \psi^\alpha = -\chi^\alpha \psi_\alpha. \tag{3.26}$$

3.3 The susy algebra

We now turn to defining the susy algebra. Since $\{Q^i_\alpha, Q^j_\beta\}$ is symmetric under the exchange of (αi) with (βj), it follows that its right-hand side is either symmetric in $(\alpha\beta)$ and (ij) or antisymmetric in $(\alpha\beta)$ and (ij). In $d = 4$, we can check that

$$(C\gamma^\mu)_{\alpha\beta} = (C\gamma^\mu)_{\beta\alpha}, \ \ (C\gamma^{\mu\nu})_{\alpha\beta} = (C\gamma^{\mu\nu})_{\beta\alpha}, \ \ (C\gamma_5)_{\alpha\beta} = -(C\gamma_5)_{\beta\alpha}. \tag{3.27}$$

Note that

$$(C\gamma^\mu)_{\alpha\beta} = -(\gamma^\mu)_{\alpha\beta}, \tag{3.28}$$

where on the right-hand side the upper index is lowered with our rules.

Given the above symmetry rules, and the fact that the symmetry generators are P_μ and $J_{\mu\nu}$, the only thing we can have on the right-hand side of the susy algebra is

$$\{Q^i_\alpha, Q^j_\beta\} = m(C\gamma^\mu)_{\alpha\beta}P_\mu\delta^{ij} + n(C\gamma^{\mu\nu})_{\alpha\beta}J_{\mu\nu}\delta^{ij} + C_{\alpha\beta}U^{ij} + (C\gamma_5)_{\alpha\beta}V^{ij}, \tag{3.29}$$

where $U^{ij} = -U^{ji}$ and $V^{ij} = -V^{ij}$ are antisymmetric functions of the i, j indices that commute with all the generators (appear only on the right-hand side of $\{Q^i_\alpha, Q^j_\beta\}$), so are called *central charges*. Here m and n are constants. Note that $C\gamma^\mu\gamma_5$ is antisymmetric in $(\alpha\beta)$, but there is no other generator to multiply it. It could have been a central charge with μ and $[ij]$ indices (note that the complete set of 4×4 matrices for the $\alpha\beta$ indices is $(\mathbb{1}, \gamma_\mu, \gamma_5, \gamma_\mu\gamma_5, \gamma_{\mu\nu})$), but such a thing does not exist.

The Jacobi identities then fix $n = 0$, and one can normalize the generators such as to have $m = 2$, so finally

$$\{Q^i_\alpha, Q^j_\beta\} = 2(C\gamma^\mu)_{\alpha\beta}P_\mu\delta^{ij} + C_{\alpha\beta}U^{ij} + (C\gamma_5)_{\alpha\beta}V^{ij}. \tag{3.30}$$

In higher dimensions, we can have more "central charges" on the right-hand side, that is, objects that commute with the rest of the generators.

From the Jacobi identities, we also find

$$[Q^i_\alpha, P_\mu] = 0, \tag{3.31}$$

whereas the commutator of Q_α with the internal symmetry generators give a representation of $T_r, (V_r)^i{}_j$, so

$$[Q^i_\alpha, T_r] = (V_r)^i{}_j Q^j_\alpha. \tag{3.32}$$

3.4 Two-dimensional Wess–Zumino model and on-shell susy

To understand supersymmetry, we will start with the simplest model, which happens for $d = 2$ dimensions. Here, a Dirac spinor has $2^{[d/2]} = 2$ complex components. A Majorana fermion has then two real components. But the Dirac equation $\not\partial\psi = 0$ is a matrix equation that relates half of the components of the spinor with the other half, so on-shell a Majorana fermion has just one degree of freedom.

Now a symmetry in general must map one degree of freedom to another. That is not immediately obvious for bosonic symmetries, when usually it is the same kind of field on both sides of a variation (something like $\delta\phi^i = M^i{}_j\phi^j$), but for supersymmetry that becomes an important condition on the theory that can be supersymmetric, since one side is bosonic, and the other is fermionic.

This means that we must match the number of bosonic and fermionic degrees of freedom. If the matching is valid on-shell, we say we can have an on-shell susy, and if it is valid off-shell, we say we can have off-shell susy.

Then, in $d = 2$, on-shell, we have one degree of freedom for a Majorana fermion, so we must have one real scalar (with one degree of freedom on-shell) in order to be able to have supersymmetry. This is called the Wess–Zumino (WZ) model.

In the case of no interactions, the (free version of the) WZ model is

$$S = -\frac{1}{2}\int d^2x[(\partial_\mu\phi)^2 + \bar{\psi}\slashed{\partial}\psi].$$ (3.33)

Note that here is one half for the fermionic action, since we have Majorana fermions, which means that ψ and $\bar{\psi}$ are not independent, like in the Dirac fermion case.

From the above action, we see that the mass dimensions of the fields are $[\phi] = 0$ and $[\psi] = 1/2$ (since $[d^2x] = -2$ and the action should have mass dimension zero).

To find the supersymmetry transformations, we start with the variation of the boson into the fermion times the susy parameter ϵ, with the index contraction fixed by Lorentz invariance,

$$\delta\phi = \bar{\epsilon}\psi = \bar{\epsilon}_\alpha\psi^\alpha = \epsilon^\beta C_{\beta\alpha}\psi^\alpha.$$ (3.34)

This is a definition, but is the simplest, since we need both ϵ and ψ on the right-hand side. Matching mass dimensions on both sides of this susy transformation, we find that $[\epsilon] = -1/2$. Note that we have defined the order of indices as

$$\bar{\chi}\psi = \bar{\chi}_\alpha\psi^\alpha = \chi^\beta C_{\beta\alpha}\psi^\alpha.$$ (3.35)

Next, we define the variation of the fermion ψ, which again should go, at least, into the boson ϕ times the parameter ϵ. But $[\psi] = 1/2$, while $[\phi\epsilon] = -1/2$, so we need an extra object of mass dimension one, with no vector indices. There is a unique such object, namely $\slashed{\partial}$, so we must have

$$\delta\psi = \slashed{\partial}\phi\epsilon.$$ (3.36)

Note that, strictly speaking, once we have fixed the normalization of ϵ by $\delta\phi = \bar{\epsilon}\psi$, there could be a constant in front of $\slashed{\partial}\phi\epsilon$ in $\delta\psi$, but we can fix that by requiring susy, and we find it equal to 1.

Before we can prove the invariance of the action under the susy rules, we must prove the following:

3.4.1 Majorana spinor identities

They are of two kinds:

Two of them are valid in both $d = 2$ and $d = 4$:

$$(1) \quad \bar{\epsilon}\chi = +\bar{\chi}\epsilon; \quad (2) \quad \bar{\epsilon}\gamma_\mu\chi = -\bar{\chi}\gamma_\mu\epsilon \tag{3.37}$$

To prove the first identity, we write (using the Majorana conjugate) $\bar{\epsilon}\chi = \epsilon^\alpha C_{\alpha\beta}\chi^\beta$, but $C_{\alpha\beta}$ is antisymmetric and ϵ and χ anticommute, being spinors, thus we get $-\chi^\beta C_{\alpha\beta}\epsilon^\alpha = +\chi^\beta C_{\beta\alpha}\epsilon^\alpha$. To prove the second identity, we use the fact that, from (3.21), $C\gamma_\mu = -\gamma_\mu^T C = \gamma_\mu^T C^T = (C\gamma_\mu)^T$, thus now $(C\gamma_\mu)$ is symmetric and the rest is the same.

Next, there are two identities that differ in $d = 2$ and $d = 4$. To write them, we first define $\gamma_3 = i\gamma_0\gamma_1$ in $d = 2$ and $\gamma_5 = i\gamma_0\gamma_1\gamma_2\gamma_3$ in $d = 4$. We then get

$$\begin{aligned} &(3) \quad &\bar{\epsilon}\gamma_3\chi = -\bar{\chi}\gamma_3\epsilon; \, (d=2) \quad &\bar{\epsilon}\gamma_5\chi = +\bar{\chi}\gamma_5\epsilon; \, (d=4)\\ &(4) \quad &\bar{\epsilon}\gamma_\mu\gamma_3\chi = -\bar{\chi}\gamma_\mu\gamma_3\epsilon; \, (d=2) \quad &\bar{\epsilon}\gamma_\mu\gamma_5\chi = +\bar{\chi}\gamma_\mu\gamma_5\epsilon; \, (d=4). \end{aligned} \tag{3.38}$$

To prove these, we need to use also the fact that $C\gamma_3 = +i\gamma_0^T\gamma_1^T C = -i(C\gamma_1\gamma_0)^T = +(C\gamma_3)^T$, whereas $C\gamma_5 = +i\gamma_0^T\gamma_1^T\gamma_2^T\gamma_3^T C = -i(C\gamma_3\gamma_2\gamma_1\gamma_0)^T = -(C\gamma_5)^T$, as well as the anticommutation relations $\{\gamma_\mu, \gamma_3\} = \{\gamma_\mu\gamma_5\} = 0$ and $\{\gamma_\mu^T, \gamma_3^T\} = -\{\gamma_\mu^T, \gamma_5^T\} = 0$.

3.4.2 Invariance of the action and susy algebra

We can now go back to the invariance of the action under susy.

The variation of the action in general gives

$$\delta S = -\int d^2x \left[-\phi\Box\delta\phi + \frac{1}{2}\delta\bar{\psi}\,\partial\!\!\!/\,\psi + \frac{1}{2}\bar{\psi}\,\partial\!\!\!/\,\delta\psi \right] = -\int d^2x[-\phi\Box\delta\phi + \bar{\psi}\,\partial\!\!\!/\,\delta\psi], \tag{3.39}$$

where in the second equality we have used partial integration together with identity (2) above to show that the two terms with $\delta\psi$ are equal.

Then substituting the susy transformation law, we get

$$\delta S = -\int d^2x[-\phi\Box\bar{\epsilon}\psi + \bar{\psi}\,\partial\!\!\!/\,\partial\!\!\!/\,\phi\epsilon] \tag{3.40}$$

But we have the following identity:

$$\partial\!\!\!/\,\partial\!\!\!/ = \partial_\mu\partial_\nu\gamma^\mu\gamma^\nu = \partial_\mu\partial_\nu\frac{1}{2}\{\gamma_\mu, \gamma_\nu\} = \partial_\mu\partial_\nu g^{\mu\nu} = \Box \tag{3.41}$$

and by using this, together with two partial integrations, we finally obtain that $\delta S = 0$.

This means then that the action is invariant *without the use of the equations of motion.* That would seem to suggest that we have off-shell susy after all?

Actually, no, because the invariance of the action is not enough. We also need that the transformation laws "close on the fields", that is, that the susy algebra is represented on the fields ϕ and ψ, forming a Lie algebra.

Since we have a single Majorana fermion, we can have at most one supersymmetry, so Q_α (no i index). Then the susy algebra must be

$$\{Q_\alpha, Q_\beta\} = 2(C\gamma^\mu)_{\alpha\beta}P_\mu, \tag{3.42}$$

and on the fields, P_μ is represented by ∂_μ.

In general, the variation of fields, with parameter ϵ^a, is defined as $\delta_\epsilon = \epsilon^a T_a$. Now, for susy, we thus have $\delta_\epsilon = \epsilon^a Q_\alpha$. By multiplying the susy algebra from the left by ϵ_1^α and from the right by ϵ_2^β, it follows that on the left-hand side we obtain

$$\epsilon_1^\alpha Q_\alpha Q_\beta \epsilon_2^\beta + \epsilon_1^\alpha Q_\beta Q_\alpha \epsilon_2^\beta = \epsilon_1^\alpha Q_\alpha Q_\beta \epsilon_2^\beta - \epsilon_2^\beta Q_\beta Q_\alpha \epsilon_1^\alpha = -[\delta_{\epsilon_1}, \delta_{\epsilon_2}], \tag{3.43}$$

and on the right-hand side we get

$$2\bar{\epsilon}_1 \gamma^\mu \epsilon_1 \partial_\mu = -(2\bar{\epsilon}_2 \gamma^\mu \epsilon_1)\partial_\mu, \tag{3.44}$$

which means that the algebra of variations that we need to represent on fields is

$$[\delta_{\epsilon_1}, \delta_{\epsilon_2}] = 2\bar{\epsilon}_2 \gamma^\mu \epsilon_1 \partial_\mu. \tag{3.45}$$

Since the fields of the on-shell WZ multiplet are ϕ and ψ, we need to have

$$[\delta_{\epsilon_{1,\alpha}}, \delta_{\epsilon_{2\beta}}] \begin{pmatrix} \phi \\ \psi \end{pmatrix} = 2\bar{\epsilon}_2 \gamma^\mu \epsilon_1 \partial_\mu \begin{pmatrix} \phi \\ \psi \end{pmatrix}. \tag{3.46}$$

On the ϕ field, we obtain

$$[\delta_{\epsilon_1}, \delta_{\epsilon_2}]\phi = \delta_{\epsilon_1}(\bar{\epsilon}_2 \psi) - 1 \leftrightarrow 2 = \bar{\epsilon}_2(\slashed{\partial}\phi)\epsilon_1 - 1 \leftrightarrow 2 = 2\bar{\epsilon}_2 \gamma^\rho \epsilon_1 \partial_\rho \phi, \tag{3.47}$$

where in the last equality we have used Majorana spinor relation (2) to show that the two terms are equal, giving the factor of 2 in the final result.

Since we have not used any equation of motion, we see that on ϕ, the algebra closes off-shell.

On the other hand, on ψ, that will not be the case. We obtain

$$[\delta_{\epsilon_1}, \delta_{\epsilon_2}]\psi = \delta_{\epsilon_1}(\slashed{\partial}\phi)\epsilon_2 - 1 \leftrightarrow 2 = (\bar{\epsilon}_1 \partial_\mu \psi)\gamma^\mu \epsilon_2 - 1 \leftrightarrow 2. \tag{3.48}$$

3.4.3 Fierz identities (Fierz recouplings)

To proceed further, we need to prove some identities about how to "recouple" spinors, known as Fierz identities. They depend on dimension, so we will show them for $d = 2$ and $d = 4$.

In $d = 2$, the basic Fierz identity is

$$M\chi(\bar{\psi}N\phi) = -\sum_j \frac{1}{2}MO_jN\phi(\bar{\psi}O_j\chi), \tag{3.49}$$

where χ, ψ, ϕ are arbitrary Majorana spinors, M and N are arbitrary matrices, the minus is a result of changing the order of the fermions ϕ and χ, and O_j is the complete set of 2×2 matrices for the spinor representation in $d = 2$,

$$O_j = \{\, \mathbb{1}, \gamma_\mu, \gamma_3 \}. \tag{3.50}$$

It is derived from the completeness relation for 2×2 matrices, which is written as

$$\delta_\alpha^\beta \delta_\gamma^\delta = \frac{1}{2}(O_i)_\alpha^\delta (O_i)_\gamma^\beta. \tag{3.51}$$

This is indeed a completeness relation, since by multiplication with $M^\gamma{}_\beta$ (an arbitrary matrix), we obtain the decomposition of the matrix M in the complete set O_i,

$$M^\delta{}_\alpha = \frac{1}{2} \text{Tr}(M O_i)(O_i)^\delta{}_\alpha. \tag{3.52}$$

We note that the factor of $1/2$ in front comes because of the 2 in $\text{Tr}[O_i O_j] = 2\delta_{ij}$: if we multiply the above with $(O_i)^\alpha{}_\delta$, we obtain an identity.

Multiplying the completeness relation (3.51) with $M^\epsilon{}_\beta$, $(N\phi)^\gamma$, $\bar\psi_\delta$, and χ^α, we obtain the needed Fierz relation.

The Fierz relation in $d = 4$ is similar, the only thing changing is the $1/2$ changes to $1/4$, since in $d = 4$, we have 4×4 matrices acting on spinors, which are decomposed into the complete set

$$O_i = \{ \mathbb{1}, \gamma_\mu, \gamma_5, i\gamma_\mu\gamma_5, i\gamma_{\mu\nu} \}, \tag{3.53}$$

and these obey the orthonormalization relation $\text{Tr}(O_i O_j) = 4\delta_{ij}$, being 4×4 matrices. So we have

$$M\chi(\bar\psi N\phi) = -\sum_j \frac{1}{4} M O_j N\phi(\bar\psi O_j \chi). \tag{3.54}$$

3.4.4 Closure of algebra on ψ

We can now go back to the closure of the susy algebra on ψ, and use the Fierz relation for $M = \gamma_\mu$, $N = \partial_\mu$, $\bar\psi = \bar\epsilon_1$, $\phi = \psi$, $\chi = \epsilon_2$, to obtain

$$\begin{aligned}
&\gamma^\mu \epsilon_2(\bar\epsilon_1 \partial_\mu \psi) - 1 \leftrightarrow 2 \\
&= -\frac{1}{2}[\gamma^\mu \mathbb{1} \partial_\mu \psi(\bar\epsilon_1 \mathbb{1}\epsilon_2) + \gamma^\mu \gamma_\nu \partial_\mu \psi(\bar\epsilon_1 \gamma^\nu \epsilon_2) + \gamma^\mu \gamma_3 \partial_\mu \psi(\bar\epsilon_1 \gamma_3 \epsilon_2)] - 1 \leftrightarrow 2 \\
&= +\gamma^\mu \gamma_\nu \partial_\mu \psi(\bar\epsilon_2 \gamma_\nu \epsilon_1) + \gamma^\mu \gamma_3 \partial_\mu \psi(\bar\epsilon_2 \gamma_3 \epsilon_2) \\
&= 2(\bar\epsilon_2 \gamma^\mu \epsilon_1)\partial_\mu \psi - \gamma^\nu(\partial\!\!\!/\psi)(\bar\epsilon_2 \gamma_\nu \epsilon_1) - \gamma_3(\partial\!\!\!/\psi)(\bar\epsilon_2 \gamma_3 \epsilon_1),
\end{aligned} \tag{3.55}$$

where in the second equality we have used Majorana spinor relations (1), (2), and (3) to find that the first term vanishes against $1 \leftrightarrow 2$ and the second and third terms double, and in the third equality, we have anticommuted the gamma matrices: $\gamma^\mu \gamma_\nu = -\gamma_\nu \gamma^\mu + 2\delta^\mu_\nu$ and $\gamma^\mu \gamma_3 = -\gamma_3 \gamma^\mu$.

We see that we can use the equation of motion of the spinor, $\partial\!\!\!/\psi = 0$, and after that we find exactly the correct susy algebra on ψ as well, so we find closure on-shell only for the algebra. This means that, indeed, we have on-shell susy for this model.

3.5 Two-dimensional WZ model: off-shell susy

We describe off-shell susy in the same $d = 2$ WZ model. Off-shell, the Majorana fermion ψ has Two degrees of freedom, whereas the real scalar ϕ stays with a single degree of freedom. This means that, in order to have off-shell susy, we must add an auxiliary degree of freedom, for a field F, with action that imposes $F = 0$ (so that on-shell we only have ϕ and ψ, as before), that is, $\int d^2x \frac{F^2}{2}$. Thus, the action for the WZ model with off-shell susy is

$$S = -\frac{1}{2} \int d^2x [(\partial_\mu \phi)^2 + \bar{\psi}\slashed{\partial}\psi - F^2]. \tag{3.56}$$

We see that F has mass dimension $[F] = 1$, and equation of motion $F = 0$. Since off-shell, the susy algebra didn't close on ψ (whereas it closed on ϕ), we need to add to $\delta\psi$ a term proportional to the equation of motion for the new boson F (so that it vanishes on-shell), which is just F itself, so we add $F\epsilon$. It has dimension $1/2$, as needed for $\delta\psi$. On the other hand, we also need to write a variation for F, and F itself is an equation of motion, so it should vary into another equation of motion, which should be fermionic, so it should vary into $\slashed{\partial}\psi\epsilon$, which has the correct mass dimension of 1.

This means that the susy transformation rules for the off-shell WZ multiplet are

$$\delta\phi = \bar{\epsilon}\psi;$$
$$\delta\psi = \slashed{\partial}\phi\epsilon + F\epsilon;$$
$$\delta F = \bar{\epsilon}\slashed{\partial}\psi. \tag{3.57}$$

Since we only had non-closure off-shell for the algebra on ψ, we would need to check only $[\delta_{\epsilon_1}, \delta_{\epsilon_2}]\psi$, but nevertheless we also check the effect of the new terms on $[\delta_{\epsilon_1}, \delta_{\epsilon_2}]\phi$. We have

$$\delta_{\epsilon_1}\delta_{\epsilon_2}\phi = \delta_{\epsilon_1}(\bar{\epsilon}_2\psi) = \bar{\epsilon}_2\slashed{\partial}\phi\epsilon_1 + \bar{\epsilon}_2\epsilon_1 F, \tag{3.58}$$

and using Majorana relations (1) and (2), when subtracting $1 \leftrightarrow 2$, the first term gives the correct result, as before, and the term with F vanishes, giving again

$$[\delta_{\epsilon_1}, \delta_{\epsilon_2}]\phi = 2(\bar{\epsilon}_2\gamma^\mu\epsilon_1)\partial_\mu\phi. \tag{3.59}$$

For the action on ψ, we obtain

$$\delta_{\epsilon_1}\delta_{\epsilon_2}\psi = \delta_{\epsilon_1}(\slashed{\partial}\phi\epsilon_2 + F\epsilon_2) = \gamma^\mu\epsilon_2(\bar{\epsilon}_1\partial_\mu\psi) + (\bar{\epsilon}_1\slashed{\partial}\psi)\epsilon_2. \tag{3.60}$$

The first term gives the contribution previously calculated, so consider the extra term only. By using the Fierz identities for $M = 1, N = \slashed{\partial}$, its contribution to the commutator $[\delta_{\epsilon_1}, \delta_{\epsilon_2}]\psi$ is

$$(\bar{\epsilon}_1\slashed{\partial}\psi)\epsilon_2 - (1 \leftrightarrow 2) = -\frac{1}{2}[1 \cdot \slashed{\partial}\psi(\bar{\epsilon}_1 1\epsilon_2) + \gamma^\mu\slashed{\partial}\psi(\bar{\epsilon}_1\gamma_\mu\epsilon_2) + \gamma_3\slashed{\partial}\psi(\bar{\epsilon}_1\gamma_3\epsilon_2)] - 1 \leftrightarrow 2$$
$$= -(\bar{\epsilon}_1\gamma_\mu\epsilon_2)\gamma^\mu\slashed{\partial}\psi - (\bar{\epsilon}_1\gamma_3\epsilon_2)\gamma_3\slashed{\partial}\psi$$
$$= (\bar{\epsilon}_2\gamma_\mu\epsilon_1)\gamma^\mu\slashed{\partial}\psi + (\bar{\epsilon}_2\gamma_3\epsilon_1)\gamma_3\slashed{\partial}\psi, \tag{3.61}$$

where in the second equality we have used Majorana spinor relations (1), (2), and (3) to find that the first term vanishes under subtraction of $(1 \leftrightarrow 2)$, and the second and third terms double.

We see that the resulting 2 extra terms exactly cancel the extra terms that we had in the on-shell case for $[\delta_{\epsilon_1}, \delta_{\epsilon_2}]\psi$, and we are left only with the correct term for closure of the algebra.

Therefore, now we didn't use the equations of motion, and the algebra (as well as the action being invariant) closes off-shell, meaning that, indeed, we have an off-shell susy invariant action.

3.6 Four dimensions: free off-shell WZ model

We can now easily extend our analysis in $d = 2$ to $d = 4$. We start with the on-shell case.

A $d = 4$ Majorana fermion has two real on-shell degrees of freedom, so now we need *two* real scalars, A and B, to match degrees of freedom. This means that the free on-shell action for the WZ model is

$$S_0 = -\frac{1}{2} \int d^4x [(\partial_\mu A)^2 + (\partial_\mu B)^2 + \bar{\psi} \partial\!\!\!/ \psi]. \tag{3.62}$$

The susy transformation rules are as before, just that we differentiate between the scalars A and B by the fact that one transforms as before, and the other with an extra $i\gamma_5$, so

$$\delta A = \bar{\epsilon}\psi, \quad \delta B = \bar{\epsilon} i\gamma_5 \psi$$
$$\delta \psi = \partial\!\!\!/(A + i\gamma_5 B)\epsilon. \tag{3.63}$$

The proof, and on-shell closure of the algebra, are left as an exercise.

For the off-shell action, the off-shell Majorana spinor has four real degrees of freedom, so we need to add two more auxiliary scalars, F and G, with the same action, so

$$S = S_0 + \int d^4x \left[\frac{F^2}{2} + \frac{G^2}{2} \right]. \tag{3.64}$$

The transformation rules differentiate again, not just between A and B, but also between F and G, by an extra $i\gamma_5$, otherwise we have the same rules as in $d = 4$, so

$$\delta A = \bar{\epsilon}\psi; \quad \delta B = \bar{\epsilon} i\gamma_5 \psi;$$
$$\delta \psi = \partial\!\!\!/(A + i\gamma_5 B)\epsilon + (F + i\gamma_5 G)\epsilon$$
$$\delta F = \bar{\epsilon} \partial\!\!\!/ \psi; \quad \delta G = \bar{\epsilon} i\gamma_5 \partial\!\!\!/ \psi. \tag{3.65}$$

We have worked with real fields, but it is perhaps more common to work with complex fields, forming $\phi = A + iB$ and $M = F + iG$, so having the multiplet (ϕ, ψ, M). The action is then

$$S = -\int d^4x \left[(\partial_\mu \phi)\partial^\mu \bar{\phi} - \bar{M}M + \frac{1}{2} \bar{\psi} \partial\!\!\!/ \psi \right], \tag{3.66}$$

and the susy transformation rules are

$$\delta \phi = \bar{\epsilon}\psi$$
$$\delta \psi = \partial\!\!\!/\phi\epsilon + M\epsilon$$
$$\delta M = \bar{\epsilon} \partial\!\!\!/ \psi. \tag{3.67}$$

Important concepts to remember

- A graded Lie algebra can contain the Poincaré algebra, internal algebra, and supersymmetry.
- The supersymmetry Q_α relates bosons and fermions.

- If the on-shell number of degrees of freedom of bosons and fermions match, we have on-shell supersymmetry, and if the off-shell number matches, we have off-shell supersymmetry.
- For off-shell supersymmetry, the supersymmetry algebra must be realized on the fields.
- The prototype for all (linear) supersymmetry is the two-dimensional WZ model, with $\delta\phi = \bar{\epsilon}\psi, \delta\psi = \partial\!\!\!/\phi\epsilon$.
- The WZ model in four dimensions has a fermion and a complex scalar on-shell. Off-shell there is also an auxiliary complex scalar.

References and further reading

For a very basic introduction to supersymmetry, see the introductory parts of [7] and [8]. Good introductory books are West [9] and Wess and Bagger [10]. An advanced book that is harder to digest but contains a lot of useful information is [11]. An advanced student might want to try also volume 3 of Weinberg [12], which is also more recent than the above, but it is harder to read and mostly uses approaches seldom used in string theory. A book with a modern approach but emphasizing phenomenology is [13]. For a good treatment of spinors in various dimensions and spinor identities (symmetries and Fierz rearrangements), see [14]. For an earlier but less detailed acount, see [5]. The original papers on supersymmetry, by Wess and Zumino, are [15, 16].

Exercises

(1) Prove that the matrix

$$C_{AB} = \begin{pmatrix} \epsilon^{\alpha\beta} & 0 \\ 0 & \epsilon_{\dot\alpha\dot\beta} \end{pmatrix} ; \epsilon^{\alpha\beta} = \epsilon^{\dot\alpha\dot\beta} = \begin{pmatrix} 0 & 1 \\ -1 & 0 \end{pmatrix} \tag{3.68}$$

is a representation of the four-dimensional C-matrix, that is, $C^T = -C, C\gamma^\mu C^{-1} = -(\gamma^\mu)^T$, if γ^μ is represented by

$$\gamma^\mu = \begin{pmatrix} 0 & \sigma^\mu \\ \bar\sigma^\mu & 0 \end{pmatrix} ; \quad (\sigma^\mu)_{\alpha\dot\alpha} = (1, \vec\sigma)_{\alpha\dot\alpha}; \quad (\bar\sigma^\mu)^{\alpha\dot\alpha} = (1, -\vec\sigma)^{\alpha\dot\alpha}. \tag{3.69}$$

(2) Show that the susy variation of the four-dimensional on-shell WZ model is zero, paralleling the two-dimensional WZ model.
(3) Using the general form of the Fierz identities, check that in four dimensions, we have

$$(\bar\lambda^a \gamma^\mu \lambda^c)(\bar\epsilon \gamma_\mu \lambda^b) f_{abc} = 0, \tag{3.70}$$

using the fact that f_{abc} is totally antisymmetric, and the identities $\gamma_\mu \gamma_\rho \gamma^\mu = -2\gamma_\rho$, $\gamma_\mu \gamma_{\rho\sigma} \gamma^\mu = 0$ (prove those as well).

(4) For the off-shell WZ model in two dimensions,

$$S = -\frac{1}{2} \int d^2x [(\partial_\mu \phi)^2 + \bar{\psi} \partial\!\!\!/ \psi - F^2],$$ (3.71)

check that

$$[\delta_{\epsilon_1}, \delta_{\epsilon_2}]F = 2(\bar{\epsilon}_2 \gamma^\mu \epsilon_1) \partial_\mu F.$$ (3.72)

4 Introduction to supersymmetry 2: Multiplets and extended supersymmetry

In this chapter, we will be focusing on four dimensional models, and derive the multiplets of extended supersymmetry and their Lagrangians and supersymmetry transformations rules.

4.1 Two-component notation for spinors: dotted and undotted indices

As such, it will be useful to use two-components notation for spinors, with dotted and undotted indices. The four-dimensional Dirac spinor index is $A, B = 1, ..., 4$ and splits into $\alpha, \dot{\alpha} = 1, 2$ indices. Thus a general Dirac spinor is written as

$$\psi = \begin{pmatrix} \psi_\alpha \\ \bar{\chi}^{\dot{\alpha}} \end{pmatrix}, \tag{4.1}$$

where $\bar{\chi}^{\dot{\alpha}} = \epsilon^{\dot{\alpha}\alpha}(\chi_\alpha)^*$. We use a representation for the four-dimensional C-matrix in terms of $\alpha, \dot{\alpha}$, as

$$C_{AB} = \begin{pmatrix} \epsilon^{\alpha\beta} & 0 \\ 0 & \epsilon_{\dot{\alpha}\dot{\beta}} \end{pmatrix}, \tag{4.2}$$

where we have defined as usual $\epsilon^{\alpha\beta} = \epsilon^{\dot{\alpha}\dot{\beta}} = +1$, and then we find also $\epsilon_{\dot{\alpha}\dot{\beta}} = -\epsilon^{\dot{\alpha}\dot{\beta}} = (\epsilon^{\dot{\alpha}\dot{\beta}})^{-1}$. For the gamma matrices we use the representation

$$\gamma^\mu = \begin{pmatrix} 0 & \sigma^\mu \\ \bar{\sigma}^\mu & 0 \end{pmatrix}, \tag{4.3}$$

where $(\sigma^\mu)_{\alpha\dot{\alpha}} = (1, \vec{\sigma})_{\alpha\dot{\alpha}}$ and $(\bar{\sigma}^\mu)^{\dot{\alpha}\alpha} = \epsilon^{\alpha\beta}\epsilon^{\dot{\alpha}\dot{\beta}}(\sigma^\mu)_{\beta\dot{\beta}} = (1, -\vec{\sigma})^{\alpha\dot{\alpha}}$.

A Majorana spinor has $\psi_\alpha = \chi_\alpha$, which means that there is a single independent two-component spinor inside the Dirac one, that is,

$$\begin{pmatrix} \psi_\alpha \\ \bar{\psi}^{\dot{\alpha}} \end{pmatrix}. \tag{4.4}$$

We will use the notation $\psi\chi \equiv \psi^\alpha\chi_\alpha$ and $\bar{\psi}\bar{\chi} \equiv \bar{\psi}_{\dot{\alpha}}\bar{\chi}^{\dot{\alpha}}$, derived from the four-dimensional notation, since

$$\bar{\psi}\chi = \psi^A C_{AB}\chi^B = \psi^\beta\chi_b + \bar{\chi}_{\dot{\beta}}\bar{\chi}^{\dot{\beta}} \equiv \psi\chi + \bar{\psi}\bar{\chi}, \tag{4.5}$$

where we have used raising and lowering of two-dimensional indices with the epsilon tensor,

$$\psi^\beta = \psi_\alpha \epsilon^{\alpha\beta}, \quad \bar{\psi}_{\dot\beta} = \bar{\psi}^{\dot\alpha} \epsilon_{\dot\alpha\dot\beta}. \tag{4.6}$$

The four-dimensional single-susy algebra (where as usual $Q_A = \begin{pmatrix} Q_\alpha \\ \bar{Q}^{\dot\alpha} \end{pmatrix}$) is

$$\{Q_A, Q_B\} = 2(C\gamma^\mu)_{AB} P_\mu, \tag{4.7}$$

where $Q_A = C_{AB} Q^B$, and becomes in two-component notation

$$\{Q_\alpha, \bar{Q}_{\dot\alpha}\} = -2(\sigma^\mu)_{\alpha\dot\alpha} P_\mu$$
$$\{Q_\alpha, Q_\beta\} = 0; \quad \{\bar{Q}_{\dot\alpha}, \bar{Q}_{\dot\beta}\} = 0. \tag{4.8}$$

In the case of $\mathcal{N}$ supersymmetries, so the $\mathcal{N}$-extended susy algebra, in the case *without central charges*, is a trivial extension of the above,

$$\{Q_A^i, Q_B^j\} = 2(C\gamma^\mu)_{AB} P_\mu, \quad i, j = 1, ..., \mathcal{N}, \tag{4.9}$$

or, in two-component notation,

$$\{Q_\alpha^i, \bar{Q}_{\dot\alpha j}\} = -2(\sigma^\mu)_{\alpha\dot\alpha} P_\mu \delta_j^i,$$
$$\{Q_\alpha^i, Q_\beta^j\} = 0; \quad \{\bar{Q}_{\dot\alpha i}, \bar{Q}_{\dot\beta j}\} = 0. \tag{4.10}$$

4.2 Massless irreducible representations

We start constructing massless irreducible representations. If $m = 0$, there is a reference frame where the momentum is along the third direction, so $P^\mu = p(1, 0, 0, 1)$, and thus $P_\mu = p(-1, 0, 0, 1)$. Then $\sigma^\mu P_\mu = p(-\mathbb{1} + \sigma_3)$, so the nontrivial part of the susy algebra reduces to

$$\{Q_\alpha^i, \bar{Q}_{j\dot\alpha}\} = 2p(\mathbb{1} - \sigma_3)\delta_j^i = 4p \begin{pmatrix} 0 & 0 \\ 0 & 1 \end{pmatrix} \delta_j^i. \tag{4.11}$$

This means that now $\{Q_1^i, \bar{Q}_{1j}\} = 0$, but the charges are complex conjugates, so we must impose on physical states $|\psi\rangle$ that they vanish, $Q_1^i |\psi\rangle = \bar{Q}_{1j} |\psi\rangle = 0$.

On the other hand, for $\alpha = \dot\alpha = 2$, we have a nontrivial anticommutator,

$$\{Q_2^i, \bar{Q}_{2j}\} = 4p\delta_j^i, \tag{4.12}$$

which is an $a, a^\dagger$ algebra, if we define

$$a^i = \frac{1}{2\sqrt{p}} Q_2^i, \quad a^{\dagger i} = \frac{1}{2\sqrt{p}} \bar{Q}_{2i}, \tag{4.13}$$

constructing a set of fermionic harmonic oscillators,

$$\{a^i, a^{\dagger j}\} = \delta^{ij}, \quad \{a^i, a^j\} = \{a^{\dagger i}, a^{\dagger j}\} = 0. \tag{4.14}$$

Once we have reduced the susy algebra to harmonic oscillators, we can use the Wigner method for constructing the irreducible representations of the algebra, by acting with the creation operators on a vacuum state.

Consider the vacuum state of helicity λ, $|\Omega_\lambda\rangle$, or $|\lambda\rangle$ for short. Helicity refers to the projection of the spin onto the momentum direction, here the third direction, so it refers to the Lorentz generator $J_3 = J_{12}$. Then

$$J_3|\Omega_\lambda\rangle = \lambda|\Omega_\lambda\rangle. \tag{4.15}$$

Remembering the commutation relation from the superalgebra, stating that Q_A is a spin 1/2 representation of the Lorentz group, namely

$$[Q_A^i, J_{\mu\nu}] = \frac{1}{2}(\gamma_{\mu\nu})_A{}^B Q_B^i, \tag{4.16}$$

we obtain for $J_3 = J_{12}$ and for $\bar{Q}_{\dot{2}i}$ that (using our representation for the gamma matrices)

$$[\bar{Q}_{\dot{2}i}, J_3] = +\frac{1}{2}\bar{Q}_{\dot{2}i}, \tag{4.17}$$

so that when acting on a helicity state of helicity j, we lower it by 1/2,

$$(a^{\dagger i}J_3 - J_3 a^{\dagger i}) = \frac{1}{2}a^{\dagger i}|j\rangle \Rightarrow$$
$$J_3\left(a^{\dagger i}|j\rangle\right) = (j - 1/2)\left(a^{\dagger i}|j\rangle\right). \tag{4.18}$$

Then, for $\mathcal{N}$ supersymmetries, we have two states (with or without $a^\dagger$) for each i, so $2^{2\mathcal{N}}$ states in the irreducible representation (irrep). We therefore obtain the following irreps, depending on the helicity λ of the "vacuum" state,

$$\mathcal{N} = 1 \quad : \quad |\lambda\rangle; |\lambda - 1/2\rangle = a^\dagger|\lambda\rangle.$$
$$\mathcal{N} = 2 \quad : \quad |\lambda\rangle; 2|\lambda - 1/2\rangle = \left(a^{\dagger 1}|\lambda\rangle, a^{\dagger 2}|\lambda\rangle\right) ; |\lambda - 1\rangle = a^{\dagger 1}a^{\dagger 2}|\lambda\rangle.$$
$$\mathcal{N} = 4 \quad : \quad |\lambda\rangle; 4|\lambda - 1/2\rangle = a^{\dagger i}|\lambda\rangle; 6|\lambda - 1\rangle = (a^{\dagger i}a^{\dagger j})|\lambda\rangle, i \neq j;$$
$$4|\lambda - 3/2\rangle = (a^{\dagger i}a^{\dagger j}a^{\dagger k})|\lambda\rangle, \ i \neq j \neq k; |\lambda - 2\rangle = a^{\dagger 1}a^{\dagger 2}a^{\dagger 3}a^{\dagger 4}|\lambda. \tag{4.19}$$

But irreps are not (necessarily) CPT invariant, as we want. Since CPT reverses λ, we want to add the CPT-conjugate representations.

At the moment, we are interested in representations of maximum helicity, so of maximum spin, of $\lambda = 1$, since these will be related to matter and gauge fields. Spin (and so helicity) of 3/2 and 2 are related to gravitino and graviton, respectively, and we will consider them later.

Then we obtain the following irreps:

$$\mathcal{N} = 1, \quad \lambda = 1/2: \quad |1/2\rangle, \ |0\rangle \ \text{ and } \ |-1/2\rangle, \ |0\rangle.$$
$$\lambda = 1: \quad |1\rangle, \ |1/2\rangle \ \text{ and } \ |-1\rangle, \ |-1/2\rangle.$$
$$\mathcal{N} = 2, \quad \lambda = 1/2: \quad |1/2\rangle, \ 2|0\rangle, \ |-1/2\rangle \ \text{ and } \ |-1/2\rangle, \ 2|0\rangle, \ |+1/2\rangle.$$
$$\lambda = 1: \quad |1\rangle, \ 2|1/2\rangle, \ |0\rangle \ \text{ and } \ |-1\rangle, \ 2|-1/2\rangle, \ |0\rangle.$$
$$\mathcal{N} = 4, \quad \lambda = 1: \quad |1\rangle, \ 4|1/2\rangle, \ 6|0\rangle, \ 4|-1/2\rangle, \ |-1\rangle. \tag{4.20}$$

Note that the unique $\mathcal{N} = 4$ representation is CPT self-conjugate already, so one doesn't need to add anything to it.

Also note that for spins ≤ 1, the helicity is at most between 1 and -1, so the number of supersymmetries is bounded by $\mathcal{N} \leq 2(1 - (-1)) = 4$, so there is nothing else available ($\mathcal{N} = 3$ susy implies $\mathcal{N} = 4$ susy).

Finally then, the multiplets are as follows:

- $\mathcal{N} = 1, \lambda = 1/2$: one Majorana spinor (with $|+1/2\rangle, |-1/2\rangle$ helicities) and two real scalars (the 2 $|0\rangle$'s). This is the Wess–Zumino (WZ), or chiral, or scalar multiplet we analyzed before. The fields are (ϕ, ψ), where ϕ was a *complex* scalar.
- $\mathcal{N} = 1, \lambda = 1$: one Majorana spinor (with $|+1/2\rangle, |-1/2\rangle$) and one vector (with $|-1\rangle, |+1\rangle$). This is the gauge (vector) multiplet, composed of the gauge field and gaugino, both in the adjoint representation of the gauge group (A_μ^a, λ^a).
- $\mathcal{N} = 2, \lambda = 1/2$: two $\mathcal{N} = 1$ WZ multiplets, (ϕ_1, ψ_1) and (ϕ_2, ψ_2), forming a *hypermultiplet*.
- $\mathcal{N} = 2, \lambda = 1$: an $\mathcal{N} = 1$ WZ multiplet and one $\mathcal{N} = 1$ vector multiplet, (ϕ, ψ) and (A_μ, λ), forming an $\mathcal{N} = 2$ vector (gauge) multiplet. Note that all fields are in the adjoint representation of the gauge group.
- $\mathcal{N} = 4, \lambda = 1$: there is a unique multiplet, the $\mathcal{N} = 4$ vector multiplet, composed of an $\mathcal{N} = 2$ vector multiplet and an $\mathcal{N} = 2$ hypermultiplet, or 3 $\mathcal{N} = 1$ WZ multiplets, (ϕ_r, ψ_r), $r = 1, 2, 3$, and one $\mathcal{N} = 1$ vector multiplet, (A_μ, ψ_4), all fields in the adjoint representation of the gauge group, fitting into $(A_\mu^a, \psi^{ai}, \phi_{[ij]}^a)$, where $i, j = 1, 2, 3, 4$ are fundamental $SU(4)$ indices, and $\phi_{[ij]}$ is complex, but satisfying the reality condition $\phi_{ij}^\dagger = \phi^{ij} \equiv \frac{1}{2}\epsilon^{ijkl}\phi_{kl}$. Note that $SU(4) \simeq SO(6)$, so the six $\phi_{[ij]}$ (complex, but with a reality condition on them) fields can be reassembled into the (real) fundamental representation of $SO(6)$, ϕ_m.

4.3 Massive representations

4.3.1 Representations without central charges

In the rest frame, the momentum is $P^\mu = M(1, 0, 0, 0)$, so $P_\mu = -M(1, 0, 0, 0)$, so $\sigma^\mu P_\mu = -M \, \mathbb{1}$, and therefore the susy algebra reduces to

$$\{Q_\alpha^i, \bar{Q}_{\dot\alpha j}\} = 2M \, \mathbb{1}\delta_{ij} = 2M \begin{pmatrix} 1 & 0 \\ 0 & 1 \end{pmatrix} \delta_j^i. \tag{4.21}$$

This means that now we have twice as many $a, a^\dagger$'s,

$$a_\alpha^i = \frac{1}{\sqrt{2M}} Q_\alpha^i, \quad a_\alpha^{\dagger i} = \frac{1}{\sqrt{2M}} \bar{Q}_{\dot\alpha i}, \quad \alpha, \dot\alpha = 1, 2, \tag{4.22}$$

satisfying the algebra

$$\{a_\alpha^i, a_\beta^{\dagger j}\} = \delta^{ij}\delta_{\alpha\beta}, \quad \{a_\alpha^i, a_\beta^j\} = \{a_\alpha^{\dagger i}, a_\beta^{\dagger j}\} = 0. \tag{4.23}$$

Then the number of states in an irrep is $2^{2\mathcal{N}}$, but the analysis of irreps leads to the same multiplets, just massive.

4.3.2   Massive representations with central charges

The more interesting case is when we have central charges. The extended supersymmetry algebra is then

$$\{Q_A^i, Q_B^j\} = 2(C\gamma^\mu)_{AB} P_\mu \delta^{ij} + C_{AB} U^{ij} + (C\gamma_5)_{AB} V^{ij}. \tag{4.24}$$

In two-component notation, this becomes

$$\{Q_\alpha^i, \bar{Q}_{\dot\alpha j}\} = -2(\sigma_\mu)_{\alpha\dot\alpha} P_\mu \delta^{ij}$$
$$\{Q_\alpha^i, Q_\beta^j\} = 2\epsilon_{\alpha\beta} Z^{ij}$$
$$\{\bar{Q}_{\dot\alpha i}, \bar{Q}_{\dot\beta j}\} = 2\epsilon_{\dot\alpha\dot\beta} Z_{ij}^*, \tag{4.25}$$

where we have defined the complex central charge $Z^{ij} = U^{ij} + iV^{ij}$.

Consider the simplest case, of $\mathcal{N} = 2$. Diagonalize Z^{ij} by the action of a global $SU(2)$ acting on i, j indices, called R-symmetry, which will be described better in Section 4.3.3. Then $Z^{ij} = Z\epsilon^{ij}$. Moreover, we can also use a global $U(1)$, also an R-symmetry, which will make Z real. If we also work in the rest frame, then as before $-2\sigma^\mu P_\mu = 2M$. Finally then, the $\mathcal{N} = 2$ susy algebra in two-component notation becomes

$$\{Q_\alpha^i, \bar{Q}_{\dot\beta j}\} = 2M\delta_{\alpha\beta}\delta^{ij}$$
$$\{Q_\alpha^i, Q_\beta^j\} = 2Z\epsilon^{ij}\epsilon_{\alpha\beta}$$
$$\{\bar{Q}_{\dot\alpha i}, \bar{Q}_{\dot\beta j}\} = 2Z\epsilon_{ij}\epsilon_{\dot\alpha\dot\beta}, \tag{4.26}$$

where $\bar{Q}_i^{\dot\alpha} = (Q_\alpha^i)^\dagger$ and $\bar{Q}_{\dot\alpha i} = \bar{Q}_i^{\dot\beta}\epsilon_{\dot\beta\dot\alpha}$.

Then we can define two sets of fermionic harmonic oscillators mixing Q's and $\bar{Q}$'s as follows:

$$a_\alpha = \frac{1}{\sqrt{2}}[Q_\alpha^1 + \epsilon_{\alpha\beta}\bar{Q}_{2\dot\beta}]$$

$$a_\alpha^\dagger = \frac{1}{\sqrt{2}}[\bar{Q}_{1\dot\alpha} + \epsilon_{\alpha\beta} Q_\beta^2]$$

$$b_\alpha = \frac{1}{\sqrt{2}}[Q_\alpha^1 - \epsilon_{\alpha\beta}\bar{Q}_{2\dot\beta}]$$

$$b_\alpha^\dagger = \frac{1}{\sqrt{2}}[\bar{Q}_{1\dot\alpha} - \epsilon_{\alpha\beta} Q_\beta^2]. \tag{4.27}$$

They satisfy the algebra

$$\{a_\alpha, a_\beta^\dagger\} = 2(M - Z)\delta_{\alpha\beta}$$
$$\{b_\alpha, b_\beta^\dagger\} = 2(M + Z)\delta_{\alpha\beta}. \tag{4.28}$$

But, because these operators can act on a vacuum and create states, which need to have positive norm (not to be ghosts), we need that the right-hand sides of Eq. (4.28) must be positive, $M \geq Z$ and $M \geq (-Z)$, so

$$M \geq |Z|, \tag{4.29}$$

known as the Bogomol'nyi–Prasad–Sommerfield, or BPS bound.

Note that the BPS bound started as a bound in the theory of normal monopoles, where in the limit of the scalar ϕ^4 coupling going to zero, we have the bound that the energy (or mass) of a field configuration is $\geq$ a topological charge, constructed from the fields. When embedding the monopole in a supersymmetric theory, there is no need to take any limit, and we always have a bound that the mass of a field configuration is bounded by a topological charge, constructed from the fields. But we see that this is nothing but a consequence of the above BPS bound, since the central charge must be represented on the fields of a multiplet, and its representation equals the topological charge of the field configuration.

But now we observe an interesting fact. At saturation of the BPS bound, so when $M = \pm Z$, we have that half of the harmonic oscillators become trivial, $\{a, a^\dagger\} = 0$ (so don't create any states), and we end up with a "short multiplet," or BPS multiplet, with the same dimension as in the case of massless irreps, $2^{\mathcal{N}}$.

4.3.3 R-symmetry

R-symmetry, which was mentioned earlier, is an internal symmetry rotating the charges Q^i_α. When representing it on the fields, it becomes a global symmetry of the action, rotating the fields, in particular the fermions. In the case of an $\mathcal{N}$-extended supersymmetry, the maximum allowed R-symmetry is $U(N)$, but in some cases it can be smaller. In the $\mathcal{N} = 2$ case considered earlier, we have $SU(2) \times U(1) \simeq U(2)$ R-symmetry.

Note that in $d = 4$, we can choose Weyl, or complex, spinors (they are equivalent to the Majorana ones) and then, for $\mathcal{N} = 2$, the $SU(2)_R$ symmetry rotates the two $\mathcal{N} = 1$ WZ multiplets in the $\mathcal{N} = 2$ hypermultiplet, and the two fermions in the $\mathcal{N} = 2$ vector multiplet. In the case of the $\mathcal{N} = 4$ SYM multiplet, the R-symmetry is $SU(4)$.

We mention also $d = 3$ (although we have not addressed it until now in this chapter), where the spinors are Majorana (real; there is no Weyl spinor in $d = 3$), so the R-symmetry is $SO(\mathcal{N})$. This means that for spins less or equal to 1, the maximal case is $\mathcal{N} = 8$, which has an $SO(8)_R$ symmetry. Another relevant case is $\mathcal{N} = 6$, which will have $SO(6)_R \simeq SU(4)_R$ symmetry.

4.4 Lagrangians for the multiplets in $d = 4$

4.4.1 Chiral (WZ) multiplet

We start with the $\mathcal{N} = 1$ chiral (WZ) multiplet. The Lagrangian for the fermions in the two-component notation is written as (using the representation for the γ_μ from before, we find two terms, equal under partial integration, canceling the $\frac{1}{2}$ in front)

$$-i\bar{\psi}\bar{\sigma}^\mu \partial_\mu \psi. \tag{4.30}$$

Then the free Lagrangian for the WZ model is (denoting the complex auxiliary field by F instead of M)

$$\mathcal{L}_0 = -\sum_i \left[\partial_\mu \phi_i \partial^\mu \phi_i^\dagger - F_i F_i^\dagger + i \bar{\psi}_i^{\dot\alpha} (\bar{\sigma}^\mu)_{\dot\alpha\alpha} \partial_\mu \psi_i^\alpha \right]. \tag{4.31}$$

But we can write a first generalization of this free case by introducing a metric on field space, called a Kähler metric, obtained from a "Kähler potential" $K(\phi, \bar{\phi})$ through

$$g^{i\bar{j}} = \frac{\partial^2 K}{\partial \phi_i \partial \phi_j^\dagger}. \tag{4.32}$$

Then the Lagrangian of the resulting *nonlinear sigma model* (a nonlinear sigma model is a scalar model with a metric on field space, denoted originally in the effective low-energy Lagrangian for QCD) is

$$\mathcal{L}_0 = \mathcal{L}_{\text{kinetic}} = -\sum_{i,j} g^{i\bar{j}} \left[\partial_\mu \phi_i \partial^\mu \phi_j^\dagger - F_i F_j^\dagger + i \bar{\psi}_i^{\dot\alpha} (\bar{\sigma}^\mu)_{\dot\alpha\alpha} \partial_\mu \psi_j^\alpha \right]. \tag{4.33}$$

The next generalization involves adding extra terms, constructed from a holomorphic function of ϕ's (but not of $\bar{\phi}$'s) called the superpotential $W(\phi)$, out of which we obtain

$$\mathcal{L}_{\text{superpotential}} = \frac{\partial W}{\partial \phi_i} F_i + \frac{\partial \bar{W}}{\partial \phi_i^\dagger} \bar{F}_i - \frac{1}{2} \frac{\partial^2 W}{\partial \phi_i \partial \phi_j} \psi_i \psi_j - \frac{1}{2} \frac{\partial^2 \bar{W}}{\partial \phi_i^\dagger \partial \phi_j^\dagger} \bar{\psi}_i \bar{\psi}_j. \tag{4.34}$$

If we are interested in fundamental theories (not effective ones), we want renormalizable quantum field theories. The most general renormalizable superpotential is up to third power,

$$W(\phi) = \lambda \phi + \frac{m}{2} \phi^2 + \frac{g}{3} \phi^3. \tag{4.35}$$

We note that now the auxiliary field has a slightly more nontrivial action, but it is still an *algebraic* action (no derivative, so no dynamics), so can be solved. By solving for it, we now obtain the scalar potential. Indeed, the terms with F in the Lagrangian are

$$\mathcal{L}(F) = \sum_i \left(F_i \bar{F} + \frac{\partial W}{\partial \phi_i} F_i + \frac{\partial \bar{W}}{\partial \phi_i^\dagger} \bar{F}_i \right). \tag{4.36}$$

Then

$$F_i = -\frac{\partial \bar{W}}{\partial \phi_i^\dagger}, \tag{4.37}$$

so we obtain a potential term,

$$\mathcal{L}(F) = -V(F) = -\sum_i |F_i|^2 = -\sum_i \left| \frac{\partial W}{\partial \phi_i} \right|^2. \tag{4.38}$$

This is called an *F-term*.

4.4.2 The $\mathcal{N} = 1$ vector (YM) multiplet

The fields of the $\mathcal{N} = 1$ vector multiplet are the gauge field A_μ^a and the gaugino λ^a, both in the adjoint representation of the gauge group.

The free action is then

$$
S = \int d^4x \, \mathrm{Tr}\left[-\frac{1}{4} F_{\mu\nu} F^{\mu\nu} - \frac{1}{2}\bar{\lambda}\slashed{D}\lambda \right]
$$
$$
= \int d^4x \left[-\frac{1}{4} F_{\mu\nu}^a F^{a\mu\nu} - i\bar{\lambda}^a \bar{\sigma}^\mu D_\mu \lambda^a \right], \tag{4.39}
$$

where we have used the normalization $\mathrm{Tr}[T^a T^b] = \delta^{ab}$.

On-shell, the Majorana spinor has two degrees of freedom, as does the vector. Off-shell, the vector has three degrees of freedom, but the Majorana spinor has four. This means that we need one real auxiliary scalar (also in the adjoint representation), D^a, in order to obtain an off-shell multiplet. Then the action we need to add for it is $\int d^4x \frac{D^a D^a}{2}$.

Moreover, we can introduce a topological theta term, which is supersymmetry invariant by itself (since it is topological).

This means that the free action for the vector multiplet off-shell is

$$
S_0 = \int d^4x \left\{ \frac{1}{g^2}\left[-\frac{1}{4} F_{\mu\nu}^a F^{a\mu\nu} - i\bar{\lambda}^a \bar{\sigma}^\mu D_\mu \lambda^a + \frac{D^a D^a}{2} \right] + \frac{\theta}{32\pi^2} F_{\mu\nu}^a \tilde{F}^{a\mu\nu} \right\}. \tag{4.40}
$$

We can easily write down the susy transformation rules for the model, by analogy with the WZ model (in Dirac spinor notation):

$$
\delta A_\mu^a = \bar{\epsilon}\gamma_\mu \lambda^a
$$
$$
\delta\lambda^a = \left[-\frac{1}{2}\gamma^{\mu\nu} F_{\mu\nu}^a + i\gamma_5 D^a \right]\epsilon
$$
$$
\delta D^a = i\bar{\epsilon}\gamma_5 \slashed{D}\lambda^a. \tag{4.41}
$$

Indeed, δA_μ is similar to $\delta\phi = \bar{\epsilon}\psi$, just matching indices (so we need to introduce the constant matrix γ_μ). δD_a is also proportional to the gaugino field equation $\slashed{D}\lambda^a$, the only nontrivial thing is the introduction of the constant matrix $i\gamma_5$, but we saw already that $i\gamma_5$ can appear in the case we considered the two real scalars instead of a complex scalar for the WZ model. In $\delta\lambda^a$, $D^a\epsilon$ is clearly needed, and the $i\gamma_5$ was again not surprising from the real scalar version of the WZ model. The fact that we have $\gamma^{\mu\nu}F_{\mu\nu}^a\epsilon$ is understood from $\slashed{\partial}\phi\epsilon$ in the WZ model, now constrained also by gauge covariance and Lorentz invariance.

In fact, we see that from Lorentz invariance and gauge invariance/covariance, we can fix all possible terms, except of course the numerical coefficients. In principle we can leave the coefficients free, and then they are fixed by requiring susy invariance of the action. We should check the susy invariance of this action; however, we will find it in Chapter 5, from the point of view of the superspace formalism, which is manifestly supersymmetric.

4.4.3 Coupling to matter (chiral multiplets)

We consider the most general coupling of gauge multiplets with chiral multiplets. First, obviously, the normal derivatives ∂_μ in the action for the chiral multiplets become covariant derivatives D_μ.

Next, we add the coupling terms

$$
-D^a \phi^{\dagger i}(T_a)_{ij}\phi^j - i\sqrt{2}\phi^{\dagger i}(T_a)_{ij}\lambda^a \psi^j + i\sqrt{2}\bar{\psi}^i(T_a)_{ij}\bar{\lambda}^a \phi^j. \tag{4.42}
$$

We see that now the scalar potential has a new contribution. Indeed, now the auxiliary fields in the gauge multiplet, D^a, also become more complicated, yet their Lagrangian is still algebraic:

$$\mathcal{L}(D^a) = +\frac{D^a D^a}{2g^2} - D^a \phi^{\dagger i}(T_a)_{ij}\phi^j, \tag{4.43}$$

so by eliminating D^a,

$$D^a = -g^2 \phi^{\dagger i}(T^a)_{ij}\phi^j, \tag{4.44}$$

we obtain

$$\mathcal{L}(D^a) = -V(D^a) = -\frac{g^2}{2}\left[\phi^{\dagger i}(T^a)_{ij}\phi^j\right]^2. \tag{4.45}$$

This is called a "D-term."

The supersymmetry transformation rules get modified to (using two-component notation):

– for the WZ model,

$$\delta\phi^i = \bar{\epsilon}\psi^i$$
$$\delta\psi^i = \epsilon F^i + i\sigma^\mu\bar{\epsilon}D_\mu\phi^i$$
$$\delta\bar{\psi}^i = \bar{\epsilon}\bar{F}^i - i\epsilon\sigma^\mu D_\mu\phi^{\dagger i}$$
$$\delta F^i = i\epsilon\bar{\sigma}^\mu D_\mu\psi^i - \sqrt{2}i(T_a)_{ij}\phi^j\epsilon\bar{\lambda}^a. \tag{4.46}$$

– for the gauge multiplet,

$$\delta A_\mu^a = -i\bar{\epsilon}\bar{\sigma}^\mu\lambda^a + i\bar{\lambda}^a\bar{\sigma}_\mu\epsilon$$
$$\delta D^a = \bar{\epsilon}\bar{\sigma}^\mu D_\mu\lambda^a + D_\mu\bar{\lambda}^a\bar{\sigma}^\mu\epsilon$$
$$\delta\lambda^a = \frac{1}{2}\sigma^{\mu\nu}\epsilon F_{\mu\nu}^a + i\epsilon D^a$$
$$\delta\bar{\lambda}^a = \frac{1}{2}\bar{\epsilon}\bar{\sigma}^{\mu\nu}F_{\mu\nu}^a - i\bar{\epsilon}D^a. \tag{4.47}$$

In the presence of coupling to chiral multiplets, we can write a more general kinetic term for the gauge multiplet, introducing also a metric in field space (depending on the WZ scalars, though in the adjoint of the gauge group) for the gauge field multiplet, defined from a function $\mathcal{F}(\phi^i)$ through the "metric"

$$\mathcal{F}_{ab}(\phi^c) \equiv \frac{\partial^2\mathcal{F}}{\partial\phi^a\partial\phi^b}, \tag{4.48}$$

giving (assuming there are both scalars in the adjoint ϕ^a and other scalars ϕ^i)

$$-\frac{1}{4}\text{Re}\left[\mathcal{F}_{ab}(\phi)F_{\mu\nu}^a F^{a\mu\nu}\right] - \frac{1}{2}\text{Re}\left[\mathcal{F}_{ab}(\phi)\bar{\lambda}^a\not{D}\lambda^b\right] + \text{Re}\left[\mathcal{F}_{ab}(\phi)\frac{D^a D^b}{2}\right]$$
$$+\frac{\partial W}{\partial\phi^i}\left(\frac{\partial\mathcal{F}_{ab}}{\partial\phi^i}\right)\bar{\lambda}^a\lambda^b, \tag{4.49}$$

where we note that the terms on the first line become the usual kinetic terms when $\mathcal{F}_{ab} = \delta_{ab}$, while the terms on the second line vanish.

4.5 $\mathcal{N} = 2$ susy models

When considering $\mathcal{N} = 2$ supersymmetry, we must have two $\mathcal{N} = 1$ multiplets in the same representation of the gauge group.

For the $\mathcal{N} = 2$ vector multiplet, we have the $\mathcal{N} = 1$ vector multiplet, plus an $\mathcal{N} = 1$ chiral multiplet in the adjoint representation, so (ϕ^a, ψ^a, F^a). Then $(T^a)_{ij}$ becomes $(T^a)_{bc} = -if^a{}_{bc}$, which means we will form commutators inside a trace.

The auxiliary field terms are then

$$\frac{1}{g^2} \operatorname{Tr} \left(\frac{1}{2} D^2 + D[\phi^\dagger, \phi] + F^\dagger F \right), \tag{4.50}$$

where $D[\phi^\dagger, \phi]$ comes from $D^a \phi^{\dagger b}(T_a)_{bc} \phi^c$. Eliminating the auxiliary fields F^a, D^a, we find the scalar potential

$$V = -\frac{1}{2g^2} \operatorname{Tr} \left([\phi^\dagger, \phi] \right)^2. \tag{4.51}$$

Thus the Lagrangian for the $\mathcal{N} = 2$ vector multiplet for the on-shell fields is

$$\mathcal{L} = \frac{1}{g^2} \operatorname{Tr} \left\{ -\frac{1}{4} F_{\mu\nu} F^{\mu\nu} + g^2 \frac{\theta}{32\pi^2} F_{\mu\nu} \tilde{F}^{\mu\nu} + (D_\mu \phi)^\dagger D^\mu \phi \right.$$
$$\left. -\frac{1}{2} [\phi^\dagger, \phi]^2 - i\lambda \sigma^\mu D_\mu \bar{\lambda} - i\bar{\psi} \bar{\sigma}^\mu D_\mu \psi - i\sqrt{2} [\lambda, \psi] \phi^\dagger - i\sqrt{2} [\bar{\lambda}, \bar{\psi}] \phi \right\}. \tag{4.52}$$

The $\mathcal{N} = 2$ hypermultiplet is composed of one $\mathcal{N} = 1$ chiral multiplet Q and one $\mathcal{N} = 1$ anti-chiral multiplet $\tilde{Q}$ (with opposite charge under the coupling to the $\mathcal{N} = 2$ vector multiplet). We will not write its action here, as it is easier to write in the superspace formulation in Chapter 5.

4.6 $\mathcal{N} = 4$ SYM

The multiplet is composed of one $\mathcal{N} = 2$ vector multiplet and one $\mathcal{N} = 2$ hypermultiplet, giving one $\mathcal{N} = 1$ vector multiplet and three $\mathcal{N} = 1$ chiral multiplets.

The action and susy transformation rules are much easier to obtain from dimensional reduction from 10 dimensions. Indeed, in 10 dimensions, the $\mathcal{N} = 4$ SYM multiplet comes from a single $\mathcal{N} = 1$ SYM multiplet, with only kinetic terms.

In 10 dimensions, we have Majorana–Weyl spinors ψ_Π, with $\Pi = 1, ..., 16$, satisfying $\Gamma_{11} \psi = \psi$ (chirality) and $\bar{\psi} = \psi^T C_{10}$ (Majorana), as well as a vector A_M, $M = 0, 1, ..., 9$. The action is

$$S_{10d, \mathcal{N}=4SYM} = \int d^{10}x \left[-\frac{1}{4} F^{aMN} F^a_{MN} - \frac{1}{2} \bar{\lambda}^a \Gamma^M D_M \lambda^a \right], \tag{4.53}$$

and the susy transformation rules are the usual ones,

$$\delta A^a_M = \bar{\epsilon} \Gamma_M \lambda^a$$

$$\delta\lambda^a = -\frac{1}{2}\Gamma^{MN}F^a_{MN}\epsilon. \tag{4.54}$$

When dimensionally reducing, the C-matrix is a tensor product (like all matrices, including the gamma matrices), $C_{10} = C_4 \otimes C_6$, so we have the condition $\bar{\psi} = \psi^T C_4 \otimes C_6$, which means that the spinor must also be a tensor product, $\psi_\Pi = \pi_{Ai}$, $A = 1, ..., 4$ (four-dimensional Majorana spinor) and $i = 1, ..., 4$ (six-dimensional Majorana–Weyl spinor).

Moreover, the 10-dimensional gauge field splits into a 4-dimensional gauge field and 6 scalars, $A_M = (A_\mu, \phi_m)$, $\mu = 0, 1, 2, 3$ and $m = 1, ..., 6$. As we saw, the six scalars (in the fundamental of $SO(6)$) can be rewritten as six scalars in the antisymmetric of $SU(4)$, with a reality condition, and the relation between the two expressions is given in terms of a Clebsch–Gordan coefficient,

$$\phi_{[ij]} \equiv \phi_m \tilde{\gamma}^m_{[ij]}, \quad \tilde{\gamma}^m_{[ij]} \equiv \frac{1}{2}(C_6\gamma_m\gamma_7)_{[ij]}, \tag{4.55}$$

where $\gamma^m_{[ij]}\gamma_n^{[ij]} = \delta^m_n$, and the resulting reality condition on $\phi_{[ij]}$ is

$$\phi^\dagger_{ij} = \frac{1}{2}\epsilon^{ijkl}\phi_{kl}. \tag{4.56}$$

Substituting the dimensional reduction into the action, we obtain (the details are left as an exercise)

$$\begin{aligned}
S_{4d,\mathcal{N}=4SYM} &= \int d^4x \, \text{Tr}\left\{-\frac{1}{4}F_{\mu\nu}F^{\mu\nu} - \frac{1}{2}\bar{\psi}_i \slashed{D}\psi^i \right. \\
&\quad \left. -g\bar{\psi}^i[\phi_{ij}, \psi^j] - \frac{g^2}{4}[\phi_{ij}, \phi_{kl}][\phi^{ij}, \phi^{kl}]\right\} \\
&= \int d^4x \, \text{Tr}\left\{-\frac{1}{4}F_{\mu\nu}F^{\mu\nu} - \frac{1}{2}\bar{\psi}_i \slashed{D}\psi^i \right. \\
&\quad \left. -g\bar{\psi}^i[\phi_n, \psi^j]\tilde{\gamma}^n_{[ij]} - \frac{g^2}{4}[\phi_m, \phi_n][\phi^m, \phi^n]\right\}, \tag{4.57}
\end{aligned}$$

where we have used the trace normalization $\text{Tr}[T^aT^b] = \delta^{ab}$.

The susy transformation rules become

$$\begin{aligned}
\delta A^a_\mu &= \bar{\epsilon}_i\gamma_\mu\psi^{ai} \\
\delta\phi^{[ij]}_a &= 2\bar{\epsilon}^{[i}\psi^{j]a} \\
\delta\lambda^{ai} &= -\frac{1}{2}\gamma^{\mu\nu}F^a_{\mu\nu}\epsilon - 2\gamma^\mu D_\mu\phi^{a[ij]}\epsilon + 2gf^a_{\ bc}(\phi^b\phi^c)^{[ij]}\epsilon, \tag{4.58}
\end{aligned}$$

where $(\phi^b\phi^c)^{[ij]} \equiv \phi^{ai}_{\ k}\phi^{bk}_{\ j}$.

The $\mathcal{N} = 4$ SYM model has $SU(4)_4 = SO(6)_R$ symmetry, as advertised.

Important concepts to remember

- Massless irreps are constructed using the Wigner method and have $2^{\mathcal{N}}$ states.
- For $\mathcal{N} = 1$ susy, we have chiral (WZ) multiplets (ϕ, ψ) and vector multiples (A^a_μ, λ^a).

- For $\mathcal{N} = 2$ susy, we have vector multiplets, one $\mathcal{N} = 1$ vector plus one $\mathcal{N} = 1$ chiral multiplets, and hypermultiplets, two $\mathcal{N} = 1$ chiral multiplets, and for $\mathcal{N} = 4$ susy, we have the vector multiplet, one $\mathcal{N} = 2$ vector + one $\mathcal{N} = 2$ hypermultiplet.
- The higher susy algebras have central charges.
- For massive representations without central charges, we have $2^{2\mathcal{N}}$ states, and for massive representations with central charges, we have the BPS bound $M \geq |Z|$: generically we have $2^{2\mathcal{N}}$ states, but for BPS multiplets, with $M = |Z|$, we have again $2^{\mathcal{N}}$ states.
- The Lagrangian for the chiral multiplets is defined by the Kähler potential $K(\phi, \phi^\dagger)$, defining the metric $g_{i\bar{j}}$ on scalar field space, and the holomorphic superpotential $W(\phi)$, defining the scalar potential.
- When the gauge multiplets are coupled to chiral multiplets, the kinetic function is defined by the function $\mathcal{F}(\phi)$, through $\mathcal{F} = \partial_a \partial_b \mathcal{F}$.
- The $\mathcal{N} = 4$ SYM model can be easily derived from the 10-dimensional $\mathcal{N} = 1$ SYM with only free kinetic terms, via dimensional reduction.

References

Same as for Chapter 3, but I followed mostly [7] and [8].

Exercises

(1) Prove that, defining $Z^{IJ} = 2\epsilon^{IJ}Z$ and making the redefinitions for the $\mathcal{N} = 2$ susy algebra

$$a_\alpha = \frac{1}{\sqrt{2}}[Q^1_\alpha + \epsilon_{\alpha\beta}(Q^2_\beta)^\dagger]$$
$$b_\alpha = \frac{1}{\sqrt{2}}[Q^1_\alpha - \epsilon_{\alpha\beta}(Q^2_\beta)^\dagger], \tag{4.59}$$

we obtain for a massive representation in the rest frame

$$\{a_\alpha, a^\dagger_\beta\} = 2(M + |Z|)\delta_{\alpha\beta}$$
$$\{b_\alpha, b^\dagger_\beta\} = 2(M - |Z|)\delta_{\alpha\beta}, \tag{4.60}$$

and that this implies the *BPS bound* $M \geq |Z|$. In the above, take Z real (though the BPS bound is valid for complex Z).

(2) Check the invariance of the $\mathcal{N} = 1$ off-shell SYM action

$$S = \int d^4x \left[-\frac{1}{4}(F^a_{\mu\nu})^2 - \frac{1}{2}\bar{\psi}^a \slashed{D}\,\psi_a + \frac{1}{2}D^a D^a \right] \tag{4.61}$$

under the supersymmetry transformations

$$\delta A^a_\mu = \bar{\epsilon}\gamma_\mu\psi^a, \quad \delta D^a = i\bar{\epsilon}\gamma_5\slashed{D}\,\psi^a$$
$$\delta\psi^a = \left(-\frac{1}{2}\gamma^{\mu\nu}F^a_{\mu\nu} + i\gamma_5 D^a\right)\epsilon. \tag{4.62}$$

(3) Check the details of the dimensional reduction of the $\mathcal{N} = 1$ SYM action in $d = 10$ down to the $\mathcal{N} = 4$ SYM action in $d = 4$.

(4) Calculate the number of off-shell degrees of freedom of the on-shell $\mathcal{N} = 4$ SYM action in the text. Propose a set of bosonic + fermionic auxiliary fields that could make the number of degrees of freedom match. Are they likely to give an off-shell formulation, and why? Will this formulation be off-shell $\mathcal{N} = 4$ supersymmetric?

5 Introduction to supersymmetry 3: Superspace formalism in $d = 4$: Perturbative susy breaking

In this chapter, we continue our analysis of supersymmetry. If in the off-shell component formalism we had manifest supersymmetry, the action and supersymmetry transformation rules needed to be guessed.

However, there is another formalism that has manifest supersymmetry built in, namely the superspace formalism.

Instead of considering fields depending only on spacetime coordinates x^μ, we consider also a dependence on a fermionic coordinate θ^A so that we have *superfields* $\phi(x, \theta)$, in which case supersymmetry is manifest.

5.1 Superspace and superfields

Since fermions anticommute, $\{\theta, \theta\} = 0$, so $\theta^2 = 0$, it means that a function of a single fermion is expanded only up to linear order, $f(\theta) = a + b\theta$.

In $d = 4$, the fermion θ^A has four components, so $f(\theta)$ is expanded up to θ^4 (such that each component can appear in the product).

The $\mathcal{N} = 1$ supersymmetry algebra is

$$\{Q_\alpha, \bar{Q}_{\dot\alpha}\} = +2(\sigma^\mu)_{\alpha\dot\alpha} P_\mu, \quad \{Q_\alpha, Q_\beta\} = \{\bar{Q}_{\dot\alpha}, \bar{Q}_{\dot\beta}\} = 0. \tag{5.1}$$

It can be represented on the superfields, written in the two-component index notation as

$$\phi(z^M) = \phi(x, \theta) = \phi(x^\mu, \theta_\alpha, \bar{\theta}^{\dot\alpha}), \tag{5.2}$$

in terms of derivative operators

$$Q_\alpha = \partial_\alpha - i(\sigma^\mu)_{\alpha\dot\alpha} \bar{\theta}^{\dot\alpha} \partial_\mu,$$
$$\bar{Q}_{\dot\alpha} = -\partial_{\dot\alpha} + i(\sigma^\mu)_{\alpha\dot\alpha} \theta^\alpha \partial_\mu,$$
$$P_\mu = i\partial_\mu. \tag{5.3}$$

Note that $\partial_\alpha \equiv \partial/\partial\theta^\alpha$ and $\partial_{\dot\alpha} \equiv \partial/\partial\theta^{\dot\alpha}$ are fermions, so they anticommute.

The supersymmetry variations with parameter $(\xi_\alpha, \bar{\xi}^{\dot\alpha})$ (remember that in general $\delta_\epsilon = \epsilon^a T_a$) acting on z^M are

$$\delta z^M = (\xi Q + \bar{\xi}\bar{Q}) z^M, \tag{5.4}$$

or explicitly

$$x^\mu \to x'^\mu = x^\mu + i\theta\sigma^\mu\bar{\xi} - i\xi\sigma^\mu\bar{\theta},$$
$$\theta \to \theta' = \theta + \xi,$$
$$\bar{\theta} \to \bar{\theta}' = \bar{\theta} + \bar{\xi}. \tag{5.5}$$

But there is another representation of the supersymmetry algebra, just that the algebra with a different sign in the anticommutator, given by the "covariant derivatives,"

$$D_\alpha = \partial_\alpha + i(\sigma^\mu)_{\alpha\dot{\alpha}}\bar{\theta}^{\dot{\alpha}}\partial_\mu,$$
$$\bar{D}_{\dot{\alpha}} = -\partial_{\dot{\alpha}} - i(\sigma^\mu)_{\alpha\dot{\alpha}}\theta^\alpha\partial_\mu, \tag{5.6}$$

which satisfy

$$\{D_\alpha, \bar{D}_{\dot{\alpha}}\} = -2i(\sigma^\mu)_{\alpha\dot{\alpha}}\partial_\mu. \tag{5.7}$$

Note the different sign in front, as we said. Moreover, these covariant derivatives anticommute with the supersymmetry generators, $\{Q, D\} = 0$, as we can easily check.

General superfields of some Lorentz spin are reducible representations of supersymmetry. In order to obtain an irrep of susy, we need to introduce constraints that don't break supersymmetry, which means that the constraints must anticommute with the Q's. But we already know that the D's anticommute with the Q's, so imposing constraints involving the covariant derivatives will leave susy intact.

The simplest superfield that we can have is the complex scalar superfield $\Phi(x,\theta)$, but it is not an irrep, so to obtain an irrep of susy, we must impose the simplest constraint, the chirality constraint,

$$\bar{D}_{\dot{\alpha}}\Phi = 0, \tag{5.8}$$

obtaining a *chiral superfield*, or WZ superfield, which is indeed an irreducible representation of supersymmetry. Imposing the conjugate condition,

$$D_\alpha\Phi = 0, \tag{5.9}$$

we obtain an *anti-chiral superfield*.

In order to solve the constraint, we must find combinations of the superspace coordinates that solve it. We construct

$$y^\mu = x^\mu + i\theta\sigma^\mu\bar{\theta}, \tag{5.10}$$

and then we can easily check that

$$\bar{D}_{\dot{\alpha}}y^\mu = 0; \quad \bar{D}_{\dot{\alpha}}\theta^\beta = 0. \tag{5.11}$$

Of course, $\bar{D}_{\dot{\alpha}}\bar{\theta}^\beta \neq 0$. Then it means that the most general solution of the chiral constraint is a superfield that depends *independently and arbitrarily* on y^μ and θ^α, but not on $\bar{\theta}^\alpha$, $\Phi(y,\theta)$. We can then expand it up to θ^2 (since θ_α has two components), as

$$\Phi = \Phi(y,\theta) = \phi(y) + \sqrt{2}\theta\psi(y) + \theta\theta F(y). \tag{5.12}$$

Here we have defined $\theta^2 = \theta\theta = \theta^\alpha\theta_\alpha$, $\bar{\theta}^2 = \bar{\theta}\bar{\theta} = \bar{\theta}_{\dot{\alpha}}\bar{\theta}^{\dot{\alpha}}$.

We see that ϕ and ψ_α are the complex scalar and the arbitrary components of the Majorana spinor of the WZ model, and F is the auxiliary field of the same.

Since

$$\epsilon^{\alpha\beta} \frac{\partial}{\partial\theta^\alpha} \frac{\partial}{\partial\theta^\beta} \theta\theta = -4 \Rightarrow D^2\theta^2\big|_{\theta=\bar\theta=0}, \tag{5.13}$$

we find that we can obtain the component fields of the superfield through the action of covariant derivatives at zero thetas:

$$\phi(x) = \Phi\big|_{\theta=\bar\theta=0},$$

$$\psi_\alpha(x) = \frac{1}{\sqrt{2}} D_\alpha \Phi\big|_{\theta=\bar\theta=0},$$

$$F(x) = -\frac{1}{4} D^2\Phi\big|_{\theta=\bar\theta=0}. \tag{5.14}$$

We can now expand also the y's in the θ's and obtain

$$\Phi = \phi(x) + \sqrt{2}\theta\psi(x) + \theta^2 F(x)$$
$$+ i\theta\sigma^\mu\bar\theta\partial_\mu\phi(x) - \frac{i}{2}\theta^2(\partial_\mu\psi\sigma^\mu\bar\theta) - \frac{1}{4}\theta^2\bar\theta^2\partial^2\phi(x), \tag{5.15}$$

where we have expanded $\phi(y)$ to quadratic order in θ's, $\psi(y)$ only to linear order, since it was multiplied by a θ already, and we haven't expanded $F(y)$, since it was multiplied by θ^2 already.

5.2 Actions in terms of superfields: chiral multiplets

We remember that the fermionic integral acts like a derivative, since

$$\int d\theta\, 1 = 0; \qquad \int d\theta\,\theta = 1, \tag{5.16}$$

and define

$$d^2\theta = -\frac{1}{4}d\theta^\alpha d\theta^\beta \epsilon_{\alpha\beta}, \tag{5.17}$$

such that $\int d^2\theta\,(\theta\theta) = 1$.

For the calculation of actions, we need the following identities, for $d^2\theta$ and $d^2\bar\theta$ inside the spatial integral $\int d^4x$:

$$\int d^4x \int d^2\theta = -\frac{1}{4}\int d^4x\, D^2\big|_{\theta\theta=0} = -\frac{1}{4}\int d^4x\, D^\alpha D_\alpha\big|_{\theta=\bar\theta=0}$$

$$\int d^4x \int d^2\bar\theta = -\frac{1}{4}\int d^4x\, \bar D^2\big|_{\theta\bar\theta=0} = -\frac{1}{4}\int d^4x\, \bar D^\alpha \bar D_\alpha\big|_{\theta=\bar\theta=0}. \tag{5.18}$$

Note that in $\int d^4\theta = \int d^2\theta \int d^2\bar\theta$, the order of the D's and the $\bar D$'s matters, since they do not anticommute.

We can now write the most general action for the chiral superfield Φ. We can write an arbitrary function $K(\Phi, \Phi^\dagger)$, which must be integrated over the whole superspace $\int d^4\theta$, as it depends on both θ and $\bar{\theta}$. This will be called the *Kähler potential*. We will see later on in the book that the name is chosen because we obtain a certain type of complex geometry, that in mathematics is named Kähler geometry.

We can also write an arbitrary function of just Φ, but not $\Phi^\dagger$, $W(\Phi)$, which is integrated only over $\int d^2\theta$. Indeed, we can shift the integration over y^μ to integration over x^μ, leaving inside $\int d^4x \int d^2\theta\, W(\Phi)$ no $\bar{\theta}$'s to be integrated over. For reality of the action, we must add the hermitian conjugated term. Here $W(\Phi)$ is called the *superpotential*.

Then, the most general action for Φ is

$$\mathcal{L}_{\text{most general}} = \int d^4\theta\, K(\Phi, \Phi^\dagger) + \int d^2\theta\, W(\Phi) + \int d^2\bar{\theta}\, \bar{W}(\Phi^\dagger). \tag{5.19}$$

The Kähler potential term gives kinetic terms, since $\int d^4\theta$ will have a component with $\Box$ in it on the scalar, or with ∂_μ on the fermions (where $d^2\theta$ will be used to generate the fermions, and the others to obtain ∂_μ), and the superpotential will give nonderivative interactions.

In an effective $\mathcal{N} = 1$ supersymmetric field theory, below a UV cutoff, we can have anything for K and W. For a fundamental theory, however, we must have a renormalizable theory, which restricts a lot the possible terms. Then only possibilities are

$$K = \Phi^\dagger \Phi,$$
$$W = \lambda \Phi + \frac{m}{2}\Phi^2 + \frac{g}{3}\Phi^3. \tag{5.20}$$

Indeed, for a renormalizable coupling, the mass dimension must be positive or zero (otherwise, the coupling term blows up at large energies). But $[\phi] = 1$ and $[\psi] = 3/2$, which means that the superfield has $[\Phi] = 1$, and $[\theta] = -1/2$. Then $[\int d^2\theta] = [\partial/\partial\theta] = +1/2$. Finally, that means $[K] = 2$ and $[W] = 3$. Since $[K] = 2$ and it must be made from at least one Φ and one $\Phi^\dagger$, $K = \Phi^\dagger\Phi$ is unique. Moreover, then, for Φ^m with $m > 3$, we would have its coupling with negative mass dimension.

Note that the superpotential will give the $g\phi^4$, $m^2\phi^2$, and $g\phi\psi\psi$ (Yukawa) terms, as well as a cosmological constant (λ) term.

We now calculate the action. We have

$$\int d^4x \int d^2\theta \left(\lambda\Phi + \frac{m}{2}\Phi^2 + \frac{g}{3}\Phi^3\right) = -\frac{1}{4}\int d^4x\, D^2 \left(\lambda\Phi + \frac{m}{2}\Phi^2 + \frac{g}{3}\Phi^3\right)\Big|_{\theta=\bar{\theta}=0}. \tag{5.21}$$

Moreover, from

$$D^2(\Phi^2)\big|_{\theta=\bar{\theta}=0} = 2\, (D^2\Phi)\big|_{\theta=\bar{\theta}=0}\, \Phi\big|_{\theta=\bar{\theta}=0} + 2\, (D^\alpha\Phi)\big|_{\theta=\bar{\theta}=0}\, (D_\alpha\Phi)\big|_{\theta=\bar{\theta}=0}$$
$$D^2(\Phi^3)\big|_{\theta=\bar{\theta}=0} = 3\, (D^2\Phi)\big|_{\theta=\bar{\theta}=0}\, \Phi\big|_{\theta=\bar{\theta}=0}\, \Phi\big|_{\theta=\bar{\theta}=0}$$
$$+ 6\, (D^\alpha\Phi)\big|_{\theta=\bar{\theta}=0}\, (D_\alpha\Phi)\big|_{\theta=\bar{\theta}=0}\, \Phi\big|_{\theta=\bar{\theta}=0}, \tag{5.22}$$

and the definitions of the fields in terms of covariant derivatives, (5.14), we obtain

$$\int d^4x \int d^2\theta\, W(\Phi) = -\frac{1}{4}\int d^4x [2m\psi\psi + 4g\phi\psi\psi - 4F(\lambda + m\phi + g\phi^2)]. \tag{5.23}$$

For the Kähler potential term, we use the fact that (left as an exercise to prove) for a chiral superfield,

$$\bar{D}^2 D^2 \Phi = 16\Box\Phi \Rightarrow D^2\bar{D}^2\Phi^\dagger = 16\Box\Phi^\dagger. \tag{5.24}$$

We also remember that

$$\{D_\alpha, \bar{D}_{\dot\alpha}\} = -2i(\sigma^\mu)_{\alpha\dot\alpha}\partial_\mu, \tag{5.25}$$

from which we find

$$D^2\bar{D}^2 = \bar{D}^2 D^2 + 8i(\sigma^\mu)_{\alpha\dot\alpha}\partial_\mu\bar{D}^{\dot\alpha}D^\alpha + 16\Box. \tag{5.26}$$

Then we obtain for the Kähler potential term

$$\begin{aligned}
\frac{1}{16}\int d^4x D^2\bar{D}^2 \left.(\Phi^\dagger\Phi)\right|_{\theta=\bar\theta=0} = \frac{1}{16}\int d^4x \Big[&\left.(D^2\bar{D}^2\Phi^\dagger)\right|_{\theta=\bar\theta=0} \left.\Phi\right|_{\theta=\bar\theta=0} \\
&+ \left.(\bar{D}^2\Phi^\dagger)\right|_{\theta=\bar\theta=0} \left.(D^2\Phi)\right|_{\theta=\bar\theta=0} \\
&+ 8i(\sigma^\mu)^{\alpha\dot\alpha}\left.(\partial_\mu\bar{D}_{\dot\alpha}\ \Phi^\dagger)\right|_{\theta=\bar\theta=0} \left.(D_\alpha\Phi)\right|_{\theta=\bar\theta=0} \Big],
\end{aligned} \tag{5.27}$$

which gives the kinetic terms

$$\int d^4x[\phi^*\Box\phi + F^*F + i(\partial_\mu\bar\psi^{\dot\alpha})(\sigma^\mu)_{\alpha\dot\alpha}\psi^\alpha]. \tag{5.28}$$

Eliminating F from the Kähler potential term and the superpotential term, through its equations of motion

$$F^* = -(\lambda + m\phi + g\phi^2) = -\frac{\partial W}{\partial\phi}, \tag{5.29}$$

and replacing it in the action, we obtain the term

$$-\int d^4x \left|\lambda + m\phi + g\phi^2\right|^2 \equiv -V, \tag{5.30}$$

giving the scalar potential.

Then the action of the most general renormalizable model is

$$\begin{aligned}
S_{\text{renorm.}} = \int d^4x \Big[&\phi^*\Box\phi - i(\partial_\mu\bar\psi)(\sigma^\mu)^T\psi - m\bar\psi\psi \\
&-2\text{Re}[g\phi\bar\psi\psi] - \left|\lambda + m\phi + g\phi^2\right|^2 \Big].
\end{aligned} \tag{5.31}$$

5.3 Vector multiplet

The vector multiplet is represented, in terms of superfields, by a *real* scalar superfield V, so subject to the constraint $V = V^\dagger$. Then it is $V = V(x, \theta, \bar\theta)$, and it can be expanded up to $\theta^2\bar\theta^2$. The coefficient of $\theta\bar\theta$ will give the gauge field: $-\theta\sigma^\mu\bar\theta A_\mu$.

Since the gauge field appears in the expansion of V, we must have a gauge invariance for V, which *in the Abelian vector case* is

$$V \to V + i\Lambda - i\Lambda^\dagger, \tag{5.32}$$

where Λ is a chiral superfield (so $\Lambda^\dagger$ is anti-chiral). Because of this gauge invariance, we can set to zero the coefficients of the $\theta, \bar\theta$ (which are Majorana spinors) and $\theta^2, \bar\theta^2$ (which are complex scalars) terms, while the coefficient of the zeroth order term is an auxiliary scalar, that can also be put to zero. This gives the *Wess–Zumino gauge*, and in it we obtain the usual off-shell vector multiplet,

$$V = -\theta\sigma^\mu\bar\theta A_\mu + i\theta^2(\bar\theta\lambda) - i\bar\theta^2(\theta\lambda) + \frac{1}{2}\theta^2\bar\theta^2 D. \tag{5.33}$$

Note that in the WZ gauge we lose *manifest* superspace supersymmetry (as it happens whenever we fix a gauge), we only retain the component off-shell supersymmetry of Chapter 4.

Also note that, while there was a superfield gauge invariance, we have not fixed all of it, we retain the usual gauge invariance, $A_\mu \to A_\mu + \partial_\mu\lambda(x)$.

We can define the Abelian superfield strength,

$$W_\alpha = -\frac{1}{4}\bar{D}^2 D_\alpha V, \tag{5.34}$$

which is gauge invariant, since $\bar{D}_{\dot\alpha}\Lambda = D_\alpha\Lambda^\dagger = 0$. It is a chiral superfield, Majorana, and satisfies a reality condition, so

$$\bar{D}_{\dot\alpha}W_\alpha = 0, \quad (W_\alpha)^\dagger = W_{\dot\alpha}, \quad D^\alpha W_\alpha = D^{\dot\alpha}W_{\dot\alpha} \leftrightarrow \mathrm{Im}(D^\alpha W_\alpha) = 0. \tag{5.35}$$

In the WZ gauge, the Abelian field strength is expanded as

$$W_\alpha = -i\lambda_\alpha(y) - \theta_\alpha D - \frac{i}{2}(\sigma^\mu\bar\sigma^\nu\theta)_\alpha F_{\mu\nu} + \theta^2(\sigma^\mu\partial_\mu\bar\lambda)_\alpha. \tag{5.36}$$

In the non-Abelian case, with $V = V^a T_a$ and $\Lambda = \Lambda^a T_a$, the gauge transformation is

$$e^{-2V} \to e^{i\Lambda^\dagger}e^{-2V}e^{-i\Lambda}. \tag{5.37}$$

The non-Abelian field strength superfield is

$$W_\alpha = \frac{1}{8}\bar{D}^2\left(e^{2V}D_\alpha e^{-2V}\right) = W_\alpha^a T_a, \tag{5.38}$$

and it transforms covariantly, so

$$W_\alpha \to e^{i\Lambda}W_\alpha e^{-i\Lambda}. \tag{5.39}$$

In the WZ gauge, it is expanded as

$$W_\alpha = T_a\left[-i\lambda_\alpha^a(y) - \theta_\alpha D^a - \frac{i}{2}(\sigma^\mu\bar\sigma^\nu\theta)_\alpha F_{\mu\nu}^a + \theta^2(\sigma^\mu\partial_\mu\bar\lambda^a)_\alpha\right]. \tag{5.40}$$

The $\mathcal{N} = 1$ SYM action in superfield form is then

$$S_{\mathcal{N}=1\mathrm{SYM}} = -\frac{1}{4}\int d^4x\left[\int d^2\theta\,\mathrm{Tr}(W^\alpha W_\alpha) + h.c.\right], \tag{5.41}$$

where the hermitian conjugate is $\int d^2\bar\theta\,\bar{W}_{\dot\alpha}\bar{W}^{\dot\alpha}$.

We can also add a topological theta term that, as noted in Chapter 4, is supersymmetric by itself. But now there is a nice formalism that includes it. Define the complex coupling

$$\tau \equiv \frac{\theta}{2\pi} + \frac{4\pi i}{g^2}, \tag{5.42}$$

then the sum of the previous term and the theta term is

$$\mathcal{L} = -\frac{1}{8\pi}\text{Im}\left[\tau \int d^2\theta \, \text{Tr}(W^\alpha W_\alpha)\right]. \tag{5.43}$$

The scalar potential, in the case of coupling with chiral multiplets, is obtained, like in the component formalism, from a D-term (besides the F term for the chiral multiplets).

Before we write it, however, note that in the case of an *Abelian* theory, we can add a *Fayet–Iliopoulos (FI) term*,

$$\mathcal{L}_{\text{FI}} = \int d^2\theta \int d^2\bar{\theta}\xi^a V^a. \tag{5.44}$$

Therefore, when solving for the auxiliary field D^a, we obtain in the most general case

$$D^a = \xi^a + \phi^{\dagger i}(T^a)_{ij}\phi^j. \tag{5.45}$$

The superspace action for the vector multiplet coupled to a chiral multiplet with canonical kinetic term (so with $K = \Phi^\dagger\Phi$ in the absence of coupling to vectors) and a superpotential $W(\Phi)$ is

$$\mathcal{L} = -\frac{1}{8\pi}\text{Im}\left[\tau \int d^2\theta \, \text{Tr}(W^\alpha W_\alpha)\right] + \int d^2\theta \int d^2\bar{\theta}\Phi^{\dagger i}\left(e^{-2V}\right)_{ij}\Phi^j + \int d^2\theta W + \int d^2\bar{\theta}\bar{W}. \tag{5.46}$$

Note that the canonical Kähler potential is contained in the middle term in the above equation, for the case that $V = 0$.

5.4 $\mathcal{N} = 2$ superspace

We can write a formalism for a superspace that is $\mathcal{N} = 2$ supersymmetric, by introducing not only θ, but another one, $\tilde{\theta}$, as well. Then the chiral superfield Ψ is chiral with respect to both θ and $\tilde{\theta}$, that is,

$$\bar{D}_{\dot{\alpha}}\Psi = 0, \quad \tilde{\bar{D}}_{\dot{\alpha}}\Psi = 0. \tag{5.47}$$

For an irreducible representation, however, we need to impose a reality condition as well.

We can write the expansion of Ψ in the $\tilde{\theta}$ in the same way as for the expansion in θ for $\mathcal{N} = 1$, but with a small modification due to the reality condition. The coefficients of the expansion are now $\mathcal{N} = 1$ superfields in θ,

$$\Psi = \Phi(\tilde{y},\theta) + \sqrt{2}\tilde{\theta}^\alpha W_\alpha(\tilde{y},\theta) + \tilde{\theta}^2 G(\tilde{y},\theta), \tag{5.48}$$

but the $\tilde{y}^\mu$ now contains, symmetrically, both θ and $\tilde{\theta}$, even though we expanded only in $\tilde{\theta}$:

$$\tilde{y}^\mu \equiv x^\mu + i\theta\sigma^\mu\bar{\theta} + i\tilde{\theta}\sigma^\mu\tilde{\bar{\theta}}. \tag{5.49}$$

We see that we obtain the independent $\mathcal{N} = 1$ superfields Φ (chiral) and W_α (vector), and the extra superfield is just a function of them,

$$G(\tilde{y},\theta) = -\frac{1}{2}\int d^2\bar{\theta}\left[\Phi(\tilde{y},\theta)\right]^\dagger e^{-2V(\tilde{y},\theta)}. \tag{5.50}$$

The most general *local* action is then an arbitrary function of Ψ, $\mathcal{F}(\Psi)$, integrated over the doubly chiral measure $\int d^2\theta \int d^2\tilde{\theta}$, so

$$
\begin{aligned}
S &= \frac{1}{16\pi} \text{Im} \int d^4x \int d^2\theta \int d^2\tilde{\theta} \mathcal{F}(\Psi) \\
&= \frac{1}{32\pi} \text{Im} \left[\int d^2\theta \mathcal{F}_{ab}(\Phi) W^{a\alpha} W^b_\alpha + 2 \int d^2\theta d^2\bar{\theta} \left(\Phi^\dagger e^{2gV} \right)^a \mathcal{F}_a(\Psi) \right],
\end{aligned}
\tag{5.51}
$$

where we have defined the derivatives

$$
\mathcal{F}_a(\Psi) \equiv \frac{\partial \mathcal{F}}{\partial \Psi^a}, \quad \mathcal{F}_{ab}(\Psi) \equiv \frac{\partial^2 \mathcal{F}}{\partial \Psi^a \partial \Psi^b}.
\tag{5.52}
$$

Then, the Kähler potential is

$$
K = \text{Im} \left(\Phi^{\dagger a} \mathcal{F}_a(\Phi) \right),
\tag{5.53}
$$

and the metric

$$
g_{ab} = \text{Im}(\partial_a \partial_b \mathcal{F})
\tag{5.54}
$$

defines a *special Kähler geometry* that will be defined later on in the book (and from where the name Kähler potential comes).

If we are interested in an effective theory, below a UV cutoff, then again $\mathcal{F}(\Psi)$ can be anything, but if we want a fundamental theory, therefore, a renormalizable and local one, there is now a unique function available:

$$
\mathcal{F}_{\text{fund.}}(\Psi) = \frac{\tau}{2} \text{Tr} \, \Psi^2.
\tag{5.55}
$$

Note that we could have written also a term

$$
\int d^4x d^4\theta d^4\bar{\theta} \mathcal{H}(\Psi),
\tag{5.56}
$$

and in fact we can write it for effective theories (which are also not at low energies), except that the term is nonlocal and/or nonrenormalizable.

Indeed, since $[\int d^4\theta \int^4 \tilde{\theta}] = 4$, the mass dimension of $\mathcal{H}$ is zero, $[\mathcal{H}] = 0$. But since $\Psi = \Phi + ... = \phi + ...$, $[\Psi] = 1$, a term with 2 Ψ's (the minimum possible) must have either a nonrenormalizable coupling g with $[g] < 0$ or a term with $1/\square$, $\phi \frac{1}{\square} \phi$, which is nonlocal.

For the most general $\mathcal{N} = 2$ preserving coupling of N_f hypermultiplets (with index i) in the fundamental representation of the gauge group G, with the vector multiplet, we cannot write an action in $\mathcal{N} = 2$ superspace, only an action in $\mathcal{N} = 1$ superspace (where the hypermultiplet splits into Q and $\tilde{Q}$, and the vector into V and Φ), which is

$$
\int d^2\theta \int d^2\bar{\theta} \left[Q^\dagger_i e^{-2V} Q_i + \tilde{Q}_i e^{2V} \tilde{Q}^\dagger_i \right] + \left[\int d^2\theta \left(\sqrt{2} \tilde{Q}_i \Phi Q_i + m_i \tilde{Q}_i Q_i \right) + h.c. \right].
\tag{5.57}
$$

The unique $\mathcal{N} = 4$ SYM theory, described by Ψ coupled to one massless hypermultiplet in the adjoint representation, contains the function $\mathcal{F}$ for a fundamental theory, $\mathcal{F}(\Psi) = \frac{\tau}{2} \Psi^2$, plus the above coupling to the hypermultiplets. We can treat all $\mathcal{N} = 1$ chiral superfields, $\Phi, Q, \tilde{Q}$, on an equal footing, writing Φ^i, with $i = 1, 2, 3$, and then the $\mathcal{N} = 1$ superpotential is

$$
W = \epsilon_{ijk} \text{Tr} \left(\Phi^i [\Phi^j, \Phi^k] \right).
\tag{5.58}
$$

Note that there is no (Lorentz invariant) formulation of an $\mathcal{N} = 4$ superspace in which to write $\mathcal{N} = 4$ SYM.

5.5 Perturbative supersymmetry breaking

We now say a few things about perturbative supersymmetry breaking. For that, we must start with some generalities about supersymmetry breaking.

5.5.1 Witten index

Witten introduced a quantity, now called the Witten index, that measures whether a theory breaks supersymmetry or not, whether we can find an explicit susy breaking mechanism or not (just from the general properties of the theory). We will now describe it.

Since the susy algebra is of the type $\{Q, Q\} \propto H$, and the vacuum should be supersymmetric by itself, so $Q|0\rangle = 0$, which means that the norm of the state is zero also, $||Q|0\rangle|| = 0$, by multiplying the susy algebra with $\langle 0|$ from the left and with $|0\rangle$ from the right, we obtain zero on the left-hand side, and on the right-hand side, $E_0 \langle 0|0\rangle = E_0$, which means that we need to have $E_0 = 0$. This is then a *necessary condition for unbroken supersymmetry*.

At nonzero energy level, we have the same number of bosonic and fermionic states, since

$$Q|b\rangle = \sqrt{E}|f\rangle, \quad Q|f\rangle = \sqrt{E}|b\rangle, \tag{5.59}$$

which indeed satisfies $\{Q, Q\} \propto H$. Note that this means that the supersymmetry invariant state is a linear combination of the $|b\rangle$ and $|f\rangle$ states. Then the above means also that for $E \neq 0$, the number of bosonic states equals the number of fermionic states,

$$n_B^{E \neq 0} = n_F^{E \neq 0}. \tag{5.60}$$

On the other hand, for the ground state, $Q|b\rangle = 0$ and $Q|f\rangle = 0$, where now $|b\rangle$ and $|f\rangle$ stand for zero modes of the vacuum, that is, $b_{\text{vac}}^\dagger |0\rangle$ and $f_{\text{vac}}^\dagger |0\rangle$, respectively. Then $n_B^{E=0} - n_F^{E=0}$ need not be zero and is invariant under small changes of parameters (for instance, if one boson state gets a small mass, so does a fermionic state, and vice versa).

This means that we can define the *Witten index*,

$$\text{Tr}(-1)^F = \langle i|(-1)^F|i\rangle = n_B^{E=0} - n_F^{E=0}, \tag{5.61}$$

where F is the fermion number (note that for the massive states n_B and n_E cancel against each other, as we said). This Witten index is independent of the parameters of the theory (is invariant under small changes of parameters) and characterizes susy breaking. Note that, since the fermions are characterized by getting a -1 under a rotation by 2π around an axis, say z, we can also write

$$\text{Tr}(-1)^F = \text{Tr}\, e^{2\pi i J_z}. \tag{5.62}$$

Then, we have the following possible situations:

- $\text{Tr}(-1)^F \neq 0$. In this case, susy is unbroken, since this can only be possible if $E_0 = 0$.
- $\text{Tr}(-1)^F = 0$, but $n_F^{E=0} = n_F^{E=0} \neq 0$. In this case, susy is still unbroken, since it means that there is an operator C such that $\text{Tr}[(-1)^F C] \neq 0$, and all that means is that the fermion number must be redefined.
- $n_B^{E=0} = n_F^{E=0} = 0$, which means that susy is broken. But this means, of course, that the true vacuum has higher energy, $E_0 > 0$.

Then for a supersymmetric Higgs gauge theory (like for MSSM, for instance), we have the following possible situations:

- If the true vacuum is at $\phi = 0$ and has $E = 0$ (and maybe a false vacuum at $\phi \neq 0, E \neq 0$), susy is unbroken, gauge symmetry is unbroken.
- If the true vacuum is at $\phi = 0$ and has $E_1 > 0$ (and maybe a false vacuum at $\phi \neq 0, E_2 > E_1$), susy is broken, gauge symmetry is unbroken.
- If the false vacuum is at $\phi = 0$ and has $E_1 > 0$, but the true vacuum is at $\phi \neq 0$ and has $E_2 = 0$, susy is unbroken, but gauge symmetry is broken.
- If the false vacuum is at $\phi = 0$ and has $E_1 > 0$ and the true vacuum is at $\phi \neq 0$ and has $0 < E_2 < E_1$, susy is broken, gauge theory is broken.

5.5.2 Tree-level susy breaking

Finally, we come to susy breaking at the tree level (in the action). As we saw, unbroken susy means that minimizing $V(\phi)$, we should find $E_{\min} = 0$. If not, susy is broken. Since the scalar potential is the sum of squares, being the F terms and the D terms, it means that for unbroken susy, we need to put both the F terms and the D terms to zero,

$$F_i = \frac{\partial W}{\partial \Phi_i} = 0, \quad D^a = \xi^a + \Phi^{\dagger i}(T^a)_{ij}\Phi^j = 0. \tag{5.63}$$

Note that the number of equations (labelled by i and a) is equal to the number of unknowns (Φ_i and ξ^a). This means that for generic $W(\Phi)$, we can find solutions, and we have unbroken susy.

- If there are no $U(1)$ factors, it is generically possible to find solutions.
- If there are $U(1)$ factors, but the FI terms $\xi^a = 0$, then we must put $F^i = D^a = 0$, and again it is generically possible to find solutions.

There are two standard cases of tree-level susy breaking: the O'Raifeartaigh model and the Fayet–Iliopoulos mechanism.

The O'Raifeartaigh model is based on a nongeneric superpotential, for which there is no minimum to W. Indeed, consider two sets of scalars Φ, X_n, and Y_i, with superpotential

$$W = \sum_i Y_i f_i(X_n). \tag{5.64}$$

Then the vanishing of the F terms means

$$\frac{\partial W}{\partial Y_i} = f_i(X_n) = 0, \quad \frac{\partial W}{\partial X_n} = \sum_i Y_i \frac{\partial f_i}{\partial X_n} = 0. \tag{5.65}$$

In the second set of equations, we can always put $Y_i = 0$ and thus solve them, but then if there are more Y_i's, than X_n's, we cannot solve the equations $f_i(X_n) = 0$ (since there are more equations than variables). Thus we have no minimum for W (no way to put the F terms to zero).

The Fayet–Iliopoulos mechanism corresponds to the case when we have a $U(1)$ factor with two chiral superfields of opposite charge with respect to it, $\pm e$, namely Φ_+ and Φ_-. Then the most general renormalizable and $U(1)$ symmetric superpotential is

$$W = m\Phi_+\Phi_-, \tag{5.66}$$

since we cannot have zero charge with one or three Φ's. The scalar potential coming from it, if we have also an FI term ξ, is

$$V(\Phi_+, \Phi_-) = m^2|\phi_+|^2 + m^2|\phi_-|^2 + (\xi + e^2|\phi_+|^2 - e^2|\phi_-|^2)^2. \tag{5.67}$$

We see that if $\xi \neq 0$, there is no minimum at $V = 0$, since if we put $\phi_+ = \phi_- = 0$ to cancel the F terms, the D term does not vanish (and if we vanish the D term, the F terms don't).

Finally, note that if we have tree-level supersymmetry breaking, one can prove that we have the "sum rules" (valid separately for each conserved charge)

$$\sum_{\text{spin}0} \text{mass}^2 - 2 \sum_{\text{spin}1/2} \text{mass}^2 + 3 \sum_{\text{spin}1} \text{mass}^2 = 0. \tag{5.68}$$

But in the case of the Standard Model (+ more), this contradicts experiment! This means that in the Standard Model, if we extend it to some supersymmetric model like MSSM, susy cannot be broken at tree level. In fact, the argument can be extended to perturbation theory.

On the other hand, since the couplings of the Standard Model are perturbative, and the only way to break the susy is non-perturbative, that led to the current model (which will be explained later on in the book in more detail), where there is a "hidden sector" where susy is broken non-perturbatively, and the susy breaking is translated to the visible sector (the Standard Model or MSSM) via some "messenger fields," which can be supergravity fields.

Important concepts to remember

- Superspace is made up of the usual space x^μ and a spinorial coordinate θ^α.
- Superfields are fields in superspace and can be expanded up to linear order in θ components, $f(\theta) = a + b\theta$, since $\theta^2 = 0$.
- Irreducible representations of susy are obtained by imposing constraints in terms of the covariant derivatives D on superfields, since the D's commute with the susy generators Q's, thus preserve susy.
- A chiral superfield is an arbitrary function $\Phi(y, \theta)$ of $y^\mu = x^\mu + i\theta\sigma^\mu\bar{\theta}$ and θ.
- Fermionic integrals and derivatives are the same.

- The action for a chiral superfield has a function $K(\Phi, \bar{\Phi})$, called Kähler potential giving kinetic terms, and a function $W(\Phi)$ called superpotential, giving potentials and Yukawas.
- To derive the component Lagrangian from the superfield one, we can either do the full θ expansion or (simpler) use the fact that $\int d^4x \int d^2\theta = -1/4 \int d^4x D^2|_{\theta=0}$ (and its c.c.) and the definitions $\phi(x) = \Phi|_{\theta=0}$, $\psi(x) = 1/\sqrt{2}D_\alpha\Phi|_{\theta=0}$, and so on, but we need to be careful with the Kähler potential.
- The $\mathcal{N} = 2$ superfields are a double expansion of the $\mathcal{N} = 1$ type.
- By imposing a double chiral condition, we obtain the $\mathcal{N} = 2$ vector superfield Ψ, made up of an $\mathcal{N} = 1$ vector and a chiral superfield, and defined by a function $\mathcal{F}(\Psi)$ (and an $\mathcal{H}(\Psi, \Psi^\dagger)$ if the theory is effective).
- The other possible $\mathcal{N} = 2$ supermultiplet is the hypermultiplet, made up of two chiral $\mathcal{N} = 1$ superfields.
- The $\mathcal{N} = 2$ vector and hypermultiplets together make up the unique $\mathcal{N} = 4$ supermultiplet of spin ≤ 1, the vector.
- The (necessary) condition for unbroken susy is $E_0 = 0$.
- The Witten index $\mathrm{Tr}(-1)^F$ decides whether susy is broken or not: if it is nonzero, susy is unbroken. If it is zero, and there are no C's such that $\mathrm{Tr}[(-1)^F C] \neq 0$, then susy is broken. Equivalently, if $E_0 > 0$, susy is broken.
- For tree-level susy breaking, we must put $F_i = 0$ and $D^a = 0$ for unbroken susy. If there are no solutions, susy is broken.
- The O'Raifeartaigh model (W has no minimum) and the Fayet–Iliopoulos mechanism (nonzero FI terms) are examples of tree-level susy breaking.
- In the Standard Model extensions, susy is not broken at tree level.

References

Same as for Chapters 3 and 4, but I followed mostly [7] and [8].

Exercises

(1) Show that the superspace coupling of the $\mathcal{N} = 1$ gauge to the $\mathcal{N} = 1$ chiral multiplet gives the component terms in the action of the type explained in Chapter 4.

(2) Check explicitly (without the use of y^μ) that $\bar{D}_{\dot{\alpha}}\Phi = 0$, where

$$\Phi = \phi(x) + \sqrt{2}\psi(x) + \theta^2 F(x) + i\theta\sigma^\mu\bar{\theta}\partial_\mu\phi(x) - \frac{i}{2}\theta^2(\partial_\mu\psi\sigma^\mu\bar{\theta}) - \frac{1}{4}\theta^2\bar{\theta}^2\partial^2\phi(x). \quad (5.69)$$

(3) Prove that for a chiral superfield

$$D^2\bar{D}^2\Phi = 16\Box\Phi. \quad (5.70)$$

(4) Consider the Lagrangian

$$\mathcal{L} = \int d^2\theta d^2\bar\theta \, \Phi_i^\dagger \Phi_i + \left(\int d^2\theta \, W(\Phi_i) + h.c. \right). \tag{5.71}$$

Do the θ integrals to obtain in components

$$\mathcal{L} = (\partial_\mu A_i)^\dagger \partial^\mu A_i - i\bar\psi_i \bar\sigma^\mu \partial_\mu \psi_i + F_i^\dagger F_i +$$
$$+ \frac{\partial W}{\partial A_i} F_i + \frac{\partial \bar W}{\partial A_i^\dagger} F_i^\dagger - \frac{1}{2} \frac{\partial^2 W}{\partial A_i \partial A_j} \psi_i \psi_j - \frac{1}{2} \frac{\partial^2 \bar W}{\partial A_i^\dagger \partial A_j^\dagger} \bar\psi_i \bar\psi_j. \tag{5.72}$$

Four-dimensional on-shell supergravity and how to count degrees of freedom

In this chapter we will begin the study of supergravity with the simplest, $\mathcal{N} = 1$, on-shell four-dimensional supergravity. Supergravity is understood both as a supersymmetric theory of gravity, as well as a theory of local supersymmetry. In practice, we will use a bit of both definitions in order to construct it. As a supersymmetric theory of gravity, it will contain the gravitational field (of spin 2), and other fields of lesser spin. But before we do that, however, we must understand how to count degrees of freedom for more general fields, since to construct a supergravity we will need (as for any supersymmetric theory) to match the number of fermionic degrees of freedom with the number of bosonic degrees of freedom, $n_F = n_B$.

6.1 Degrees of freedom counting

6.1.1 Off-shell counting

- Scalar ϕ. For a scalar, either propagating (which means with kinetic terms involving two derivatives; one can consider also kinetic terms with more than two derivatives, but those are very special cases) or auxiliary (with algebraic equation of motion), off-shell we have one degree of freedom.
- Gauge field A_μ. A gauge field has gauge invariance $\delta A_\mu = \partial_\mu \lambda$, which is a redundancy that means we can put one component of A_μ to zero, leaving $d - 1$ independent degrees of freedom.
- Graviton $g_{\mu\nu}$. It is a symmetric matrix, so with $d(d+1)/2$ components, but it also has a gauge invariance, the general coordinate invariance, with a parameter ξ^μ, so containing d components. Therefore the number of independent components is $d(d + 1)/2 - d = d(d - 1)/2$ degrees of freedom.
- Spinor of spin 1/2 λ_α. A Majorana spinor, as we are considering, has $n \equiv 2^{[d/2]}$ real components.
- Gravitino, of spin 3/2, $\psi_{\mu\alpha}$. This is the field that will be the superpartner of the graviton. Since it has one vector (μ) and one spinor (α) indices, it has $n \cdot d$ components. However, it is also subject to a "gauge invariance," it being the gauge field of local supersymmetry = supergravity. We will see that it is $\delta\psi_\mu = D_\mu\epsilon$, with a parameter ϵ_α that is a spinor of spin 1/2, with n components, so ψ_μ has $n \cdot (d - 1)$ degrees of freedom off-shell.
- Antisymmetric tensor $A_{\mu_1...\mu_r}$. It has a field strength $F_{\mu_1...\mu_{r+1}} = (r + 1)\partial_{[\mu_1}A_{\mu_2...\mu_{r+1}]}$, so it is subject to a gauge invariance

$$\delta A_{\mu_1 \dots \mu_r} = \partial_{[\mu_1} \lambda_{\mu_2 \dots \mu_r]}, \quad \lambda_{\mu_2 \dots \mu_r} \neq \partial_{[\mu_2} \epsilon_{\mu_3 \dots \mu_r]}, \tag{6.1}$$

so the parameter has $\binom{d}{r-1}$ components, but not those that can be expressed as derivatives, removing one value from d to leave $d - 1$, so all in all

$$\binom{d}{r} - \binom{d-1}{r-1} = \frac{(d-1) \dots (d-r)}{1 \cdot 2 \cdot \dots r} = \binom{d-1}{r}, \tag{6.2}$$

so for $A_{\mu_1 \dots \mu_r}$ with indices taking $d - 1$ values instead of d values.

6.1.2 On-shell counting

- Scalar ϕ. The KG equation, $(\Box - m^2)\phi = 0$, or $(p^2 + m^2)\phi = 0$, doesn't constrain anything, only the functional form of the degree of freedom, so there is still one degree of freedom for a dynamical (propagating) scalar. For an auxiliary scalar, we have no degrees of freedom.
- Gauge field A_μ. The equation of motion is $\partial^\mu(\partial_\mu A_\nu - \partial_\nu A_\mu) = 0$, and in principle we should analyze what it does to the field. But there is a shortcut: If we use the Lorenz (covariant) gauge $\partial^\mu A_\mu = 0$, the equation of motion becomes $\Box A_\nu = 0$, the KG equation, which we saw that doesn't kill degrees of freedom. However, to get to it, we used one constraint, the Lorenz gauge, which restricts the polarizations of the field, $k^\mu \epsilon_\mu(k) = 0$, hence the number of degrees of freedom on-shell is less by one, namely $d - 2$. We see that the gauge field has only transverse components (the longitudinal A_z and timelike component A_0 are not propagating), expressed by $k^\mu \epsilon_\mu(k) = 0$.
- Graviton $g_{\mu\nu}$. More precisely, we consider the fluctuation $h_{\mu\nu} = g_{\mu\nu} - \eta_{\mu\nu} \ll 1$. For these small fluctuations, the action is the Fierz–Pauli action,

$$\mathcal{L} = -\frac{1}{2} h^2_{\mu\nu,\rho} + h^2_\mu - h^\mu h_{,\mu} + \frac{1}{2} h^2_{,\mu}, \tag{6.3}$$

where $h_\mu \equiv \partial^\nu h_{\nu\mu}$, $h \equiv h^\mu_\mu$ and $h_{,\mu} = \partial_\mu h$, $h_{\mu\nu,\rho} = \partial_\rho h_{\mu\nu}$.

Again we should analyze the effect of the complicated equation of motion coming from the above action, but there is the same trick (shortcut). We can find an analog of the Lorenz gauge, called the de Donder gauge condition,

$$\partial^\mu \bar{h}_{\mu\nu} = 0, \quad \bar{h}_{\mu\nu} = h_{\mu\nu} - \eta_{\mu\nu} \frac{h}{2}, \tag{6.4}$$

and in this gauge, the equation of motion is just KG: $\Box \bar{h}_{\mu\nu} = 0$, which restricts just the functional form (restricts k^2 to 0), but not the degrees of freedom. Then, we have d conditions for the Lorenz gauge, $k^\mu \epsilon_{\mu\nu}(k) = 0$, in terms of polarizations, which means that we have

$$\frac{d(d-1)}{2} - d = \frac{(d-1)(d-2)}{2} - 1 \tag{6.5}$$

degrees of freedom, or transverse ($k^\mu \epsilon_{\mu\nu}(k) = 0$) and traceless (since $\bar{h}_{\mu\nu}$ is the KG field).

- Spinor of spin 1/2 λ_α. The Dirac equation in momentum space is

$$(\gamma^\mu p_\mu - m)u(k) = 0, \qquad (6.6)$$

  and it relates half of the components of the spinor with the other half, so it halves the number of degrees of freedom, and we are left with $n/2$.

- Gravitino $\psi_{\mu\alpha}$, of spin 3/2. Naively this looks like a spinor times a gauge field, so we would say that we have $\frac{n}{2}(d-2)$ degrees of freedom: the $n(d-1)$ degrees of freedom would be halved by a Dirac equation, but that would be obtained only by imposing a spinor constraint analogous to the Lorenz gauge $\partial^\mu \psi_\mu = 0$, so $\frac{n}{2}(d-1) - \frac{n}{2}$.

  However, there is a subtlety. $\psi_{\mu\alpha}$ is not an irreducible representation. We can see this easily, since $1 \otimes 1/2 = 3/2 \oplus 1/2$ (in terms of spin) or, in terms of multiplicities, $3 \otimes 2 = 4 \oplus 2$. That means that, in order to obtain only the spin 3/2 irrep, we must remove the spin 1/2 one, which is just the gamma-trace, $\gamma^\mu \psi_\mu$, and put it to zero. But this gives a condition on $n/2$ components, so in total we have now $\frac{n}{2}(d-3)$ components.

- Antisymmetric tensor $A_{\mu_1\ldots\mu_r}$. This is a generalization of the gauge field, so the same logic also applies now. We could analyze the full equation of motion, but it easier to take the shortcut and note that by imposing the Lorenz-like gauge condition $\partial^{\mu_1} A_{\mu_1\mu_2\ldots\mu_r} = 0$ (in the gauge condition, μ must be different from the others $r-1$ indices, which removes one value from the $d-1$ possibilities), and we are left with only the KG equation, $\Box A_{\mu_1\ldots\mu_r} = 0$, which doesn't lose degrees of freedom. Thus we have $\left(d-2//r-1\right)$ constraints, for a total of

$$\binom{d-1}{r} - \binom{d-2}{r-1} = \binom{d-2}{r} \qquad (6.7)$$

  degrees of freedom, or only transverse indices in $A_{\mu_1\ldots\mu_r}$, as defined by the transversality condition $k^\mu \epsilon_{\mu\mu_2\ldots\mu_r} = 0$.

6.2 Supergravity

Supergravity is defined either as a supersymmetric theory of gravity or as a theory of local supersymmetry, but in practice we used both definitions to construct it.

Since it is a theory of local supersymmetry, the susy parameter ϵ_α becomes local, $\epsilon_\alpha(x)$. But, when we make a global symmetry local, like for instance for a complex scalar field $-\int |\partial_\mu \phi|^2$, invariant under $\phi \to e^{i\alpha}\phi$, for which when we put $\alpha = \alpha(x)$, we need to introduce covariant derivatives with a gauge field, $D_\mu = \partial_\mu - ieA_\mu$.

Now the equivalent of "$A_\mu^\alpha(x)$" is the gravitino field $\psi_\mu^\alpha(x)$. Note that here μ is a curved spacetime index, but α is a local Lorentz, so flat, index. Of course, in flat space, $\psi_{\mu\alpha}$ (with $\gamma^\mu \psi_\mu = 0$) is a spin 3/2 field, where μ and α are the same kind of indices.

On the other hand, for a supersymmetric theory of gravity, $\delta\psi_{\mu\alpha}$ must equal Q_α(gravity), which can only mean that gravity is represented by the vielbein e_μ^a (object with a single curved index, μ). And we have fermions, so we must consider the spin connection ω_μ^{ab}, so we must use the vielbein-spin connection formulation of general relativity.

We now count degrees of freedom to see what fields there are in supergravities in three and four dimensions.

Three dimensions

- On-shell, the graviton vielbein e_μ^a has $(d-1)(d-2)/2 - 1 = 1 \cdot 2/2 - 1 = 0$, so no degrees of freedom. But also the gravitino $\psi_{\mu\alpha}$ has $n(d-3)/2 = \frac{2}{2}(3-3) = 0$ degrees of freedom, so we have a match without any other fields need, albeit having a somewhat trivial theory.
- Off-shell, it is less trivial. The graviton e_μ^a has $d(d-1)/2 = 2 \cdot 3/2 = 3$ degrees of freedom, while the gravitino $\psi_{\mu\alpha}$ has $n(d-1) = 2(3-1) = 4$ degrees of freedom. That means that we need an auxiliary scalar field, S, to match degrees of freedom, for a multiplet $\{e_\mu^a, S, \psi_{\mu\alpha}\}$.

Four dimensions

- On shell, we have the first nontrivial model. The graviton e_μ^a has $2 \cdot 3/2 - 1 = 2$ degrees of freedom, while the gravitino has $4/2(4-3) = 2$ degrees of freedom, so again we have a match without other fields being needed.
- Off-shell, the graviton e_μ^a has $4 \cdot 3/2 = 6$ degrees of freedom, while the gravitino has $4(4-1) = 12$ degrees of freedom. That means that we need still six auxiliary bosonic degrees of freedom, which will be the minimal choice. But we could also choose, for instance, 4 auxiliary fermionic degrees of freedom (an auxiliary Majorana spinor) and 10 bosonic auxiliary degrees of freedom. For the minimal choice, the minimal set of bosonic auxiliary fields are two real scalar S and P and a vector A_μ, so $\{S, P, A_\mu\}$. We could form a complex auxiliary scalar, $M = S + iP$.

6.3 $\mathcal{N} = 1$ on-shell four-dimensional supergravity

We now consider the simplest nontrivial model of on-shell supergravity, $\mathcal{N} = 1$ in four dimensions, with only fields e_μ^a and $\psi_{\mu\alpha}$.

To construct it, we find it easier to start with the susy transformation rules. Remembering that for a scalar, we had $\delta\phi = \bar{\epsilon}\psi$, and for a vector, we had $\delta A_\mu = \bar{\epsilon}\gamma_\mu\psi$, now for the vielbein transforming into the gravitino, matching indices, the only possibility is

$$\delta e_\mu^a = \frac{\kappa_N}{2}\bar{\epsilon}\gamma^a\psi_\mu, \qquad (6.8)$$

where κ_N is the Newton constant, added for dimensional reasons, and the 1/2 is a convenient normalization, but otherwise the transformation is fixed.

The gravitino is a gauge field of local susy, so it should transform into a derivative of the parameter, like $\delta A_\mu = \partial_\mu\epsilon$. Just that now we have gravity, so the derivative must be gravitationally covariant, and the parameter is a spinor, so

$$\delta\psi_\mu = \frac{1}{\kappa_N}D_\mu\epsilon, \qquad (6.9)$$

where again κ_N is for dimensional reasons and

$$D_\mu \epsilon = \partial_\mu \epsilon + \frac{1}{4} \gamma_{ab} \omega_\mu^{ab} \epsilon. \tag{6.10}$$

Those are the susy transformation rules for the $\mathcal{N} = 1$ model. Of course, for higher $\mathcal{N}$, we would have extra terms.

Next, we write the action. The action for gravity must be the Einstein–Hilbert action, and naively we would want to write

$$S_{\text{E--H}} = \frac{1}{2\kappa_N^2} \int d^d x \sqrt{-g} R(\Gamma), \tag{6.11}$$

where we could take $\Gamma = \Gamma(g)$ in a second-order formulation, or an independent Γ in the (original form of the) Palatini formalism.

However, we have spinors in the theory, so we must have ω_μ^{ab}, and we also saw the need for e_μ^a. Then the action we need is actually

$$S_{\text{EH}} = \frac{1}{2k_N^2} \int d^d x (\det e) R_{\mu\nu}^{ab}(\omega)(e^{-1})_a^\mu (e^{-1})_b^\nu, \tag{6.12}$$

which in four dimensions can be rewritten as

$$S_{\text{EH}} = \frac{1}{2k_N^2} \int d^4 x \epsilon_{abcd} \epsilon^{\mu\nu\rho\sigma} e_\mu^a e_\nu^b R_{\rho\sigma}^{cd}$$
$$\equiv \frac{1}{2k_N^2} \int d^4 x \epsilon_{abcd} e^a \wedge e^b \wedge R^{cd}(\omega), \tag{6.13}$$

with an obvious extension to higher dimensions, where we have $\epsilon_{a_1 \ldots a_d} R^{a_1 a_2} \wedge e^{a_3} \ldots \wedge e^{a_d}$ (in form language) inside the integral.

From now on we will drop the (-1) index on the inverse vielbein, since it is clear from the position of the μ (curved) index whether it is a vielbein or an inverse one. The position of the flat index a doesn't matter, since it is raised or lowered with η_{ab}.

As usual (in form language)

$$R^{ab} = d\omega^{ab} + \omega^{ac} \wedge \omega^{cb}. \tag{6.14}$$

Since in the case of Yang–Mills, the commutator of two covariant derivatives gives the field strength,

$$[D_\mu, D_\nu] = F_{\mu\nu} \equiv F_{\mu\nu}^a T_a, \tag{6.15}$$

we now write a similar relation for the local Lorentz ($SO(1, d-1)$) gauge field ω_μ^{ab},

$$[D_\mu(\omega), D_\nu(\omega)] = R_{\mu\nu}^{rs} \frac{1}{4} \gamma_{rs}. \tag{6.16}$$

Defining a covariant derivative with flat indices,

$$D_a \equiv e_a^\mu D_\mu, \tag{6.17}$$

we obtain for its commutator

$$[D_a, D_b] = (2e_a^\mu e_b^\nu D_{[\mu} e_{\nu]}^c) D_c + (e_a^\mu e_b^\nu R_{\mu\nu}^{rs}(\omega)) \frac{1}{4} \gamma_{rs}$$
$$\equiv T_{ab}^c D_c + R_{ab}^{rs} M_{rs}, \tag{6.18}$$

where we have defined the torsion and curvature with flat indices as

$$T^c_{ab} \equiv e^\mu_a e^\nu_b T^c_{\mu\nu} = 2e^\mu_a e^\nu_b D_{[\mu} e^c_{\nu]}$$
$$R^{cd}_{ab} = e^\mu_a e^\nu_b R^{cd}_{\mu\nu}(\omega). \tag{6.19}$$

In general, we will define the torsion as the object that multiplies a covariant derivative D_a in the commutator of covariant derivatives, and the curvature as the object multiplying the generators T_A on the right-hand side of the commutator of covariant derivatives. In this way, we will generalize the notions of torsion and curvature to superspace and (super-) Yang–Mills theories, to be studied later on in the book.

The action for the gravitino, as a free spin 3/2 field in flat space, was written by Rarita and Schwinger,

$$S_{\mathrm{RS}} = -\frac{1}{2} \int d^d x \bar{\psi}_\mu \gamma^{\mu\nu\rho} \partial_\nu \psi_\rho. \tag{6.20}$$

In four dimensions, we can rewrite it as

$$S_{\mathrm{RS}} = \frac{i}{2} \int d^4 x \epsilon^{\mu\nu\rho\sigma} \bar{\psi}_\mu \gamma_5 \gamma_\nu \partial_\rho \psi_\sigma \tag{6.21}$$

by using the relations

$$i\epsilon^{\mu\nu\rho\sigma} \gamma_5 \gamma_\nu = -\gamma^{\mu\rho\sigma}, \quad \gamma_5 = i\gamma_0\gamma_1\gamma_2\gamma_3. \tag{6.22}$$

In curved space, we just replace the normal derivative with the gravitationally covariant derivative, and add the integration measure, to write

$$S_{\mathrm{RS}} = -\frac{1}{2} \int d^d x (\det e) \bar{\psi}_\mu \gamma^{\mu\nu\rho} D_\nu \psi_\rho \tag{6.23}$$

in a general dimension, and in four dimensions also

$$S_{\mathrm{RS}} = \frac{i}{2} \int d^4 x \epsilon^{\mu\nu\rho\sigma} \bar{\psi}_\mu \gamma_5 \gamma_\nu D_\rho \psi_\sigma, \tag{6.24}$$

where (note that there is no need for a Christoffel symbol term because of the antisymmetry of the vector indices ρ, σ)

$$D_{[\rho} \psi_{\sigma]} = \partial_{[\rho} \psi_{\sigma]} + \frac{1}{4} \omega^{ab}_{[\rho} \gamma_{ab} \psi_{\sigma]}. \tag{6.25}$$

Then the action of the on-shell $\mathcal{N} = 1$ supergravity in four dimensions is just the sum

$$S_{\mathcal{N}=1} = S_{\mathrm{EH}}(\omega, e) + S_{\mathrm{RS}}(\psi_\mu), \tag{6.26}$$

and it obeys the susy transformation rules

$$\delta e^a_\mu = \frac{\kappa_N}{2} \bar{\epsilon} \gamma^a \psi, \quad \delta \psi_\mu = \frac{1}{\kappa_N} D_\mu \epsilon. \tag{6.27}$$

But it is not enough to write the above, we must also specify the formalism we use.

- **Second-order formalism.** Here we have independent e^a_μ, $\psi_{\mu\alpha}$, but ω^{ab}_μ is dependent. However, note that since we have dynamical fermions (the gravitinos), we have a nonzero torsion $T^a_{\mu\nu} \neq 0$, so $\omega \neq \omega(e)$, but rather

$$\omega^{ab}_\mu = \omega^{ab}_\mu(e, \psi) = \omega^{ab}_\mu(e) + \psi\psi \text{ terms}, \tag{6.28}$$

and this is found by varying the action with respect to ω_μ^{ab}, as if it is an independent field,

$$\frac{\delta S_{\mathcal{N}=1}}{\delta \omega_\mu^{ab}} = 0 \Rightarrow \omega_\mu^{ab}(e, \psi). \tag{6.29}$$

- **First-order formalism.** Here we have independent e_μ^a, $\psi_{\mu\alpha}$ and ω_μ^{ab}. Now, however, besides the transformation rules for e_μ^a and $\psi_{\mu\alpha}$, we must also add transformation rules for ω_μ^{ab},

$$\delta \omega_\mu^{ab}(\text{first order}) = -\frac{1}{4}\bar{\epsilon}\gamma_5 \gamma_\mu \tilde{\psi}^{ab} + \frac{1}{8}\bar{\epsilon}\gamma_5(\gamma^\lambda \tilde{\psi}_\lambda^b e_\mu^a - \gamma^\lambda \tilde{\psi}_\lambda^a e_\mu^b)$$

$$\tilde{\psi}^{ab} \equiv \epsilon^{abcd}\psi_{cd}; \quad \psi_{ab} \equiv e_a{}^\mu e_b{}^\nu (D_\mu \psi_\nu - D_\nu \psi_\mu). \tag{6.30}$$

On shell for ω_μ^{ab}, we go back to the second-order formalism.

Note that the equation of motion for ψ_μ is $\epsilon^{\mu\nu\rho\sigma}\gamma_5 \gamma_\nu D_\rho \psi_\sigma = 0$, and by multiplying it with γ_5, we obtain

$$\gamma^\lambda \tilde{\psi}_{\lambda\mu} = 0. \tag{6.31}$$

- **1.5-order formalism.** This is a very simple, but very powerful observation. We use the second-order formalism, but in $S(e, \psi, \omega(e, \psi))$, when we vary the fields we don't vary $\omega(e, \psi)$ by the chain rule, since it is anyway multiplied by $\delta S/\delta \omega$, which vanishes in the second-order formalism. More precisely, we have

$$\delta S = \frac{\delta S}{\delta e}\delta e + \frac{\delta S}{\delta \psi}\delta \psi + \frac{\delta S}{\delta \omega}\left(\frac{\delta \omega}{\delta e}\delta e + \frac{\delta \omega}{\delta \psi}\delta \psi\right), \tag{6.32}$$

and the last two terms vanish because of the prefactor. Of course, that means that in the action we cannot substitute $\omega(e, \psi)$ in terms of e and ψ, we must leave it as it is.

The transformation laws of the fields under the other symmetries are as follows.

Under Einstein (or general coordinate) transformations, we have

$$\delta_E e_\mu^a = \xi^\nu \partial_\nu e_\mu^a + (\partial_\mu \xi^\nu)e_\nu^a$$

$$\delta_E \omega_\mu^{ab} = \xi^\nu \partial_\nu \omega_\mu^{ab} + (\partial_\mu \xi^\nu)\omega_\nu^{ab}$$

$$\delta_E \psi_\mu = \xi^\nu \partial_\nu \psi_\mu + (\partial_\mu \xi^\nu)\psi_\nu, \tag{6.33}$$

and we see that all the fields transform in a similar manner, since all are vectors under Einstein transformation, with the index μ.

Under local Lorentz ($SO(1, d - 1)$) transformations, we have

$$\delta_{\text{IL}} e_\mu^a = \lambda^{ab} e_\mu^b$$

$$\delta_{\text{IL}} \omega_\mu^{ab} = D_\mu \lambda^{ab} = \partial_\mu \lambda^{ab} + \omega_\mu^{ac}\lambda^{cb} - \omega_\mu^{bc}\lambda^{ca}$$

$$\delta_{\text{IL}} \psi_\mu = -\lambda^{ab}\frac{1}{4}\gamma_{ab}\psi_\mu \tag{6.34}$$

6.3.1 Susy invariance of the $S_{\mathcal{N}=1}$ action in 1.5-order formalism

We now prove the invariance of the supergravity action under the susy transformation rules, in the 1.5-order formalism.

The Lagrangian is

$$\mathcal{L} = -\frac{1}{2} e e_a^\mu e_b^\nu R_{\mu\nu}^{ab}(\omega) - \frac{1}{2} \epsilon^{\mu\nu\rho\sigma} \bar\psi_\mu \gamma_5 \gamma_\nu D_\rho(\omega) \psi_\sigma. \tag{6.35}$$

Since we are in the 1.5-order formalism, we don't vary ω, so also don't vary $R_{\mu\nu}^{ab}(\omega)$ and $D_\rho(\omega)$. We put $\kappa_N = 1$ for simplicity, so we have

$$\delta e_\mu^a = \frac{1}{2} \bar\epsilon \gamma^a \psi_\mu. \tag{6.36}$$

Since $e_{\mu a} e^{\nu a} = \delta_\mu^\nu$, $\delta(e_{\mu a} e^{\nu a}) = 0$, so we obtain

$$\delta e^{\mu a} = -\frac{1}{2} \bar\epsilon \gamma^\mu \psi^a. \tag{6.37}$$

Then the variation of the spin 2 part (gravity) of the supergravity action is (we need to vary only e, e_a^μ and e_b^ν, equivalent to the variation with respect to $g_{\mu\nu}$ of the Einstein–Hilbert action, which gives the Einstein tensor times the variation of the metric)

$$\delta\mathcal{L}^{(\text{spin 2})} = \delta e \frac{\delta\mathcal{L}^{(\text{spin 2})}}{\delta e} = \frac{1}{2} \bar\epsilon \gamma^\mu \psi^a G_{\mu a}, \tag{6.38}$$

where

$$G_{\mu a} = R_{\mu a} - \frac{1}{2} e_{\mu a} R \tag{6.39}$$

is the Einstein tensor with one flattened index.

Next we vary the spin 3/2 part (gravitino) of the action. We can vary $\bar\psi_\mu$ and ψ_σ, as well as $\gamma_\nu = e_{\nu a}\gamma^a$. We leave the vielbein variation for later, and concentrate on the gravitino variations:

$$\delta\mathcal{L}_{\psi_\mu\text{ terms}}^{(\text{spin 2})} = -\frac{1}{2}\epsilon^{\mu\nu\rho\sigma}\left(\delta\bar\psi_\mu \gamma_5 \gamma_\nu D_\rho \psi_\sigma + \bar\psi_\mu \gamma_5 \gamma_\nu D_\rho \delta\psi_\sigma\right). \tag{6.40}$$

Since

$$\epsilon^{\mu\nu\rho\sigma} D_\rho \psi_\sigma = \frac{1}{2}\epsilon^{\mu\nu\rho\sigma}[D_\rho, D_\sigma]\epsilon = \epsilon^{\mu\nu\rho\sigma} R_{\rho\sigma}^{ab}(\omega)\frac{1}{4}\gamma_{ab}\epsilon, \tag{6.41}$$

and in $\epsilon^{\mu\nu\rho\sigma}\delta\bar\psi_\mu... = \epsilon^{\mu\nu\rho\sigma}D_\mu\bar\epsilon...$ we can partially integrate the D_μ to act on $\gamma_\nu D_\rho \psi_\sigma$, we can form the same $\frac{1}{2}[D_\mu, D_\rho]\psi_\sigma = R_{\mu\rho}^{ab}\frac{1}{4}\gamma_{ab}\psi_\sigma$, and we get an *extra term*, to be treated later (and called "extra"), where D_μ is commuted past γ_ν.

Then

$$\delta\mathcal{L}_{\psi_\mu\text{ terms, no extra}}^{(\text{spin 2})} = -\frac{1}{8}\epsilon^{\mu\nu\rho\sigma}\left[\bar\psi_\mu \gamma_5 \gamma_\nu \gamma_{cd}\epsilon - \bar\epsilon \gamma_5 \gamma_\nu \gamma_{cd} \psi_\mu\right] R_{\rho\sigma}^{cd}. \tag{6.42}$$

But using the Majorana spinor identity

$$\bar\epsilon \gamma_5 \gamma_\nu \gamma_{cd} \psi_\mu = -\bar\psi_\mu \gamma_5 \gamma_{cd} \gamma_\nu \epsilon \tag{6.43}$$

and the gamma matrix identities

$$\gamma_\nu \gamma_{cd} + \gamma_{cd}\gamma_\nu = \epsilon_{cdab}\gamma_5\gamma^b e_\nu^a$$

$$e_\mu^a \epsilon^{\mu\nu\rho\sigma} \epsilon_{abcd} = 6e\delta_{bcd}^{\nu\rho\sigma}, \qquad (6.44)$$

which are easy to prove (the first by using the commutator of gamma matrices, and then giving values to the indices, the second by using the definition of the determinant for e_μ^a), we finally find

$$\delta\mathcal{L}^{(\text{spin }2)} = \frac{e}{2}(\bar{\psi}^\mu \gamma^a \epsilon)G_{\mu a} + \delta\mathcal{L}^{(\text{spin }3/2)}_{\gamma_\nu \text{ term}} + \delta\mathcal{L}^{(\text{spin }3/2)}_{\text{extra}}. \qquad (6.45)$$

But there is a Majorana spin identity $\bar{\epsilon}\gamma^a\psi^\mu = -\bar{\psi}^\mu\gamma^a\epsilon$ so, together with the fact that the Einstein tensor $G_{\mu a}$ is symmetric and we can redefine the sum over μ as sum over a and vice versa, so $\gamma^\mu\psi^a G_{\mu a} = \gamma^a\psi^\mu G_{\mu a}$, we see that the terms with $G_{\mu a}$, in the variation of the spin 2 part, and the variation of the spin 3/2 part, cancel.

The extra term, where the D_μ is commuted past γ_ν in the partial integration, gives

$$+\frac{1}{2}\epsilon^{\mu\nu\rho\sigma}\bar{\epsilon}\gamma_5(D_\mu e_\nu^a)\gamma_a D_\rho \psi_\sigma. \qquad (6.46)$$

On the other hand, the variation of $\gamma_\nu = e_\nu^a \gamma_a$ term gives

$$-\frac{1}{2}\epsilon^{\mu\nu\rho\sigma}(\bar{\psi}_\mu\gamma_5\gamma_a D_\rho\psi_\sigma)\left(\frac{1}{2}\bar{\epsilon}\gamma^a\psi_\nu\right). \qquad (6.47)$$

Then, the remaining terms (which should cancel) are

$$-\frac{1}{2}\epsilon^{\mu\nu\rho\sigma}(\bar{\psi}_\mu\gamma_5\gamma_a D_\rho\psi_\sigma)\left(\frac{1}{2}\bar{\epsilon}\gamma^a\psi_\nu\right) + \frac{1}{2}\epsilon^{\mu\nu\rho\sigma}(\bar{\epsilon}\gamma_5\gamma_a D_\rho\psi_\sigma)D_\mu e_\nu^a, \qquad (6.48)$$

and in the last term, we can use the fact that the torsion is purely fermionic, so $D_{[\mu}e_{\nu]}^a = \frac{1}{4}\bar{\psi}_{[\mu}\gamma^a\psi_{\nu]}$.

But in the first term, we use the Fierz identity

$$\epsilon^{\mu\nu\rho\sigma}\bar{\psi}_\mu (\bar{\epsilon}\gamma^a\psi_\nu) = -\frac{1}{4}\left(\epsilon^{\mu\nu\rho\sigma}\bar{\psi}_\mu O_j\psi_\nu\right)\frac{1}{2}\bar{\epsilon}\gamma^a O_j, \qquad (6.49)$$

where O_j are the complete set of 4×4 matrices $\{1, \gamma_5, \gamma_\lambda, \gamma_5\gamma_\lambda, \gamma_{\lambda\tau}\}$. But, due to Majorana spinor identities in four dimensions, and the fact that $\bar{\psi}_\mu O_j\psi_\nu$ is antisymmetric in μ, ν, only the γ_λ survives.

Moreover, then, $\gamma^a\gamma_\lambda\gamma_5\gamma_a = 2\gamma_\lambda\gamma_5$, so the first term becomes

$$\frac{1}{8}(\bar{\psi}_\mu\gamma_\lambda\psi_\nu)\epsilon^{\mu\nu\rho\sigma}\bar{\epsilon}\gamma^\lambda\gamma_5 D_\rho\psi_\sigma. \qquad (6.50)$$

Finally using the anticommutation $\gamma^\lambda\gamma_5 = -\gamma_5\gamma^\lambda$, the two terms now cancel, thus proving the supersymmetry of the $\mathcal{N} = 1$ supergravity action.

Important concepts to remember

- Supergravity is a supersymmetric theory of gravity and a theory of local supersymmetry.
- The gauge field of local supersymmetry and superpartner of the vielbein (graviton) is the gravitino ψ_μ.
- Supergravity (local supersymmetry) is of the type $\delta e_\mu^a = (k/2)\bar{\epsilon}\gamma^a\psi_\mu + ...$, $\delta\psi_\mu = (D_\mu\epsilon)/k + ...$

- The action for gravity is the Einstein–Hilbert action in the vielbein-spin connection formulation.
- Torsion and curvature are defined respectively as the terms proportional to D_c and T_m on the right-hand side of $[D_a, D_b]$.
- The action for the gravitino is the Rarita–Schwinger action.
- The most useful formulation is the 1.5-order formalism: second-order formalism, but don't vary $\omega(e, \psi)$ by the chain rule.
- In three dimensions on-shell, for e_μ^a and ψ_μ^α there are no degrees of freedom for the $\mathcal{N} = 1$ supergravity, whereas in four dimensions there are two bosonic and two fermionic degrees of freedom.
- In three dimensions off-shell, we need an auxiliary scalar S, for a multiplet $(e_\mu^a, S, \psi_{\mu\alpha})$.
- In four dimensions off-shell, we need six bosonic auxiliary degrees of freedom more than the fermionic auxiliary degrees of freedom. Choosing just the bosonic auxiliary fields (A_μ, S, P) is the minimal set.

References and further reading

An introduction to supergravity, but one that might be hard to follow for the beginning student, is found in West [9] and Wess and Bagger [10]. A good supergravity course, which starts at an introductory level and reaches quite far, is [5]. In this chapter, I followed mostly [5] (you can find more details in Sections 1.2–1.6 of the reference). You can also follow the book [17]. The original paper on supergravity is [18].

Exercises

(1) Find $\omega_\mu^{ab}(e, \psi) - \omega_\mu^{ab}(e)$ in the second-order formalism for $\mathcal{N} = 1$ supergravity.
(2) Calculate the number of off-shell bosonic and fermionic degrees of freedom of $\mathcal{N} = 8$ on-shell supergravity in four dimensions, with field content $\{(2, 3/2) + 7 \times (3/2, 1) + 21 \times (1, 1/2) + 35 \times (1/2, 0)$, specifically $\{e_\mu^a, \psi_\mu^i, A_\mu^{[IJ]}, \chi_{[IJK]}, \nu\}$, where $i, j, k = 1, ..., 8; I, J = 1, ..., 8$; and ν= matrix of 70 real scalars (the scalar in the WZ multiplet $(1/2, 0)$ is complex).
(3) Consider the spinors η^I satisfying the "Killing spinor equation,"

$$D_\mu \eta^I = \pm \frac{i}{2} \gamma_\mu \eta^I. \tag{6.51}$$

Prove that they live on a space of constant positive curvature (a sphere), by computing the curvature of the space.
(4) Write down explicitly the variation of the $\mathcal{N} = 1$ four-dimensional supergravity action in 1.5-order formalism, as a function of δe and $\delta \psi$.

7 Three-dimensional $\mathcal{N} = 1$ off-shell supergravity

We now consider the first more nontrivial example of supergravity, namely of off-shell supergravity. The simplest such model is $\mathcal{N} = 1$ off-shell supergravity in three dimensions.

We saw that on-shell we don't have any degrees of freedom, since for e_μ^a, we have $(d-1)(d-2)/2 - 1 = 0$ degrees of freedom, and for ψ_μ, we have $2[d/2](d-3)/2 = 0$ degrees of freedom, so the model is somewhat trivial.

Off-shell, however, it is more interesting: for e_μ^a, we have $d(d-1)/2 = 3$ degrees of freedom, and for ψ_μ, we have $(d-1)2^{[d/2]} = 4$ degrees of freedom. This means that we need to add an auxiliary scalar S so that we have an extra bosonic degree of freedom.

7.1 Action and symmetries

In this chapter, we will use the normalization of the EH gravity action as (replacing $\kappa \to 2\kappa$)

$$S_{\text{EH}} = -\frac{1}{8\kappa^2} \int d^3x \, eR, \tag{7.1}$$

where as usual, we have

$$R = R_{\mu\nu}^{mn}(\omega)e_m^\mu e_n^\nu,$$
$$R_{\mu\nu}^{mn}(\omega) = \partial_\mu \omega_\nu^{mn} - \partial_\nu \omega_\mu^{mn} + \omega_\mu^{mp}\omega_\nu^{pn} - \omega_\nu^{mp}\omega_\mu^{pn}. \tag{7.2}$$

Note that in three dimensions, the mass dimension of κ is $[\kappa] = -1/2$ (in four dimensions, it has the mass dimension of -1).

We make an aside, to note that in three dimensions, the action for gravity in the vielbein–spin connection formulation is a true gauge theory, at least classically. In higher dimensions, as we already noted, only if we are half-on-shell, by having $\omega = \omega(e)$, and if we identify curved and flat indices, do we have a gauge theory.

But in three dimensions, we have a true gauge, of the CS (rather than Yang–Mills) type. Indeed, the action can be rewritten in form language as

$$S_{\text{EH}} = \frac{1}{16\pi G_N} \int_{M_3=\partial M_4} d^3x \, \epsilon_{abc} R^{ab}(\omega) \wedge e^c$$
$$= \frac{1}{16\pi G_N} \int_{M_4} d^4x \, \epsilon_{abc} R^{ab}(\omega) \wedge T^c, \tag{7.3}$$

where in the second line we used the Stokes relation $\int_{\partial M} A = \int_M dA$, together with the fact that some extra terms involving ω vanish, since $T^a = De^a = de^e + \omega^{ab} \wedge e^b$.

This action is of the general form of the CS action, which can be rewritten in four dimensions (using the Stokes relation) as

$$S_{CS} = \int_{M_4} d_{AB} F^A \wedge F^B, \tag{7.4}$$

where A, B are group indices in the adjoint representation of some group, F^A is a field strength ("curvature"), and d_{AB} is a (symmetric) group invariant.

Here $R^{ab}(\omega)$ and $T^c(e)$ are the curvatures of the Poincaré group $ISO(2, 1)$, corresponding to the generators J_{ab} and P_a, and ϵ_{abc} is the (symmetric in the interchange of $A = a$ and $B = (bc)$) group invariant d_{AB}.

This means that the three-dimensional gravity action S_{EH} is gauge invariant, though it is of CS, not YM, form.

The gravitino action is

$$S_{RS} = -\frac{1}{2} \int d^3x \, e \, \bar{\psi}_\mu \gamma^{\mu\nu\rho} D_\nu(\omega) \psi_\rho, \tag{7.5}$$

where as usual the gravitational covariant derivative involves only the spin connection,

$$D_\nu \psi_\rho = \partial_\nu \psi_\rho + \frac{1}{4} \omega_\nu^{mn} \gamma_{mn} \psi_\rho. \tag{7.6}$$

However, in three dimensions, we have $\gamma^{mnp} = -\epsilon^{mnp}$ (for flat indices), which means that $e\gamma^{\mu\nu\rho} = -\epsilon^{\mu\nu\rho}$ (with curved indices; the relation is obtained by using the definition of the determinant of e_μ^a, which is e), so we can rewrite the gravitino action as

$$S_{RS} = +\frac{1}{2} \int d^3x \, \epsilon^{\mu\nu\rho} \bar{\psi}_\mu D_\nu \psi_\rho = \frac{1}{2} \int \bar{\psi} \wedge D\psi, \tag{7.7}$$

in form language.

An observation on how the Rarita–Schwinger action is obtained is as follows. In flat space, the most general action for ψ_μ is

$$\mathcal{L}_{3/2} = \bar{\psi}_\mu \mathcal{O}^{\mu\nu\rho} \partial_\nu \psi_\rho, \tag{7.8}$$

where the relation to a current is $\mathcal{O}^{\mu\nu\rho} \partial_\nu \psi_\rho = J_\mu$.

This means that in the effective action S_{eff}, we have a term $\frac{1}{2} J^\mu P_{\mu\nu} J^\nu$, where $P_{\mu\nu} = [\mathcal{O}^{\mu\nu\rho} \partial_\nu]^{-1}$. But tree-level unitarity implies that the complex residues at the physical $k^2 = 0$ poles are positive. Imposing this condition gives the unique result $\mathcal{O}^{\mu\nu\rho} = +\gamma^{\mu\nu\rho}$. In this case, the action has a gauge invariance $\delta \psi_\sigma = \partial_\sigma \epsilon$, as we know. The RS action is generalized to curved space in the usual manner.

The transformation rules under Einstein (general coordinate) transformations are the same as in any dimension,

$$\delta_E e_\mu^a = \xi^\nu \partial_\nu e_\mu^a + (\partial_\mu \xi^\nu) e_\nu^a,$$
$$\delta_E \omega_\mu^{ab} = \xi^\nu \partial_\nu \omega_\mu^{ab} + (\partial_\mu \xi^\nu) \omega_\nu^{ab},$$
$$\delta_E \psi_\mu = \xi^\nu \partial_\nu \psi_\mu + (\partial_\mu \xi^\nu) \psi_\nu, \tag{7.9}$$

whereas the transformation rules for local Lorentz symmetry are (also as in any dimension)

$$\delta_{l.L.} e_\mu^a = \lambda^{ab} e_\mu^b,$$
$$\delta_{l.L.} \omega_\mu^{ab} = D_\mu \lambda^{ab} = \partial_\mu \lambda^{ab} + \omega_\mu^{ac} \lambda^{cb} - \omega_\mu^{bc} \lambda^{ca},$$

$$\delta_{l.L.}\psi_\mu = -\lambda^{ab}\frac{1}{4}\gamma_{ab}\psi_\mu, \tag{7.10}$$

where the transformation rule for ω is obtained because ω_μ^{ab} is the gauge field for local Lorentz transformations (now this is a precise statement, unlike in four dimensions).

The action for the auxiliary field S is

$$S_S = -\frac{1}{2}\int d^3x \, eS^2. \tag{7.11}$$

We should observe the sign in front, which is the opposite to the rigid supersymmetry (say, the WZ or vector multiplet) case, and it comes from imposing local supersymmetry of the action. The reason that the sign is different is that in the rigid case, we had a free action for the auxiliary scalar, whereas now the formerly free action interacts with gravity (through the determinant factor e), so S^2 will contribute to the cosmological constant (just like in the free case F^2 or D^aD^a contributed to the scalar potential, or the vacuum energy, at the minimum).

Since we have a contribution to the cosmological constant with the opposite sign from the "matter," that is, WZ or vector multiplet, there is a possibility that the two contributions will cancel against each other. Indeed, in supergravity plus matter, we can have zero cosmological constant.

The transformation rules for the scalar under the Einstein and local Lorentz symmetries are obvious. The S field is a scalar, so it only transforms by translation under Einstein, and it doesn't transform under local Lorentz, so

$$\delta_E S = \xi^\nu \partial_\nu S,$$
$$\delta_{l.L.} S = 0. \tag{7.12}$$

As always, we have a choice of formalisms for gravity:

- in the first-order formalism, ω is independent.
- in the second-order formalism, $\omega = \omega(e, \psi)$, and this solution satisfies the equation of motion of the first-order formalism (as opposed to just the vielbein postulate, which leads to $\omega = \omega(e)$ only).
- in the 1.5-order formalism, which is used in the following, we use the second-order formalism, but since $\delta\omega = \delta\omega/\delta e \delta e + \delta\omega/\delta\psi \delta\psi$ is multiplied by $\delta S/\delta\omega$, which is zero in the second-order formalism, we keep $\omega(e, \psi)$ as it is (without replacing e and ψ in it) and we don't vary it.

In order to have off-shell susy, we must in fact use the 1.5 (or second)-order formalism, since in the first-order formalism, we have too many bosonic degrees of freedom, so we don't have matching of bosonic and fermionic degrees of freedom.

7.2 Supersymmetry transformation rules

The free action for bosons (scalars) plus fermions in three dimensions is of the type $\int d^3x[(\partial_\mu\phi)^2 + \bar\psi\slashed\partial\psi]$, which means that the mass dimensions are $[\phi] = 1/2$ and $[\psi] = 1$.

But under rigid susy, $\delta\phi \sim \bar{\epsilon}\psi$, which means that a $[\epsilon] = -1/2$. Then $\delta\psi_\mu \sim D_\mu\epsilon$ doesn't match mass dimensions, unless we have a $1/\kappa$ in front (since the gravity action has $\frac{1}{\kappa^2}\int d^3xR$, so $[\kappa] = -1/2$). Finally then

$$\delta\psi_\mu = \frac{1}{\kappa}D_\mu\epsilon = \frac{1}{\kappa}\left(\partial_\mu\epsilon + \frac{1}{4}\omega_\mu^{mn}\gamma_{mn}\right)\epsilon. \tag{7.13}$$

Here $\omega = \omega(e, \psi)$. Equation (7.13) was for on-shell supergravity. For off-shell supergravity, we must add a term with S in the transformation rule for the gravitino.

We saw in the case of rigid supersymmetry that it is not enough for the susy transformation rules to leave the action invariant. We must also represent the susy algebra on the fields. In the case of rigid $\mathcal{N} = 1$ susy, group theory considerations gave us the unique susy algebra in any dimension,

$$\{Q_\alpha, Q_\beta\} = 2(C\gamma^\mu)_{\alpha\beta}P_\mu, \tag{7.14}$$

and from it we obtained (without any ambiguity) the algebra of susy transformations, by multiplying with $\bar{\epsilon}_2$ from the left and ϵ_1 from the right, and forming the variation $\delta_\epsilon = \bar{\epsilon}Q$ (in general $\delta_\epsilon = \epsilon^a Q_a$),

$$[\delta_{\epsilon_1}, \delta_{\epsilon_2}] = \xi^\mu\partial_\mu,$$
$$\xi^\mu \equiv 2\bar{\epsilon}_2\gamma^\mu\epsilon_1. \tag{7.15}$$

In the local susy case, translations P_μ become general coordinate transformations, since as we saw, Einstein transformations are the local version of translations: $\delta_E(\xi)(...) = \xi^\nu\partial_\nu(...) + (\partial_\mu\xi^\nu)(...)$ terms. On each field, the extra terms differ. Moreover, the local susy algebra depends on dimension, and the parameters depend on the particular fields the algebra is represented on (on the susy representation). We will see that in this case, of $\mathcal{N} = 1$ off-shell susy, we will find the local algebra

$$[\delta_Q(\epsilon_1), \delta_Q(\epsilon_2)] = \delta_{g.c.}(\xi^\mu) + \delta_{l.L.}(\xi^\mu\omega_\mu^{mn}) + \delta_Q(-\xi^\mu\psi_\mu)$$
$$\xi^\mu \equiv 2\bar{\epsilon}_2\gamma^\mu\epsilon_1. \tag{7.16}$$

It is worth emphasizing that this local version of the susy algebra cannot be derived from group theory alone, like in the rigid case, and it depends on the representation.

Since we don't know the algebra beforehand, what we require first is the *closure of the algebra* on the fields, that is, that the commutator of two transformations is a sum of the various symmetry transformations of the theory. We thus find the algebra by requiring closure on one of the fields, and then we impose that it is satisfied on the other fields as well.

We saw that in the rigid case, the algebra closes even on-shell on the dynamical scalar (in that case a scalar), and it doesn't on the fermion. Therefore we assume (and indeed find) that the same happens in the local case and find the local algebra by requiring closure of the algebra on the graviton, that is, vielbein e_μ^a. Once we do, we find that the algebra is realized on all of the fields.

Since we are using the 1.5-order formalism, $R_{\mu\nu}^{ab}(\omega)$ is not varied, since it is only a function of ω. Then the variation of the Einstein–Hilbert action gives

$$\delta S_{\text{EH}} = -\frac{1}{8\kappa^2}\int d^3x\, R_{\mu\nu}^{mn}(\omega)\delta[ee_m^\mu e_n^\nu]$$

$$= -\frac{1}{4\kappa^2} \int d^3x \, e \left[R^m_\mu - \frac{1}{2} e^m_\mu R \right] \delta e^\mu_m, \tag{7.17}$$

where in the second line we have used $\delta e = e e^\mu_m \delta e^m_\mu = -e e^m_\mu \delta e^\mu_m$. In the second line, in the square bracket [...], we have the Einstein tensor $R_{\mu\nu} - 1/2 g_{\mu\nu} R$ with flattened indices.

We will see that we get the same structure from varying the Rarita–Schwinger action S_{RS}. Using the fact that $\delta\psi_\mu = \frac{1}{\kappa} D_\mu \epsilon$, the fact that the variation of $\bar\psi$ and ψ gives the same term (because of the Majorana spinor relation $\bar\chi \gamma_m \psi = -\bar\psi \gamma_m \chi$ applied to the two resulting terms), and the fact that

$$[D_\mu(\omega), D_\nu(\omega)] = \frac{1}{4} R^{mn}_{\mu\nu} \gamma_{mn}, \tag{7.18}$$

we obtain

$$\delta S_{\mathrm{RS}} = \frac{1}{\kappa} \int d^3x \epsilon^{\mu\nu\rho} \bar\psi_\mu D_\nu D_\rho \epsilon$$
$$= \frac{1}{8\kappa} \int d^3x \epsilon^{\mu\nu\rho} R^{mn}_{\nu\rho} \bar\psi_\mu \gamma_{mn} \epsilon. \tag{7.19}$$

We use the relations

$$\gamma_{mn} = -\epsilon_{mnr} \gamma^r,$$
$$\epsilon^{\mu\nu\rho} \epsilon_{mnr} = -6 e e^{[\mu}_m e^\nu_n e^{\rho]}_r. \tag{7.20}$$

The first one is obtained from $\gamma_{mnr} = -\epsilon_{mnr}$, and the second one from $e = \epsilon^{\mu\nu\rho} \epsilon_{mnr} e^m_\mu e^p_\nu e^r_\rho$. We then rewrite the following expression appearing in δS_{RS}:

$$\epsilon^{\mu\nu\rho} R^{mn}_{\nu\rho} \gamma_{mn} = -\epsilon^{\mu\nu\rho} \epsilon_{mnr} R^{mn}_{\nu\rho} \gamma^r$$
$$= 6 e R^{mn}_{\nu\rho} e^{[\mu}_m e^\nu_n e^{\rho]}_r \gamma^r$$
$$= 4 e e^\mu_m \left(R^m_\rho - \frac{1}{2} R e^m_\rho \right) \gamma^r e^\rho_r, \tag{7.21}$$

where all antisymmetrizations are done with "strength one," that is, =(sum of terms)/(number of terms). We finally obtain

$$\delta S_{\mathrm{RS}} = \frac{1}{2\kappa} \int d^3x \, e \left(R^m_\rho - \frac{1}{2} R e^m_\rho \right) \bar\psi_m \gamma^\rho \epsilon. \tag{7.22}$$

We see that in order to cancel the variation of the gravity term against the variation of the gravitino term, $\delta S_{\mathrm{EH}} + \delta S_{\mathrm{RS}} = 0$, we need the variation of the inverse vielbein to be

$$\delta e^\mu_m = 2k \bar\psi_m \gamma^\mu \epsilon. \tag{7.23}$$

Using $\delta(e^n_\mu e^\mu_m) = e^n_\mu \delta e^\mu_m + e^\mu_m \delta e^n_\mu = 0$, and the Majorana spinor relations, which in three dimensions are

$$\bar\psi \chi = \bar\chi \psi,$$
$$\bar\psi \gamma_m \chi = -\bar\chi \gamma_m \psi, \tag{7.24}$$

we obtain the variation of the vielbein as

$$\delta e^m_\mu = 2\kappa \bar\epsilon \gamma^m \psi_\mu. \tag{7.25}$$

7.3　The susy algebra

Since we have found the susy transformation rules, we now impose the closure of the susy algebra (the commutator of two supersymmetries) on the graviton, that is, to write everything on the right-hand side as a sum of invariances, in order to find this local susy algebra. We obtain

$$
[\delta_1, \delta_2]e_\mu^m = 2\kappa\bar{\epsilon}_2\gamma^m\left(\frac{1}{\kappa}D_\mu\epsilon_1\right) - (1 \leftrightarrow 2)
$$
$$
= 2\partial_\mu(\bar{\epsilon}_2\gamma^m\epsilon_1) + 2\left[\frac{1}{4}\omega_\mu^{rs}\bar{\epsilon}_2\gamma^m\gamma_r\gamma_s\epsilon_1 - (1 \leftrightarrow 2)\right],
\tag{7.26}
$$

where we have used the Majorana spinor relations (7.24) to add up the two terms with derivatives into a single total derivative term. Defining as before $\xi^\mu = 2\bar{\epsilon}_2\gamma^\mu\epsilon_1$ and $\xi^m = \xi^\mu e_\mu^m$, we find

$$
[\delta_1, \delta_2]e_\mu^m = (\partial_\mu\xi^\nu)e_\nu^m + \xi^\nu(\partial_\mu e_\nu^m) + 2\left[\frac{1}{4}\omega_\mu^{rs}\bar{\epsilon}_2\gamma^m\gamma_r\gamma_s\epsilon_1 - (1 \leftrightarrow 2)\right]
$$
$$
= (\partial_\mu\xi^\nu)e_\nu^m + \xi^\nu\partial_\nu e_\mu^m + \xi^\nu(\partial_\mu e_\nu^m - \partial_\nu e_\mu^m) + 2\left[\frac{1}{4}\omega_\mu^{rs}\bar{\epsilon}_2\gamma^m\gamma_r\gamma_s\epsilon_1 - (1 \leftrightarrow 2)\right].
\tag{7.27}
$$

We transform the first bracket, $(\partial_\mu e_\nu^m - \partial_\nu e_\mu^m)$, using the vielbein constraint,

$$
\partial_\mu e_\nu^m - \partial_\nu e_\mu^m + \omega_\mu^{mn}(e)e_\nu^n - \omega_\nu^{mn}(e)e_\mu^n = 0,
\tag{7.28}
$$

which is true since $\omega(e)$ is the solution of $D_{[\mu}e_{\nu]}^m = 0$. We then get

$$
[\delta_1, \delta_2]e_\mu^m = \delta_E(\xi^\nu)e_\mu^m + \xi^\nu(-\omega_\mu^{mn}(e)e_\nu^n + \omega_\nu^{mn}(e)e_\mu^n) + 2\left[\frac{1}{4}\omega_\mu^{rs}\bar{\epsilon}_2\gamma^m\gamma_r\gamma_s\epsilon_1 - (1 \leftrightarrow 2)\right].
\tag{7.29}
$$

Note that in the first bracket, we have $\omega(e)$ and not our $\omega(e,\psi)$!

Next, we decompose $\gamma^m\gamma^r\gamma^s$ in the basis elements γ^m, γ^{mn}, and γ^{mnp} of the complete set of 2×2 matrices (for space of spinors in three dimensions) as

$$
\gamma^m\gamma^r\gamma^s = \gamma^{mrs} + \eta^{mr}\gamma^s + \eta^{rs}\gamma^m - \eta^{ms}\gamma^r.
\tag{7.30}
$$

This is proven as follows. First, we cannot have γ^{mr} terms, since we would be left with a single Lorentz index, and there is no invariant with a single index we can multiply it with. So we write the above sum with arbitrary coefficients, and then fix the coefficients by taking particular cases: Taking $m = 0, r = 1, s = 2$, we fix the coefficient of γ^{mrs}. Because different gamma matrices anticommute, and γ^{mrs} is antisymmetrized with strength one, it follows that $\gamma^{012} = \gamma^0\gamma^1\gamma^2$. Next, the coefficient of $\eta^{mr}\gamma^s$, for instance, is found by taking, for example, $m = r = 1, s = 2$, and the fact that $\gamma^1\gamma^1\gamma^2 = \gamma^2$.

Since in (7.29) we have $\bar{\epsilon}_2\gamma^m\gamma_r\gamma_s\epsilon_1$, we use $\gamma^{mrs} = -\epsilon^{mrs}$ and the Majorana spinor relations (7.24) to find for the square bracket [...] in (7.29).

$$
2\omega_\mu^{rs}(e,\psi)\bar{\epsilon}_2\gamma_s\epsilon_1 = \omega_\mu^{rs}(e,\psi)\xi_s.
\tag{7.31}
$$

We then obtain

$$[\delta_1, \delta_2]e_\mu^m = \delta_E(\xi^\nu)e_\mu^m + [\xi^\nu\omega_\nu^{mn}(e)]e_{\mu n} + [\omega_\mu^{ms}(e, \psi) - \omega_\mu^{ms}(e)]\xi_s. \tag{7.32}$$

But the difference in the spin connections is given by

$$\omega_\mu^{mn}(e, \psi) - \omega_\mu^{mn}(e) = \omega_\mu^{mn}(\psi),$$
$$\omega_{mn\,\mu}(\psi) = \kappa^2(\bar\psi_\mu\gamma_m\psi_n - \bar\psi_\mu\gamma_n\psi_m + \bar\psi_m\gamma_\mu\psi_n). \tag{7.33}$$

Then, adding and subtracting a term so as to form $\omega(e, \psi)$ instead of $\omega(e)$ in the second term, we obtain

$$[\delta_1, \delta_2]e_\mu^m = \delta_E(\xi^\nu)e_\mu^m + [\xi^\nu\omega_\nu^{mn}(e, \psi)]e_{\mu n} + [\omega_{ms\,\mu}(\psi) - \omega_{m\mu\,s}(\psi)]\xi_s. \tag{7.34}$$

Using the Majorana spinor relations (7.24), we obtain for the term with the last bracket

$$[\omega_{ms\,\mu}(\psi) - \omega_{m\mu\,s}(\psi)]\xi_s = 2\kappa^2\bar\psi_\mu\gamma^m\psi_s\xi^s = \delta_Q(-k\xi^s\psi_s). \tag{7.35}$$

We have therefore finally obtained the local susy algebra, from its representation on the vielbein e_μ^m, as

$$[\delta_1, \delta_2]e_\mu^m = \delta_E(\xi^\nu) + \delta_{l.L.}(\xi^\nu\omega_\nu^{mn}(e, \psi)) + \delta_Q(-k\xi^\nu\psi_\nu). \tag{7.36}$$

Now it remains to introduce the auxiliary field, since we are interested in off-shell susy. We must find its contribution to the susy transformation rules. The variation of the auxiliary field action is

$$\delta\int d^3x\left(-\frac{1}{2}eS^2\right) = \int d^3x(-\kappa\bar\epsilon\gamma^\mu\psi_\mu S^2 - eS\delta S). \tag{7.37}$$

By analogy with the rigid susy case (WZ), we add a new term to the variation of the gravitino. It has to be proportional to S so as to be zero on-shell, and then by Lorentz invariance (and matching dimensions) it can only be

$$\delta_S\psi_\mu = cS\gamma_\mu\epsilon. \tag{7.38}$$

Under this new term in the transformation of the gravitino, the variation of the RS action gets the new term

$$\delta_S I_{RS} = c\int d^3x S\epsilon^{\mu\nu\rho}[\bar\epsilon\gamma_\mu D_\nu(\omega)\psi_\rho]. \tag{7.39}$$

Requiring cancellation of (7.37) against (7.39) implies

$$\delta S = \kappa\bar\epsilon\gamma^\mu\psi_\mu S + \frac{c}{e}\epsilon^{\mu\nu\rho}\bar\epsilon\gamma_\mu D_\nu(\omega)\psi_\rho. \tag{7.40}$$

But now we also get an extra term in the commutator we have just computed, of two supersymmetries acting on the vielbein,

$$[\delta_1, \delta_2]e_\mu^m|_{extra} = 2c\kappa S\bar\epsilon_2\gamma^m\gamma_\mu\epsilon_1 - (1 \leftrightarrow 2) = 4c\kappa S\bar\epsilon_2\gamma^{mn}\epsilon_1 e_{\mu n}, \tag{7.41}$$

where in the second line we have used the Majorana spinor relations (7.24) and $\gamma_{mn} = -\epsilon_{mnr}\gamma_r$. Finally, this term is written as *an extra contribution to the susy algebra*,

$$\delta_{l.L.}(4c\kappa S\bar\epsilon_s\gamma^m_{\ n}\epsilon_1) = \delta_{l.L.}(-2c\kappa S\epsilon^m_{\ ns}\xi^s). \tag{7.42}$$

Thus indeed, as we said, the local susy algebra depends on the representation (on the fields) it acts on.

On the gravitino, we have

$$[\delta_1, \delta_2]\psi_\mu - \frac{1}{4\kappa}[\delta_1 \omega_\mu^{mn}(e, \psi)]\gamma_{mn}\epsilon_2 + c(\delta_1 S)\gamma_\mu\epsilon_2 - (1 \leftrightarrow 2). \qquad (7.43)$$

It can be proven that the local susy algebra is represented on ψ_μ as well. The same is true on S, and the representation on S (as on ψ_μ) fixes $c = 1$. We see then that, indeed, imposing the closure of the algebra fixes a priori free coefficients in the susy transformation rules.

Important concepts to remember

- In three dimensions on-shell, we only need to add an auxiliary field, with the action of the opposite sign from the rigid susy case.
- To get an off-shell susy representation, we need to represent the local susy algebra on the fields.
- The local susy algebra cannot be obtained from group theory alone, and it depends on dimension, and its parameters depend on the fields of the representation.
- The algebra is found by requiring closure on fields, that is, on the right-hand side of the commutator $[\delta_{\epsilon_1}, \delta_{\epsilon_2}]$, we need to get a sum of invariances of the theory.
- We find the local susy algebra from imposing closure on the graviton, and then the closure on the other fields should follow, and fix any free coefficients.
- The vielbein variation is found by canceling the variation of S_{RS} against the variation of S_{EH}, given $\delta\psi$.
- We fix the term added to $\delta\psi$ by invariance to $cS\gamma_\mu\epsilon$. Then δS follows from cancellation of the extra terms, and the value of $c = 1$ is found by closure of the algebra on S, $[\delta_1, \delta_2]S$.

References and further reading

For more details on $\mathcal{N} = 1$ off-shell supergravity in three dimensions, see [19].

Exercises

(1) Prove the closure of the general coordinate transformation

$$[\delta_{g.c.}(\eta^\mu), \delta_{g.c.}(\xi^\mu)] = \delta_{g.c.}(\xi^\mu \partial_\mu \eta^\nu - \eta^\mu \partial_\mu \xi^\nu), \qquad (7.44)$$

when acting on e_μ^a and ω_μ^{ab}.

(2) Check that

$$D_\mu e_\nu^a = \partial_\mu e_\nu^a + \omega_\mu^{ab} e_\nu^b - \Gamma_{\mu\nu}^\rho(g)e_\rho^a \qquad (7.45)$$

is Einstein and Lorentz covariant, by substituting the Einstein and Lorentz transformations of e, ω and $\Gamma(g)$.

(3) Write down explicitly $[\delta_1, \delta_2]S$ for three-dimensional $\mathcal{N} = 1$ supergravity.

(4) Check that in three dimensions, the EH gravity action in first-order formalism (for e and ω) is gauge invariant (up to global issues), for instance, by writing it as a CS theory,

$$S = \int d^3x \ \text{Tr}\left(dA \wedge A + \frac{2}{3}A \wedge A \wedge A\right) \tag{7.46}$$

for the gauge field $A_\mu = e_\mu^a P_a + \omega_\mu^{ab} J_{ab}$, with the bilinear form $\text{Tr}(P_a J_{bc}) = \epsilon_{abc}$ and the rest zero.

Coset theory and rigid superspace

In this chapter, we will describe coset theory, and then use it to construct rigid superspace as a coset manifold.

8.1 Coset theory

Consider a group G and a subgroup H of G. Then the coset G/H is the group G modulo the equivalence under multiplication by an element of H. So,

$$G/H = \{g \in G | g \sim g \cdot h, \ \forall h \in H\}. \tag{8.1}$$

If G and H are continuous groups, then G/H is a manifold, known as the coset manifold.

Considering coordinates $z^\alpha \in \mathbb{R}^d$ on the coset manifold, a coset representative for the Lie group is an element $L(z) \in G$.

To understand the concepts, we consider the simplest example of coset, the sphere,

$$S^n = \frac{SO(n+1)}{SO(n)}. \tag{8.2}$$

Here $SO(n+1)$ is the invariance group of S^n, and the $SO(n)$ is the group of local rotations of S^n (the "local Lorentz group"), which leave a point of S^n invariant. $SO(n)$ is then a linear coordinate change, or change of reference frame, that leaves the point invariant.

The easiest to understand is the usual sphere, $S^2 = SO(3)/SO(2)$. The $SO(3)$ invariance of the sphere is the rotation group in Euclidean three-dimensional space, and it takes us between *different* points on S^2. On the other hand, the $SO(2) = U(1)$ rotations correspond to rotations around an axis of the sphere going through the point, thus leaving it invariant. Therefore S^2 is generated by rotations in $SO(3)$, but not in $SO(2)$, that is, in $SO(3)/SO(2)$. This generalizes trivially to S^n.

Since the multiplication by an element $g \in SO(3)$ moves us on the coset (on the S^2), the action of g on the coset representative should move us on the coset.

To understand better this fact, consider the example of the S^n in stereographic projection. This means that we consider S^n embedded in $\mathbb{R}^{n+1}$, with coordinates x^μ, and we consider a plane that bisects the sphere S^n, passing through its origin. The coordinates on the plane are $z^\alpha \in \mathbb{R}^n$, and the sphere is projected onto the plane by a line from the north pole to the plane, which passes through a unique point on the sphere.

Consider a sphere of radius 1, so $x^\mu x^\mu = 1$, where $x^\mu = (x^\alpha, x^{n+1})$ are the coordinates of a point on the sphere. Then, from the similarity of the triangles formed by the vertical, the projecting line and the line in the plane (vs. the line parallel to it, from the point on the sphere), we obtain

$$1 - x^{n+1} = \frac{x^\alpha}{z^\alpha}. \tag{8.3}$$

Moreover, we have

$$x^\alpha = \frac{2z^\alpha}{1+z^2} \Rightarrow x^{n+1} = 1 - \frac{x^\alpha}{z^\alpha} = \frac{z^2 - 1}{z^2 + 1}. \tag{8.4}$$

In that case, since the coset representative, which is an element $L(z) \in G$, should move us on the coset (sphere), consider a coset representative $L(z) \in SO(n+1)$ that takes us from the north pole, with coordinates $x_N^\mu = (\vec{0}, 1)$, to the arbitrary point of coordinates x^μ.

We can check that, if we choose the coset representative

$$L(z) = \begin{pmatrix} \delta^{\alpha\beta} - \frac{2z^\alpha z^\beta}{1+z^2} & \frac{2z^\alpha}{1+z^2} \\ \frac{2z^\alpha}{1+z^2} & \frac{z^2-1}{z^2+1} \end{pmatrix} \in SO(n+1), \tag{8.5}$$

(which is an $(n+1) \times (n+1)$ orthogonal matrix of unit determinant, as we can easily check), then we have

$$L(z) \cdot (x_N^\mu) = \begin{pmatrix} \frac{2z^\alpha}{1+z^2} & \frac{z^2-1}{z^2+1} \end{pmatrix} = x^\mu. \tag{8.6}$$

Next, we define the coset manifold in detail. Consider the Lie algebra of the continuous group G, with generators T_a, satisfying

$$[T_a, T_b] = f_{ab}{}^c T_c. \tag{8.7}$$

Since we will want to apply the formalism to superspace, that is, to graded Lie algebras as well, consider the generalization

$$[T_a, T_b\} = f_{ab}{}^c T_c, \tag{8.8}$$

where the notation means, as usual, commutator or anticommutator, depending on the $\mathbb{Z}_2$ grading of the generators (bosonic or fermionic),

$$[T_a, T_b\}: \quad [B, B]; \quad \{F, F\}; \quad [B, F]. \tag{8.9}$$

Since we want to consider a subgroup H and a coset G/H, split the generators as $T_a = \{H_i, K_\alpha\}$, where $H_i \in H$ forms a subalgebra (by an abuse of notation we will write the same letters G and H for the group and the algebra, hoping that there will be no confusion), and $K_\alpha \in G/H$.

In the following, we will usually consider a *reductive algebra*,

$$[H_i, H_j] = f_{ij}{}^k H_k$$
$$[H_i, K_\alpha] = f_{i\alpha}{}^\beta K_\beta$$
$$[K_\alpha, K_\beta] = f_{\alpha\beta}{}^i H_i + f_{\alpha\beta}{}^\gamma K_\gamma, \tag{8.10}$$

where the first relation is the subgroup relation, the second is due to the reductive algebra, and the last is general. (if we also have $f_{\alpha\beta}{}^\gamma = 0$, we call the algebra symmetric, but we will not need this here).

Note that the *Killing metric* of the Lie algebra is defined as

$$\gamma_{ab} = f_{ad}{}^e f_{be}{}^d. \tag{8.11}$$

Because of the equivalence relation defining the coset (modulo elements $h \in H$), a coset element is

$$e^{z^\alpha K_\alpha} h, \quad \forall h. \tag{8.12}$$

We say that we have a *coset representative*, when we have the above for any fixed h (for instance, $h = 1$). Consider z^α coordinates on the coset, defining it as a *manifold*. We also write

$$h = e^{y^i H_i}. \tag{8.13}$$

As seen for S^n, a general group element $g \in G$ induces a motion on the coset, since

$$g e^{z^\alpha K_\alpha} = e^{z'^\alpha K_\alpha} h(z, g). \tag{8.14}$$

Here $e^{z^\alpha K_\alpha} = L(z)$ is a coset representative, so we can write the above also as

$$gL(z) = L(z')h(z, g). \tag{8.15}$$

The notation of H_i, K_α is usual in group theory, but since we will apply the coset formalism to general relativity and superspace, we will instead use α, μ (for curved) and m (for flat) indices. We write $z \cdot K \equiv z^\mu K_\mu$ and $dz \cdot K \equiv dz^m K_m$ (for 1-forms).

In order to construct the coset as a manifold, we define the following objects:

- (inverse) vielbein $e_m^\mu(z)$
- (spin) H-connection $\omega_\mu^i(z)$
- Lie vector (or Killing vector) $f_a^\mu(z)$
- H-compensator $\Omega_a^i(z)$,

by multiplying the coset representative with infinitesimal group elements from the left and from the right,

$$e^{z \cdot K} e^{dz \cdot K} = e^{z \cdot K + dz^m e_m^\mu(z) K_\mu} \times e^{dz^m e_m^\mu(x) \omega_\mu^i(z) H_i} + \mathcal{O}(dz^2) \tag{8.16}$$

$$e^{dg^a T_a} e^{z \cdot K} = e^{z \cdot K + dg^a f_a^\mu(z) K_\mu} \times e^{-dg^a \Omega_a^i(z) H_i} + \mathcal{O}(dg^2). \tag{8.17}$$

Here as usual, μ is a "curved index" and m is a "flat index," though the way we defined them seems, for now, different from the usual general relativity definition.

An equivalent definition for the vielbein (e_μ^m, instead of the inverse vielbein e_m^μ) and the H-connection ω_μ^i in (8.16) is given by

$$L^{-1}(z)\partial_\mu L(z) = e_\mu^m(z) K_m + \omega_\mu^i(z) H_i, \tag{8.18}$$

or in form language

$$L^{-1}dL = e \cdot K + \omega \cdot H, \tag{8.19}$$

where $L(z) = e^{-z^\alpha K_\alpha}$ is a coset element.

To show the equivalence to (8.16), we multiply the latter by $L(z)$ from the left, and denoting

$$dz^m e_m^\mu(z) \equiv \Delta z^\mu, \tag{8.20}$$

we obtain

$$e^{-\Delta z^\mu e_\mu^m K_m} = L(z)L^{-1}(z + \Delta z)e^{\Delta z^\mu \omega_\mu^i H_i}, \tag{8.21}$$

and then we expand in Δz to obtain the equivalent form. We leave the details as an exercise.

On the other hand, (8.17) is a rewriting of the relation

$$gL(z) = L(z')h(z, g), \tag{8.22}$$

argued for before, in the case of infinitesimal g.

From it, we derive

$$L^{-1}(z')\partial_\mu' L(z') - L^{-1}(z)\partial_\mu L(z) = \left[e_\mu^m(z')K_m + \omega_\mu^i(z')H_i\right] - \left[e_\mu^m(z)K_m + \omega_\mu^i(z)H_i\right]$$
$$= hL^{-1}(z)\partial_\mu'(L(z)h^{-1}) - L^{-1}(z)\partial_\mu L(z), \tag{8.23}$$

since the constant g matrix canceled with g^{-1} in the form in the second equality; the first equality uses the definition of the vielbein and the H-connection.

Since we also have, by the definition of z',

$$z'^\mu = z^\mu + dg^a f_a^\mu, \tag{8.24}$$

and thus

$$L(z') - L(z) = dg^a f_a^\mu(z)\partial_\mu L(z)$$
$$= dg^a(T_a L(z) + L(z)\Omega_a^i H_i), \tag{8.25}$$

where the second equality is a consequence of the definition (8.17), and where

$$h(z, g) \simeq 1 - dg^a \Omega_a^i H_i, \tag{8.26}$$

and since (8.24) gives

$$\partial_\mu' = \frac{\partial}{\partial z'^\mu} = -(\partial_\mu f_a^\nu(z)dg^a)\partial_\nu, \tag{8.27}$$

we can calculate $\partial_\nu e_\mu^m$ and $\partial_\nu \omega_\mu^i$ from the coefficients of K_m and H_i, respectively, in the first equality in (8.23).

But, we can define the *Lie derivative* along K_a from the general definition of a Lie derivative as a general coordinate transformation with parameter $\xi^\mu = f_a^\mu$, which on e_μ^m and ω_μ^i gives

$$l_{K_a} e_\mu^m \equiv f_a^\nu \partial_\nu e_\mu^m + (\partial_\mu f_a^\nu)e_\nu^m$$
$$l_{K_a} \omega_\mu^m \equiv f_a^\nu \partial_\nu \omega_\mu^i + (\partial_\mu f_a^\nu)\omega_\nu^i. \tag{8.28}$$

Then, from the calculation above (whose details we leave as exercise), we find

$$l_{K_a} e_\mu^m = e_\mu^n(z)f_{ni}{}^m \Omega_a^i(z)$$
$$l_{K_a} \omega_\mu^i(z) = \partial_\mu \Omega_a^i(z) + \omega_\mu^j(z)f_{jk}{}^i \Omega_a^k(z). \tag{8.29}$$

In form language, this becomes

$$l_{K_a} e^m(z) = e^n(z) f_{ni}{}^m \Omega_a^i(z)$$
$$l_{K_a} \omega^i(z) = d\Omega_a^i + \omega^j(z) f_{jk}{}^i \Omega_a^k(z). \tag{8.30}$$

Then the first observation is that the vielbein is only invariant under the motion along the Killing vector if we also add a compensating H transformation, which is why $\Omega_a^i(z)$ is called the H-compensator.

The second observation is that the H-connection ω_μ^i "transforms," if we consider the Lie derivative like the analog of a gauge transformation (it is a derivative, but by multiplying with a differential, it is indeed the general coordinate transformation along the Killing vector, as we already observed, and the latter is a sort of gauge transformation), as a Yang–Mills gauge field (hence the name of "connection," or gauge field in math terms):

$$l_{K_a} \omega^i = d\Omega^i + \omega \cdot f \cdot \Omega^i \tag{8.31}$$

is the analog of

$$\delta A^i = d\epsilon^i + A \cdot f \cdot \epsilon^i. \tag{8.32}$$

Of course, now Ω_a^i is a fixed function of the group, but that is because the Lie derivative is also a fixed function (it is a derivative, so not multiplied by an arbitrary differential, like for the gauge transformation).

Note that the Lie derivative without curved indices is

$$dg^a l_{K_a} = dg^a f_a^\mu \partial_\mu. \tag{8.33}$$

8.2   Parallel transport and general relativity on the coset manifold

On a manifold, we have a notion of parallel transport of a vector, from $V^\mu(x)$ to $V^\mu(x+dx)$. In flat space, we always keep the angle α of the vector V^μ with respect to a fixed straight line along which we transport fixed. From x^μ to $x^\mu + dx^\mu$, the line is defined by dx^μ, so the angle α between $V^\mu(x)$ and dx^μ is the same as between $V^\mu(x+dx)$ and dx^μ, which means that $V^\mu(x+dx)$ is parallel with $V(x)$, hence the name parallel transport.

In curved space, the procedure is analogous, just that the equivalent of the straight line is a geodesic in the curved space, so parallel transport (keeping the angle of a – contravariant – vector with respect to the geodesic fixed) is defined by the Christoffel symbol, via

$$V^\mu(x+dx) = V^\mu(x) - dx^\nu \Gamma_{\nu\rho}^\mu V^\rho(x). \tag{8.34}$$

But on a coset manifold, we can define a notion of parallel transport via the motion induced by a group element $g \in G$. If the two ways of defining parallel transport (as a manifold, via the Christoffel symbol, and as a coset, via multiplication by a group element) are compatible, we say that we have a *group invariant connection* (the connection on the manifold is group invariant).

On the coset, we first define the flat vector as

$$V^m(z) = V^\mu(z) e_\mu^m(z), \tag{8.35}$$

and then we define parallel transport of $V^m(z)$ via the group element, as

$$V^m(z+dz) = V^m(z) - dz^\nu \omega_\nu{}^m{}_n(z) V^n(z), \tag{8.36}$$

where we define $\omega_\nu{}^m{}_n$ as follows.

For fields $\Phi^I(z)$ in some representation $D^I{}_J$ of H, we define

$$\omega_\mu{}^I{}_J = \omega_\mu^i(z)(H_i)^I{}_J. \tag{8.37}$$

Then the covariant derivative with flat indices of the field Φ^I is

$$D_m \Phi^I(z) = e_m^\mu(z) \left[\partial_\mu \Phi^I(z) + \omega_\mu{}^I{}_J(z) \Phi^J(z) \right]. \tag{8.38}$$

In particular, for a flat vector,

$$D_\mu v^m = \partial_\mu v^m + \omega_\mu{}^m{}_n(z) v^n, \tag{8.39}$$

where we have defined

$$\omega_\mu{}^m{}_n(z) = \omega_\mu^i(z)(H_i)^m{}_n. \tag{8.40}$$

If the two ways of defining parallel transport (manifold, with Christoffel symbol, and group motion) are compatible, that is, if we have a group invariant connection, then the un-symmetrized vielbein postulate is valid,

$$D_\mu e_\nu^m = \partial_\mu e_\nu^m - \Gamma_{\mu\nu}^\rho(z) e_\rho^m + \omega_\mu{}^m{}_n(z) e_\nu^n = 0, \tag{8.41}$$

which as we know is solved by $\Gamma = \Gamma(e)$ (from the symmetric part) and $\omega_\mu{}^m{}_n = \omega_\mu{}^m{}_n(e)$ (from the antisymmetric part).

On a *reductive* coset manifold, if parallel transport is compatible with group action, we have

$$\omega_\mu{}^m{}_n(z) = e_\mu^r(z) \omega_r{}^m{}_n(0) + \omega_\mu^i(z) f_{in}{}^m, \tag{8.42}$$

where $\omega_r{}^m{}_n(0)$ is an H-invariant tensor (an invariant tensor of the subgroup H) and $f_{in}{}^m$ are structure constants of the Lie algebra.

The *Maurer–Cartan equations* come from the simple observation (since $d(L \cdot L^{-1}) = 0$, so $dL^{-1} = -L^{-1} \cdot dL \cdot L^{-1}$) that

$$d(L^{-1}dL) = -(L^{-1}dL) \wedge (L^{-1}dL), \tag{8.43}$$

by substituting the explicit form of $L^{-1}dL$ and calculating.

In our case,

$$L^{-1}dL = e^m K_m + \omega^i H_i, \tag{8.44}$$

so by substituting, calculating, and identifying the coefficients of K_m and H_i on the left-hand side and the right-hand side, we obtain

$$de^m + \frac{1}{2} f_{np}{}^m e^n \wedge e^p + f_{im}{}^n \omega^i \wedge e^m = 0$$

$$d\omega^i + \frac{1}{2} f_{jk}{}^i \omega^j \wedge \omega^k + \frac{1}{2} f_{mn}{}^i e^m \wedge e^n = 0. \tag{8.45}$$

From these Maurer–Cartan equations, we can find that the spin connection is

$$\omega_\mu{}^m{}_n = \frac{1}{2}e_\mu^r \bar{f}_{rn}{}^m + \omega_\mu^i f_{in}{}^m, \tag{8.46}$$

where

$$\omega_p{}^m{}_n(0) \equiv \frac{1}{2}\bar{f}_{pn}{}^m = \frac{1}{2}\left[f_{pn}{}^m + \delta^{mm'}(f_{m'p}{}^{n'}\delta_{n'n} + f_{m'n}{}^{n'}\delta_{n'p})\right] \tag{8.47}$$

is an H-invariant tensor, so we obtain indeed the form (8.42), as advertised.

Then, after a calculation, we find that the Riemann tensor with flat indices is constant,

$$R_{pq}{}^m{}_n = f_{ni}{}^m f_{pq}{}^i - \frac{1}{2}(\bar{f}_{rn}{}^m f_{pq}{}^r)$$
$$+ \frac{1}{4}(\bar{f}_{qn}{}^r \bar{f}_{pr}{}^m - \bar{f}_{pn}{}^r \bar{f}_{qr}{}^m). \tag{8.48}$$

We note that for a *group manifold* (there are also coset manifolds that are group manifolds as well: for instance $S^3 = SO(4)/SO(3) \simeq SO(3)$, since $SO(4) \simeq SO(3) \times SO(3)$), one finds instead

$$R_{pq}{}^m{}_n = -\frac{1}{4}f_{pq}{}^r f_{rn}{}^m. \tag{8.49}$$

8.3 *H*-covariant Lie derivatives

We have seen that the Lie derivative of the vielbein is not zero, as expected for the Lie derivative along a Killing vector, due to the H-compensator.

But we can define a modified Lie derivative, an *H-covariant Lie derivative* that does vanish on the vielbein, as

$$\mathcal{L}_{H,K_a}e^m \equiv l_{K_a}e^m - \Omega_a^i f_i{}^{mn}e^n = 0. \tag{8.50}$$

This definition is generalized to an arbitrary field Φ^I in some representation as

$$\mathcal{L}_{H,K_a}\Phi^I(z) = l_{K_a}\Phi^I(z) + \Omega_a^i(z)(H_i)^I{}_J\Phi^J(z). \tag{8.51}$$

In the case of the vielbein, $\Phi^I \to e^m$, thus for an adjoint representation of the coset (e^m is in the adjoint representation of the coset, coupling to K_m), we have

$$(H_i)^I{}_J \to -f_i{}^{mn}. \tag{8.52}$$

In the case of a representation of $H = SO(n)$, we have

$$(H_i)^I{}_J = (-f_i{}^{mn})(t_{mn})^I{}_J, \tag{8.53}$$

and in the particular case of the vector (fundamental) representation of H, we have

$$(t_{mn})^I{}_J = \frac{1}{2}(\delta^{mI}\delta_J^n - \delta^{nJ}\delta_J^m). \tag{8.54}$$

The H-covariant Lie derivative obeys a number of useful commutation identities, which we will use to construct KK spherical harmonics from the coset theory in a future chapter, and which we state here without proof:

1) $[\mathcal{L}_{H,K_a}, \mathcal{L}_{H,K_b}] = -f_{ab}{}^c \mathcal{L}_{H,K_c}$

$$2) \quad [D_m, \mathcal{L}_{H,K_a}] = 0$$

$$3) \quad [dz^\mu D_\mu, \mathcal{L}_{H,K_a}] = 0$$

$$4) \quad [\mathcal{L}_{H,K_a}, \tau^m D_m] = 0$$

$$5) \quad [\mathcal{L}_{H,K_a}, D_\mu^2] = 0, \tag{8.55}$$

where τ^m are Pauli matrices, so the H-covariant Lie derivative obeys the Lie algebra, commutes with the flat covariant derivative, with the differential operator, the Dirac operator, and the Box operator.

Finally, the group-invariant integration measure on the coset manifold is exactly as we expect, that is, in

$$\int_M \mu(z) f(z) d^n z, \tag{8.56}$$

we obtain

$$\mu(z) = [\det e_\mu^m(z)] \mu(0), \tag{8.57}$$

from the Jacobian of the transformation $z \to z'$ on the coset.

8.4   Rigid superspace from the coset formalism

We now apply the coset formalism to the construction of rigid superspace. Superspace is invariant under the super-Poincaré group, and a Lorentz transformation doesn't change the superspace point, so G is the super-Poincaré group and H is the Lorentz group, so we can define superspace as the coset

$$\frac{\text{super-Poincaré}}{\text{Lorentz}}. \tag{8.58}$$

Therefore $H_i = \{M_{mn}\}$ and $K_\alpha = \{P_\mu, Q_A, Q_{\dot{A}}\}$ (note that we use the four-dimensional dotted and undotted indices notation; in other dimensions, we have corresponding modifications to the notation: for instance, in three dimensions, considered Chapter 9, we just drop the $\dot{A}$ space).

The general coset element is then

$$e^{z^\alpha K_\alpha} e^{y^i H_i} = e^{\xi^\mu P_\mu + \epsilon^A Q_A + \epsilon^{\dot{A}} Q_{\dot{A}}} e^{\lambda^{mn} M_{mn}}. \tag{8.59}$$

Correspondingly, the superspace will be denoted as before by $\{x^\mu, \theta^A, \bar{\theta}^{\dot{A}}\}$.

We decompose as usual

$$Q_\alpha = \begin{pmatrix} Q^A \\ \bar{Q}_{\dot{A}} \end{pmatrix}, \tag{8.60}$$

and the gamma matrix representation is, also as usual,

$$(\gamma^m)^\alpha{}_\beta = \begin{pmatrix} 0 & -i(\sigma^m)^{A\dot{B}} \\ i(\bar{\sigma}^m)_{\dot{A}B} & 0 \end{pmatrix}, \tag{8.61}$$

where $\sigma^m = (\vec{\sigma}, \mathbb{1})$ and $\bar{\sigma}^m = (\vec{\sigma}, -\mathbb{1})$.

The transformation law, that is, the action of the G group on the coset, in our case the super-Poincaré group on superspace, is found, as we saw, by the action of a general group element g on the coset representative, $ge^{z^\alpha K_\alpha} = e^{z'^\alpha K_\alpha} h(z, g)$.

Therefore, the supersymmetry transformation law is

$$e^{\epsilon^A Q_A + \bar{\epsilon}^{\dot{A}} Q_{\dot{A}} + \xi^\mu P_\mu + \frac{1}{2}\lambda_{mn} M^{mn}} e^{\bar{\theta}Q + x^\mu P_\mu} = e^{\bar{\theta}'Q + x'^\mu P_\mu} h. \tag{8.62}$$

We will describe better in Chapter 9 how we extract from this the standard transformation rules, but suffice it to say, they are the usual,

$$x'^\mu = x^\mu + \xi^\mu + \frac{1}{2}\bar{\theta}^{\dot{B}}\epsilon^A(-2i(\sigma^\mu)_{A\dot{B}}) + \frac{1}{2}\theta^B\bar{\epsilon}^{\dot{A}}(-2i(\sigma^\mu)_{B\dot{A}}) + \lambda^\mu{}_\nu x^\nu$$

$$\theta'^A = \theta^A + \epsilon^A + \frac{1}{4}\lambda^{mn}(\sigma_{mn})^A{}_B \theta^B$$

$$\bar{\theta}'^{\dot{A}} = \bar{\theta}^{\dot{A}} + \bar{\epsilon}^{\dot{A}} - \frac{1}{4}\lambda^{mn}(\bar{\sigma}_{mn})^{\dot{A}}{}_{\dot{B}} \bar{\theta}^{\dot{B}}. \tag{8.63}$$

We continue with the coset formalism, to define the supervielbein.

For the super-Poincaré algebra, we can easily check that in $[K, K]$ there is no H piece, only K (M in on the right-hand side of only the $[M, M]$ commutator, not any other). But that in turn means that in (8.16) and (8.17), we cannot have any H terms on the right-hand side, since now both $[K, K] \sim K$ and $[H, K] \sim K$. That is,

$$\omega^i_\Lambda = \Omega^i_\Lambda = 0. \tag{8.64}$$

Then the relation defining the supervielbein becomes simply (our notation is: $\Lambda = (\mu, A, \dot{A})$ are curved indices in superspace, and $M = (m, V, \dot{V})$ are flat indices in superspace)

$$e^{z^\Lambda K_\Lambda} e^{dz^M K_M} = e^{z^\Lambda K_\Lambda + dz^M E^\Lambda_M K_\Lambda} + \mathcal{O}(dz^2), \tag{8.65}$$

(since $\omega^i_\Lambda = 0$), and then we find

$$E^\Lambda_M = \begin{pmatrix} \delta^\mu_m & 0 & 0 \\ -i\sigma^\mu_{A\dot{B}}\theta^{\dot{B}} & \delta^B_A & 0 \\ -i\sigma^\mu_{B\dot{A}}\theta^B & 0 & \delta^{\dot{B}}_{\dot{A}} \end{pmatrix}. \tag{8.66}$$

We define flat covariant derivatives as in the general relativity case, with the supervielbein, as

$$D_M \equiv (D_m, D_A, D_{\dot{A}})$$
$$= E^\Lambda_M(\partial_\Lambda + \omega^i_\Lambda T_i) = E^\Lambda_M \partial_\Lambda. \tag{8.67}$$

Substituting the supervielbein, we get

$$D_m = \partial_m$$
$$D_A = \partial_A - i\sigma^\mu_{A\dot{B}}\bar{\theta}^{\dot{B}}\partial_\mu$$
$$D_{\dot{A}} = \partial_{\dot{A}} - i\sigma^\mu_{B\dot{A}}\theta^B\partial_\mu. \tag{8.68}$$

Then we can define torsions and curvatures as before, by the commutator of two covariant derivatives with flat indices, namely by

$$[D_M, D_N\} = T^P_{MN}D_P + R^i_{MN}T_i. \tag{8.69}$$

Here we have written $R^i_{MN}T_i$ to emphasize the general case, which we will use later, but at this moment we have $T_i = M_{rs}$, so the curvature terms are actually $R^{rs}_{MN}M_{rs}$.

We have also defined, as usual,

$$R^i_{MN} = e^\Lambda_M e^\Sigma_N R^i_{\Lambda\Sigma}$$
$$R^i_{\Lambda\Sigma} = \partial_\Lambda \omega^i_\Sigma - \partial_\Sigma \omega^i_\Lambda + f_{jk}{}^i \omega^j_\Lambda \Omega^k_\Sigma$$
$$T^P_{MN} = e^\Lambda_M (D_\Lambda e^\Sigma_N)e^P_\Sigma - M \leftrightarrow N$$
$$D_\Lambda e^\Sigma_N = \partial_\Lambda e^\Sigma_N + \omega^i_\Lambda f_{Ni}{}^P e^\Sigma_P. \tag{8.70}$$

But since $\omega^i_\Lambda = 0$ in rigid superspace, we have no curvatures (R's), only torsions (T's), and specifically the only nonzero torsion is

$$T^m_{A\dot B} = f_{A\dot B}{}^m, \tag{8.71}$$

which means that the only nontrivial commutator is

$$\{D_A, D_{\dot B}\} = T^m_{A\dot B}D_m. \tag{8.72}$$

For a transformation of coordinates on superspace,

$$\begin{pmatrix} x' \\ \theta' \end{pmatrix} = \begin{pmatrix} A & B \\ C & D \end{pmatrix}\begin{pmatrix} x \\ \theta \end{pmatrix} \equiv M \begin{pmatrix} x \\ \theta \end{pmatrix}, \tag{8.73}$$

we have to define a superjacobian. A superjacobian is obtained as a superdeterminant, that is, a determinant on superspace. If we calculate the effect of a quadratic bosonic Lagrangian on a path integral, the result is $(\det A)$, whereas for fermions the result is $(1/\det D)$. For a matrix that mixes bosons and fermions, like M, result is

$$\text{sdet } M \equiv \frac{\det(A - BD^{-1}C)}{\det D}. \tag{8.74}$$

Thus for the change of coordinates above, the superjacobian is $J = \text{sdet } M$.

The integration measure is, like in the general coset case,

$$\int d^4x\, d^4\theta\, \mu(x,\theta)f(x,\theta) = \int d^4x\, d^4\theta\, J(x,\theta)\mu(0)f(x,\theta), \tag{8.75}$$

where the measure is the (super, in this case) determinant of the (super)vielbein,

$$\mu = \text{sdet } E^M_\Lambda. \tag{8.76}$$

But for rigid superspace, we can calculate from (8.66) that

$$\text{sdet } E^M_\Lambda = 1/\text{sdet } E^\Lambda_M = 1, \tag{8.77}$$

so the measure on rigid superspace is trivial. But in the case of local superspace, which will be analyzed in Chapter 9, we will find the same [sdet E^M_Λ] measure, which is why we have emphasized it here.

In order to find irreducible representations, we must impose supersymmetry-preserving constraints on superfields, as we saw in previous chapters. Since D_M's commute with the Q's, we write constraints in terms of D_M's = "covariant constraints."

In the case of rigid superspace, the torsions and curvatures are fixed and almost trivial, but in general, torsions and curvatures contain information, and as we saw, arise on the right-hand side of commutators $[D_M, D_N\}$, so we will in fact impose constraints using torsions and curvatures.

In principle, we should also treat the covariant formulation of SYM in superspace in the same rigid superspace treatment, but we will treat it in Chapter 9, when discussing local superspace, since the local superspace will be using a formalism similar to it.

Important concepts to remember

- A coset G/H is the reduction of the group G under the equivalence relation generated by a subgroup H, and for a continuous group it is a manifold.
- A reductive algebra has $[H, K] \sim K$.
- A general group element g generates a motion on the coset by $g e^{z^\alpha K_\alpha} = e^{z'^\alpha K_\alpha} h(z, g)$.
- On the coset we can define a vielbein $e_m^\mu(x)$, H-connection $\omega_\mu^i(x)$, Lie derivative $f_a^\mu(x)$, and H-compensator $\Omega_a^i(x)$.
- The vielbein is invariant under a Lie derivative only if we add a compensating H transformation, and the H-connection "transforms" under the Lie derivative as a YM connection under gauge transformation, with $\epsilon^i = \Omega^i$.
- The compatibility of parallel transport with group motion fixes $\Gamma_{\nu\rho}^\mu$ in terms of e_μ^m and ω_μ^i.
- For a compatible spin connection on the coset manifold, the Riemann tensor with flat indices is constant.
- The H-covariant Lie derivative vanishes on the vielbein and satisfies commutation relations.
- Rigid superspace is the coset super-Poincaré/Lorentz.
- In rigid superspace, $\omega = \Omega = 0$ and $l = \mathcal{L}_H$.
- Rigid superspace has no curvatures, and only one torsion.
- The measure on superspace is trivial, since sdet $E_\Lambda^M = 1$.

References and further reading

For more details on coset theory see, for example, section 5.3 of [5], [20], or [11], though the most complete discussion is in the lectures [14] and [21].

Exercises

(1) Prove that from parallel transport defined on the coset manifold as

$$V^m(x + dx) = V^m(x) - dx^\nu \omega_{\nu\,n}^{\,m} V^n(x), \tag{8.78}$$

we obtain the unsymmetrized vielbein postulate.

(2) Prove that

$$[D_m, \mathcal{L}_H] = 0, \tag{8.79}$$

using the fact that $\mathcal{L}_H e_m^\mu = 0$ and $\mathcal{L}_H \omega_\mu^i = 0$.

(3) Prove that the definition

$$e^{x\cdot K} e^{dx\cdot K} = e^{x\cdot K + dx^m e_m^\mu(x) K_\mu} \times e^{dx^m e_m^\mu(x)\omega_\mu^i(x)H_i} + \mathcal{O}(dx^2) \tag{8.80}$$

is equivalent to the definition

$$L^{-1}\partial_\mu L = e_\mu^m(x)K_m + \omega_\mu^i(x)H_i$$
$$L \equiv e^{-x^\alpha K_\alpha}, \tag{8.81}$$

by expanding to first order in dx.

(4) Use

$$e^{\epsilon^A Q_A + \bar\epsilon^{\dot A} Q_{\dot A} + \xi^\mu P_\mu + \frac{1}{2}\lambda_{mn}M^{mn}} e^{\bar\theta Q + x^\mu P_\mu} = e^{\bar\theta' Q + x'^\mu P_\mu} h \tag{8.82}$$

to prove the transformation laws of x^μ and θ's,

$$x'^\mu = x^\mu + \xi^\mu + \frac{1}{2}\bar\theta^{\dot B}\epsilon^A(-2i(\sigma^\mu)_{A\dot B}) + \frac{1}{2}\theta^B\bar\epsilon^{\dot A}(-2i(\sigma^\mu)_{B\dot A}) + \lambda^\mu{}_\nu x^\nu$$
$$\theta'^A = \theta^A + \epsilon^A + \frac{1}{4}\lambda^{mn}(\sigma_{mn})^A{}_B\theta^B$$
$$\bar\theta'^{\dot A} = \bar\theta^{\dot A} + \bar\epsilon^{\dot A} - \frac{1}{4}\lambda^{mn}(\bar\sigma_{mn})^{\dot A}{}_{\dot B}\bar\theta^{\dot B}. \tag{8.83}$$

(5) Show the details of the derivation of

$$l_{K_a}e_\mu^m = e_\mu^n(z)f_{ni}{}^m\Omega_a^i(z)$$
$$l_{K_a}\omega_\mu^i(z) = \partial_\mu\Omega_a^i(z) + \omega_\mu^j(z)f_{jk}{}^i\Omega_a^k(z). \tag{8.84}$$

9

Covariant formulation of YM in rigid superspace and local superspace formalisms

In this chapter, we will apply the coset formalism to construct local superspace, specifically in three dimensions. But before we do that, we will continue with the rigid superspace, and construct a covariant formulation of YM in rigid superspace, that we will then use to define local superspace in the coset formalism.

9.1 Rigid superspace as a coset: transformation rules and constraints

The transformation rules for fields under a general transformation for the coset are defined as follows. Define $D_I{}^J(h)$ as a representation of $h \in H$ (the equivalence group, or "Lorentz" group), so

$$D_I{}^J(h_1)D_J{}^K(h_2) = D_I{}^K(h_1 \cdot h_2). \tag{9.1}$$

Then $u(g)$ is in a representation of $g \in G$, the invariance group, acting on the fields $\Phi^I(x)$ defined on the coset. In general, we have that the motion of the invariance group on the coset shifts $x \to x'$ and adds an H motion, so

$$ge^{x \cdot K} = e^{\tau(g^{-1}x) \cdot K}h(\tau(g^{-1}x), g), \tag{9.2}$$

where $x' = \tau(g^{-1})x$. This becomes on the fields

$$[u(g)\Phi^I](x) = D_I{}^J\left(h(\tau(g^{-1}), g)\right)\Phi_J(\tau(g^{-1}x)), \tag{9.3}$$

or, in infinitesimal form, using the definitions of the Lie vector (Killing vector) $f_a^\mu(x)$ and the H-compensator $\Omega_a^i(x)$ from Chapter 8,

$$\delta\Phi^I = [u(g) - \mathbb{1}]\Phi_I = -dg^a\left[f_a^\mu(x)\partial_\mu\Phi_I + \Omega_a^i(x)(H_i)_I{}^J\Phi_J\right]. \tag{9.4}$$

Applying to supersymmetry, where $H = \{M_{mn}\}$ (Lorentz group), since in the commutators of K's and of K's and H's ([K, K] and [H, K]) there are no H's, from the definitions (8.16) and (8.17) (so acting with K's on K's cannot give H's) we find that

$$\omega_\mu^i = \Omega_a^i = 0, \tag{9.5}$$

so that from (8.17) applied to the susy case, now

$$e^{\xi^\mu P_\mu + \epsilon^A Q_A + \bar{\epsilon}^{\dot{A}} \bar{Q}_{\dot{A}} + \frac{1}{2}\lambda^{mn}M_{mn}} \cdot e^{x^\mu P_\mu + \theta^A Q_A + \bar{\theta}^{\dot{A}} \bar{Q}_{\dot{A}}} = e^{x'^\mu P_\mu + \theta'^A Q_A + \bar{\theta}'^{\dot{A}} \bar{Q}_{\dot{A}}}h(x, g), \tag{9.6}$$

we find the transformation rules (that include the general coordinate, or Einstein, transformations with parameter ξ^μ, local Lorentz with parameter $\lambda^\mu{}_\nu$, and supersymmetry with parameters ϵ^A and $\bar\epsilon^{\dot A}$)

$$x'^\mu = x^\mu + \xi^\mu + \frac{1}{2}\bar\theta^{\dot B}\epsilon^A(-2i(\sigma^\mu)_{A\dot B}) + \frac{1}{2}\theta^B\bar\epsilon^{\dot A}(-2i(\sigma^\mu)_{B\dot A}) + \lambda^\mu{}_\nu x^\nu,$$

$$\theta'^A = \theta^A + \epsilon^A + \frac{1}{4}\lambda^{mn}(\sigma_{mn})^A{}_B\theta^B,$$

$$\bar\theta'^{\dot A} = \bar\theta^{\dot A} + \bar\epsilon^{\dot A} - \frac{1}{4}\lambda^{mn}(\bar\sigma_{mn})^{\dot A}{}_{\dot B}\bar\theta^{\dot B}. \tag{9.7}$$

But, since $\Omega^i_a = 0$, in the infinitesimal case, we have $\delta\Phi^I = -dg^a f^\mu_a \partial_\mu\Phi^I$, or more precisely on superspace, $\delta\Phi^I = -dg^\Lambda f^\Lambda_a \partial_\Lambda\Phi^I$, where $\Lambda = (\mu, A, \dot A)$. Specializing to supersymmetry transformations, with parameters ϵ^A and $\bar\epsilon^{\dot A}$, and in this case renaming the Lie (or Killing) vectors f^Λ_a as l^Λ_A, we have

$$\delta x^\Lambda \partial_\Lambda = (\epsilon^A l^\Lambda_A + \bar\epsilon^{\dot A} l^\Lambda_{\dot A})\partial_\Lambda, \tag{9.8}$$

or, with $l_A \equiv l^\Lambda_A \partial_\Lambda$,

$$l_A = \frac{\partial}{\partial\theta^A} + i(\sigma^\mu)_{A\dot B}\bar\theta^{\dot B}\partial_\mu,$$

$$l_{\dot A} = \frac{\partial}{\partial\theta^{\dot A}} + i(\sigma^\mu)_{B\dot A}\theta^B\partial_\mu. \tag{9.9}$$

Then, the supersymmetry transformation of a superfield is

$$\delta\Phi(x,\theta) = (\epsilon^A l_A + \epsilon^{\dot A} l_{\dot A})\Phi(x,\theta), \tag{9.10}$$

and we find that $(-l_A)$ satisfies a representation of the susy algebra

$$\{D_A, D_{\dot B}\} = -2i(\sigma^m)_{A\dot B}D_m, \tag{9.11}$$

that is

$$\{-l_A, -l_B\} = -2i(\sigma^m)_{A\dot B}(-l_m). \tag{9.12}$$

But, in general, as we said, the torsions and curvatures are defined from the commutators of covariant derivatives,

$$[D_m, D_n] = T^p_{mn}(x)D_p + R^i_{mn}(x)T_i,$$

$$T^p_{mn} = e^\mu_m e^p_\nu(D_\mu e^\nu_n) - (m \leftrightarrow n),$$

$$D_\mu e^\nu_n = \partial_\mu e^\nu_n + \omega^i_\mu f^p_{ni} e^\nu_p,$$

$$R^i_{mn} = e^\mu_m e^\nu_n[\partial_\mu\omega^i_\nu - \partial_\nu\omega^i_\mu + f^i_{jk}\omega^j_\mu\omega^k_\nu]. \tag{9.13}$$

Since for supersymmetry $\omega^i_\mu = 0$, there are no curvatures, and only torsions. Moreover, the only nonzero torsion is

$$T^m_{A\dot B} = f_{A\dot B}{}^m. \tag{9.14}$$

We also have the fact that the H-covariant Lie derivatives vanish on both the vielbein and the spin connections, in the general case (not just for superspace)

$$\mathcal{L}e^\mu_m = \mathcal{L}\omega^i_\mu = 0, \tag{9.15}$$

and that the Maurer–Cartan equations give

$$\partial_\mu e_\nu^s - \omega_\mu^i f_{mi}{}^s e_\nu^m = 0,$$
$$R_{\mu\nu}^i(\omega) + f^i{}_{mn} e_\mu^m e_\nu^n = 0. \tag{9.16}$$

In the case of superspace, we have to impose susy preserving constraints on the super-fields, which as we saw must be in terms of the covariant derivatives (which commute with the supercharges, represented by the Lie derivatives).

But we consider the possible constraints always in increasing order of mass dimension (so in increasing order of irrelevance at low energies). Since $[\theta] = -1/2$ and, under $\int d^4x$, $\int d^2\theta = -\frac{1}{2}D^2$, it follows that $[d\theta] = [\partial/\partial\theta] = +1/2$.

Then the lowest order constraint, at dimension 1/2, is the chiral constraint,

$$D_A\phi = 0. \tag{9.17}$$

Note that imposing both chiral and antichiral constraints, $D_A\phi = \bar{D}_{\dot{A}}\phi = 0$, implies ϕ is constant, so it is trivial. Imposing only the chiral constraint gives the chiral superfield.

The next order constraint in terms of mass dimension is the dimension one constraint

$$D^2\phi = 0 = \bar{D}^2\phi, \tag{9.18}$$

where the conjugate constraint comes from the reality of ϕ. This gives the real linear superfield.

On the other hand, the YM multiplet in the previous, "prepotential," formalism is a real linear superfield, specifically the vector superfield $V(x,\theta)$, that contains A_μ. It has the gauge invariance

$$\delta V = -\Lambda - \Lambda^\dagger. \tag{9.19}$$

Its super field strength is the gauge invariant and chiral superfield

$$W_A = \bar{D}^2 D_A V, \tag{9.20}$$

which is also Majorana and obeys the Bianchi identity, specifically

$$(W_A)^\dagger = W_{\dot{A}} \quad D^A W_A + D^{\dot{A}} W_{\dot{A}} = 0. \tag{9.21}$$

A good action for this is

$$I = \mathrm{Tr}\int d^4x \int d^2\theta\, W^A W_A. \tag{9.22}$$

9.2 Covariant formulation of four-dimensional YM in rigid superspace

In this section, we will use the notation: curved indices $M = \{\mu, \alpha\}$ (bosonic and fermionic) and flat indices $A = \{m, a\}$ (bosonic and fermionic). Note that in three dimensions, to which we will later apply the formalism, there are no dotted fermionic indices. We need this notation in order to deal with curved and flat indices, since for the latter we usually use A.

In YM theory, we have a gauge field $A_\mu^{\tilde{a}}(x)$, and we define covariant derivatives $D_\mu = \partial_\mu + A_\mu^{\tilde{a}} T_{\tilde{a}}$. We extend this concept of covariant derivatives to superspace: We first define super-gauge fields in superspace, $A_M^{\tilde{a}}(x, \theta)$, and then super-gauge-covariant derivatives

$$\mathcal{D}_M \equiv \partial_M + A_M^{\tilde{a}} T_{\tilde{a}}. \tag{9.23}$$

We then define covariant derivatives with flat indices

$$\mathcal{D}_A \equiv E_A^M \mathcal{D}_M = E_A^M \partial_M + E_A^M A_M = D_A + A_A, \tag{9.24}$$

where in rigid superspace $D_A = E_A^M \partial_M$ (as we saw in Chapter 8), and we have defined $A_A = A_M E_A^M$.

We define torsions and curvatures in the usual manner, except that: (a) we have a graded Lie algebra, so we must use the graded commutator of covariant derivatives and (b) we have not only gravitational curvatures, extended to superspace as R_{AB}^{rs}, but now also YM curvatures (super field strengths of the super-gauge field) $F_{AB}^{\tilde{a}}$, so

$$\{\mathcal{D}_A, \mathcal{D}_B\} \equiv T_{AB}^C \mathcal{D}_C + \frac{1}{2} R_{AB}^{rs} M_{rs} + F_{AB}^{\tilde{a}} T_{\tilde{a}}, \tag{9.25}$$

and the YM curvature is also defined using the graded commutators and derivatives,

$$F_{AB} = D_A A_B - (-)^{AB}(A \leftrightarrow B) + [A_A, A_B] - T_{AB}^C A_C. \tag{9.26}$$

Moreover, we see that we needed to subtract $T_{AB}^C A_C$ in Eq. (9.26), so as to remain only with $T_{AB}^C D_C$, and not $T_{AB}^C \mathcal{D}_C$ as it was without it, on the right-hand side of the graded commutator.

The above definitions are general. But since we are in rigid superspace, we have no curvatures, that is, $R_{AB}^{rs} = 0$, and we have fixed torsion only, $T_{A\dot{B}}^m = f_{A\dot{B}}^m$, and the rest are zero.

In order to obtain a good multiplet, we have to impose constraints on the superfields. As we mentioned in Chapter 8, we can impose constraints involving covariant derivatives (since they commute with the supercharges, so preserve supersymmetry). But since the commutators of two covariant derivatives define torsions and curvatures, we can impose constraints on the torsions and curvatures. However, in rigid superspace, we already saw that the torsions and gravitational curvatures are fixed. That leaves the YM curvatures (or super field strengths) as objects on which we can impose constraints.

We can impose three types of constraints:

– 1. **Representation preserving constraints**, which are needed in order to find a good (irreducible) representation. For SYM, these are

$$F_{\alpha\beta} = F_{\dot{\alpha}\dot{\beta}} = 0, \tag{9.27}$$

or, in the previous notation,

$$F_{AB} = F_{\dot{A}\dot{B}} = 0. \tag{9.28}$$

- 2. **Conventional constraints** (or *optional constraints*). These are constraints that we may or may not impose, depending on convenience. For SYM, these are

$$F_{\alpha\dot\beta} = 0, \tag{9.29}$$

or, using the previous notation, $F_{A\dot B} = 0$.

To understand them, a good analogy, or rather an example of the general idea is the no-torsion constraint, or vielbein postulate, of pure general relativity: $T^a_{\mu\nu} = D_{[\mu} e^a_{\nu]} = 0$. We can impose it or not, obtaining the first- or the second-order formalism, with independent ω, or with $\omega = \omega(e)$, respectively.

- 3. On top of the constraints, we have also to solve **Bianchi identities**, which arise from the super-Jacobi identities

$$[\mathcal{D}_A, [\mathcal{D}_B, \mathcal{D}_C\}\} + (-)^{A(B+C)}[\mathcal{D}_B, [\mathcal{D}_C, \mathcal{D}_A\}\} + (-)^{C(A+B)}[\mathcal{D}_C, [\mathcal{D}_A, \mathcal{D}_B\}\} = 0. \tag{9.30}$$

These are of course identities, that is, by expanding the commutators, we get $0 = 0$. But, as always, we substitute the commutators with their expressions in terms of torsions and curvatures. Then, we must check whether these definitions are consistent (which is what solving them means).

The Bianchi identities are then of the type $\sim DR + DT + DF = 0$. The simplest case of Bianchi identity is of course the Maxwell case, when $F_{\mu\nu} = \partial_\mu A_\nu - \partial_\nu A_\mu$, and the Bianchi identity is

$$\partial_{[\mu} F_{\nu\rho]} = 0, \tag{9.31}$$

which is one of Maxwell's equations, and follows identically from the definition of F in terms of the gauge field A, that is, the solution of the Bianchi identity is F as a function of A, $F = dA$.

In general, we can start by solving either the constraints or the Bianchi identities, but in the end we have to satisfy both. Moreover, as always, we consider the constraints and/or Bianchi identities in order of increasing mass dimension, solving the lower dimension ones and then substituting in the higher dimension ones.

9.3 Solving the constraints and Bianchi identity and relation to prepotential formalism

In this section, we go back to denoting the flat spinor indices by $A, \dot A$, since we don't need to talk about superspace flat indices anymore.

Since $[D_\alpha] = 1/2$ and $[\partial_\mu] = 1$, so also $[A_\mu] = 1$ and $[A_\alpha] = 1/2$, we have $[F_{\alpha\beta}] = 1$, $[F_{\alpha\mu}] = 3/2$, and $[F_{\mu\nu}] = 2$. As usual, we can exchange a vector index for a bi-spinor index, so

$$F_{A\dot C, B} \equiv (\sigma^m)_{A\dot C} F_{mB} \quad F_{mn} = (\sigma^m)^{A\dot A} (\sigma^n)^{B\dot B} F_{A\dot A, B\dot B}. \tag{9.32}$$

We solve the Bianchi identities first. The lowest dimension identity comes from (A, B, C) (in the previous notation) and is actually $(A\dot CB)$ (in the current notation). Since the only nonzero torsion is $T^m_{A\dot{B}} = 2i(\sigma^m)_{A\dot{B}}$, multiplying $\mathcal{D}_m$ on the right-hand side of a commutator, when we do the second commutator, we obtain again the curvatures, and we focus on the nontrivial one, the YM curvature. But since it is multiplied by $-i(\sigma^m)_{A\dot{B}}$, we get the Bianchi identity of lowest dimension, 3/2,

$$F_{A\dot{C},B} + F_{B\dot{C},A} = 0, \tag{9.33}$$

together with the similar one, obtained from indices $(C\dot{A}\dot{B})$ in the Jacobi identity,

$$F_{C\dot{A},\dot{B}} + F_{C\dot{B},\dot{A}} = 0. \tag{9.34}$$

The solution to the first is

$$F_{A\dot{C},B} = \epsilon_{AB}\left(-\frac{i}{4}w_{\dot{C}}\right), \tag{9.35}$$

which is a definition for the quantity in brackets, and for the second is

$$F_{C\dot{A},\dot{B}} = \epsilon_{\dot{A}\dot{B}}\left(-\frac{i}{4}w_C\right). \tag{9.36}$$

From conjugation, we obtain that $(w_c)^\dagger = -w_{\dot{C}}$, so w is an (anti)Majorana spinor.

The next in mass dimension is dimension 2, for which the Bianchi identity for indices (ABn), with n replaced by $C\dot{C}$, is

$$\mathcal{D}_A F_{\dot{B},C\dot{C}} + \mathcal{D}_B F_{A,C\dot{C}} = 0, \tag{9.37}$$

and replacing $F_{B,C\dot{C}}$ from the solution to the previous Bianchi identity, we obtain the condition

$$\mathcal{D}_A w_{\dot{B}} = 0 \Rightarrow \mathcal{D}_{\dot{A}}w_B = 0, \tag{9.38}$$

so w is a chiral superfield.

Also at mass dimension 2, the Bianchi identity for indices $(A\dot{B}n)$, with n replaced by $C\dot{C}$, gives

$$\mathcal{D}_A F_{\dot{B},C\dot{C}} + \mathcal{D}_{\dot{B}} F_{A,C\dot{C}} + 2iF_{A\dot{B},C\dot{C}} = 0, \tag{9.39}$$

which splits into an antisymmetric part giving

$$\mathcal{D}^A w_A + \mathcal{D}^{\dot{A}} w_{\dot{A}} = 0 \Rightarrow \mathcal{D}^A w_A \in \mathbb{R}, \tag{9.40}$$

and the symmetric part giving

$$\mathcal{D}_{(A} w_{C)} + 4(\sigma^{mn})_{AC} F_{mn} = 0. \tag{9.41}$$

At mass dimension 5/2, we have the Bianchi identity for indices (Amn), giving

$$\mathcal{D}_A F_{mn} + \mathcal{D}_n F_{Am} + \mathcal{D}_m F_{nA} = 0. \tag{9.42}$$

Using $F_{A\dot{A},B\dot{B}}$ instead of F_{mn}, because of the antisymmetry in $[mn]$, we can write

$$F_{A\dot{A},B\dot{B}} = \epsilon_{AB}f_{\dot{A}\dot{B}} + \epsilon_{\dot{A}\dot{B}}f_{AB} \quad (f_{AB})^\dagger = f_{\dot{A}\dot{B}}. \tag{9.43}$$

Then the Bianchi identity relates f_{AB}, so F_{mn}, in terms of w_B.

Finally, at mass dimension 3, the Bianchi identity for indices (mnp) is identically true, since this is the usual YM Bianchi identity, and we have expressed F_{mn} in terms of A_m.

The action is

$$S = \int d^4x \int d^2\theta \, \text{Tr}[w^A w_A].$$ (9.44)

Then finally all the F_{MN}'s are written in terms of the field w_A, which is chiral, Majorana, and satisfying $\mathcal{D}^A w_A + \mathcal{D}^{\dot{A}} w_{\dot{A}} = 0$. This is similar to the W_A in the prepotential formalism, except now we have defined the constraints of the superfield strength with $\mathcal{D}_A$, not with D_A.

It remains to solve the constraints, and find the relation to the prepotential formalism.

The conventional constraint

$$F_{A\dot{B}} = 0$$ (9.45)

is solved by (remember that now F is defined by subtraction of a torsion term, which is only nonzero for $F_{A\dot{B}}$, and is $T^m_{A\dot{B}} A_m = 2i(\sigma^m)_{A\dot{B}} A_m$)

$$A_m = \frac{i}{4}(\sigma_m)^{A\dot{B}}(D_A A_{\dot{B}} + D_{\dot{B}} A_A + \{A_A, A_{\dot{B}}\}).$$ (9.46)

Then the representation preserving constraint $F_{AB} = 0$ is solved by

$$\mathcal{D}_A = D_A + A_A = e^{-\Omega} D_A e^{\Omega},$$ (9.47)

which is a definition for Ω (note that here D_A acts both on e^Ω and on whatever it is on the right-hand side).

Then, solving for A_A, we find

$$A_A = e^{-\Omega}(D_A e^\Omega),$$ (9.48)

where now D_A only acts on e^Ω, and similarly the constraint $F_{\dot{A}\dot{B}} = 0$ is solved by

$$A_{\dot{A}} = e^{-\bar{\Omega}}(D_{\dot{A}} e^{\bar{\Omega}}).$$ (9.49)

The gauge invariance on A_A is then given by

$$A'_A = e^K A_A e^{-K},$$ (9.50)

whereas the gauge invariance on the prepotential was

$$e^{V'} = e^{\bar{\Lambda}} e^V e^{-\Lambda},$$ (9.51)

so that the gauge invariance of Ω is

$$e^{\Omega'} = e^{\bar{\Lambda}} e^{\Omega} e^{-K},$$ (9.52)

or in infinitesimal form

$$\Delta\Omega = \bar{\Lambda} - K + \text{more}.$$ (9.53)

Then we can relate the prepotential with the covariant approach by

$$e^V = e^{\Omega} e^{-\bar{\Omega}},$$ (9.54)

where the right-hand side gives A_A.

We can define a new type of covariant derivative, the *chiral representation of derivatives* by

$$\mathcal{D}_{0M} = e^{\bar{\Omega}} \mathcal{D}_M e^{-\bar{\Omega}}, \tag{9.55}$$

so that

$$\mathcal{D}_{0\dot{A}} = D_{\dot{A}}$$
$$\mathcal{D}_{0A} = e^{\bar{\Omega}}(e^{-\Omega} D_A e^{\Omega})e^{-\bar{\Omega}} = e^{-V} D_A e^V. \tag{9.56}$$

Then, using $\mathcal{D}_{0M}$, we can define w_{0A} in the same way that we have defined w_A from $\mathcal{D}_M$, and therefore find that the relation between the super field strength in the covariant approach and in the prepotential approach is

$$w_{0A} = W_A, \tag{9.57}$$

and, because of how w_{0A} is defined, it is related to the previously defined w_A by

$$w_{0A} = e^{\bar{\Omega}} w_A e^{-\bar{\Omega}}, \tag{9.58}$$

so the relation between w_A and the prepotential W_A is

$$W_A = e^{\bar{\Omega}} w_A e^{-\bar{\Omega}}. \tag{9.59}$$

9.4 Coset approach to three-dimensional supergravity

We will use rigid superspace, with the covariant formulation of SYM, gauging the super-Poincaré Lie algebra. Therefore we have a "YM theory of super-Poincaré on rigid superspace."

The group is then

$$T_I = \{P_\mu, Q_\alpha, M_{rs}\}, \tag{9.60}$$

and correspondingly we define gauge fields H_A^I. Therefore the index I splits as $I = \{M, rs\} = \{\mu, \alpha, rs\}$, so into curved plus Lorentz. As before, we have the usual coset construction

$$\frac{\text{Super-Poincaré}}{\text{Lorentz}} = \frac{\{T_I\}}{\{M_{rs}\}}. \tag{9.61}$$

As in the covariant SYM formulation, we define covariant derivatives (now called ∇_A instead of $\mathcal{D}_A$ to emphasize that we do local superspace in this way, so introduce gravity)

$$\nabla_A \equiv D_A + H_A^I T_I, \tag{9.62}$$

where D_A are the rigid superspace covariant derivatives.

We represent T_I by the Lie derivatives $-\mathcal{L}_I$, which are $\mathcal{L}_I = (\mathcal{L}_M, \mathcal{L}_{rs})$, specifically

$$\mathcal{L}_\mu = i\partial_\mu,$$
$$\mathcal{L}_\alpha = \partial_\alpha - i\theta^\beta (\gamma^\mu)_{\beta\alpha} \partial_\mu,$$
$$\frac{1}{2} L^{rs} \mathcal{L}_{rs} = L^\mu{}_\nu x^\nu \partial_\mu + \frac{1}{4} L^{rs} (\gamma_{rs})^\alpha{}_\beta \theta^\beta \partial_\alpha + \frac{1}{2} L^{rs} M_{rs}. \tag{9.63}$$

In the last equation, we have shown the multiplication with a parameter L^{rs}, emphasizing that when it multiplies $x^\mu \partial_\nu$, the indices are curved, $L^\mu{}_\nu$. More explicitly, we would write

$$\mathcal{L}^{rs} = (\delta^\mu_r \delta^\nu_s - \delta^\nu_r \delta^\mu_s) x^\nu \partial_\mu + \frac{1}{4} (\gamma_{rs})^\alpha{}_\beta \theta^\beta \partial_\alpha + M_{rs}. \tag{9.64}$$

As we saw before, $T_I = -\mathcal{L}_I$ commute with the D_M's in rigid superspace,

$$[T_I = -\mathcal{L}_I, D_M\} = 0. \tag{9.65}$$

We now look for another basis in T_I, specifically rewriting the $\mathcal{L}_M$'s as linear combinations of the D_M's. This means that we have also another basis for H^I_A. Of course, the M_{rs} part is unchanged by this procedure. We then redefine

$$H^I_A T_I \equiv h^M_A D_M + \frac{1}{2} \phi^{rs}_A M_{rs}. \tag{9.66}$$

Then we obtain

$$\nabla_A = \delta^M_A D_M + h^M_A D_M + \frac{1}{2} \phi^{rs}_A M_{rs}$$
$$\equiv E^M_A D_M + \frac{1}{2} \phi^{rs}_A M_{rs}, \tag{9.67}$$

where we have defined the *local supervielbein* by

$$E^M_A = \delta^M_A + h^M_A, \tag{9.68}$$

and D_M is the covariant derivative of rigid superspace.

As before, we define torsions and curvatures by

$$[\nabla_A, \nabla_B\} = T^C_{AB} \nabla_C + \frac{1}{2} R^{rs}_{AB} M_{rs} \tag{9.69}$$

just that now, due to the fact that the supervielbein is nontrivial (contains degrees of freedom through h^M_A), the torsions and curvatures are also nontrivial. Therefore now *we need to* impose constraints on torsions and curvatures.

In the following, we will substitute vector indices for bi-spinor indices (as in the case of YM constraints considered before), via

$$v_{ab} = (\gamma_\mu)_{ab} v^\mu,$$
$$v_\mu = -\frac{1}{2} (\gamma_\mu)^{ab} v_{ab}. \tag{9.70}$$

Then we impose the following *conventional constraints* (with the same meaning as in the covariant YM approach, just that now we have constraints involving not just the F's, but also the T's and the R's):

$$\{\nabla_a, \nabla_b\} = 2i \nabla_{ab},$$
$$T^{de}_{a,bc} = 0. \tag{9.71}$$

The first one is implicit, being equivalent to

$$T^{cd}_{a,b} = 2i \delta^{(c}_a \delta^{d)}_b; \quad T^c_{a,b} = 0; \quad R^{rs}_{a,b} = 0. \tag{9.72}$$

We will not present here the derivation, but solving the first set of constraints and the Bianchi identities, we can express everything in terms of E^M_a and ϕ^{rs}_a, and solving the second set of constraints, we can also express ϕ_a^{rs} in terms of $E_a^M(x,\theta)$. The only remaining independent field is then $E_a^M(x,\theta)$.

A part of the solution is defined as follows. We first write

$$[\nabla_a, \nabla_{bc}] = \frac{1}{2}\epsilon_{ab}W_c + \frac{1}{2}\epsilon_{ac}W_b, \tag{9.73}$$

then expand W_a in T_I as

$$W_a = W_a{}^b\nabla_b + \hat{W}_a{}^{bc}\nabla_{bc} + \frac{1}{2}W_a^{rs}M_{rs}, \tag{9.74}$$

with $D^aW_a = 0$, then obtain

$$W_{ab} = \epsilon_{ab}R,$$
$$\hat{W}_a^{bc} = 0,$$
$$W_{a,bc} \equiv \frac{1}{4}W_a^{rs}(\gamma_{rs})_{bc} = G_{abc} - \frac{1}{3}\epsilon_{ab}\nabla_c R - \frac{1}{3}\nabla_b R,$$
$$\nabla^a G_{abc} = \frac{2i}{3}\nabla_{bc}R, \tag{9.75}$$

where R is a real superfield and G_{abc} is a real, totally symmetric superfield. Moreover, we have

$$[\nabla_{ab}, \nabla_{cd}] = \epsilon_{bc}f_{ad} + \epsilon_{ad}f_{bc},$$
$$[\nabla_{(a}W_{b)} = 2if_{ab}. \tag{9.76}$$

Both R and G_{abc} can be expressed in terms of the independent components $E_a^M(x,\theta)$. This means that, finally, $E_a^M(x,\theta)$ is the only independent superfield, also called a "prepotential," though not in the SYM sense.

Next, we use the symmetries to simplify further the result.

Since we have a supervielbein, we can say without construction an action that we should have the usual symmetries, just extended to superspace. On the vielbein, we can act with Einstein (general coordinate) transformations with parameter $\xi^\mu(x)$ and local Lorentz transformations with parameter $\lambda^m{}_n(x)$. Therefore on the supervielbein, we can act with super-Einstein transformations with parameter $k^M(x,\theta)$ and super-local Lorentz transformations with parameter $\lambda^A{}_B(x,\theta)$.

Since we have now only E_a^M as independent fields, it means that the super-local Lorentz transformations with mixed indices (Bose–Fermi), $\lambda^m{}_a$, are not invariances anymore, and on the other hand, $\lambda^m{}_n(x,\theta)$ will act only on the dependent fields (on E_m^M only), so that leaves us with $\lambda^a{}_b(x,\theta)$, $k^\alpha(x,\theta)$, and $k^m(x,\theta) \leftrightarrow k^{\alpha\beta}(x,\theta)$ nontrivial invariances, that we can use.

Another way of saying the above is that the super-Einstein transformations come from the super-gauge transformations, since $E_A^M = \delta_A^M + h_A^M$, and h_A^M are gauge fields, so super-Einstein transformations have parameters $k^M(x,\theta)$, and $\lambda^a{}_b(x,\theta)$ come from the super-H-transformations, with parameter L^{rs}, transforming into $L^{rs}(\gamma_{rs})^\alpha{}_\beta$. This will remain so after solving all the constraints in terms of $E_a^M(x,\theta)$.

Now we use these symmetries to fix some gauges:

- We can use the local Lorentz transformation with L^{rs} or rather with $\lambda^a{}_b = L^{rs}(\gamma_{rs})^a{}_b$ to fix $E_a^\alpha = \delta_a^\alpha\psi(x,\theta)$.

- We can use the fermionic super-Einstein transformation with parameter $k^\alpha(x, \theta)$ to fix $E_a^{\alpha\beta}\delta_\alpha^a = 0$, after which we are left only with $E^{(a\alpha\beta)}(x, \theta)$ (the totally symmetric part, since we can decompose $E_a^{(\alpha\beta)}$ into a totally symmetric part and a trace).
- We can use the bosonic super-Einstein transformation with parameter $k^{\alpha\beta}(x, \theta)$ to fix some more components. But first, we decompose the superfield parameter in parameters in xA space (expand the superfield into component fields) as

$$k^{\alpha\beta}(x, \theta) = \xi^{\alpha\beta}(x) + i\theta^{(\alpha}\epsilon^{\beta)}(x) + i\theta_\gamma\eta^{(\gamma\alpha\beta)}(x) + i\theta^2\zeta^{\alpha\beta}(x). \tag{9.77}$$

Note that above we have used the same decomposition of an object with three spinor indices (two symmetrized) into a totally symmetric part and a trace, namely $\theta_\gamma\sigma^{\gamma,(\alpha\beta)} = \theta_\gamma\left[\delta^{\gamma\alpha}\delta_{\delta\epsilon}\sigma^{\delta(\epsilon\beta)} + \sigma^{(\gamma\alpha\beta)}\right]$.

We recognize ξ^{ab} as just the general coordinate parameter $\xi^\mu(x)$ and $\epsilon^\alpha(x)$ as the supersymmetry parameter. This means that $\eta^{(\alpha\beta\gamma)}(x)$ and $\zeta^{(\alpha\beta)}(x)$ are extra symmetries that we could use to fix more components, or not. The purpose of fixing more components is to get to the off-shell supergravity multiplet, but the formulation we have now is also good.

The remaining independent components are expanded as

$$\begin{aligned} \psi(x, \theta) &= h(x) + i\theta^\alpha\lambda_\alpha(x) + i\theta^2 S \\ &= e_{m\mu}(x)\delta^{\mu m} + i\theta^\alpha(\gamma^\mu\psi_\mu)_\alpha(x) + i\theta^2 S, \\ E^{(a\alpha\beta)}(x, \theta) &= \chi^{(a\alpha\beta)} + \theta^{(\alpha}X^{\alpha\beta)} + \delta_{bc}^{\alpha\beta}(\theta_d h^{(abcd)} + i\theta^2\psi^{(abc)}). \end{aligned} \tag{9.78}$$

Note that for the terms in $E^{(a\alpha\beta)}$ linear in θ, we have used the same decomposition of the object with three symmetrized indices and an independent one into a totally symmetric one and a trace.

We then obtain the fields:

- $h(x) = e_{\mu m}\delta^{\mu m}$ is the trace of the graviton.
- $\lambda_\alpha = (\gamma^\mu\psi_\mu)_\alpha$ is the gamma-trace of the gravitino.
- S is the off-shell supergravity auxiliary field.
- $\psi^{(abc)}$ is the gamma-traceless part of the gravitino $\psi_{\mu a}$.
- $h^{(abcd)}$ is the traceless part of the symmetrized graviton $e_{\mu m}$.

The first three fields, from $\psi(x, \theta)$, are all fields that vanish on-shell, whereas the fields in $E^{(a\alpha\beta)}$ are fields that are nonzero on-shell.

We see that we still have the fields $\chi^{(a\alpha\beta)}$ and $X^{(\alpha\beta)}$ left. But we can now use the $\eta^{(\gamma\alpha\beta)}$ and $\zeta^{(\alpha\beta)}$ transformations to fix a "Wess–Zumino (WZ)-like gauge" where we have just the off-shell supergravity multiplet $\{e_{\mu m}, \psi_\mu, S\}$. But this is a choice, which is not required.

We finally turn to finding the action for supergravity in superspace.

The first step is finding the measure for integration. It is of the same functional form as the one in rigid superspace, for the same reason:

$$\int d^3x d^2\theta \; \text{sdet} \, E_M^A(x, \theta), \tag{9.79}$$

except, of course, now the supervielbein E_M^A is not trivial anymore, but contains degrees of freedom. This has an obvious generalization to higher dimensions. In fact, in four dimensions, we will see that the action is just this integration measure over local superspace. In three dimensions, however, we need more fields.

We notice however a problem with finding an action, even if we know the form of the equations of motion that it should reproduce. Since we have expressed a large part of $E_M^A(x, \theta)$ in terms of independent components, when we vary such an action, we have a potential problem. It can be resolved by taking three different possible paths:

- Write the action in terms of unconstrained superfields $E_a^M(x, \theta)$ and vary them independently.
- Choose a gauge as discussed on page 109, where the independent fields are ψ and $E^{(a\alpha\beta)}$, and vary them. We should add compensating transformations to stay in the gauge, except if the action is gauge invariant.
- Write the action in terms of E_M^A and ϕ_M^{rs}, but find their independent variations and only allow those in the action.

In four dimensions, we will use the last one (the action is simpler and it is possible), but in three dimensions, we use the second combined with the third.

The action is found by finding the unique candidate possible given dimension and symmetries.

We can find (by symmetry and dimension considerations) that the equations of motion, which reproduce the component supergravity equations of motion, are (with the addition of a cosmological constant Λ for completeness)

$$R = \Lambda; \quad G_{abc} = 0. \tag{9.80}$$

Then we can also find the action, through similar symmetry and dimension considerations,

$$S = \frac{1}{k^2} \int d^3x d^2\theta \ \text{sdet} \, E_A^M (R + \Lambda), \tag{9.81}$$

where R must be expressed in terms of $E_a^M(x, \theta)$.

Specifically, note first that in three dimensions, $[\psi] = 1$, $[D_\mu] = 1$, $[e] = 0$. Then the field equation for the gravitino ψ_μ, $D_{[\mu}\psi_{\nu]} = 0$, has dimension 2. On the other hand, the vielbein e field equation, the no-torsion constraint $D_{[\mu}e_{\nu]}^a = 0$, has dimension 1. From $\partial_\mu\partial_\nu e$, the only covariant object one can form is $R_{\mu\nu}^{rs}$, which has dimension 2. Finally, it then follows that on-shell, all T's and R's of dimension less than 2 should vanish.

The dimensions of the T's and R's are obtained as

$$[T] = \left[\frac{\nabla \nabla}{\nabla}\right] \quad [R] = [\nabla_A, \nabla_B]. \tag{9.82}$$

Then, again in increasing order of dimension, we start at dimension 0, where the unique object is $T_{ab}^m \sim \gamma_{ab}^m = \text{constant}$, so non-dynamical. At dimension 1/2, we have the constraints $T_{ab}^c = 0$ and $T_{am}^n = 0$. At dimension 1, we have the constraint $R_{ab}^{rs} = 0$, and we have exhausted all constraints. This means that T_{am}^b and T_{mn}^r, which also have dimension 1, should vanish on-shell. At dimension 3/2, we have T_{mn}^a and R_{am}^{rs}, which are not constraints, so also should vanish on-shell.

Finally, we arrive at dimension 2, where we have R^{rs}_{mn}, but we did not need to impose anything. However, from the Bianchi identities, we find that its single contraction, the Ricci tensor, needs to be zero, which is as it should be, since this is the vacuum Einstein equation.

Finally then, considering our solution of the Bianchis and constraints, in terms of W_a, which is itself expressed in terms of R and G_{abc}, we find that we need to have $R = 0$ and $G_{abc} = 0$ on-shell. This then, as we can check from the expression for $[\nabla_{ab}, \nabla_{cd}]$, also implies that the Ricci tensor coming from R^{rs}_{mn} should vanish.

Thus indeed, the equations of motion must be $R = 0$ (extended to $R = \Lambda$ in the presence of a cosmological constant) and $G_{abc} = 0$.

Then note that in the action (9.81), the integration is done using $d^2\theta = \nabla^a \nabla_a$, and that we then get the object

$$\nabla^a \nabla_a R(x, \theta) = R^{rs}_{mn}(x, \theta)\delta^r_m \delta^2_n, \tag{9.83}$$

whose $\theta = 0$ component is the Ricci scalar $\mathcal{R}$ appearing in the Einstein–Hilbert action.

9.5 Super-geometric approach

Finally, we come to the super-geometric approach, which is the easiest to explain, but is less formalized as the coset formalism. It is the approach that will easily generalize to any dimension, and we will use in four dimensions.

The idea is to generalize the description of general relativity in terms of vielbeins and spin connection to superspace. That is, we write now supervielbeins $E^A_M(x, \theta)$ and super-spin connections $\Omega^{AB}_M(x, \theta)$ in superspace. Then define

$$[D_A, D_B] = T^C_{AB}D_C + \frac{1}{2}R^{rs}_{AB}M_{rs}, \tag{9.84}$$

where now the covariant derivatives are defined using the supervielbein and super-spin connection as in general relativity. Then we restrict the independent components using invariances, physical input, and constraints on torsions and curvatures.

Note that unlike the coset approach, now we have also a super-spin connection Ω^{AB}_M. In the coset case, the supervielbein was a derived notion, $E^M_A = \delta^M_A + h^M_A$. Of course, we had a ϕ^{rs}_A, but it was defined as one of the gauge fields.

Important concepts to remember

- The susy transformation rules are obtained, in the coset formulation, from the action of the group on the coset.
- In the covariant formulation of SYM in rigid superspace, we write super-gauge fields $A^{\tilde{a}}_M(x, \theta)$ and covariant derivatives $\mathcal{D}_M = \partial_M + A^{\tilde{a}}_M T_{\tilde{a}}$.
- We write with flat indices $\mathcal{D}_A = D_A + A_A$ and define super-torsion, super-curvature, and super field strengths using the graded Lie commutator of $\mathcal{D}_A$.
- The YM representation preserving constraints are $F_{\alpha\beta} = F_{\dot\alpha\dot\beta} = 0$ and the conventional constraints (optional) are $F_{\alpha\dot\beta} = 0$.

- The Bianchi identities are identities (come from super-Jacobi identities), but because of the way we define torsions and curvatures, they become consistency conditions that need to be solved together with the constraints.
- Solving the Bianchis and constraints for the covariant YM formulation, we obtain fields w_A defined by the same constraints as W_A in the prepotential formulation, but with the covariant derivatives $\mathcal{D}_A$ instead of the D_A ones.
- We can define also a chiral representation of derivatives, by $\mathcal{D}_{0M} = e^{\bar{\Omega}}\mathcal{D}_M e^{-\bar{\Omega}}$, such that $\mathcal{D}_{0\dot{A}} = D_{\dot{A}}$ and $\mathcal{D}_{0A} = e^{-V}D_A e^V$, and then $w_{0A} = W_A$. Thus $W_A = e^{\bar{\Omega}}w_A e^{-\bar{\Omega}}$.
- The coset approach to three-dimensional supergravity is the covariant formulation of the YM theory of the super-Poincaré group on rigid superspace.
- The gauge fields corresponding to the coset are redefined linearly by $H_A^I T_I = h_A^M D_M + 1/2\phi_A^{rs}M_{rs}$ and we define the vielbein as $E_A^M = \delta_A^M + h_A^M$, obtaining the independent fields E_A^M and ϕ_A^{rs}, to be subject to constraints.
- The constraints in three dimensions are $\{\nabla_a, \nabla_b\} = 2i\nabla_{ab}$ and $T_{a,bc}^{de} = 0$.
- The solution of the Bianchis and constraints gives everything in terms of $E_a^M(x,\theta)$.
- Using invariances, we are left with the superfields $\psi(x,\theta)$ in $E_a^\alpha = \delta_a^\alpha\psi$ and $E^{(\alpha\alpha\beta)}(x,\theta)$. In the WZ-like gauge, we find the off-shell supergravity multiplet $\{e_{\mu m}, \psi_\mu, S\}$.
- In the super-geometric approach, we generalize general relativity to superspace, writing supervielbeins $E_M^A(x,\theta)$ and super-spin connections $\Omega^{AB}(x,\theta)$ on superspace and using invariances, physical input, and constraints to define the system.

References and further reading

For more details, see [19].

Exercises

(1) Prove that only
$$T_{A\dot{B}}^m = f_{A\dot{B}}{}^m \tag{9.85}$$
is nonzero in rigid superspace.

(2) If we start with $E_a^\alpha\delta_\beta^a \neq 0$ and $\chi^{(\alpha\alpha\beta)}, X^{\alpha\beta} \neq 0$, calculate $k^\alpha, \eta^{\alpha\beta\gamma}, \zeta^{\alpha\beta}$ that bring them to zero.

(3) Check that the solution of the constraints satisfies the Bianchi identity
$$[\nabla_a, \{\nabla_b, \nabla_c\}] + \text{super} - \text{cyclic} = 0. \tag{9.86}$$

(4) Calculate $T_{a,b}^c, R_{a,b}^{rs}$ in terms of explicit components of E_A^M and ϕ_A^{rs}.

(5) Calculate $T_{a,bc}{}^d$ and $R_{a,bc}{}^{de}$ using the solutions of the Bianchi identities and constraints, and then check that the constraints
$$T_{am}{}^b = T_{mn}{}^r = T_m{}^a = R_{am}{}^{rs} = 0 \tag{9.87}$$
give the equations of motion $R = 0, G_{abc} = 0$.

$\mathcal{N} = 1$ Four-dimensional off-shell supergravity

In this chapter, we apply the off-shell supergravity formalism to the four-dimensional $\mathcal{N} = 1$ case. The logic will follow the three-dimensional case, though the details are different.

As we mentioned, off-shell the four-dimensional graviton e_μ^a has $d(d-1)/2 = 4 \cdot 3/2 = 6$ degrees of freedom, the gravitino ψ_μ has $2^{[d/2]}(d-1) = 2^{[4/2]} \cdot 3 = 12$ degrees of freedom. This means that for a good on-shell representation, we need six bosonic auxiliary degrees of freedom more than the fermionic auxiliary degrees of freedom. We could have several choices, but the minimal set of auxiliary fields is S, P, and A_μ, where S is scalar and P is pseudoscalar, so we can compose a complex scalar $M = S + iP$.

10.1 Supersymmetry transformation rules

We start by writing the supersymmetry transformation rules, just like in the on-shell case, and we will write the action later.

In practice, once we have some unperturbed ("kinetic") supersymmetric model, we can introduce interactions order by order in nonlinearity (i.e., linear, quadratic, etc. in the fields), to both the transformation rules and the action. We write the most general terms (to both) allowed by symmetries and physical input at each order, and then fix the coefficients in both the action and the transformation rules, such that we have supersymmetry.

Of course, in the case of supergravity we can guess (use physical input) to simplify the procedure. For instance, the Lagrangian will always have $e = \det e_\mu^m$ in front, which is a quartic term in the field e_μ^m, but must always be there.

In the case of off-shell supergravity, we start with the on-shell action and transformation rules, and add terms to them: start with adding terms linear in the (auxiliary) fields to the transformation rules, and terms quadratic in the fields in the action. Then continue with quadratic terms in the transformation rules, and so on.

In any case, the result for the susy transformation rules is

$$\delta e_\mu^m = \frac{\kappa_N}{2}\bar{\epsilon}\gamma^m\psi_\mu$$

$$\delta\psi_\mu = \frac{1}{\kappa_N}\left(D_\mu + \frac{i\kappa_N}{2}A_\mu\gamma_5\right)\epsilon - \frac{1}{2}\gamma_\mu\eta\epsilon$$

$$\delta S = \frac{1}{4}\bar{\epsilon}\gamma^\mu R_\mu^{\mathrm{cov}}$$

$$\delta P = -\frac{i}{4}\bar{\epsilon}\gamma_5\gamma^\mu R_\mu^{\text{cov}}$$

$$\delta A_m = \frac{3i}{4}\bar{\epsilon}\gamma_5\left(R_m^{cov} - \frac{1}{3}\gamma_m\gamma^\mu R_\mu^{\text{cov}}\right), \tag{10.1}$$

where

$$\eta \equiv -\frac{1}{3}(S - i\gamma_5 P - iA_\rho\gamma^\rho\gamma_5) \tag{10.2}$$

and

$$R^\mu = \epsilon^{\mu\nu\rho\sigma}\gamma_5\gamma_\nu D_\rho\psi_\sigma \tag{10.3}$$

is the gravitino field equation, but with *supercovariant derivatives*, that is, their variation doesn't have $\partial_\mu\epsilon$ terms. That is,

$$R^{\mu,\text{cov}} = \epsilon^{\mu\nu\rho\sigma}\gamma_5\gamma_\nu\left[D_\rho\psi_\sigma - \frac{i}{2}A_\rho\gamma_5\psi_\sigma + \frac{1}{2}\gamma_\rho\eta\psi_\sigma\right]$$

$$\equiv \epsilon^{\mu\nu\rho\sigma}\gamma_5\gamma_\nu\psi_{\rho\sigma}^{\text{cov}}. \tag{10.4}$$

It is left as an exercise to check that indeed, the variation of $R^{\mu,\text{cov}}$ does not contain any $\partial_\mu\epsilon$ terms.

A few comments are in order about the rules in (10.1). First, of course, when $S = P = A_m = 0$, that is, on-shell, the rules reduce to the on-shell rules we already wrote. As before, in the variation of the vielbein e_μ^m we don't add anything, since we don't have fermionic auxiliary fields, which would be zero on-shell and could appear in the variation of a boson. In the variation of the gravitino, we add the (bosonic) auxiliary fields, adding γ matrices to fix the indices, and the coefficients are found by requiring invariance of the action.

The variation of the bosonic auxiliary fields must be proportional to something which is zero on-shell, so it can only be proportional to the gravitino field equation. Then again we add gamma matrices to fix indices, and the coefficients are found by requiring invariance of the action.

Also note that the covariantization of the gravitino field equation R^μ introduces nonlinear terms in the auxiliary fields, $A_\rho\psi_\sigma$ and $\eta\psi_\sigma$, which appear on the right-hand side of the auxiliary field transformation rules. Correspondingly, there will be interacting (higher than quadratic) terms in the auxiliary fields in the action.

10.2 Action

The Lagrangian is

$$\mathcal{L} = -\frac{e}{2}R(e,\omega) - \frac{1}{2}\epsilon^{\mu\nu\rho\sigma}\bar{\psi}_\mu\gamma_5\gamma_\nu D_\rho\psi_\sigma$$

$$-\frac{e}{3}(S^2 + P^2 - A_\mu^2) \tag{10.5}$$

where the first line is the on-shell Lagrangian we already wrote, and the second are the auxiliary field terms.

We observe again that, while in rigid supersymmetry, the auxiliary, fields are truly auxiliary, that is, their action is free $(+F^2/2)$, in the case of supergravity the auxiliary fields couple to gravity, that is, to the vielbein, through $e = \det e^m_\mu$, so the action is indeed interacting from the point of view of the auxiliary fields, as we noted it should be when talking about transformation rules. Thus scalar field VEVs ($\langle S \rangle$, $\langle P \rangle$, or $\langle A_\mu \rangle$) would give a cosmological constant for gravity, $\int d^4x \, e \, \Lambda$.

Also note that, while the vector A_μ has action with the same sign as the auxiliary field action in rigid supersymmetry $(+e A_\mu A^\mu /3)$, the auxiliary scalars have the opposite sign $(-(S^2 + P^2)/3$. Moreover, the coefficients are also $1/3$ and not $1/2$ (though that is a normalization). All of these facts come from requiring invariance of the action under the susy rules, which, as we said, fixes the coefficients of the interacting terms in both the transformation rules and the action.

One more observation is that, from the action, the mass dimensions of the auxiliary scalars are $[S] = [P] = [A_\mu] = 2$. Since we know that the susy transformation parameter has $[\epsilon] = -1/2$, and we see from the action that the gravitino equation of motion has dimension $[R^{\text{cov}}_\mu] = 5/2$, this means the new terms added to the susy rules had indeed the right dimension, since $\delta[S, P, A_m u] \sim \bar{\epsilon} R^{\text{cov}}_\mu$.

The supergravity equations of motion are

$$0 = -2\frac{\delta I}{\delta e_{av}} = e\left(R^{av} - \frac{1}{2}e^{av}R\right) - \frac{1}{4}\bar{\psi}_\lambda \gamma_5 \gamma^a \tilde{\psi}^{\lambda v} - \frac{e}{3}e^{av}(S^2 + P^2 - A^2_m)$$

$$\frac{\delta I}{\delta\bar{\psi}_\mu} = R^\mu = \epsilon^{\mu v \rho \sigma}\gamma_5 \gamma_v D_\rho \psi_\sigma$$

$$S = P = A_m = 0, \tag{10.6}$$

where

$$\tilde{\psi}^{\lambda v} = 2\epsilon^{\lambda v \rho \sigma} D_\rho \psi_\sigma. \tag{10.7}$$

10.3 Susy algebra closure

As we saw before, in order to have off-shell supersymmetry, it is not enough for the susy rules to leave the Lagrangian invariant, but we must also have a representation of the susy algebra on the fields. In the rigid case, the susy algebra and its representation in terms of commutators are completely determined from group theory considerations; however, in the local case, it is not enough. The local algebra depends on dimension and on the fields. In order to find it, we must impose *closure of the algebra*, that is, the commutator of two transformations must be a linear combination of the other invariances of the theory. For simplicity, we must do it on a field on which the algebra closes on-shell already. We know from previous examples that this happens on the vielbein.

We also have the same comment as in three dimensions. We can't use the first-order formalism for ω, since an independent spin connection will give extra degrees of freedom off-shell, and then we would not have matching of off-shell degrees of freedom anymore.

With the auxiliary fields put to zero (i.e., on-shell), we find

$$[\delta_{\epsilon_1}, \delta_{\epsilon_2}]e_\mu^m = \frac{1}{2}\bar{\epsilon}_2\gamma^m D_\mu\epsilon_1 - 1 \leftrightarrow 2$$
$$= \delta_E(\xi^\mu)e_\mu^m + \delta_{l.l.}(\xi^\mu\omega_\mu^{mn}(e,\psi))e_\mu^m + \delta_Q(-k\xi^\mu\psi_\mu)e_\mu^m, \tag{10.8}$$

where the second line is exactly the same calculation as in three dimensions, so we will not repeat it here (it is left as an exercise). Also, here

$$\xi^\mu = \frac{1}{2}\bar{\epsilon}_2\gamma^\mu\epsilon_1, \tag{10.9}$$

as before.

On the gravitino, we can start with (doing the first variation of the gravitino)

$$[\delta_{\epsilon_1}, \delta_{\epsilon_2}]\psi_\mu = \frac{1}{2}\sigma_{mn}\epsilon_2\delta_{\epsilon_1}\omega_\mu^{mn} + \frac{i}{2}\delta_{\epsilon_1}A_\mu\gamma_5\epsilon_2 - \frac{1}{2}\gamma_\mu\delta_{\epsilon_1}\eta\epsilon_2, \tag{10.10}$$

so we also need the variation of the spin connection in second-order formalism (same as in the 1.5-order formalism we mostly use). It is given by

$$\delta\omega_{\mu ab} = \frac{1}{4}\bar{\epsilon}(\gamma_b\psi_{\mu a}^{cov} - \gamma_a\psi_{\mu b}^{cov} - \gamma_\mu\psi_{ab}^{cov}) + \frac{1}{2}\bar{\epsilon}(\sigma_{ab}\eta + \eta\sigma_{ab})\psi_\mu. \tag{10.11}$$

We notice that again, for the bosonic auxiliary fields (S, P, A_μ) set to zero, we get the on-shell variation (in second-order formalism).

However, we will not continue the calculation on the gravitino and will focus just on the vielbein. We get

$$[\delta_{\epsilon_1}, \delta_{\epsilon_2}]e_\mu^m = \text{previous} + \frac{\kappa_N}{4}\bar{\epsilon}_2\gamma^m\left(iA_\mu\gamma_5 + \frac{1}{3}\gamma_\mu(S - i\gamma_5 P - iA_\rho\gamma^\rho\gamma_5)\right)\epsilon_1 - 1 \leftrightarrow 2. \tag{10.12}$$

We separate the S and P pieces in the extra terms,

$$\frac{\kappa_N}{12}\bar{\epsilon}_2(\delta^{mn} + \gamma^{mn})(S - i\gamma_5 P)\epsilon_1 e_\mu^n - 1 \leftrightarrow 2, \tag{10.13}$$

(Here we wrote $\gamma_\mu = \gamma^n e_\mu^n$ and $\gamma^m\gamma^n = 1/2\{\gamma^m, \gamma^n\} + 1/2[\gamma^m, \gamma^n] = \delta^{mn} + \gamma^{mn}$) and the A_μ pieces

$$\frac{i\kappa_N}{4}\left[A_\mu\bar{\epsilon}_2\gamma^m\gamma_5\epsilon_1 - \frac{1}{3}A_\rho\bar{\epsilon}_2\gamma^m\gamma_\mu\gamma_\rho\gamma_5\epsilon_1\right] - 1 \leftrightarrow 2. \tag{10.14}$$

To continue, we write Majorana spinor relations and gamma matrix decompositions. We already know that

$$\bar{\epsilon}\chi = +\bar{\chi}\epsilon$$
$$\bar{\epsilon}\gamma_\mu\chi = -\bar{\chi}\gamma_\mu\epsilon. \tag{10.15}$$

Using $C\gamma_\mu = -\gamma_\mu^T C$, $\gamma_5 = i\gamma_0\gamma_1\gamma_2\gamma_3$, and $C^T = -C$, we similarly find (as in the previous cases)

$$\bar{\epsilon}\gamma^{mn}\chi = -\bar{\chi}\gamma^{mn}\epsilon$$
$$\bar{\epsilon}\gamma_5\chi = +\bar{\chi}\gamma_5\epsilon$$
$$\bar{\epsilon}\gamma^m\gamma_5\chi = +\bar{\chi}\gamma^m\gamma_5\epsilon$$
$$\bar{\epsilon}\gamma^{mn}\gamma_5\chi = -\bar{\chi}\gamma^{mn}\gamma_5\epsilon. \tag{10.16}$$

We decompose

$$\gamma^m \gamma^\mu \gamma^\rho = \gamma^{m\mu\rho} + \eta^{m\mu}\gamma^\rho - \gamma^\mu \eta^{m\rho} + \gamma^m \eta^{\mu\rho}. \qquad (10.17)$$

The decomposition is in terms of the only possible Lorentz structures involving the 4×4 gamma matrix basis elements ($\mathcal{O}_I = \{ \mathbb{1}, \gamma_\mu, \gamma_5, \gamma_\mu \gamma_5, \gamma_{\mu\nu} \}$), and the coefficients are found by taking different index values $((m\mu\rho) = (123), (112), (211), (121))$ and identifying the left- and right-hand sides (considering that for instance, $\gamma^{123} = \gamma^1 \gamma^2 \gamma^3$, $\gamma^{121} = 0$, etc.).

We next find

$$\gamma_d \gamma_5 = \frac{i}{6} \epsilon_{abcd} \gamma^{abc}, \qquad (10.18)$$

for instance by taking $d = 0$, $(abc) = (123)$, considering that there are six permutations in the sum over the indices $(abc) = (123)$, and identifying the left- and right-hand sides. Then by multiplying with $\epsilon^{a'b'c'd}\gamma_5$ from the right, we find

$$\epsilon^{abcd}\gamma_d = i\gamma^{abc}\gamma_5. \qquad (10.19)$$

Then we see that the δ^{mn} terms in (10.13) are symmetric, so they vanish under $1 \leftrightarrow 2$, while the γ^{mn} terms are antisymmetric, so they remain. The first terms, with $\gamma^m \gamma_5$, in (10.14) are symmetric, so they vanish under $1 \leftrightarrow 2$, while the terms with $\gamma^m \gamma_\mu \gamma^\rho \gamma_5$ decompose using (10.17) and (10.19) into terms with γ_a and $\gamma_a \gamma_5$. The terms with $\gamma_a \gamma_5$ are symmetric, so they cancel under $1 \leftrightarrow 2$, while the terms with γ_a are antisymmetric, so they survive. We finally get

$$\frac{\kappa_N}{6}\bar\epsilon_2 \gamma^{mn}(S - i\gamma_5 P)\epsilon_1 e^n_\mu - \frac{i\kappa_N}{6}A_p \bar\epsilon_2 \gamma^{mnp}\epsilon_1 e^n_\mu. \qquad (10.20)$$

Finally, we obtain for the full algebra

$$[\delta_{\epsilon_1}, \delta_{\epsilon_2}] = \delta_E(\xi^\mu) + \delta_Q(-\xi^\mu \psi_\mu) + \delta_{l.L.}\left[\xi^\mu \hat\omega^{mn}_\mu + \frac{1}{3}\bar\epsilon_2 \sigma^{mn}(S - i\gamma_5 P)\epsilon_1\right],$$

$$\hat\omega^{mn}_\mu = \omega^{mn}_\mu - \frac{i}{3}\epsilon_\mu{}^{mnc}A_c,$$

$$\xi^\mu = \frac{1}{2}\bar\epsilon_2 \gamma^\mu \epsilon_1. \qquad (10.21)$$

Now that we have obtained the algebra; we can check that it is represented on fields. We can also prove that the algebra is realized on the auxiliary fields S, P, and A_μ and on the gravitino ψ_μ, but it is a long calculation, which we will skip. We will only make a few comments about the closure on the gravitino.

The algebra on the gravitino is

$$[\delta_{\epsilon_1}, \delta_{\epsilon_2}]\psi_\mu = \frac{1}{\kappa_N}\left[\frac{1}{4}\delta_{\epsilon_1}\omega^{ab}_\mu(e, \psi)\gamma_{ab} - \frac{3\kappa_N}{4}\bar\epsilon_1 \gamma_5 \left(R^{cov}_\mu - \frac{1}{3}\gamma_\mu \gamma^\nu R^{cov}_\nu\right)\right]\epsilon_2$$

$$- \frac{\kappa_N}{2}(\bar\epsilon_1 \gamma^m \psi_\mu)\gamma_n \eta \epsilon_2 - \frac{1}{2}\gamma_\mu \delta_{\epsilon_1}\eta \epsilon_2 + \frac{i\kappa_N}{2}\bar\epsilon_1 \gamma^m \psi_\mu \gamma_5 \epsilon_2 A_m - (1 \leftrightarrow 2), \qquad (10.22)$$

where $\delta_{\epsilon_1}\eta$ has also many terms involving R^{cov}_μ. So all the terms not coming from $\delta\omega^{ab}_\mu(e, \psi)$ (the only on-shell term) are proportional to some R^{cov}_λ.

In the absence of S, P, and A_μ, that is, on-shell, we have simply

$$[\delta_{\epsilon_1}, \delta_{\epsilon_2}]\psi_\mu = \frac{1}{4\kappa_N}\delta_{\epsilon_1}\omega_\mu^{ab}(e, \psi)\gamma_{ab}\epsilon_2. \tag{10.23}$$

Fierzing, we obtain

$$-\frac{1}{32}\bar{\epsilon}_1\mathcal{O}_I\epsilon_2\left[2\sigma_{ab}\mathcal{O}_I\gamma_b\psi_{\mu a} + \sigma_{ab}\mathcal{O}_I\gamma_\mu\psi_{ba}\right] - (1 \leftrightarrow 2). \tag{10.24}$$

But we have that

$$\gamma_b\psi_{\mu a} - \gamma_a\psi_{\mu b} = -\gamma_\mu\psi_{ab} + R_{\text{cov}}^\lambda \text{ terms.} \tag{10.25}$$

Finally, after a calculation, one finds that on-shell (in the case of no S, P, A_μ), one has

$$[\delta_{\epsilon_1}, \delta_{\epsilon_2}]\psi_\mu = \frac{1}{2}(\bar{\epsilon}_2\gamma^\lambda\epsilon_1)(D_\lambda\psi_\mu - D_\mu\psi_\lambda)$$
$$+ \frac{1}{4}(\bar{\epsilon}^1\gamma^\alpha\epsilon_2)T_{\mu\alpha\beta}^{(1)}R^\beta + \frac{1}{4}(\bar{\epsilon}^1\gamma^{\rho\sigma}\epsilon_2)T_{\mu\rho\sigma\tau}^{(2)}R^\tau, \tag{10.26}$$

where

$$eT_{\mu\alpha\beta}^{(1)} = \frac{1}{4}g_{\mu\alpha}\gamma_\beta + \frac{1}{2}e\gamma_{\mu\alpha\beta\tau}\gamma_5\gamma^\tau$$
$$eT_{\mu\rho\sigma\tau}^{(2)} = g_{\mu\rho}g_{\sigma\tau} + \frac{1}{2}g_{\mu\tau}\sigma_{\rho\sigma} - \frac{1}{2}e\epsilon_{\mu\rho\sigma\tau}\gamma_5. \tag{10.27}$$

Then, noting that

$$R_\mu^{\text{cov}} = R_\mu + A_\mu \text{ terms,} \tag{10.28}$$

we can rewrite everything in terms of R_λ^{cov} terms plus A_μ terms. All the other terms involved such R_λ^{cov} terms, which will cancel.

We will be left with A_μ terms, from the above, and also with S and P terms, that come from extending the variation of ω_μ^{ab} to the off-shell case,

$$\delta\omega_{\mu ab} = \frac{1}{4}\bar{\epsilon}\left(\gamma_b\psi_{\mu a}^{\text{cov}} - \gamma_a\psi_{\mu b}^{\text{cov}} - \gamma_\mu\psi_{ab}^{\text{cov}}\right) + \frac{1}{2}\bar{\epsilon}(\sigma_{ab}\eta + \eta\sigma_{ab})\psi_\mu, \tag{10.29}$$

where

$$\psi_{\rho\sigma}^{\text{cov}} = D_\rho\psi_\sigma - \frac{i}{2}A_\sigma\gamma_5\psi_\rho$$
$$R_{\text{cov}}^\mu = \epsilon^{\mu\nu\rho\sigma}\gamma_5\gamma_\nu\psi_{\rho\sigma}^{\text{cov}}. \tag{10.30}$$

Important concepts to remember

- In four dimensions, the minimal set of auxiliary fields is S, P, A_μ.
- The variation of the auxiliary fields involves the gravitino equation of motion, with supercovariant derivatives, that is, such that their susy variations contain no $\partial_\mu\epsilon$ terms.
- The local susy algebra of the four-dimensional $\mathcal{N} = 1$ supergravity is found by requiring closure of the susy commutator on the vielbein. Then it is realized also on the gravitino and auxiliary fields.

References and further reading

For more details, see sections 1.9 and 1.10 of [5].

Exercises

(1) Redo $[\delta_{\epsilon_1}, \delta_{\epsilon_2}]e_\mu^m$ exactly as in three dimensions and show that we get the right on-shell algebra (without the auxiliary fields).

(2) Check that

$$[\delta_Q(\epsilon), \delta_{g.c.}(\xi^\mu)] = \delta_Q(\xi^\mu \partial_\mu \epsilon). \tag{10.31}$$

(3) Check that

$$\psi_{\rho\sigma}^{cov} \equiv D_\rho \psi_\sigma^{cov} \equiv D_\rho \psi_\sigma - \frac{i}{2}A_\sigma \gamma_5 \psi_\rho + \frac{1}{2}\gamma_\sigma \eta \psi_\rho \tag{10.32}$$

is supercovariant, that is, its susy variation has no $\partial_\mu \epsilon$ terms.

(4) Check that the extra terms coming from the off-shell in $\delta_Q S$ cancel under the given susy laws, using that R_μ is the gravitino field equation, that is,

$$\delta S_\psi = \int R^\mu \delta \psi_\mu. \tag{10.33}$$

$\mathcal{N} = 1$ Four-dimensional supergravity in superspace

In this chapter, we will construct $\mathcal{N} = 1$, four-dimensional supergravity in superspace. Unlike the three-dimensional case, now we will construct local superspace using the super-geometric approach. This means that we will generalize general relativity to superspace, but we will have to use some physical input and constraints. In that sense, it is less well defined as the coset approach (which is more algorithmic), but it is easier to generalize.

11.1 Super-geometric approach: invariances, gauge choices, and fields

In this chapter, we will denote the flat indices as $M = (m, a)$, where m is bosonic and $a = (A, \dot{A})$ is fermionic, and curved indices by $\Lambda = (\mu, \alpha)$.

Since we generalize general relativity, we write a supervielbein $E_\Lambda^M(x, \theta)$ in superspace, as well as a super-spin connection Ω_Λ^{MN}, also in superspace. Since the indices on the fields are superspace indices, the symmetry transformations that act on them must also be generalized to superspace:

- super-Einstein transformations $\xi^\Lambda(x, \theta)$, splitting into bosonic ξ^μ, which contains the usual Einstein transformation parameter as its $\theta = 0$ component, $\xi^\mu(x, \theta = 0) = \xi^\mu(x)$, and fermionic ξ^α, which contains the local supersymmetry transformation parameter as its $\theta = 0$ component, $\xi(\theta = 0) = \epsilon^\alpha$.
- super-local Lorentz $\Lambda^{MN}(x, \theta)$ transformations. But here we must use some physical input. We don't want these transformations to mix bosons and fermions, since bose or fermi is related to Lorentz spin, which we want to be preserved by the super version of the Lorentz transformations. So the matrix of transformations has to be diagonal.

Moreover, the number of Lorentz generators should not be increased in superspace, which would mean extra symmetries, so all the components should be parametrized by the same Λ^{mn}. This means that we must finally have

$$\Lambda^{MN} = \begin{pmatrix} \Lambda^{mn} & 0 & 0 \\ 0 & -\frac{1}{4}(\sigma_{mn})_{AB}\Lambda^{mn} & 0 \\ 0 & 0 & \frac{1}{4}(\sigma_{mn})_{\dot{A}\dot{B}}\Lambda^{mn} \end{pmatrix}. \tag{11.1}$$

But since Ω_Λ^{MN} is a connection (gauge field) for the Λ^{MN} transformations, it follows that the same form must be true for the super-spin connection, that is,

$$\Omega_\Lambda^{MN} = \begin{pmatrix} \Omega_\Lambda^{mn} & 0 & 0 \\ 0 & -\frac{1}{4}(\sigma_{mn})_{AB}\Omega_\Lambda^{mn} & 0 \\ 0 & 0 & \frac{1}{4}(\sigma_{mn})_{\dot{A}\dot{B}}\Omega_\Lambda^{mn} \end{pmatrix}. \tag{11.2}$$

We then define super-GR-covariant derivatives in the usual way, by

$$D_\Lambda = \partial_\Lambda + \frac{1}{2}\Omega_\Lambda^{mn}M_{mn}, \tag{11.3}$$

and covariant derivatives with flat indices also as before, by

$$D_M = E_M^\Lambda D_\Lambda, \tag{11.4}$$

and finally torsions and curvatures from the graded commutator,

$$[D_M, D_N\} = T_{MN}^P D_P + \frac{1}{2}R_{MN}^{mn}M_{mn}. \tag{11.5}$$

Again, these torsions and curvatures will satisfy as consistency conditions the Bianchi identities, obtained from the Jacobi identities,

$$[D_M, [D_N, D_P\}\} + \text{supercyclic} = 0, \tag{11.6}$$

by substituting $[D_N, D_P\}$ in terms of R's and T's. (As we said, the Bianchi identities follow from the Jacobi identities, so they are of type 0=0, but once we define torsions and curvatures from the commutator of derivatives, they become consistency conditions for this definition.)

The rigid superspace limit of the local superspace is given by $E_M^\Lambda \to E_M^{(0)\Lambda}$ (where the rigid inverse supervielbein $E_M^{(0)\Lambda}$ was defined before) and $\Omega_\Lambda^{mn} = 0$.

Guided by the above limit, we now take the following gauge choice, which fixes some of the extra components in the superspace transformations.

$$\begin{aligned} E_\mu^m(x, \theta = 0) &= e_\mu^m \\ E_\mu^a(x, \theta = 0) &= \psi_\mu^a \\ \Omega_\mu^{mn}(x, \theta = 0) &= \omega_\mu^{mn}. \end{aligned} \tag{11.7}$$

We note that this is different from the three-dimensional coset approach, where we used the extra invariances to fix $E_a^\alpha = \delta_a^\alpha \psi$ (the fermi–fermi component) and $E^{(\alpha\beta)}$ as the other independent field (the symmetrization of flat fermi, curved bose indices), quite different from this choice. The moral is that in general, the gauge choice and its relation to physical x-space fields depends on dimension and on theory, there is no general prescription.

11.2 Constraints

We now impose constraints on the system. We have now three types of constraints. The first two types are like in the case of the coset approach, or more precisely the covariant YM formulation on the coset superspace. The third type, super-conformal choice, is new, and comes because we need to avoid having extra symmetries in the action. Unlike super-Einstein and super-local Lorentz transformations, these extra transformations cannot be described before we write down the action, so we will describe them later.

The three sets of constraints are then:

– Conventional constraints,

$$T^p_{mn} = 0$$
$$T^C_{AB} = 0; \quad T^{\dot C}_{A\dot B} = 0$$
$$T^m_{A\dot B} + 2i(\sigma^m)_{A\dot B} = 0$$
$$T^{\dot C}_{A\dot B} - \frac{1}{4}T^n_{Am}(\bar\sigma^m{}_n)_{\dot B}{}^{\dot C} = 0, \tag{11.8}$$

as well as

$$T_{A(\dot B}{}^{\dot C)} = 0$$
$$T_{\dot A(B}{}^{C)} = 0, \tag{11.9}$$

or, equivalently,

$$T^n_{Am}(\bar\sigma^m{}_n)_{\dot B}{}^{\dot C} = 0$$
$$T^n_{Am}(\sigma^m{}_n)_B{}^C = 0. \tag{11.10}$$

These are optional constraints, and we only choose them in order to find the required x-space multiplet, but they are a priori not required. The analog, or more precisely example, since we will shortly see that they in fact generalize it to superspace, is the no-torsion constraint (or vielbein postulate) of general relativity, which takes us from the first-order formalism to the second-order formalism.

– Representation preserving (or consistency) constraints,

$$T^{\dot C}_{AB} = T^m_{AB} = 0. \tag{11.11}$$

– Super-conformal choice constraints,

$$T^m_{A\,m} = 0. \tag{11.12}$$

The representation preserving, or consistency, constraints arise from the consistency condition in the presence of chiral superfields. (Anti-)Chiral superfields ϕ will be defined by $D_A\phi = 0$. But this in turn means that

$$\{D_A, D_B\}\phi = 0 = T^N_{AB}D_N\phi = T^C_{AB}D_C\phi + T^{\dot C}_{AB}D_{\dot C}\phi + T^m_{AB}D_m\phi, \tag{11.13}$$

which means that we must have

$$T^{\dot C}_{AB} = T^m_{AB} = 0, \tag{11.14}$$

since otherwise $D_A\phi = 0$ will imply $D_m\phi = 0, D_{\dot A}\phi = 0$ as well, which we don't want.

11.3   Solution of the constraints and Bianchis

We now move on to solving the constraints. We first consider the conventional constraints, which we will see are a generalization to superspace of the vielbein constraint, whose solution is known. The first constraint, $T^r_{mn} = 0$, is the usual bosonic no-torsion constraint (vielbein postulate), just that now with summed indices being also fermionic, as well as for superfields instead of regular fields. So we can define as usual the quantity

$$C_{MN}{}^P = E^\Lambda_M(\partial_\Lambda E^\Pi_N)E^P_\Pi - (-)^{MN}(M \leftrightarrow N), \tag{11.15}$$

and in terms of it, the solution of $T^r_{mn} = 0$ has the same form as $\omega = \omega(e)$ in general relativity, that is,

$$\Omega_{mnr} = -\frac{1}{2}(C_{mnr} + C_{rnm} - C_{nrm}). \tag{11.16}$$

Similarly then, the solution of the constraint with all fermionic undotted indices, $T^C_{AB} = 0$, is the fermionic version of the same, namely,

$$\Omega_{ABC} = -\frac{1}{2}(C_{ABC} + C_{CBA} - C_{BCA}). \tag{11.17}$$

Similarly, the solution of the constraint with all dotted indices, $T^{\dot{C}}_{\dot{A}\dot{B}} = 0$, is

$$\Omega_{\dot{A}\dot{B}\dot{C}} = -\frac{1}{2}(C_{\dot{A}\dot{B}\dot{C}} + C_{\dot{C}\dot{B}\dot{A}} - C_{\dot{B}\dot{C}\dot{A}}). \tag{11.18}$$

These solutions were algebraic and solved for Ω_{mnr}, Ω_{ABC}, and $\Omega^{\dot{C}}_{\dot{A}\dot{B}}$ in terms of E^Λ_M. We can also use

$$T_{A(\dot{B}}{}^{\dot{C})} = 0$$
$$T_{\dot{A}(B}{}^{C)} = 0, \tag{11.19}$$

or, equivalently,

$$T^n_{Am}(\bar{\sigma}^m{}_n)_{\dot{B}}{}^{\dot{C}} = 0$$
$$T^n_{\dot{A}m}(\sigma^m{}_n)_B{}^C = 0, \tag{11.20}$$

to solve for $\Omega_{A\dot{B}}{}^{\dot{C}}$, $\Omega_{\dot{A}B}{}^C$ in terms of E^Λ_M.

The rest of the constraints, the remaining conventional ones,

$$T^m_{A\dot{B}} + 2i(\sigma^m)_{A\dot{B}} = 0$$
$$T^{\dot{C}}_{A\dot{B}} - \frac{1}{4}T^n_{Am}(\bar{\sigma}^m{}_n)_{\dot{B}}{}^{\dot{C}} = 0, \tag{11.21}$$

fixing E^Λ_m in terms of E^Λ_A and $E^\Lambda_{\dot{A}}$, the representation preserving ones

$$T^{\dot{C}}_{AB} = T^m_{AB} = 0, \tag{11.22}$$

the super-conformal choice $T^m_{Am} = 0$, and the Bianchi identities solve the components of E^Λ_M in terms of independent fields.

We could write this solution for E^Λ_M in terms of independent fields, and then from the form of R's and T's in terms of E^Λ_M, derive the form of R's and T's in terms of independent fields, but it is more useful to write directly R's and T's in terms of independent fields. We will do so after writing the action.

We therefore write the action for supergravity in superspace. We have already written the measure in superspace for the three-dimensional case, albeit that was in the coset formalism. But the invariant measure is independent of the formalism used. Therefore the first try for the action is the invariant supermeasure,

$$S = \frac{1}{2\kappa_N^2} \int d^4x \, d^4\theta \quad \text{sdet } E_\Lambda^M. \tag{11.23}$$

This has the right dimension, since $[d^4\theta] = 2$ and $[E] = 0$ and, after putting the required $1/2\kappa_N^2$ in front, the integrand for $\int d^4x$ must have dimension 2. In principle, we could have some other function of E_Λ^M, like it happened in three dimensions in the coset formalism, but in this case, we actually don't need anything else, and this action is enough to reproduce pure $\mathcal{N} = 1$ supergravity. In fact, we don't have any other scalar function of dimension zero we could put there, so the choice is unique.

Now we can finally explain the appearance of the the super-conformal choice constraint. In its absence, both the above action *and the rest of the constraints and Bianchis* would be invariant under superconformal transformations,

$$E_A^\Lambda \to e^L E_A^\Lambda; \quad E_{\dot{A}}^\Lambda \to e^{L^*} E_A^\Lambda. \tag{11.24}$$

Note that a conformal transformation would be a rescaling of the vielbein, hence the above is called a superconformal transformation. But since both the action and the constraints are invariant, we could parametrize $E_A^\Lambda = \psi \bar{E}_A^\Lambda$, where $\bar{E}_A^\Lambda$ satisfies the same constraints, and then ψ would be an independent variable. Varying the action with respect to the independent variable ψ, we would get the equation of motion sdet $E_\Lambda^M = 0$, which is impossible. This means, we must break the invariance, and taking the super-conformal choice constraint $T_{A\,m}^m = 0$ achieves that.

The full analysis of the Bianchis and constraints is involved, so we will not reproduce it here, just (parts of) the final result. We can express all torsions and curvatures in terms of three chiral superfields, $R, G_{A\dot{b}}$, and W_{ABC}. They can be defined for instance from the following piece of the solution:

$$R_{\dot{A}\dot{B}\dot{C}\dot{D}} = \frac{1}{6}(\epsilon_{\dot{D}\dot{B}}\epsilon_{\dot{C}\dot{A}} + \epsilon_{\dot{C}\dot{B}}\epsilon_{\dot{D}\dot{A}})R^*$$

$$T_{C\dot{C}\dot{A}D} = \frac{i}{12}\epsilon_{CD}\epsilon_{\dot{C}\dot{A}}R^*$$

$$T_{C\dot{C}DE} = \frac{1}{4}(\epsilon_{CE}G_{D\dot{C}} + 3\epsilon_{CD}G_{E\dot{C}} - 3\epsilon_{DE}G_{C\dot{C}})$$

$$T_{A\dot{A}B\dot{B}C} = \epsilon_{AB}\left(W_{\dot{A}\dot{B}\dot{C}} - \frac{1}{2}\epsilon_{\dot{A}\dot{C}}D^E G_{EB} - \frac{1}{2}\epsilon_{\dot{B}\dot{C}}\right) + \epsilon_{\dot{A}\dot{B}}D_{(B}G_{C)\dot{A}}, \tag{11.25}$$

where

$$T_{AB\dot{C}D\dot{D}} = T_{Am}{}^n(\sigma^m)_{B\dot{C}}(\sigma_n)_{D\dot{D}}$$

$$T_{AB\dot{C}D} = T_{m\dot{C}D}(\sigma^m)_{A\dot{B}}, \tag{11.26}$$

and so on.

The fields are chiral ($\bar{D}_{\dot{A}} R^* = \bar{D}_{\dot{A}} W_{BCD} = 0$), and $W_{(ABC)}$ is totally symmetric. They also satisfy

$$D^A G_{A\dot{A}} = -\frac{1}{24} \bar{D}_{\dot{A}} R$$
$$D^A W_{(ABC)} = -i \left(D_B{}^{\dot{D}} G_{C\dot{D}} + D_C{}^{\dot{D}} G_{B\dot{D}} \right), \qquad (11.27)$$

and $G_{A\dot{B}}$ satisfies the reality condition $G_{A\dot{B}} = (G_{B\dot{A}})^*$, $G_{A\dot{B}} = (\sigma^m)_{A\dot{B}} G_n$, with G_n real.

We now derive the equations of motion of the action. For this, we need to deal with the independent variations of the action. As we saw in three dimensions, we have a priori three choices:

- Write the action in terms of some unconstrained superfields and vary them independently.
- Choose a gauge where the independent fields are the ones of off-shell supergravity, write the action in terms of them, and vary those independently.
- Write the action in terms of E_Λ^M, but only allow independent variations.

In four dimensions, the last path is usually chosen. It is a bit complicated, but one can prove that *on the constraints*, the independent variation of the action is written in the form (note that in the bosonic case, we would have $\delta \int d^4x \det e_\mu^m = \int d^4x \det e_\mu^m \; e_m^{-1\mu} \delta e_\mu^m$, so the variation of the measure is proportional to the measure itself, and to the arbitrary variation; the same happens below)

$$\delta S = \int d^4x d^4\theta \, (\text{sdet} \, E_\Lambda^M) [v^m G_m - RU - R^* U^*], \qquad (11.28)$$

where v^m and U are arbitrary superfields (the independent variations) and as usual $G_m = G_{A\dot{B}} (\sigma_m)^{A\dot{B}}$.

Then it follows that the equations of motion are

$$R = G_m = 0. \qquad (11.29)$$

So they encode the off-shell equations of motion of $\mathcal{N} = 1$, four-dimensional supergravity. Since part of these equations are the equations of motion of the auxiliary fields $M = S + iP$ and A_m, setting those to zero, it is obvious that we must have

$$R(x, \theta = 0) = M = S + iP$$
$$G_m(x, \theta = 0) = A_m. \qquad (11.30)$$

The rest of the supergravity equations of motion in the off-shell formalism are obtained from higher-order terms in the θ expansion of the R and G_m.

Important concepts to remember

- In the four-dimensional super-geometric approach, we begin with $E_\Lambda^M(x, \theta)$ and $\Omega_\Lambda^{MN}(x, \theta)$.
- Since we must fix only diagonal super-local Lorentz transformations Λ^{MN} and write them in terms of only Λ^{mn}, it means that Ω_Λ^{MN} is also diagonal and has only Ω_Λ^{mn} independent components.

- The usual gauge choices in four dimensions are $E_\mu^m(x, \theta = 0) = e_\mu^m$, $E_\mu^a(x, \theta = 0) = \psi_\mu^a$ and $\Omega_\mu^{mn}(x, \theta = 0) = \omega_\mu^{mn}$.
- In four dimensions, we have conventional constraints, representation preserving constraints, which come from the consistency of defining chiral superfields, and super-conformal choice, which is required in order to avoid super-conformal invariance and a trivial action.
- The action in four dimensions is just the super-invariant measure on superspace, $\int d^4x d^4\theta \, \mathrm{sdet} \, E_\Lambda^M$.
- The solution of the Bianchi identities and constraints expresses everything in terms of chiral superfields R, $G_{A\dot{B}}$, and W_{ABC}.
- The equations of motion are $R = G_m \equiv G_{A\dot{B}}(\sigma_m)^{A\dot{B}} = 0$, whose $\theta = 0$ components are the auxiliary field equations of motion, $R(x, \theta = 0) = M$, $G_m(x, \theta = 0) = A_m$.

References and further reading

For more details, see chapter 16 in [9] and chapters 14–18 in [10].

Exercises

(1) Write down explicitly all the Bianchi identities,

$$[D_M, [D_N, D_P\}\} + \text{supercyclic} = 0, \tag{11.31}$$

arising for $M = A$, $N = B$, $P = m$ in terms of torsion and curvature components.

(2) Calculate $R_{ABC}{}^D$ and $R_{ABC}{}^{\dot{D}}$ in terms of $\Omega_{\Lambda m}{}^n$.

(3) Denote by $I_{\dot{A}\dot{B}m}{}^n$ the D_n component of $[D_{\dot{A}}, [D_{\dot{B}}, D_m\}\}+$ supercyclic $= 0$. Show that the form of $R_{\dot{A}\dot{B}\dot{C}\dot{D}}$ and $T_{C\dot{C}\dot{A}D}$ in the text, together with $R_{\dot{A}\dot{B}CD} = 0$ is enough to satisfy the Bianchi identity $I_{\dot{A}\dot{B}m}{}^n = 0$, if we take into account the constraints.

(4) Show that, using the constraints, the equation

$$(\sigma^m)_{A\dot{B}} T_{mn}^{\dot{B}}(\theta = 0) = 0, \tag{11.32}$$

(which follows from the Bianchi identity $I_{nB\dot{D}}{}^{\dot{C}} = 0$ together with the equations $R = G_{A\dot{B}} = 0$) gives the supergravity equations of motion.

12 Superspace actions and coupling supergravity with matter

In Chapter 11, we saw how to write four-dimensional supergravity in superspace. In this chapter, we will see how to couple it to matter, first using the same superspace formalism.

We saw that the constraints imposed on four-dimensional supergravity in superspace were such that the chiral superfield constraint was consistent. The covariant derivative D_A has the correct rigid superspace limit, so we can define chiral superfields in the same way, via

$$\bar{D}_{\dot{A}}\Phi = 0. \tag{12.1}$$

The independent fields $R, G_{A\dot{A}}$ and W_{ABC}, defining supergravity in superspace in the super-geometric approach, were chiral.

If we define superfields in the coset formalism, they are defined by their H-representations, meaning local Lorentz representations for superspace, which also means that the superfield indices are flat (since the flat indices are indices in local Lorentz representations). Also, the covariant derivatives, like D_A above, used to define irreducible representations and to define components of superfields, also have flat indices, that is, local Lorentz.

The matter fields we will be interested in coupling to supergravity are the ones appearing in the Minimal Supersymmetric Standard Model or the minimal supersymmetrization of the Standard Model (MSSM), namely chiral superfields (the Standard Model has chiral fermions) and vector superfields.

Then, for instance, for a chiral superfield, we have the same formulas as in the case of rigid superspace,

$$\Phi = \Phi(x, \theta) = \phi(y) + \sqrt{2}\psi(y) + \theta^2 F(y), \tag{12.2}$$

where

$$y^\mu = x^\mu + i\theta\sigma^\mu\bar{\theta}, \tag{12.3}$$

and we then have the component fields in terms of superfields and covariant derivatives,

$$\phi(x) = \Phi|_{\theta=\bar{\theta}=0}; \quad \psi(x) = \frac{D_A\Phi|_{\theta=\bar{\theta}=0}|}{\sqrt{2}}; \quad F(x) = -4D^2\Phi|_{\theta=\bar{\theta}=0}. \tag{12.4}$$

12.1 Review of YM superfields in rigid superspace

An abelian gauge field is part of a gauge superfield V, together with the fermion (called gaugino) λ, and the auxiliary field D. The gauge superfield V is real, $V = V^\dagger$, and satisfies an abelian super-gauge symmetry

$$V \to V + i\Lambda - i\Lambda^\dagger, \tag{12.5}$$

where Λ is a chiral superfield, $\bar{D}_{\dot{A}}\Lambda = 0$. We can use parts of that gauge symmetry (leaving untouched the normal gauge symmetry of the gauge field A_μ) to fix a gauge, called the Wess-Zumino gauge, where V has only the off-shell multiplet fields, namely where

$$V = -\theta\sigma^\mu\bar{\theta}A_\mu + i\theta^2(\bar{\theta}\bar{\lambda}) - i\bar{\theta}^2(\theta\lambda) + \frac{\theta^2\bar{\theta}^2}{2}D, \tag{12.6}$$

so A_μ is among its components. We can also construct a gauge superfield strength, which will contain $F_{\mu\nu}$ among its components. The correct formula is

$$W_A = -\frac{1}{4}\bar{D}^2 D_A V. \tag{12.7}$$

It satisfies a reality condition

$$D^A W_A = D^{\dot{A}} W_{\dot{A}} \quad (\mathrm{Im}(D^A W_A) = 0) \tag{12.8}$$

and W_A is obviously chiral, $\bar{D}_{\dot{B}} W_A = 0$, since $(\bar{D})^3 = 0$.

Reversely, a chiral field Φ can always be written as $1/4\bar{D}^2 U$, where U is an arbitrary superfield.

The generalization to YM (or nonabelian) gauge superfields is done by exponentiating some results. The super-gauge invariance is now

$$e^{-V} \to e^{i\Lambda^\dagger} e^{-V} e^{-i\Lambda}, \tag{12.9}$$

and the gauge superfield strength is

$$W_A = \frac{1}{4}\bar{D}^2 e^V D_A e^{-V}. \tag{12.10}$$

The action (which is gauge invariant, as we can easily check) is

$$S = -\frac{1}{4}\int d^4x d^2\theta \, \mathrm{Tr}(W_A W^A + \mathrm{h.c.})$$
$$= -\frac{1}{4}\int d^4x F_{\mu\nu}^a F^{\mu\nu a} + ..., \tag{12.11}$$

so it contains the usual YM action, plus supersymmetric terms.

The coupling of the vector multiplet to matter (which in this case means chiral superfields) is given by

$$S_{\mathrm{matter}} = \frac{1}{4g^2}\int d^4x d^2\theta d^2\bar{\theta} \, \mathrm{Tr}(\Phi^\dagger e^V \Phi). \tag{12.12}$$

The F and D auxiliary fields are given by (solving their equations of motion)

$$F_i = \frac{\partial W}{\partial \phi^i}$$
$$D^a = \Phi^\dagger T^a \Phi \equiv \phi^{\dagger i}(T^a)_{ij}\Phi^j. \tag{12.13}$$

12.2 YM superfields in curved superspace

Considering now the generalization of the above to curved superspace, we again start with a real superfield V, that is, $V = V^\dagger$, with the same nonabelian gauge transformation

$$e^{-V} \rightarrow e^{i\Lambda^\dagger} e^{-V} e^{-i\Lambda}, \tag{12.14}$$

where Λ is chiral, $\bar{D}_{\dot{A}} \Lambda = 0$.

However, now we must modify the definition of the invariant field strength, since now the integration measure in terms of covariant derivatives (which is used to calculate the invariant field strength in actions) is written differently,

$$\int d^4x d^2\bar{\theta} = \int d^4x \frac{1}{4} \left(\bar{D}^2 - \frac{1}{3}R \right) |_{\theta=\bar{\theta}=0}, \tag{12.15}$$

that is, the formula is modified by the addition of the R term. Note that this formula is correctly chiral, since both terms are chiral, $\bar{D}_{\dot{A}} \bar{D}^2 = 0$ and $\bar{D}_{\dot{A}} R = 0$. Then the correct invariant field strength is

$$W_A = \frac{1}{4} \left(\bar{D}^2 - \frac{1}{3}R \right) e^V D_A e^{-V}. \tag{12.16}$$

12.3 Invariant measures

In order to write actions, we need to find invariant measures.

We already found the integration measure for the full superspace,

$$\int d^4x d^4\theta E, \tag{12.17}$$

where $E = \text{sdet } E_\Lambda^M$.

We can use this measure to generalize the Kähler potential term of rigid superspace to supergravity, by

$$\int d^4x d^4\theta \ E \ K(\Phi, \Phi^\dagger). \tag{12.18}$$

But in order to generalize the superpotential term as well, we must generalize the chiral measure, that is, the measure of integration for chiral superspace. We must find the equivalent of E for the chiral superspace, namely the chiral density $\mathcal{E}$ on curved superspace. It must be chiral, that is, $\bar{D}_{\dot{A}} \mathcal{E} = 0$.

It is found to be

$$\mathcal{E} = e[1 + i\theta\sigma^m \bar{\psi}_m - \theta^2(M^* + \bar{\psi}_m(\bar{\sigma}^m\sigma^n - \bar{\sigma}^n\sigma^m)\bar{\psi}_n)], \tag{12.19}$$

and moreover it can be written also as

$$\mathcal{E} = \frac{1}{4} \frac{\bar{D}^2 E}{R}, \tag{12.20}$$

where $E =$ sdet E_Λ^M and R can be put inside or outside the $\bar{D}^2$, since it is chiral.

The correct invariance of the last equation is proved as follows.

Since $W(\Phi)$ is chiral, it can be written as (as for any chiral superfield)

$$W = \int d^2\bar{\theta} U = \left(\bar{D}^2 - \frac{1}{3}R\right) U, \tag{12.21}$$

for some general U, that depends on both θ and $\bar{\theta}$, whose form will not be important. Then, we can write the invariant (obtained by integrating $U(\theta, \bar{\theta})$ with the measure for the full superspace, and then using the above relation)

$$-\frac{1}{3}\int d^4x d^4\theta EU = -\int d^4x d^4\theta \frac{E}{R}\bar{D}^2 U + \int d^4x d^4\theta \frac{E}{R}W. \tag{12.22}$$

But the first term can be rewritten as

$$\int d^4x d^4\theta E\bar{D}_{\dot{A}}\left(\frac{\bar{D}^{\dot{A}}U}{R}\right), \tag{12.23}$$

that is, as a divergence of a vector on superspace, for a superspace vector with components $V^N = (V^{\dot{A}} = \bar{D}^{\dot{A}}U/R, V^A = 0, V^m = 0)$.

But then we have

$$\int d^4x d^4\theta ED_N V^N(-)^N = 0, \tag{12.24}$$

which we can prove as follows.

We first write $D_N V^N = E_N^\Lambda \partial_\Lambda V^N$, and then partial integrate to obtain

$$\int d^4x d^4\theta V^N[-E_N^\Lambda \partial_\Lambda E - (-)^{\Lambda(\Lambda+N)}E\partial_\Lambda E_N^\Lambda]. \tag{12.25}$$

But then, using

$$\partial_\Lambda E = E(\partial_\Lambda E_M^\Pi)E_\Pi^M(-)^M$$
$$(-)^{\Lambda(\Lambda+N)}E\partial_\Lambda E_N^\Lambda = (-)^{NM}EE_M^\Lambda \partial_\Lambda E_N^\Pi E_\Pi^M(-)^M, \tag{12.26}$$

and, given that $T_{mn}^p = e_m^\mu e_n^\nu D_{[\mu}e_{\nu]}^p$, and its super-generalization, after a calculation, we obtain

$$\int d^4x d^4\theta ED_N V^N(-)^N = \int d^4x d^4\theta EV^N T_{NM}^M(-)^M, \tag{12.27}$$

which is zero by $T_{\dot{A}m}^m = 0$ (super-conformal choice constraint) and $T_{\dot{A}B}^B = 0$ and $T_{\dot{A}B}{}^{\dot{B}} = 0$ (conventional constraints).

We have therefore finally proved that

$$-\frac{1}{3}\int d^4x d^4\theta EU = \int d^4x d^4\theta \frac{E}{R}W, \tag{12.28}$$

and since the left-hand side is invariant, and on the right-hand side, we have

$$\int d^4x d^2\theta \left(d^2\bar{\theta}\frac{E}{R}\right)W, \tag{12.29}$$

we see that indeed $\bar{D}^2(E/R)$ has the right invariance to be the chiral measure of integration over $W(\Phi)$, which we called $\mathcal{E}$. The correct normalization to obtain $\mathcal{E}$ in fact follows.

12.4 Supergravity actions

We can in fact check explicitly that we can get the correct Einstein–Hilbert action only by integrating over the above chiral measure.

Write a superpotential term with $W = \int d^2\bar\theta\, U = (\bar D^2 - 1/3R)U$, for $U = 1$. We then get that

$$-\frac{1}{3}\int d^4x d^2\theta\, \mathcal{E}R = -\frac{1}{3}\int d^4x d^2\theta d^2\bar\theta\, E, \tag{12.30}$$

which in fact equals the Einstein–Hilbert action, which in superspace is

$$\int d^4x d^4\theta\, E. \tag{12.31}$$

We can also deduce it from the fact that the left-hand side of (12.30) is invariant, being the integration with the chiral measure of a chiral superfield, but the right-hand side is integrated over the whole superspace, so must be integrated with the full measure E.

In fact, the expansion in θ of the chiral superfield R is known to be

$$R = M + \theta(\sigma^m\bar\sigma^n\psi_{mn} - i\sigma^m\bar\psi_m M + i\psi_m A^m)$$
$$+ \theta^2\Big[-\frac{1}{2}R + i\bar\psi^m\sigma^n\psi_{pq} + \frac{2}{3}MM^* + \frac{A_m^2}{3} - ie_m^\mu D_\mu A^m$$
$$+ \frac{\bar\psi\psi}{2}M - \frac{1}{2}\psi_m\sigma^m\bar\psi_n\sigma^n + \frac{1}{8}(\bar\psi_m\bar\sigma_n\psi_{pq} + \psi_m\sigma_n\bar\psi_{pq})\Big] \tag{12.32}$$

so we could check explicitly the invariance properties above.

Finally, the most general $\mathcal{N} = 1$ invariant Lagrangian for supergravity coupled to matter, specifically chiral superfields and gauge superfields with canonical kinetic terms, is

$$S = \int d^4x d^4\theta\, E[K(\Phi, \Phi^\dagger) + \Phi^\dagger e^V \Phi]$$
$$+ \int d^4x d^2\theta\, \mathcal{E}[W(\Phi) + \text{Tr}(W^A W_A)] + h.c. \tag{12.33}$$

We can rewrite the first line as an integral over chiral superspace also, that is, as

$$S = \int d^4x d^2\theta\, \mathcal{E}\left[\bar D^2 - \frac{1}{3}R\right][K(\Phi, \Phi^\dagger) + \Phi^\dagger e^V \Phi]$$
$$+ \int d^4x d^2\theta\, \mathcal{E}\,[W(\Phi) + \text{Tr}(W^A W_A)] + h.c. \tag{12.34}$$

We can further generalize to the case of general kinetic terms for the gauge superfields by writing

$$\int d^4x d^4\theta\, E\, \Phi^\dagger e^V \Phi + \int d^4x d^2\theta\, \mathcal{E}\, \text{Tr}[W_A W^A] + h.c. \rightarrow$$
$$\rightarrow \int d^4x d^4\theta\, E\, (\Phi^\dagger e^V)^a F_a(\Phi) + \int d^4x d^2\theta\, \mathcal{E}\, [F_{ab}(\phi)W_A^a W^{Ab}] + h.c., \tag{12.35}$$

where $F_a = \partial F/\partial\phi^a$ and $F_{ab} = \partial^2 F/\partial\phi^a\partial\phi^b$, and F is an arbitrary function of Φ.

A few observations are in order.

- A constant term W_0 in W now is nontrivial, as is a constant term in K, or more precisely a term of the type $K = a + c\phi^\dagger$. In rigid superspace, a constant term in W drops out of the action when integrating over $d\theta$'s (which act as covariant derivatives), and similarly for $K = a + c\phi^\dagger$.

- Now, however, both constants couple to the supergravity multiplet. A constant in W corresponds to a cosmological constant, since now, after using $\int d^2\theta = D^2 - R^*/3$ on $\mathcal{E}W_0$, with $\mathcal{E}$ in (12.19), we obtain a term of the type (we isolate the θ^2 component with the derivatives) $\sim MW_0$+h.c. On the other hand, the kinetic term for supergravity contains $-MM^*$. Solving for M between the above W term and the supergravity term and substituting back, we get a term in the action $+|W_0|^2$ (with a *negative* cosmological constant).

- A constant term in the Kähler potential gives just the Einstein action, that is, the kinetic term for gravity. More precisely, consider $K = a + \Phi^\dagger\Phi$, that is, a constant term, plus the usual Kähler potential from rigid superspace. We obtain

$$-\frac{1}{3} \int d^4x d^2\theta \mathcal{E}R[a + \phi^\dagger(x)\phi(x)] + ..., \tag{12.36}$$

where we just wrote the terms where we don't act with the $\int d^2\theta$ on $K(\Phi, \Phi^\dagger)$, so we can replace with $K(\phi, \phi^\dagger)$ instead. Note that this is the *Brans–Dicke parametrization for gravity*.

That is, the first term is just the usual EH action, but the second contains a variation of the Newton constant. This was found a long time ago by Brans and Dicke, who considered the fact that the Newton constant in front of the EH action, $1/\kappa_N^2$, could in principle vary in spacetime, thus making it a scalar field. But it was soon realized that one can perform a change of metric, or "metric frame," that removes the scalar field term in front of the Einstein action. For an action $\int d^4x\sqrt{-g}R[g]\ C(\phi)$, we can choose an appropriate $A(\phi)$ such that the field redefinition

$$g_{\mu\nu} = A(\phi)\tilde{g}_{\mu\nu} \tag{12.37}$$

takes us to the standard form of the Einstein action,

$$S = \int d^4x\sqrt{-\tilde{g}}(R[\tilde{g}] + (...)(\partial\phi)^2), \tag{12.38}$$

plus scalar kinetic terms. (There are only terms with two derivatives on the scalar, since $R \sim \partial\Gamma + \Gamma\Gamma$ and $\Gamma \sim g^{-1}\partial g$, so the Einstein action has two derivatives acting on metrics.) Here $g_{\mu\nu}$ is called a "Jordan frame" metric, and $\tilde{g}_{\mu\nu}$ an "Einstein frame" metric. These are physically different metrics; the transformation between them is not like a general coordinate transformation, which does not affect the physics (physics looks the same in all systems of coordinates). Rather, it is a field redefinition (a change of field variables), which however changes the way physics looks in the two frames. Of course, we describe the same physics, but from two different perspectives. Jordan frame and Einstein frame descriptions have each its advantages and disadvantages. We are more familiar with Einstein frame, so we will use that.

We redefine $K \to K + a$ to isolate the constant term, thus obtaining

$$-\frac{1}{3}\int d^4x d^2\theta \mathcal{E} R[a + K(\phi(x), \phi^\dagger(x))].$$ (12.39)

We thus see that for $a = -3$ we get the usual Einstein action, and we obtain

$$\int d^4x d^2\theta \mathcal{E} R\left[1 - \frac{K}{3}\right].$$ (12.40)

We then redefine

$$1 - \frac{K(\Phi, \Phi^\dagger)}{3} = e^{-\frac{k(\Phi, \Phi^\dagger)}{3}}$$ (12.41)

where we call $k(\Phi, \Phi^\dagger)$ the modified Kahler potential. Of course, at the linear level, there is no difference between K and k, but at the nonlinear level, there is.

12.5 Supergravity coupled to matter: terms in the component action

The scalar potential in the absence of the supergravity coupling (in rigid superspace) was

$$V = \sum_i |F_i|^2 + \frac{g^2}{2} D^a D^a,$$ (12.42)

which is modified, in the presence of the supergravity coupling, to

$$V = \sum_i |F_i|^2 + \frac{g^2}{2} D^a D^a - \frac{1}{3}(|M|^2 + A_m^2)e^{-k/3}.$$ (12.43)

To get some idea of the final result, we observe that by a generalization of the $W = W_0$ case above, now we have a coupling $MW(\phi(x))$, and now we also have another term of a similar type, $\sim M(\phi/3)(\partial W/\partial\phi)$. From the Kahler potential term, we get $M(\partial K/\partial\phi)F$ terms. Solving the equation of motion for M, we will get

$$M \sim \phi\frac{dW}{d\phi} - 3W + F\frac{\partial K}{\partial\phi}.$$ (12.44)

After doing the "Weyl rescaling" to the Einstein frame from (12.43) with the auxiliary fields substituted with their solutions, we obtain the Einstein-frame potential

$$V = e^k\left[\sum_{i,\bar{j}}(g^{-1})^{i\bar{j}}\left(\frac{\partial W}{\partial\phi^i} + W\frac{\partial k}{\partial\phi^i}\right)\left(\frac{\partial W}{\partial\phi^j} + W\frac{\partial k}{\partial\phi^j}\right)^* - 3|W|^2\right]$$
$$+ \frac{1}{2}(F^{-1})^{ab}\left(\frac{\partial k}{\partial\phi^i}(T_a)_{ij}\phi^j\right)\left(\frac{\partial k}{\partial\phi^j}(T_b)_{kl}\phi_l\right)^*,$$ (12.45)

where

$$g_{i\bar{j}} = \frac{\partial^2 k}{\partial\phi^i\partial\bar{\phi}^{\bar{j}}}$$ (12.46)

is the *metric on scalar field space*.

Indeed, the kinetic terms for the scalars and the corresponding fermions are

$$\int d^4x\sqrt{-g}\left[g_{i\bar{j}}D_\mu\phi^i(D^\mu\phi)^{*\bar{j}} + g_{i\bar{j}}\psi^i\slashed{D}\bar{\psi}^{\bar{j}}\right]. \tag{12.47}$$

This is a metric on the scalar space since we have something like $ds^2 = g_{i\bar{j}}d\phi^i d\bar{\phi}^{\bar{j}}$. In the case of zero potential, for instance, if $W = 0$, the scalar space is called a *moduli space*, since then the fields are *moduli*, that is, their VEVs are arbitrary (it doesn't cost energy to change them).

The gauge fields kinetic terms are

$$-\frac{1}{4}\mathrm{Re}[F_{ab}(\phi)F^a_{\mu\nu}F^{b\mu\nu}]. \tag{12.48}$$

One usually defines the "Kähler-covariant derivative"

$$D_i = \frac{\partial}{\partial\phi^i} + \frac{\partial k}{\partial\phi^i}, \tag{12.49}$$

so that the scalar field at zero D terms is

$$V = e^k\left[\sum_{i\bar{j}}(g^{-1})^{i\bar{j}}D_iW(D_jW)^* - 3|W|^2\right]. \tag{12.50}$$

Finally, the gaugino action is

$$-\frac{1}{2}\mathrm{Re}[F_{ab}(\phi)\bar{\lambda}^a\slashed{D}\lambda^b] + \frac{1}{2}e^{k/2}\mathrm{Re}\sum_{i\bar{j}}(g^{-1})^{i\bar{j}}D_iW\left(\frac{\partial F_{ab}}{\partial\phi^j}\right)^*(\bar{\lambda}^a\lambda^b). \tag{12.51}$$

Important concepts to remember

- When generalizing YM fields to curved superspace, we change $\int d^2\bar{\theta} = 1/4\bar{D}^2$ to $\int d^2\theta = 1/4(\bar{D}^2 - 1/3R)$.
- On the full superspace, we have the $\int d^4x d^4\theta E$ measure.
- On chiral superspace, we have the chiral measure $\mathcal{E} = 1/4\bar{D}^2(E/R)$ and the Einstein action in terms of it is $\int d^4x d^2\theta\mathcal{E}R$.
- A constant term in K gives the pure supergravity action, a constant term in W then gives a cosmological constant.
- We naturally get the Einstein action in Brans-Dicke parametrization, so we must perform a Weyl rescaling to the Einstein frame metric.
- The Kähler potential is redefined by $1 - K/3 = e^{-k/3}$.
- The scalar field metric is $g_{i\bar{j}} = \partial_i\partial_{\bar{j}}k$, the gauge field metric is $F_{ab} = \partial_a\partial_b F$, and the scalar potential is written in terms of the Kähler-covariant derivative $D_i = \partial_i + \partial k/\partial\phi^i$.

References and further reading

For more details, see chapter 16 of [9], chapters 19–25 of [10] and chapter 31 of [12].

Exercises

(1) Check that the most general $\mathcal{N} = 1$ Lagrangian for supergravity plus matter is (nonabelian) gauge invariant.

(2) Calculate the scalar potential for the case that the modified Kähler potential $k(\rho, \bar{\rho})$ and the superpotential $W(\rho)$ are

$$k = -3\log(i(\rho - \bar{\rho}))$$
$$W(\rho) = W_0 + Ae^{-ia\rho} + Be^{ib\rho}, \tag{12.52}$$

where a and b are real and positive.

(3) Check explicitly that

$$\mathcal{E} = \frac{1}{4}\bar{D}^2\left(\frac{E}{R^*}\right), \tag{12.53}$$

using the explicit formulas for $R, \mathcal{E}$ in the text. Note: for this, it is enough to prove that the action in terms of $\mathcal{E}$ gives the correct off-shell sugra action. Why?

(4) Check that the "Jordan-frame" four-dimensional gravity action

$$\int d^4x\sqrt{-g}R[g]f(\phi) \tag{12.54}$$

transforms to the "Einstein frame" action,

$$\int d^4x\sqrt{-\tilde{g}}(R[\tilde{g}] + (\partial\phi)^2h(\phi)), \tag{12.55}$$

under the metric frame transformation

$$g_{\mu\nu} = 1/f(\phi)\tilde{g}_{\mu\nu}. \tag{12.56}$$

13   Kaluza–Klein (KK)-dimensional reduction and examples

Until now we talked about four-dimensional and three-dimensional supergravities, but in the second part of the course (Applications), we will consider supergravities in higher dimensions. For instance, string theory, which has supergravity as its low energy limit, lives in 10 dimensions.

Also, the maximal dimension for supergravities is 11, and there is a unique 11-dimensional supergravity theory. This comes about as follows. Interacting theories with a finite number of spins larger than 2 in flat space do not exist (have not been found). Of course, string theory is equivalent to an interacting theory of an infinite number of fields of arbitrary spin, and can be defined in flat space. Also, free theories of any spin can be defined, as can interacting theories in AdS (or dS) background, the so-called Vassiliev higher spin theories. But allowing for a finite number of fields in flat space, we must restrict ourselves to spins ≤ 2. Then, since constructing the representation from the action of $a^\dagger$'s constructed from the susy charges modifies the helicity by 1/2, we can have at most eight supercharges in four dimensions, which can be put into a single supercharge in 11 dimensions.

The uniqueness of the 11-dimensional supergravity means that it was thought for a while to be a good candidate for a fundamental theory. Then, when it was found that there is a potential divergence at seven loops in it, string theory took the role of possible fundamental theory. But string theory at large coupling g_s is equivalent to an 11-dimensional "M theory" on a circle of radius $R_{11} = g_s l_s$, whose low energy theory is the unique 11-dimensional supergravity.

So, either way, it seems that we are forced to consider the idea of supergravity in higher dimensions, and then the question is, what to do with the extra dimensions? The most common solution is called Kaluza–Klein (KK) theory, and it will be described in this chapter.

13.1   Kaluza–Klein compactification

The idea is an old one, going back to Theodor Kaluza (1921) and Oskar Klein (1926), which is to consider that the space is a direct product space, $M_D = M_4 \times K_n$, where M_4 is a four-dimensional space, usually our flat Minkowksi space, and K_n is a compact space (there are some things we can do with some noncompact spaces, but that is a very special case). Standard examples are spheres S^n and tori $T^n = (S^1)^n$. Then we assume that the reason why we "feel" (or "see," which for particle physics means that we scatter very light or massless particles; after all, an electron microscope functions by scattering electrons off a target) only four dimensions is that the size of K_n is very small, perhaps (as we will see shortly) comparable with the Planck scale, so we cannot probe it. The resulting theory is generally known as Kaluza–Klein (KK) theory.

Kaluza used this idea for the unification of fields, in particular, gravity ($g_{\mu\nu}$) and electromagnetism (A_μ), through the existence of higher dimensions. When he sent the paper to Einstein for refereeing, famously Einstein kept it for more than a year, thinking there must be something wrong with the idea. In the end, he accepted it, and in fact, in the later part of his life, Einstein tried to find a unified theory using Kaluza's basic idea as a starting point.

The theory went soon out of fashion, for reasons we will explain toward the end of the chapter, but then became popular again in the 1970s, after supergravity became popular, and it became clear in the 1980s that string theory needs it, being a unified theory in 10 dimensions.

Let us consider how small the extra dimensions of K_n must be.

(1) First, from a simplistic *experimental* point of view, in other words, in greatest generality (without specific mechanisms in mind), we can say that, since we don't see anything at the current accelerators, which probe energies up to about 10 TeV (and, as we said, probing with energies amounts to "seeing" the extra dimensions, if they exist), we must have $R \lesssim (10\,\text{TeV})^{-1}$.

On the other hand, there are special cases, such as the Large Extra Dimensions (LEDs) scenario, where the above bound is invalid. In particular, Arkani–Hamed, Dimopoulos, and Dvali (ADD) thought up the idea that particle physics can be restricted to a four-dimensional "wall" or "3-brane," and only gravity feels the rest, leading to some large dimensions, and then Randall and Sundrum (RS) came up with the idea of "warped" extra dimensions (where the four-dimensional metric is multiplied with a function of the extra dimension), where the extra dimensions can be even larger, or even infinite. But we will not deal with these special cases, since they fall outside the strict KK scenario.

(2) From a *theoretical* perspective, the extra dimensions modify the geometry, associated with gravity. So, if, besides the four-dimensional infinite dimensions, we have some extra dimensions that stay small due to some quantum mechanical reason, it follows that the natural scale for the extra dimensions is the *quantum gravity* Planck scale $l_{\text{Pl}} \sim G_N^{1/2} \sim (10^{19}\,\text{GeV})^{-1}$, which is the only length scale made from G_N (for gravity), $\hbar$ (for quantum mechanics), and c (for relativity).

Note that we wrote a direct product space $M_4 \times K_n$, but rather, in most generality, we should have $K_n(x^\mu)$, where x^μ are the coordinates on M_4, meaning that at each point in spacetime we have a K_n of a priori different size. Then the total spacetime is really defined by coordinates (x^μ, y^m), which is a more useful way to think about it. We will then consider fields $\phi(x^\mu, y^m)$.

In the simplest case, of a single coordinate y on an S^1 (circle), we have the Fourier theorem, which reads

$$\phi(x^\mu, y) = \sum_{n=0}^{\infty} \phi_n(x^\mu) e^{i\frac{ny}{R}}, \tag{13.1}$$

where the sum is over fields $\phi_n(x^\mu)$ from the (3+1)-dimensional point of view, replacing the (4+1)-dimensional field.

13.2 Metrics in KK theory

Defining KK theory, we note that there are three metrics that sometimes go by the name of KK metric, so we should distinguish between them:

- **The KK background metric**. The fact that the space is $M_4 \times K_n$ means that the *background* is a solution of the equations of motion, which is of direct product type, meaning block diagonal, and where the blocks depend only on the coordinate corresponding to the block,

$$g_{MN}(\vec{x}, \vec{y}) = \begin{pmatrix} g^{(0)}_{\mu\nu}(\vec{x}) & 0 \\ 0 & g^{(0)}_{mn}(\vec{y}) \end{pmatrix}. \tag{13.2}$$

Here $g^{(0)}_{\mu\nu}$ is a background metric in four dimensions, usually Minkowski, de Sitter, or Anti-de Sitter, and $g^{(0)}_{mn}$ is the metric on the compact space K_n.

Note that the metric is itself one of the fields of the theory, so it is a variable, so when we write $M_4 \times K_n$ we only mean the background, not the full fluctuating metric. For small (linearized) fluctuations around the background, this may be a pedantic point, but we will also consider the case of nonlinear (large) fluctuations in later chapters, and then it really matters.

Also note that in general, the background has to be a solution of the supergravity equations of motion, however sometimes one considers the case when it isn't. In that case, however, we must be careful about the conclusions that we derive.

- **The KK expansion**. This is an exact decomposition, the generalization of the Fourier expansion on a circle, or the spherical harmonic expansion on the 2-sphere. In the case of the Fourier expansion, the Fourier theorem says that we can always expand

$$\phi(\vec{x}, y) = \sum_n \phi_n(\vec{x}) e^{\frac{iny}{R}}, \tag{13.3}$$

if y is on a circle of radius R.

On a 2-sphere, the expansion is familiar from electromagnetism or quantum mechanics, and we can similarly always write (the expansion is again a theorem)

$$\phi(\vec{x}, \theta, \phi) = \sum_{lm} \phi_{lm}(\vec{x}) Y_{lm}(\theta, \phi), \tag{13.4}$$

where $Y_{lm}(\theta, \phi)$ are called the "spherical harmonics on the S^2." The nomenclature will be generalized to K_n.

Here the functions in which we expand are eigenfunctions of the Laplacian, since

$$\partial_y^2 e^{\frac{iny}{R}} = -\left(\frac{n}{R}\right)^2 e^{\frac{iny}{R}}$$

$$\Delta_2 Y_{lm}(\theta, \phi) = -\frac{l(l+1)}{R^2} Y_{lm}(\theta, \phi), \tag{13.5}$$

where in the second line, we put a radius R for a 2-sphere of radius R for generality, even though Y_{lm} is really defined for $R = 1$.

Similarly, in the general *spinless* case (we will consider Lorentz spin later), we can always write

$$\phi(\vec{x}, \vec{y}) = \sum_{q, I_q} \phi_q^{I_q}(\vec{x}) Y_q^{I_q}(\vec{y}), \tag{13.6}$$

where $Y_q^{I_q}(\vec{y})$ is also called a *spherical harmonic*, like in the 2-sphere case. Here q is an index that measures the eigenvalue of the Laplacian, like l for S^2, and I_q is an index in some representation of the symmetry group (like m for S^2, which takes values in a representation of the $SO(3) = SU(2)$ invariance group of S^2, namely a spin l representation). Note that for $S^n = SO(n+1)/SO(n)$, the invariance group is $SO(n+1)$.

The $Y_q^{I_q}$ are also eigenfunctions of the Laplacian on K_n, that is,

$$\Delta_n Y_q^{I_q}(\vec{y}) = -m_q^2 Y_q^{I_q}(\vec{y}). \tag{13.7}$$

Here we call the eigenvalue $-m_q^2$ since it acts as a mass term from the point of view of four dimenssions. Indeed, from the four-dimensional point of view, because of the split (due to the direct product structure of the space) into four-dimensional d'Alembertian and n-dimensional Laplacian,

$$\Box_{(D)} = \Box_{(4)} + \Delta_{(n)}, \tag{13.8}$$

we get for $\phi(\vec{x}, \vec{y}) = \phi_q^{I_q}(\vec{x}) Y_q^{I_q}(\vec{y})$ that

$$\Box_D \phi(\vec{x}, \vec{y}) = (\Box_4 + \Delta_n)\phi(\vec{x}, \vec{y}) = (\Box_4 - m_q^2)\phi(\vec{x}, \vec{y}), \tag{13.9}$$

and so if $\phi(\vec{x}, \vec{y})$ is D-dimensional massless, the above is zero, which now looks like four-dimensional massive with mass m_q.

This is the statement that in order to see structure on K_n, we must use some energy, at least m_q if we want to see information at the level of the $Y_q^{I_q}$ spherical harmonic.

Thus the KK expansion is a mathematical equality and contains no information other than the metric of the background we expand around.

- **The KK reduction ansatz.** This is an *ansatz*, which means it is a guess (in German), it is not guaranteed to work. Since we want to say that the compact space has a very small size, and we cannot probe it, we must find an effective four-dimensional description that does not see the K_n. This is the dimensional reduction ansatz, which is we keep only fields in the $n = 0$ representation, that is, "independent of y," though in general there is a given y-dependence, namely of $Y_0(\vec{y})$, but it is the simplest we can have.

Also, in general it is not necessarily the first representation that is kept for all fields, but rather it could be $n = 1$ or $n = 2$ for some fields. In the case of supergravity, the relevant factor is that we need to keep a four-dimensional supermultiplet. Also note that for M_4 being AdS for example, m_q is not necessarily zero, we could have fields that are a bit tachyonic, namely $m_q^2 < 0$, but still above some bound (the Breitenlohner–Freedman or BF bound), or even massive. The relevant fact is still that we keep the lowest supergravity supermultiplet. The BF bound arises from the condition that small perturbations are stable at infinity in the space, which is also what is required in the case of flat space (though in that case we can understand $m^2 \geq 0$ as the condition for a potential that doesn't go down to $-\infty$, $V''(\phi) \geq 0$), and the resulting BF bound depends on the spin of the field.

Thus in the KK-dimensional reduction ansatz, we keep generically speaking

$$\phi(\vec{x}, \vec{y}) = \phi_0(\vec{x}) Y_0(\vec{y}). \tag{13.10}$$

Note that the KK ansatz is an ansatz, so it could be that it doesn't work, that is, it doesn't solve the equations of motion (for instance, if $\Box \phi_n = \phi_0^2$, we cannot put ϕ_n to 0 if we keep ϕ_0 nonzero). We will deal with consistent and inconsistent truncations later in the book.

In summary then, the KK background metric is a solution, the KK expansion is a parametrization, and the KK reduction ansatz is an ansatz.

13.3 Fields with (Lorentz) spin in KK theory

The next case to consider is the case of KK theory for fields with Lorentz spin, which is, after all, why the KK theory was introduced. In this case, various fields in the lower dimensions with various spins are unified into a single higher-dimensional one.

So various components of the higher-dimensional field act as different fields (in different four-dimensional Lorentz spin representations) in four dimensions.

We consider now relevant examples.

Example 1 The simplest relevant case is the D-dimensional vector (electromagnetism) $A_M(\vec{x}, \vec{y})$, splitting into $M = (\mu, m)$ with $\mu = 0, 1, 2, 3$ and $m = 4, ..., 3 + n$. The fields are then

$$A_M(\vec{x}, \vec{y}) = (A_\mu(\vec{x}, \vec{y}), A_m(\vec{x}, \vec{y})). \tag{13.11}$$

Here $A_\mu(\vec{x}, \vec{y})$ transforms as a four-dimensional vector, and $A_m(\vec{x}, \vec{y})$ transform as four-dimensional scalars. That is, under a (3+1)-dimensional Lorentz transformation with parameter $\Lambda_\mu{}^\nu \in SO(3, 1)$, we have

$$A'_\mu(\vec{x}', \vec{y}) = \Lambda_\mu{}^\nu A_\nu(\vec{x}, \vec{y})$$
$$A'_m(\vec{x}', \vec{y}) = A_m(\vec{x}, \vec{y}). \tag{13.12}$$

We see then that the four-dimensional vector and scalars are unified into a single D-dimensional field.

The original (D-dimensional) theory is invariant under $SO(1, 3 + n)$, but the $M_4 \times K_n$ background metric breaks the invariance down to $SO(1, 3) \times SO(n)$.

We see then that $SO(n)$ acts on $A_m(\vec{x}, \vec{y})$ as an *internal* symmetry, transforming only the $\vec{y}$ coordinates,

$$A'_m(\vec{x}, \vec{y}') = \Lambda_m{}^n A_n(\vec{x}, \vec{y}), \tag{13.13}$$

so $SO(n)$ is an internal symmetry from the four-dimensional point of view, like for instance, the $SU(3) \times SU(2) \times U(1)$ of the Standard Model, not like a spacetime symmetry. This means that KK theory has *geometrized the internal symmetry*.

We can thus summarize the effects of KK theory as:

– it unifies the various fields into a single higher-dimensional field.
– it geometrizes the internal symmetry.

The KK expansion for the vector is as follows (μ sits on the four-dimensional field and m on the spherical harmonic, as is natural):

$$A_\mu(\vec{x}, \vec{y}) = \sum_{q, I_q} A_\mu^{q, I_q}(\vec{x}) Y_q^{I_q}(\vec{y})$$

$$A_m(\vec{x}, \vec{y}) = \sum_{q, I_q} A^{q, I_q}(\vec{x}) Y_m^{q, I_q}(\vec{y}). \tag{13.14}$$

Here I_q is an index in the isometry group of K_n, becoming the internal symmetry group of the four-dimensional field, and m is an index in the local Lorentz group of K_n (for the tangent space at any point).

For instance, for $S^n = SO(n+1)/SO(n)$, $SO(n+1)$ is the isometry group, with index I_q, and $SO(n)$ is the local Lorentz group, with index m. In this *particular* case, $SO(n) \subset SO(n+1)$, so we can embed the m index into I_q, but this is not what happens in general.

Another important particular case is $T^n = (S^1)^n$ (flat n-dimensional space with some identifications), where the isometry group = internal symmetry group and the local Lorentz group of K_n coincide, both being $SO(n)$. That is why, in this case, we have "$\delta_m^{I_q}$."

For the breaking of $SO(1, 3 + n)$ into $SO(1, 3) \times SO(n)$, the split of the vector representation as $M = (\mu, m)$ is described as

$$R_V = r_V \otimes \mathbb{1} \oplus \mathbb{1} \otimes r_V^{(n)}, \tag{13.15}$$

where R and r stand for representation and V for vector. Equivalently, using the multiplicity of the representations, we say that

$$\underline{(4+n)}_V = \underline{4}_V \otimes \mathbb{1} \oplus \mathbb{1} \otimes \underline{n}_V. \tag{13.16}$$

Example 2   Next, we consider the gravitational field (symmetric tensor) under KK theory. It splits as

$$g_{MN}(\vec{x}, \vec{y}) = \begin{pmatrix} g_{\mu\nu}(\vec{x}, \vec{y}) & g_{\mu m}(\vec{x}, \vec{y}) \\ g_{m\mu}(\vec{x}, \vec{y}) = g_{\mu m} & g_{mn}(\vec{x}, \vec{y}) \end{pmatrix}, \tag{13.17}$$

where $g_{\mu\nu}$ is a four-dimensional gravitational field (symmetric tensor), $g_{\mu m} = g_{m\mu}$ are vectors, and g_{mn} are scalars.

This means that under the Lorentz transformation $\Lambda_\mu^{\ \nu} \in SO(1, 3)$, we have the field transformations

$$g'_{\mu m}(\vec{x}', \vec{y}) = \Lambda_\mu^{\ \nu} g_{\nu m}(\vec{x}, \vec{y})$$

$$g'_{mn}(\vec{x}', \vec{y}) = g_{mn}(\vec{x}, \vec{y}). \tag{13.18}$$

Under the KK expansion, the fields are expanded in general as

$$g_{\mu\nu}(\vec{x}, \vec{y}) = \sum_{q, I_q} g_{\mu\nu}^{q, I_q}(\vec{x}) Y^{q, Y_q}(\vec{y})$$

$$g_{\mu m}(\vec{x}, \vec{y}) = \sum_{q, I_q} V_{\mu}^{q, I_q}(\vec{x}) Y_m^{q, I_q}(\vec{y})$$

$$g_{mn}(\vec{x}, \vec{y}) = \sum_{q, I_q} \phi^{q, I_q}(\vec{x}) Y_{mn}^{q, I_q}(\vec{y}). \tag{13.19}$$

As representations, under the breaking $SO(1, 3 + n) \to SO(1, 3) \times SO(n)$, we have

$$R_S = r_S \otimes \mathbb{1} \oplus r_V \otimes r_V^{(n)} \oplus \mathbb{1} \otimes r_S^{(n)}, \tag{13.20}$$

where S stands for symmetric tensor, or equivalently, in terms of multiplicities,

$$\underline{((4 + n)(5 + n)/2)}_S = \underline{10}_S \otimes \mathbb{1} \oplus \underline{4}_V \otimes \underline{n}_V \oplus \mathbb{1} \otimes \underline{(n(n + 1)/2)}_S. \tag{13.21}$$

Example 3 Antisymmetric tensors, that is, $p + 1$-forms $A_{M_1 \ldots M_{p+1}}$. Under KK theory, they split as a series of forms,

$$\left(A_{\mu_1 \ldots \mu_{p+1}}, \ldots, A_{\mu_1 \ldots \mu_k m_{k+1} \ldots m_{p+1}}, \ldots, A_{\mu_1 \ldots \mu_{p-n+1}, m_{p-n+2} \ldots m_{p+1}} \right), \tag{13.22}$$

which are a $p + 1$-form, ..., k-forms, ..., until $p - n + 1$-forms.

As usual, these forms have field strengths

$$F_{M_1 \ldots M_{p+2}} = (p + 2) \partial_{[M_1} A_{M_2 \ldots M_{p+2}]}, \tag{13.23}$$

and have gauge invariance

$$\delta A_{M_1 \ldots M_{p+1}} = \partial_{[M_1} \Lambda_{M_2 \ldots M_{p+1}]}. \tag{13.24}$$

Example 4 Fermions, which are in a spinor representation. ψ_Ω, where Ω is in a spinor representation in D dimensions. The spinor index Ω now splits into a *product* of an $SO(1, 3)$ index α and a spinor index z for $SO(n)$, so $\Omega = (\alpha z)$.

We then obtain a set of Lorentz spinors for $SO(1, 3)$ with a label = a spinor index for $SO(n)$,

$$\psi_\Omega(\vec{x}, \vec{y}) = \psi_{\alpha z}(\vec{x}, \vec{y}). \tag{13.25}$$

Under the KK expansion, we have

$$\psi_{\alpha z}(\vec{x}, \vec{y}) = \sum_{q, I_q} \psi_\alpha^{q, I_q}(\vec{x}) \eta_z^{q, I_q}(\vec{y}), \tag{13.26}$$

where η_z are Killing spinors, about which we will say more next chapter, but are fermionic versions of Killing vectors, representing isometries, so basically are supersymmetric isometries, or sort of a square root of Killing vectors.

One observation is that, since $\psi_{\alpha z}(\vec{x}, \vec{y})$ must be anticommuting, as they represent D-dimensional fermions, but we also want to have ψ_α^{q, I_q} anticommuting, as they represent four-dimensional fermions, this means that η_z^{q, I_q} must be *commuting spinors*.

In terms of representations under the symmetry breaking $SO(1, 3 + n) \to SO(1, 3) \times SO(n)$, we have

$$R_{\text{sp}} = r_{\text{sp}} \otimes r_{\text{sp}}^{(n)}. \tag{13.27}$$

13.4 The original KK theory: compactification on S^1

The original idea of Kaluza and Klein was to unify gravity ($g_{\mu\nu}$) and electromagnetism, described by a vector B_μ, into a single field: a five-dimensional metric g_{MN}.

Then the KK background metric is $Mink_4 \times S^1$, a solution of the five-dimensional Einstein–Hilbert action for gravity, so

$$g_{MN}^{(0)}(\vec{x}, y) = \begin{pmatrix} \eta_{\mu\nu} & \mathbf{0} \\ \mathbf{0} & g_{55} = 1 \end{pmatrix}. \tag{13.28}$$

The KK expansion, or parametrization, is the Fourier expansion on a circle, so

$$g_{MN}(\vec{x}, \vec{y}) = \begin{pmatrix} \eta_{\mu\nu} + \sum_{n\geq 0} h_{\mu\nu}^{(n)}(\vec{x})e^{i\frac{ny}{R}}, & g_{\mu 5} = g_{5\mu} \\ g_{5\mu}(\vec{x}, y) = \sum_{n\geq 0} h_{\mu 5}^{(n)}(\vec{x})e^{i\frac{ny}{R}}, & g_{55} = 1 + \sum_{n\geq 0} \phi^{(n)}(\vec{x})e^{i\frac{ny}{R}} \end{pmatrix}. \tag{13.29}$$

The KK reduction ansatz means putting all fields with $n \neq 0$ to zero, so

$$g_{MN}(\vec{x}, \vec{y}) = \begin{pmatrix} \eta_{\mu\nu} + h_{\mu\nu}^{(0)}(\vec{x}) \equiv g_{\mu\nu}(\vec{x}), & g_{\mu 5} = g_{5\mu} \\ g_{5\mu}(\vec{x}, y) = h_{5\mu}^{(0)}(\vec{x}) \equiv B_\mu, & g_{55} = 1 + \phi^{(0)}(\vec{x}) = \varphi(\vec{x}) \end{pmatrix}. \tag{13.30}$$

This ansatz is always consistent, since we are on a circle, and we kept all the zero modes, so it seems fine. But it means that, besides gravity and electromagnetism, we would also have a *massless* scalar field $\varphi(\vec{x})$, which contradicts experiment. This would create a long-range fifth force that we don't observe, so it would mess up the Keplerian motion of the planets. This is the Brans–Dicke theory from Chapter 12 and, like it was noted there, we can exchange the massless field of this, Einstein metric frame, for a spacetime-dependent Newton coupling $\kappa_N(\vec{x})$, equally contradicting experiment, in the Jordan metric frame.

An apparent solution to this problem, which Kaluza and Klein wanted to use, would be to set the background value to $\varphi = 1$, or $\phi^{(0)} = 0$, but this is inconsistent, that is, it doesn't solve the Einstein equations of motion in five dimensions. We say we have an inconsistent truncation.

So we cannot unify gravity and electromagnetism in this simple way. We need to keep φ, in which case we have a consistent ansatz, that is, theoretically valid, just that it does not agree with experiments, since we don't see φ. But in this case, even though the reduction ansatz is consistent, we still need to write nonlinear modifications in order to obtain from the KK reduction both the action for gravity in the standard Einstein–Hilbert form, and the action for electromagnetism in the standard Maxwell form. Finally then, this nonlinear KK reduction ansatz is

$$g_{MN} = \begin{pmatrix} g_{\mu\nu}(\vec{x})\varphi^{-1/2}(\vec{x}) & B_\mu(\vec{x})\varphi(\vec{x}) \\ B_\mu(\vec{x})\varphi(\vec{x}) & \varphi(\vec{x}) \end{pmatrix}, \tag{13.31}$$

which we can rewrite as

$$g_{MN} = \Phi^{(-1/3)}(\vec{x}) \begin{pmatrix} g_{\mu\nu}(\vec{x}) & B_\mu(\vec{x})\Phi(\vec{x}) \\ B_\mu(\vec{x})\Phi(\vec{x}) & \Phi(\vec{x}) \end{pmatrix}, \tag{13.32}$$

where we have redefined the scalar field as $\varphi = \Phi^{2/3}$.

Generalization to torus $T^n = (S^1)^n$

The torus is obtained by periodic identifications of $\mathbb{R}^n$, so it is flat, therefore, $g_{mn}^{(0)} = \delta_{mn}$, so we now have the KK background metric

$$g_{MN}^{(0)}(\vec{x}, \vec{y}) = \begin{pmatrix} \eta_{\mu\nu} & \mathbf{0} \\ \mathbf{0} & \delta_{mn} \end{pmatrix}. \tag{13.33}$$

The KK expansion is just a product of the Fourier expansions on the S^1's, so we have

$$g_{MN}(\vec{x}, \vec{y}) =$$
$$\begin{pmatrix} \eta_{\mu\nu} + \sum_{\{n_i\}} h_{\mu\nu}^{\{n_i\}}(\vec{x}) \exp\left(i \sum_i \frac{n_i y_i}{R_i}\right), & g_{\mu m} = g_{m\mu} \\ g_{m\mu}(\vec{x}, \vec{y}) = \sum_{\{n_i\}} B_\mu^{m\{n_i\}}(\vec{x}) \exp\left(i \sum_i \frac{n_i y_i}{R_i}\right), & g_{mn} = \delta_{mn} + \sum_{\{n_i\}} h_{mn}^{\{n_i\}}(\vec{x}) \exp\left(i \sum_i \frac{n_i y_i}{R_i}\right) \end{pmatrix}. \tag{13.34}$$

We see that the scalar spherical harmonics are just products of Fourier modes,

$$Y_{\{n_i\}}(\vec{y}) = \prod_i \exp\left(i \frac{n_i y_i}{R_i}\right). \tag{13.35}$$

The Y_{mn} (symmetric tensor) spherical harmonics are the same, just with some delta functions. Indeed, as we said, this is because in this specific case, we have the isometry and local Lorentz groups being equal, both being $SO(n)$. Then we obtain

$$Y_{mn\{n_i\}}^{pr} = \delta_m^p \delta_n^r Y_{\{n_i\}}. \tag{13.36}$$

We note that (qI_q) in our case becomes $(pr\{n_i\})$, and we have

$$h_{mn\{n_i\}}^{\{n_i\}} Y_{\{n_i\}} = h^{pr\{n_i\}} Y_{mn}^{pr\{n_i\}}, \tag{13.37}$$

as in the general case.

13.5 Some general properties of KK reductions

On a general compact space, the linearized KK ansatz for the off-diagonal metric is

$$g_{\mu m}(\vec{x}, \vec{y}) = B_\mu^{AB}(\vec{x}) V_m^{AB}(\vec{y}), \tag{13.38}$$

where $V_m^{AB}(\vec{y})$ is called a Killing vector, and it has an index in an adjoint of the gauge group of symmetries of the compact space. We have written this in the case of a sphere, where for $SO(n+1)$, the adjoint index is an antisymmetric representation in the fundamental indices (AB), and here B_μ^{AB} is a gauge field. This means that in general, for each independent Killing vector, we will get one corresponding gauge field.

In order to get the correct d-dimensional (generalizing from $d = 4$) Einstein–Hilbert action by dimensional reduction, we must consider a nonlinear ansatz for the metric $g_{\mu\nu}$, generalizing the case of the usual KK theory (for S^1), described earlier. The necessary rescaling by the scalar fields is

$$g_{\mu\nu}(\vec{x}, \vec{y}) = g_{\mu\nu}(\vec{x}) \left[\frac{\det g_{mn}(\vec{x}, \vec{y})}{\det g_{mn}^{(0)}(\vec{y})} \right]^{-\frac{2}{d-2}}. \tag{13.39}$$

The full proof of the fact that we get the correct d-dimensional Einstein–Hilbert action from the D-dimensional Einstein–Hilbert action is a bit involved, but we can give here a simple check. Since the Christoffel symbol is of the type $\Gamma \sim g^{-1}\partial g$, and the Riemann tensor and its Ricci tensor contraction is of the type $R \sim \partial\Gamma + \Gamma\Gamma$, it follows that under a *constant* rescaling $g_{\mu\nu} \to \lambda$, we have $R_{\mu\nu}$ invariant (under a non-constant λ, all that happens is that we get extra terms with derivatives on λ). But the D-dimensional Einstein–Hilbert action is $\int d^D x \sqrt{-g^{(D)}} R^{(D)}$, and $R^{(D)} = R_{MN}^{(D)} g^{(D)MN}$. This means that $R^{(D)} = R_{\mu\nu}^{(D)} g^{(D)\mu\nu} + ...$ (other terms, not of Einstein–Hilbert type, but rather scalar and vector kinetic terms).

Then under $g_{\mu\nu} \to \lambda g_{\mu\nu}$, we obtain

$$\sqrt{-g^{(D)}} R^{(D)} = \sqrt{g^{(d)}} \sqrt{\det g_{mn}} \lambda^{\frac{d}{2}-1} R^{(d)} + ... \tag{13.40}$$

We see, therefore, that we need to take

$$\lambda = \left[\frac{\det g_{mn}(\vec{x}, \vec{y})}{\det g_{mn}^{(0)}(\vec{y})} \right]^{-\frac{2}{d-2}}. \tag{13.41}$$

Finally, the notions of Killing vector and Killing spinor, which will be explained better in the next chapter (see also Chapter 8 for coset spaces), are defined as follows.

A Killing vector (named after Wilhelm Killing) satisfies the Killing equation

$$D_{(m}^{(0)} V_{n)}^{AB} = 0, \tag{13.42}$$

and is associated with an isometry. Here the gravitationally covariant derivative is with respect to the background metric on K_n.

On the other hand, the Killing spinor is a bit harder to define precisely.

– On an Einstein space, it satisfies the equation:

$$D_m^{(0)} \eta_\alpha^I = c(\gamma_m \eta^I)_\alpha. \tag{13.43}$$

– In general, is defined by the fact that the susy variation of the D-dimensional spinors vanishes,

$$\delta_Q \lambda_\Omega(\vec{x}, \vec{y}) = 0. \tag{13.44}$$

This means that, in a sense, Killing spinors are a square root of a Killing spinor, since $\{Q, Q\} \sim H$ means that $[\delta_Q, \delta_Q] \sim H$, and a Killing vector has zero energy variation (isometry) as its alternative definition. On Einstein spaces, the fact that Killing spinors are square root of Killing vectors becomes more precise, as we will see in the next chapter.

Important concepts to remember

- In KK reduction, we consider a product space $M_D = M_d \times K_n$.
- There are three KK metrics: the background metric, the KK expansion, and the KK reduction ansatz.

- The background metric is a solution of the product space type, the KK expansion is a generalization of the Fourier expansion, which is always valid, and the KK reduction ansatz is a priori valid only at the linearized level.
- The KK expansion is in terms of spherical harmonics, which are eigenfunctions of the Laplacian on the compact space.
- On the torus, the spherical harmonics are just products of Fourier mode exponentials, and the fields split into fields of different d-dimensional spin, according to the split $\Lambda = (\mu, m)$.
- The truncation to the zero modes (KK reduction) is a priori inconsistent at the nonlinear level, that is, it could be that it does not satisfy the D-dimensional equations of motion.
- Sometimes, a nonlinear redefinition of fields, or equivalently a nonlinear KK reduction ansatz from the beginning, will make a reduction ansatz consistent.
- In the original KK ansatz, the truncation $\phi = 1$ is inconsistent.
- To get the EH action in d dimensions, we need to redefine $g_{\mu\nu}$ by $[\det g_{mn}/ \det g_{mn}^{(0)}]^{-1/(d-2)}$.
- The off-diagonal metric gives a gauge field for each Killing vector, $g_{\mu m} = B_\mu^{AB} V_m^{AB}$.
- Spinors are expanded into d-dimensional spinors times Killing spinors $\psi_\Omega = \psi_\alpha^{q,I_q} \eta_z^{q,I_q}$.
- Killing spinors preserve some susy.

References and further reading

For the KK approach to supergravity, see [22]. For more details, see for instance [20] and references therein.

Exercises

(1) Write down the form of the general spherical harmonic expansion for a gravitino.
(2) For the original KK metric,

$$g_{MN} = \Phi^{-1/3} \begin{pmatrix} g_{\mu\nu} & B_\mu \Phi \\ B_\mu \Phi & \Phi \end{pmatrix}, \qquad (13.45)$$

prove that $g_{\mu\nu}$ is the metric in Einstein frame.
(3) Prove that if $g_{\mu m}(x, y) = B_\mu^{AB}(x) V_m^{AB}(y)$ and we choose the general coordinate transformation with parameter

$$\xi_m(x, y) = \lambda^{AB}(x) V_m^{AB}(y), \qquad (13.46)$$

then the transformation with parameter $\lambda^{AB}(x)$ is the nonabelian gauge transformation of B_μ^{AB}. Note: Use the fact that $V^{AB} = V^{mAB} \partial_m$ satisfies the nonabelian algebra.
(4) Explain why you get the YM action from the expansion of the Einstein–Hilbert action in the case in exercise 3, with a nonabelian gauge field multiplying Killing vectors in the metric.

14 Spherical harmonics and the KK expansion on sphere, coset, and group spaces

We saw in Chapter 13 that spherical harmonics are generalizations of the Fourier modes $e^{iny/R}$ for the Fourier expansion on S^1 and of the $Y_{lm}(\theta, \phi)$ for the spherical harmonics expansion on S^2 to an arbitrary compact space K_n: we KK expand in terms of them.

In this chapter, we will consider the construction and properties of the various spherical harmonics.

14.1 Spherical harmonics

Spherical harmonics are fields on the compact space K_n, with a certain spin = Lorentz representation for the tangent space (= local Lorentz group $SO(n)$) and in a certain representation of the symmetry group G. Then the relevant question, in order to define the KK expansion, is to find the spectrum of the kinetic operators as a function of the representations of G and $SO(n)$.

For instance:

– For the *scalars*, the kinetic operator is the d'Alembertian (or Laplacian, rather), defined as

$$\Box = D_a D^a, \tag{14.1}$$

where D_a is the gravitationally covariant derivative. The spectrum is thus defined by the eigenfunction–eigenvalue equation

$$\Box Y_q^{I_q} = -m^2 Y_q^{I_q}. \tag{14.2}$$

– The case of *antisymmetric tensors (p-forms)* is a bit more complicated, so we will take a bit of time to explain it. For p-forms, we have the *Hodge–de Rham decomposition*, which in form language is

$$p = dq + \delta r + p_h, \tag{14.3}$$

where dq is an exact part (so it is also closed, $d(dq) = 0$, since $d^2 = 0$), δr is a co-exact part, since $\delta = *d*$, and p_h is a harmonic part, $\Delta p_h = 0$.

The Hodge–de Rham operator (or Laplacian) Δ is defined recursively, by

$$\Delta D_{[\mu_1} A_{\mu_2 \dots \mu_k]} = D_{[\mu_1} \Delta A_{\mu_2 \dots \mu_k]}, \tag{14.4}$$

for which the first few forms are the action on 0-forms (scalars), 1-forms (vectors), and 2-forms,

$$\Delta\phi = \Box\phi,$$
$$\Delta A_\mu = \Box A_\mu + R_{\mu\nu}A^\nu,$$
$$\Delta A_{\mu\nu} = \Box A_{\mu\nu} + R_{\mu\sigma}A^\sigma{}_\nu + R_{\nu\sigma}A_\mu{}^\sigma + R_{\mu\nu}{}^{\rho\sigma}A_{\rho\sigma}, \dots \qquad (14.5)$$

and this Hodge–de Rham Laplacian is also written as

$$\Delta = d * \delta + \delta * d \qquad (14.6)$$

and sometimes it is also written as

$$\Delta = (d + d*)^2 = d * d + dd*, \qquad (14.7)$$

since $d^2 = (d*)^2 = 0$, though of course, d and $d*$ act on different p-forms, so it is only a formal relation.

On S^n, we can do more to describe the spherical harmonic p-forms. There, the only harmonic forms are the constant (0-form) and the epsilon tensor (also a constant, n-form), so we can ignore p_h in the Hodge-de Rham decomposition. Then we write

$$p = dq = \delta r \equiv p_e + p_c, \qquad (14.8)$$

the sum of the exact (thus closed) and co-exact (thus co-closed) terms. Then the eigenfunction relation becomes

$$\Delta(p_e + p_c) = \lambda(p_e + p_c) \Rightarrow (\Delta p_e - \lambda p_e) = -(\Delta p_c - \lambda p_c), \qquad (14.9)$$

and taking d on both sides, commuting d with Δ through the defining relation (14.4), and using the fact that p_e is closed and p_c is co-closed, we obtain that we have independently

$$d(\Delta p_e - \lambda p_e) = -d(\Delta p_c - \lambda p_c) = 0, \qquad (14.10)$$

but remembering that there are no non-trivial closed and co-closed p-forms on S^n (it is topologically trivial), we obtain that we have the same eigenfunction relation independently for both p_e and p_c,

$$\Delta p_e = \lambda p_e, \quad \Delta p_c = \lambda p_c. \qquad (14.11)$$

For the exact (thus closed) part of the p-form,

$$\partial_{[\mu_1} u_{\mu_2\dots\mu_{p+1}]} = 0, \qquad (14.12)$$

one finds (we will not show the proof) the spectrum of eigenvalues

$$\Delta u_{\mu_1\dots\mu_p} = -(k + p - 1)(k + n - p)u_{\mu_1\dots\mu_p}, \qquad (14.13)$$

for $k = 1, 2, \dots$, while for the co-exact (thus co-closed) part,

$$D^{\mu_1} u_{\mu_1\mu_2\dots\mu_{p+1}} = 0, \qquad (14.14)$$

one finds

$$\Delta u_{\mu_1\dots\mu_p} = -(k + n - p - 1)(k + p)u_{\mu_1\dots\mu_p}, \qquad (14.15)$$

for $k = 1, 2, \dots$

– Finally, the *Dirac operator* for spinors can be understood as "square root of the Hodge–de Rham Laplacian" and, given the formal form $\Delta = (d + d*)^2$, it is in a certain sense

$$S = d + d*,$$ (14.16)

though it is better understood as the usual

$$S = D_\mu \gamma^\mu = D_a \gamma^a.$$ (14.17)

14.2 Coset theory

We will mostly be interested in coset spaces, like is the case for S^n, so in that case, there is a general construction, though it has to be made concrete for every specific case of representations.

We remember from Chapter 8 the following.

The definition of the vielbein e_μ^m and H-connection ω_μ^i, Killing vector f_a^μ, and H-compensator Ω_a^i comes either from left- and right-multiplication with an infinitesimal group element or, in the case of the first two objects, also from the coset representative $L(z)$ (which was, as we remember, an element of the group parametrized by the coordinate on the coset z), via

$$L^{-1}(z)\partial_\mu L(z) = e_\mu^m(z)K_m + \omega_\mu^i(z)H_i$$
$$e^{dg^a T_a} e^{z^\mu K_\mu} = e^{z^\mu K_\mu + dg^a f_a^\mu(z) K_\mu} e^{-dg^a \Omega_a^i(z) H_i + \mathcal{O}(dg^2)}.$$ (14.18)

The Lie derivative in direction K_a is a general coordinate transformation with parameter f_a^μ, and it was also proven to be written in terms of the H-compensator Ω_a^i, as

$$l_{K_a} e_\mu^m(z) = f_a^\nu \partial_\nu e_\mu^m + (\partial_\mu f_a^\nu) e_\nu^m$$
$$= e_\mu^n(z) f_{ni}{}^m \Omega_a^i(z).$$ (14.19)

Defining the H-connection in a representation R by

$$\Omega_\mu{}^I{}_J = \omega_\mu^i(z)(H_i)^I{}_J,$$ (14.20)

we find the covariant derivative on a general field in an H representation R as

$$D_m \Phi^I(z) = e_m^\mu(z) \left[\partial_\mu \Phi^I(z) + \omega_\mu{}^I{}_J(z) \Phi^J(z) \right].$$ (14.21)

The gravitational spin connection (responsible for parallel transport of the fermions on the manifold) is given by a constant piece times the vielbein, plus the H-connection,

$$\omega(\text{spin})_\mu{}^m{}_n(z) = \frac{1}{2} e_\mu^r(z) \bar{f}_{rn}{}^m + \omega_\mu^i(z) f_{in}{}^m,$$ (14.22)

where

$$\frac{1}{2} \bar{f}_{pn}{}^m = \frac{1}{2} \left[f_{pn}{}^m + \delta^{mm'} \left(f_{m'p}{}^{n'} \delta_{n'}^m + n \leftrightarrow p \right) \right].$$ (14.23)

Then, the gravitational Riemann tensor with flat indices derived from the above spin connection is constant,

$$R_{pq}{}^m{}_n = f_{ni}{}^m f_{pq}{}^i - \frac{1}{2}\left(\bar{f}_{rn}{}^m f_{pq}{}^r\right) + \frac{1}{4}\left(\bar{f}_{qn}{}^r \bar{f}_{pr}{}^m - \bar{f}_{pn}{}^r \bar{f}_{qr}{}^m\right). \tag{14.24}$$

The H-covariant Lie derivative is defined such that its action on the vielbein gives zero,

$$\mathcal{L}_{K_a} e^m = 0, \tag{14.25}$$

which generalizes to the action on a field Φ^I in some representation R of H to

$$\mathcal{L}_{HK_a}\Phi^I(z) = l_{K_a}\Phi^I(z) + \Omega_a^i(z)(H_i)^I{}_J\Phi^J(z). \tag{14.26}$$

One has the following useful theorems for the calculation of the spectra of kinetic operators:

– Theorem 1: The H-covariant Lie derivatives form the Lie algebra of H,

$$[\mathcal{L}_{HK_a}, \mathcal{L}_{HK_b}] = -f_{ab}{}^c \mathcal{L}_{HK_c}. \tag{14.27}$$

– Theorem 2: The H-covariant Lie derivatives commute with the gravitational covariant derivative,

$$[D_m, \mathcal{L}_{HK_a}] = 0 \Leftrightarrow [dz^\mu D_\mu, \mathcal{L}_{HK_a}] = 0. \tag{14.28}$$

– Theorem 3: The H-covariant Lie derivatives commute with the Dirac operator,

$$[\mathcal{L}_{HK_a}, \gamma^m D_m] = 0, \tag{14.29}$$

which also implies
– Theorem 4: The H-covariant Lie derivative commute with the Box operator,

$$[\mathcal{L}_{HK_a}, D_\mu^2] = 0. \tag{14.30}$$

Finally, we are ready to define spherical harmonics Y on coset spaces, by the relation

$$\mathcal{L}_{HK_a}Y = -YX_a, \tag{14.31}$$

where X_a is the generator K_a in the corresponding representation.

Then, if the coset representative is $L(z)$, we have that the spherical harmonic is its inverse, in the corresponding representation,

$$Y(z) = L(z)^{-1}. \tag{14.32}$$

14.3 Examples of spherical harmonics

The classic example is the *Killing vector* V_μ^A, defined as an isometry of the metric, by the relation

$$D_{(\mu}V_{\nu)} = 0, \tag{14.33}$$

since this implies that the general coordinate transformation of the metric with parameter V_μ^A vanishes,

$$\delta_{g.c.,V_\mu^A} g_{\mu\nu} = 0. \tag{14.34}$$

In terms of the vielbein, the same general coordinate transformation can now give a local Lorentz transformation,

$$\delta_{g.c.,V_\mu} e_\mu^m = \text{local Lorentz } e_\mu^m, \tag{14.35}$$

that is,

$$V^\mu \partial_\mu e_\nu^m + (\partial_\nu V^\mu) e_\mu^m = \text{local Lorentz } e_\mu^m. \tag{14.36}$$

The Killing spinor is considered as the fermionic superpartner of the Killing vector. Construct the super-general coordinate transformation in superspace with superfield parameter $\xi^\mu = (V^\mu, \epsilon^\alpha)$, where V^μ is the Killing vector and ϵ^α will be shown to give the Killing spinor.

Then the condition to have a super-Killing vector is

$$\xi^\Lambda \partial_\Lambda E_\Pi^M + (\partial_\Pi \xi^\Lambda) E_\Lambda^M = \text{local Lorentz } E_\Lambda^M. \tag{14.37}$$

Consider the fermionic Killing vector, obeying the condition with M being the flat fermionic index a, which becomes

$$\epsilon^\alpha \partial_\alpha E_\Lambda^a + (\partial_\Lambda \epsilon^\alpha) E_\alpha^a = \lambda^{mn} (\Gamma_{mn})^a{}_b E_\Lambda^b, \tag{14.38}$$

and fix a gauge where we only have the condition with $\Lambda = \beta$. Also restrict to the $\theta = 0$ component of the condition. We obtain

$$\left(\epsilon^\alpha \partial_\alpha E_\beta^a + (\partial_\beta \epsilon^\alpha) E_\alpha^a \right)\big|_{\theta=0} = 0, \tag{14.39}$$

which turns out to be the same as the condition of supersymmetry invariance of the spinor (gravitino),

$$\delta_{susy} \psi_\beta = 0. \tag{14.40}$$

The fact that the susy transformation of the fermion vanishes means that susy is preserved by the vacuum. Indeed, in the vacuum, the VEVs of the fermions vanish, and since the variation of the bosons is fermionic, it is automatically 0, while the variation of the fermions is bosonic, with a priori nonzero VEVs, so their vanishing is nontrivial.

Thus the Killing spinor condition is the condition of susy invariance of the vacuum, which means that the number of Killing spinors equals the number of unbroken susies of the vacuum.

Moreover, since $\{Q, Q\} \sim H$, which in terms of variations becomes $[\delta_Q, \delta_Q] \sim \delta_t$ or, more generally, add to δ_t the other isometries (zero energy bosonic transformations) defined by the Killing vectors, then the Killing spinors are sort of square roots of the Killing vectors. This will be made more precise in particular cases.

Indeed, a different definition of the Killing spinor, valid only on Einstein spaces, as we will prove shortly, is by the Killing spinor equation

$$D_\alpha \eta = ic\gamma_\alpha \eta, \tag{14.41}$$

where D_α are gravitationally covariant derivatives (with spin connection in them), c is a constant, and γ_α are $SO(n)$ gamma matrices.

But we can apply the definition twice and take the commutator, to obtain

$$[D_\alpha, D_\beta]\eta = -2c^2 \gamma_{\alpha\beta}\eta, \qquad (14.42)$$

where $\gamma_{\alpha\beta} = \frac{1}{2}[\gamma_\alpha, \gamma_\beta]$. On the other hand, the commutator of two gravitational covariant derivatives gives the Riemann tensor times the Lorentz generator in the corresponding representation, to the above should also equal

$$\frac{1}{4}R_{\alpha\beta}{}^{ab}\gamma_{ab}\eta. \qquad (14.43)$$

Equating the two relations, multiplying by γ^α, using the gamma matrix relation (obtained, as usual, by writing on the right-hand side all the possible Lorentz structures with the complete system of antisymmetric gammas, and finding the coefficients by giving special values to the indices)

$$\gamma^\alpha \gamma_{ab} = \gamma^\alpha{}_{ab} + e^\alpha_a \gamma_b - e^\alpha_b \gamma_a, \qquad (14.44)$$

and using the antisymmetry relation for the Riemann tensor $R_{a[bcd]} = 0$, we find

$$\left[R_{\alpha\beta} - 4c^2(n-1)g_{\alpha\beta}\right]\gamma^\beta\eta = 0. \qquad (14.45)$$

We can multiply by $\bar\eta\gamma^\alpha$, use the relation

$$\bar\eta\gamma^\alpha\gamma^\beta\eta = g^{\alpha\beta}\bar\eta\eta, \qquad (14.46)$$

obtained from a Majorana spinor relation and $\{\gamma^\alpha, \gamma^\beta\} = 2g^{\alpha\beta}$, and obtain

$$R_{\alpha\beta} = 4c^2(n-1)g_{\alpha\beta}, \qquad (14.47)$$

which means that the n-dimensional compact space K_n is an Einstein space.

Note that the case $c = 0$ is also included in the analysis. In this case, $R_{\alpha\beta} = 0$ means the space is Ricci flat, and the Killing spinor equation reduces to $D_\alpha\eta = 0$, the condition for the spinor to be covariantly constant. Both are true, for instance, in the case of CY_n (Calabi–Yau) spaces, which will be studied toward the end of the book, in which case there is one unbroken supersymmetry after the KK reduction.

The Riemann tensor can be decomposed into the Weyl tensor $C_{\alpha\beta}{}^{ab}$ (which is a part that is invariant under Weyl, local scale, or conformal, transformations; it vanishes for de Sitter or anti-de Sitter spaces), plus a trace part,

$$R_{\alpha\beta}{}^{ab} = C_{\alpha\beta}{}^{ab} + \delta^a_{[\alpha}\delta^b_{\beta]}\frac{2}{n(n-1)}R, \qquad (14.48)$$

which means that for the Einstein space, we obtain

$$C_{\alpha\beta}{}^{ab}\gamma_{ab}\eta = 0. \qquad (14.49)$$

But we saw that on coset manifolds, the Riemann tensor with flat indices, $R_{ab}{}^{cd}$, was constant, which means (from its decomposition into Weyl plus Ricci scalar) that the same is true for Weyl tensor with flat indices, $C_{ab}{}^{cd}$.

This means that $C_{ab}{}^{cd}\gamma_{cd}$ is a constant local Lorentz matrix acting on the Killing spinors, so it is a *holonomy*. Holonomy refers to the parallel transport of a spinor around a closed curve in a manifold, and the normal holonomy is an element in the local Lorentz group $SO(n)$, defined (through the commutator of two derivatives, or the fact that the curvature of the spin connection is the Riemann tensor) by the Riemann tensor.

However, the above holonomy $C_{ab}{}^{cd}\gamma_{cd}$ is not the usual holonomy, but rather a "de Sitter holonomy," which takes values in $SO(n+1)$ instead of just $SO(n)$, with connection that joins together the $SO(n)$ spin connection and the vielbein into a $SO(n+1)$ generator, (or rather, $SO(1,n)$) $\Omega^{AB} = \{\omega(\text{spin})^{ab}, \Omega^{a,n+1} = e^a\}$.

Finally, if we have N Killing spinors η^A, $A = 1, ..., N$, the Killing vectors are obtained from them as a quadratic form (vindicating the advertised fact of the Killing spinor being a "square root of the Killing vector")

$$V_\mu^{AB} = \bar{\eta}^A \gamma_\mu \eta^B. \tag{14.50}$$

Besides the Killing spinor and the Killing vector, other important examples of spherical harmonics are modifications of them, obtained by modifying their defining relations as follows.

- The *conformal Killing vector*, satisfying the equation

$$D_\alpha V_\beta + D_\beta V_\alpha = \frac{2}{n} g_{\alpha\beta} D^\gamma V_\gamma. \tag{14.51}$$

- The *conformal Killing spinor*, satisfying the equation

$$D_\alpha \chi = \frac{1}{n} \gamma_\alpha \slashed{D} \chi. \tag{14.52}$$

One can prove (though I will not show it here) the following theorems.

- Theorem 1. The Killing vectors V_α^{AB} generate the group $SO(N)$ in the case that we have N Killing spinors, namely we have

$$[V^{AB}, V^{CD}] = 4c(\delta^{BC}\delta^{AD} + 3 \text{ more terms}). \tag{14.53}$$

- Theorem 2. The H-covariant Lie derivative equals the usual Lie derivative plus a term that is proportional to the generator t_{ab} in the corresponding representation (on which the derivative acts),

$$\mathcal{L}_V = l_V + \left(D(\omega(e))^a V^b + V^\alpha \omega(\text{spin})_\alpha{}^{ab}\right) t_{ab}. \tag{14.54}$$

- Theorem 3. The H-covariant Lie derivative on the Killing spinor generates an $SO(N)$ rotation,

$$\mathcal{L}_{V_{AB}} \eta_C = 4c(\delta_{AC}\eta_B - \delta_{BC}\eta_A). \tag{14.55}$$

- Theorem 4. The H-covariant Lie derivative in the maximal isometry group G of the coset is an $SO(N)$ rotation,

$$\mathcal{L}_{K_G} \eta_A = SO(N) \text{ rotation.} \tag{14.56}$$

- Theorem 5. $G = G^1 \otimes SO(N)$. For S^n, G' is trivial ($= \{\mathbb{1}\}$).
- Theorem 6. If the cosmological constant is nonzero, then all the $V_\alpha^{AB} \neq 0$, while if the cosmological constant is zero, that is, we are in the Ricci flat case, we have no Killing vectors V_α^{AB}.

Then, from

$$\mathcal{L}_{K_{G'}}\eta = 0$$
$$\mathcal{L}_{V_{AB}}\eta_C = 4c(\delta_{AC}\eta_B - \delta_{BC}\eta_A),\qquad(14.57)$$

we obtain

$$\mathcal{L}_{K_G}\eta_A = -\eta_B(X_G)^B{}_A,\qquad(14.58)$$

where $(X_G)^B{}_A$ are the generators of $G = G' \otimes SO(N)$ in the identity representation of G' and the vector of $SO(N)$.

Going back to the general case of spherical harmonics on cosets, both the Killing vector and the Killing spinor are examples of

$$\mathcal{L}_{HK_a}Y = -YX_a,\qquad(14.59)$$

with $Y = L^{-1}(z)$, so

$$L^{-1}dL = YdY^{-1} = -Y^{-1}dY.\qquad(14.60)$$

Multiplying from the left with Y and using the defining relation of the vielbein and H-connection, we get

$$dY = -(e^a K_a + \omega^i H_i)Y.\qquad(14.61)$$

Then the gravitational covariant derivative is

$$DY = \left(d + \omega(\text{spin})^{ab}\frac{1}{4}\gamma_{ab}\right)Y = -e^d\left(K_d + \frac{1}{2}\bar{f}_d{}^{ab}\frac{1}{4}\gamma_{ab}\right)Y.\qquad(14.62)$$

14.4 KK decomposition

To proceed further and calculate the spectrum, we need to understand the KK decomposition.

Consider a representation T of the four-dimensional Local Lorentz $SO(3, 1)$ group, with generic index μ (for instance, vector, spinor, and gravitino) and a representation R of the local Lorentz group of K_n, $SO(n)$, with index m. This will be, in general, decomposed into H-irreps as

$$\Phi_{Rm} = \sum_i \Phi_{R_i m_i}.\qquad(14.63)$$

In turn, the right-hand side will be decomposed into G-irreps I as

$$\Phi_{R_i m_i}(z) = \sum_I C_I Y^I_{R_i m_i}.\qquad(14.64)$$

Then, finally, the KK decomposition of the field $\varphi_{T_\mu, R_m}(x, z)$ is written as

$$\varphi_{T_\mu, R_m}(x, z) = \sum_i \sum_{I_i} \varphi_{T_\mu}(x) Y^{I_i}_{R_i m_i}(z).\qquad(14.65)$$

Note that with respect to the previous notation (from the previous chapter), we just had q for i, so I_q for I_i, otherwise it is basically the same

The spherical harmonics were defined via the action of the H-covariant Lie derivative, so

$$\mathcal{L}_{K_M} Y^{I_j}_{R_i m_i} = -Y^{I_{j'}}_{R_i m_i} (S_M)_{j'}{}^j, \tag{14.66}$$

where S_M is in a representation of G. We see then that only the G representations I_j's that contain the representation R_i of H contribute to the sum.

To calculate the spectrum of $\Box = D_a D^a$ (or the Dirac operator $D_a \gamma^a$) on the Y's, we use the action of the gravitational covariant derivative on it (14.62), knowing also the H-covariant Lie derivative on it (14.66), and the previous theorems for their interrelations.

Of course, for Killing spinors, we have used Einstein spaces, but not all coset spaces are Einstein. We thus must *rescale the vielbein such that one has an Einstein metric*. Starting with the vielbein whose H-covariant Lie derivative vanishes,

$$\mathcal{L}_{K_M} e^a = 0 \Rightarrow l_{K_M} e^a = \Omega^i_M f_i{}^{ab} e^b, \tag{14.67}$$

we rescale

$$e^a = \lambda^a \bar{e}^a, \tag{14.68}$$

obtaining

$$l_{K_M} \bar{e}^a = (\Omega^i_M f_i{}^{ab}) \bar{e}^b \left(\frac{\lambda^b}{\lambda^a} \right). \tag{14.69}$$

For the Maurer–Cartan equations, for $\bar{e}^a$ we obtain

$$d\bar{e}^a + \omega^i f_{ib}{}^a \bar{e}^b + \frac{1}{2} \hat{f}_{bd}{}^a \bar{e}^b \wedge e^d = 0, \tag{14.70}$$

where

$$\hat{f}_{ba}{}^d = f_{bd}{}^a \frac{\lambda^b \lambda^d}{\lambda^a}. \tag{14.71}$$

The metric is written in terms of the coset vielbein e^a_μ via the H-invariant tensor m_{ab} (η_{ab} for $SO(3,1)$),

$$g_{\mu\nu} = e^a_\mu m_{ab} e^b_\nu. \tag{14.72}$$

Then, the Lie derivative on the metric is, as usual,

$$l_{K_M} g_{\alpha\beta} = D_\alpha K_\beta + D_\beta K_\alpha. \tag{14.73}$$

For a general field Φ^I, the gravitational covariant derivative for the rescaled vielbeins is

$$D_a \Phi^I = -\left(\lambda^a K_a + \hat{\hat{f}}_a{}^{bc} \frac{1}{4} \gamma_{bc} \right) \Phi^I. \tag{14.74}$$

The first, and most relevant, fields are:

– the scalars $Y(z)$, for which we have

$$\Box Y(z) = \delta^{ab} D_a D_b Y(z). \tag{14.75}$$

– spinors, in the spin representation of H,

$$\delta_{H_i}\eta = (H_i^{\text{spin}}) \times \eta, \tag{14.76}$$

for which the relevant operator is the Dirac operator $D_a\gamma^a$ acting on η.

– vectors, which can be conserved, V_μ such that $\partial^\mu V_\mu = 0$, in which case we have

$$\Box V_a = \delta^{bc} D_b D_c V_a, \tag{14.77}$$

or pure gauge, $V_\mu = \partial_\mu \phi$.

14.5 Spherical harmonics on particular coset manifolds

14.5.1 Group spaces

Another important case is the case of group spaces.

Note that we can have a case where the space is not only a coset G/H, but also a group space, so it is $= \tilde{G}$. This is the case for instance for the three-sphere, $S^3 = SO(4)/SO(3)$, since we have $SO(4) \simeq SO(3) \times SO(3)$, so $S^3 \simeq SO(3)$. In this case, the left and right cosets are equivalent.

On groups, we can define the left-invariant vielbeins e_μ^a and the right-invariant vielbeins f_μ^a. They are defined for multiplication from the left and from the right, so

$$e^{z^\alpha K_\alpha} e^{dz^a K_a} = e^{z^\alpha K_\alpha + dz^a e_a^\alpha(z) K_\alpha} e^{-dz^a e_a^\alpha \omega_\alpha^i(z) H_i + \mathcal{O}(dz^2)}$$

$$e^{dz^a K_a} e^{z^\alpha K_\alpha} = e^{z^\alpha K_\alpha + dz^a f_a^\alpha(z) K_\alpha} e^{-dz^a f_a^\alpha \omega_\alpha^i(z) H_i + \mathcal{O}(dz^2)}. \tag{14.78}$$

We note that on cosets we defined the Lie vector f_a^α and H-compensator Ω_μ^i for the multiplication from the right (with a different expansion in the exponentials on the right-hand side), though now e have an equivalent set of vielbeins, denoted by the same symbol as the Lie vector (and the same expansion in the exponentials on the right-hand side).

Now, from the Maurer–Cartan equations, we find two sets of the same Lie algebra (just with a minus sign) from the two sets of vielbeins,

$$[e_a, e_b] = f_{ab}^{\ c} e_c$$
$$[f_a, f_b] = -f_{ab}^{\ c} f_c$$
$$[f_a, e_b] = (f_a^\mu \partial_\mu e_b^\nu - e_b^\mu \partial_\mu f_a^\nu)\partial_\nu = (l_{f_a} e_b^\mu)\partial_\mu$$
$$= 0, \tag{14.79}$$

where for the commutation of f_a and e_b we used that $l_{f_a} e_b^\mu = 0$.

The reason that now we get two commuting sets of vielbeins is that, while on coset manifolds G/H the isometry group is $G \times N(H)/H$, where $N(H)$ is the normalizer group (the largest subgroup in which H is a normal group), on group manifolds G the isometry group is $G \times G$, so we get the left- and right-Killing vectors or vielbeins.

Spheres S^n

We now apply the formalism to spheres, and write the resulting formulas. For S^n, $H = SO(n)$, with generators

$$H = \frac{1}{4}\gamma_{mn}, \tag{14.80}$$

and the group $G = SO(n+1)$, so the coset elements are generated by $\gamma_{m,d+1}$, that is,

$$K_m = -ic\gamma_m. \tag{14.81}$$

The Killing spinors, as a spherical harmonic, are obtained from the one at zero by the action of the coset representative,

$$\eta^I(x) = L^{-1}(x)\eta^I(0). \tag{14.82}$$

The Killing spinor equation is

$$D_\mu \eta^I(x) = \left(\partial_\mu + \frac{1}{4}\omega_\mu^{mn}(x)\gamma_{mn}\right)\eta^I(x) = ce_\mu^m(x)(i\gamma_m)\eta^I(x), \tag{14.83}$$

as for a general Einstein space.

If the radius of S^n is m, then we find

$$c = \pm\frac{1}{2}m. \tag{14.84}$$

Indeed, for the sphere, we have the Riemann tensor

$$R_{\mu\nu}{}^{mn} = m^2\left(e_\mu^m(x)\epsilon_\nu^n(x) - e_\mu^n(x)e_\nu^m(x)\right), \tag{14.85}$$

which means by the action of two covariant derivatives, obtaining the Riemann tensor times the generator in the spinor representation, we get

$$\begin{aligned}
[D_\mu, D_\nu]\eta &= \frac{1}{4}R_{\mu\nu}{}^{mn}\gamma_{mn}\eta = \frac{m^2}{2}\gamma_{\mu\nu}\eta \\
&= -2c^2\gamma_{\mu\nu}\eta.
\end{aligned} \tag{14.86}$$

In terms of the stereographic coordinates (obtained by intersecting a line from the North Pole of the sphere with the sphere at a point p and then with the $\mathbb{R}^n$ plane that contains the sphere center at a point z), we have for the vielbein and spin connection

$$e_\mu^m(z) = \frac{\delta_\mu^m}{1+z^2}, \quad \omega_\mu^{mn}(z) = \frac{-2\delta_\mu^m z^n + 2\delta_\mu^n z^m}{1+z^2}, \tag{14.87}$$

while the Killing spinors (of two possible chiralities in even dimensions) are related to the ones at zero by

$$\eta_\pm(z) = \frac{1 \pm i\gamma_m\delta_\mu^m z^\mu}{\sqrt{1+z^2}}\eta_\pm(0). \tag{14.88}$$

The conjugate spinor is defined as usual, for Majorana spinors, $\bar{\eta} = \eta^T C$ (or perhaps with the Ω^{IJ} matrix, if we have modified Majorana spinors), where the C matrix satisfies, as usual,

$$\gamma_\mu^T = \lambda C\gamma_\mu C^{-1}, \quad C^T = \epsilon C, \tag{14.89}$$

where $\lambda = \pm 1, \epsilon = \pm 1$. If $\lambda = -1$, then

$$\bar{\eta}_\pm^I(z)\eta_\pm^J(z) = \bar{\eta}_\pm^J(0)\eta_\pm^I(0). \tag{14.90}$$

The Killing spinors are normalized to either $\Omega^{IJ} = \begin{pmatrix} 0 & \mathbb{1} \\ -\mathbb{1} & 0 \end{pmatrix}$ or to δ^{IJ}. Moreover,

$$\bar{\eta}^I\eta^J = (\bar{\eta}^I\eta^J)^T = \epsilon(\bar{\eta}^J\eta^I). \tag{14.91}$$

The Cartesian coordinates Y^A on the sphere, $Y^A Y^A = 1$, form a vector of $SO(n+1)$, and are related to the Killing spinors as

$$Y^A = \Gamma_{IJ}^A \bar{\eta}_+^I \eta_-^J. \tag{14.92}$$

Here, in even dimensions, $\Gamma^A = \{i\gamma_m\gamma_5, \gamma_5\}$ in terms of the $SO(n)$ matrices γ_m, while in odd dimensions Γ^A are a chiral representation of $SO(n+1)$.

There is also a completeness relation for Killing spinors that, however, depends on dimension. For S^4, for instance, we have

$$\eta_J^\alpha \bar{\eta}_\beta^J = -\delta_\beta^\alpha, \tag{14.93}$$

where

$$\eta_J^\alpha \equiv \eta^{\alpha I}\Omega_{IJ}. \tag{14.94}$$

However, a similar completeness relation holds for all S^n, since then there are the same number of Killing spinors as values for the spinor index. But, obviously, this doesn't hold in a general coset space.

All spherical harmonics can be built from the Killing spinors for S^n, but the formulas depend on dimension also. We give, therefore, the formulas for S^4 as an example.

– the scalars, in the $\underline{5}$ representation

$$\phi_5^{IJ} = \bar{\eta}^I\gamma_5\eta^J, \quad \phi_5^{IJ}\Omega_{IJ} = 0, \tag{14.95}$$

or, equivalently

$$Y^A = \frac{1}{4}(\gamma^A)_{IJ}\phi_5^{IJ}. \tag{14.96}$$

– the conformal Killing vectors in another $\underline{5}$ representation,

$$C_\mu^{IJ} = \bar{\eta}^I\gamma_\mu\gamma_5\eta^J, \quad C_\mu^{IJ}\Omega_{IJ} = 0, \tag{14.97}$$

or equivalently,

$$C_\mu^A = \frac{i}{4}C_\mu^{IJ}(\gamma^A)_{IJ} = D_\mu Y^A = \partial_\mu Y^A. \tag{14.98}$$

– the Killing vectors, in a $\underline{10}$ representation,

$$V_\mu^{IJ} = \bar{\eta}^I\gamma_\mu\eta^J, \tag{14.99}$$

or, equivalently,

$$V_\mu^{AB} = -V_\mu^{BA} = -\frac{i}{8}(\gamma^{AB})_{IJ}V_\mu^{IJ} = Y^{[A}D_\mu^{(0)}Y^{B]}. \tag{14.100}$$

The inverse relation is

$$V_\mu^{IJ} = iV_\mu^{AB}(\gamma_{AB})^{IJ}. \tag{14.101}$$

– the symmetric tensor spherical harmonic, for $g_{\mu\nu}$, is actually the of two spherical harmonics with the same symmetry (Young tableau in the form of a 2×2 box), but different eigenvalues of $\Box$,

$$\eta_{\mu\nu}^{IJKL} = \eta_{\mu\nu}^{IJKL}(-2) - \frac{1}{3}\eta_{\mu\nu}^{IJKL}(-10). \tag{14.102}$$

The first is traceless, while the second is a pure trace, and the value in parentheses represents the eigenvalue of $\Box$, so

$$\Box\eta_{\mu\nu}^{IJKL}(-2) = -2\eta_{\mu\nu}^{IJKL}(-2)$$
$$\Box\eta_{\mu\nu}^{IJKL}(-10) = -10\eta_{\mu\nu}^{IJKL}(-10). \tag{14.103}$$

In terms of previously defined objects, which were in turn defined from the Killing spinors, we have

$$\eta_{\mu\nu}^{IJKL}(-2) = C_{(\mu}^{IJ}C_{\nu)}^{KL} - \frac{1}{4}g_{\mu\nu}^{(0)}C_{\lambda}^{IJ}C^{\lambda KL}$$
$$\eta_{\mu\nu}^{IJKL}(-10) = g_{\mu\nu}^{(0)}\left(\phi_5^{IJ}\phi_5^{KL} + \frac{1}{4}C_{\lambda}^{IJ}C^{\lambda KL}\right). \tag{14.104}$$

– the vector spinor spherical harmonic is also a sum, for the same symmetry (Young tableau in the form of a gun), but different values for the kinetic operator, now the Dirac operator,

$$\eta_{\mu}^{JKL} = \eta_{\mu}^{JKL}(-2) + \eta_{\mu}^{JKL}(-6), \tag{14.105}$$

where the first is a gamma-traceless piece and the second is a gamma-trace, and again the value in the parentheses is the eigenvalue of the kinetic operator (now the Dirac operator),

$$\gamma^{\nu}D_{\nu}^{(0)}\eta_{\mu}^{JKL}(-2) = -2\eta_{\mu}^{JKL}(-2)$$
$$\gamma^{\nu}D_{\nu}^{(0)}\eta_{\mu}^{JKL}(-6) = -6\eta_{\mu}^{JKL}(-6). \tag{14.106}$$

In terms of previously defined objects, we have

$$\eta_{\mu}^{JKL}(-2) = 3\left(\eta^J C_{\mu}^{KL} - \frac{1}{4}\gamma_{\mu}\gamma^{\nu}\eta^J C_{\nu}^{KL}\right)$$
$$\eta_{\mu}^{JKL}(-6) = \gamma_{\mu}\left(\eta^J \phi_5^{KL} - \frac{1}{4}\gamma^{\nu}\eta^J C_{\nu}^{KL}\right). \tag{14.107}$$

Important concepts to remember

- Spherical harmonics are eigenfunctions of the kinetic operator, usually $\Box$ or Dirac.
- For antisymmetric tensors, the Hodge–de Rham decomposition is $p = dq + \delta r + p_h$, with $\delta = *d*$ and $\Delta p_h = 0$.
- The Hodge–de Rham Laplacian is $\Delta = d*\delta + \delta*d$ or $\Delta = (d+d*)^2 = d*d + dd*$.
- Spherical harmonics are defined by $\mathcal{L}_{HK_a}Y = -YX_a$, and one has $Y(z) = L(z)^{-1}$.
- Killing vectors are defined by $D_{(\mu}V_{\nu)} = 0$ and generate isometries, and Killing spinors are superpartners of Killing vectors, generate supersymmetries, $\delta_{susy}\psi_{\beta} = 0$, and are the "square root of Killing vectors."
- On Einstein spaces, Killing spinors satisfy $D_{\alpha}\eta = ic\gamma_{\alpha}\eta$.

- For N Killing spinors η^A, we have $V_\mu^{AB} = \bar{\eta}^A \gamma_\mu \eta^B$.
- To calculate the spectrum of spherical harmonics, one rescales the vielbeins to Einstein space vielbeins, then writes the gravitational covariant derivatives in terms of constant objects, for obtaining the kinetic operators in terms of them. One also uses the theorems for the relations to the H-covariant Lie derivative.
- On group spaces, we have two copies of vielbeins (left- and right-invariant), for the symmetry $G \times G$, where G is the group space.
- For S^n, all the spherical harmonics are built from the Killing spinors, and Y^A, the Cartesian coordinates of the sphere, are scalar spherical harmonics.

References and further reading

For the Kaluza–Klein approach to supergravity, see [22]. For more details, see for instance [20] and references therein.

Exercises

(1) For a four-sphere, the Euclidean embedding coordinates Y^A are scalar spherical harmonics, satisfying $Y^A Y_A = 1$ (and so $Y^A D_\mu^{(0)} Y^A = 0$, $D_\mu^{(0)} Y^A D_\nu^{(0)} Y^A = g_{\mu\nu}^{(0)}$.) Prove then that

$$\epsilon_{A_1 \ldots A_5} dY^{A_1} \wedge dY^{A_2} = 3\sqrt{g^{(0)}} \epsilon_{\mu\nu\rho\sigma} dx^\mu \wedge dx^\nu \partial^\rho Y^{[A_3} \partial^\sigma Y^{A_4} Y^{A_5]}. \tag{14.108}$$

(2) Prove explicitly that for the sphere (Einstein space) we have theorem 3,

$$\mathcal{L}_{V_{AB}} \eta_C = 4c(\delta_{AC}\eta_B - \delta_{BC}\eta_A). \tag{14.109}$$

(3) Show explicitly that, for the sphere S^n, the Cartesian coordinates are obtained from the Killing spinors as

$$Y^A = \Gamma_{IJ}^A \bar{\eta}_+^I \eta_-^J, \tag{14.110}$$

using an explicit parametrization for the sphere.

(4) Let Y^A be 6 cartesian coordinates for the five-sphere S^5. Then Y^A is a vector spherical harmonic and $Y^{A_1 \ldots A_n} = Y^{(A_1} \ldots Y^{A_n)} - traces$ is a totally symmetric traceless spherical harmonic (i.e., $Y^{A_1 \ldots A_n} \delta_{A_m A_p} = 0$, $\forall\ 1 \leq m, p \leq n$). Check that, as polynomials in six dimensions, $Y^{A_1 \ldots A_n}$ satisfy $\Box_{6d} Y^{A_1 \ldots A_n} = 0$. Expressing $\Box_{6d}$ in terms of $\Box_{S^5}$ and ∂_r (where $Y^A Y^A \equiv r^2$), check that $Y^{A_1 \ldots A_n}$ are eigenfunctions with eigenvalues $-k(k + 5 - 1)/r^2$.

15 $\mathcal{N} = 2$ sugra in 4 dimensions, general sugra theories, and $\mathcal{N} = 1$ sugra in 11 dimensions

In this chapter, we will analyze more general supergravity theories: $\mathcal{N} = 2$ in 4 dimensions, then mention $\mathcal{N} > 2$ or $d > 4$ (which will be described in a bit more detail in Chapter 16), and then the unique $\mathcal{N} = 1$ supergravity in 11 dimensions.

15.1 $\mathcal{N} = 2$ supergravity and special geometry

In four dimensions, $\mathcal{N} = 2$ supergravity is obtained by coupling the $\mathcal{N} = 1$ supergravity multiplet $(2, 3/2)$ (graviton plus gravitino) to the $\mathcal{N} = 1$ gravitino multiplet $(3/2, 1)$, that is, gravitino plus (abelian) vector, for a total of: graviton, two gravitini, and an abelian scalar. In general, the number of gravitinos equals the number of supersymmetries, since each different supersymmetry must vary the unique graviton into another gravitino. Here we will analyze the bosonic Lagrangian of $\mathcal{N} = 2$ supergravity coupled to matter. For that, we will first look at the rigid case, in order to understand it better.

15.1.1 $\mathcal{N} = 2$ rigid supersymmetry

For $\mathcal{N} = 2$ rigid supersymmetry, we have the $\mathcal{N} = 2$ vector multiplet, made up of the $\mathcal{N} = 1$ vector W_α, $(1, 1/2)$ (vector plus spinor) plus the $\mathcal{N} = 1$ chiral multiplet Φ, $(1/2, 0)$ (spinor plus scalar). We can also have the $\mathcal{N} = 2$ hypermultiplet, made up of two chiral multiplets, Q and $\tilde{Q}$.

For the n $\mathcal{N} = 2$ vector multiplets, Ψ^A, with $A = 1, ..., n$, we can write in $\mathcal{N} = 2$ superspace the action in terms of a prepotential $F(\Psi^A)$,

$$S = \frac{1}{16\pi} \mathrm{Im} \int d^4 x d^2\theta d^2\tilde{\theta} F(\Psi^A), \tag{15.1}$$

which can be rewritten in $\mathcal{N} = 1$ language as

$$\mathrm{Im} \int d^4 x \left[\int d^2\theta F_{AB}(\Phi) W^{A\alpha} W_\alpha^B + \int d^2\theta d^2\bar{\theta} \left(\Phi^\dagger e^{-2gV} \right)^A F_A(\Phi) \right], \tag{15.2}$$

where

$$F_A(\Phi) = \frac{\partial F}{\partial \Phi^A}; \quad F_{AB} = \frac{\partial^2 F}{\partial \Phi^A \partial \Phi^B}. \tag{15.3}$$

To this we add a coupling to m hypermultiplets, $i = 1, ..., m$ with standard kinetic terms, in $\mathcal{N} = 1$ superspace language

$$\int d^2\theta d^2\bar{\theta} \left((Q^\dagger e^{-2gV})_i Q_i + (\tilde{Q}e^{2gV})_i \tilde{Q}_i\right) + \int d^2\theta \left(\sqrt{2}(\tilde{Q}\Phi)_i Q_i + m_i \tilde{Q}_i Q_i\right) + h.c., \quad (15.4)$$

where the interaction terms between the hypers and the vectors must respect the global invariances as well, but we did not write explicitly how is that realized. In general, the Q_i and $\tilde{Q}_i$ could also have a general kinetic term, coming from a Kahler potential of their own, but we will not write it here.

The kinetic terms in the Lagrangian for the $\mathcal{N} = 2$ vector multiplets Ψ^A, written in its components $((X^A, \lambda^{iA}, A^A_\mu)$ for the scalar X, 2 gauginos λ^i for $i = 1, 2$, and gauge field A_μ), is

$$\mathcal{L} = g_{A\bar{B}} \partial_\mu X^A \partial^\mu \bar{X}^{\bar{B}} + g_{A\bar{B}} \bar{\lambda}^{iA} \slashed{\partial} \lambda_i^{\bar{B}} + \text{Im}(F_{AB} \mathcal{F}^{-A}_{\mu\nu} \mathcal{F}^{-B}_{\mu\nu}), \quad (15.5)$$

where

$$g_{A\bar{B}} = \partial_A \partial_{\bar{B}} K,$$
$$K(X, \bar{X}) = i(\bar{F}_A(\bar{X}) X^A - F_A(X) \bar{X}^A),$$
$$F_A(X) = \partial_A F(X); \quad F_{AB} = \partial_A \partial_B F(X), \quad (15.6)$$

and $A, B = 1, ..., n$.

15.1.2 Special geometry

We will now couple the $\mathcal{N} = 2$ supergravity multiplet with n $\mathcal{N} = 2$ vector multiplets and m hypermultiplets. The bosonic fields here are the graviton, $n + 1$ vectors, n scalars from the vector multiplets and m vectors from the hypermultiplets. The resulting geometry on the space of scalars is called **special geometry**.

More precisely, the scalars in the vector multiplets form **special Kähler geometry**, and the scalars in the hypermultiplets form **hyper-Kähler or quaternionic geometry**.

We first write the bosonic Lagrangian, and then explain the various objects in it,

$$\frac{\mathcal{L}_{\text{sugra}}}{\sqrt{-g}} = R[g] + g_{i\bar{j}}(z, \bar{z}) \nabla^\mu z^i \nabla_\mu \bar{z}^{\bar{j}} - 2\lambda h_{uv}(q) \nabla^\mu q^u \nabla_\mu q^v$$
$$+ i\left(\bar{\mathcal{N}}_{IJ} \mathcal{F}^{-I}_{\mu\nu} \mathcal{F}^{-J\mu\nu} - \mathcal{N}_{IJ} \mathcal{F}^{+I}_{\mu\nu} \mathcal{F}^{+J\mu\nu}\right) - g^2 V, \quad (15.7)$$

where $i, j = 1, ..., n$ (previously denoted by A, B, though we will come back to the notation with $A, B = 1, ..., n$, but reserved for a special situation, as we will see), $I, J = 1, ..., n + 1$, $u, v = 1, ..., m$ (previously called i, j, but that is now taken) and

$$V = \bar{X}^I \left(4k^u_I k^v_J h_{uv} + k^i_I k^{\bar{j}}_J g_{i\bar{j}}\right) X^J + \left(U^{IJ} - 3\bar{X}^I X^J\right) \mathcal{P}^x_I \mathcal{P}^x_J \quad (15.8)$$

is the potential, and U^{IJ} and the gauge-covariant derivatives of z^i and q^u are defined by

$$U^{IJ} = -\frac{1}{2}(\text{Im}\mathcal{N})^{-1\,IJ} - \bar{X}^I X^J,$$
$$\nabla_\mu z^i = \partial_\mu z^i + g A^I_\mu k^i_I(z),$$
$$\nabla_\mu q^u = \partial_\mu q^u + g A^I_\mu k^u_I(z), \quad (15.9)$$

while the quantities (k_I^i, k_I^u) = Killing vectors and the triholomorphic momentum map $\mathcal{P}_I^x$ will be defined after some preliminaries about geometry.

Here, as we noticed before, the scalars z^i live in a geometry called Kähler geometry, since the kinetic term gives $ds^2 = g_{i\bar{j}}dz^i d\bar{z}^j$, so in the presence of only this kinetic term, motion with arbitrary initial conditions for the scalars is geodesic motion on this space. We have

$$g_{i\bar{j}} = \partial_i \partial_{\bar{j}} K(z, \bar{z}), \tag{15.10}$$

where K is the Kähler potential for the scalars.

Kähler geometry is a particular type of complex geometry. Complex geometry (technically, almost complex geometry; complex geometry has the additional vanishing of a so-called Nijenhuis tensor, but that is irrelevant here) is defined by the existence of a matrix J that locally can be diagonalized on this space, giving $J^2 = -1$ (a generalization of i from the complex plane to a complex manifold).

Then we can write a 2-form called Kähler form,

$$K = g_{i\bar{j}}dz^i \wedge dz^{\bar{j}}. \tag{15.11}$$

If this form is closed, that is, $dK = 0$, we call the space a Kähler space, and then we can write locally (at least on patches, globally there can be differences)

$$g_{i\bar{j}} = \partial_i \partial_{\bar{j}} K(z, \bar{z}), \tag{15.12}$$

for some K.

In fact, the geometry that we have for the z^i's is of a special type, called special Kähler, which we will describe after some more definitions about hyper-Kähler geometry.

For the scalars coming from the hypermultiplets, q^u, we have a hyper-Kähler or quaternionic geometry, defined as follows. We can define not only a complex structure J, but actually three of them, J^x for $x = 1, 2, 3$, satisfying the quaternionic algebra,

$$J^x J^y = -1 + \epsilon^{xyz} J^z, \tag{15.13}$$

which is a generalization of the $(J^x)^2 = -1$ relation defining each complex structure.

Then we can define a triplet of 2-forms

$$\begin{aligned} K_{uv}^x &= h_{uw}(J^x)^w{}_v, \\ K^x &= K_{uv}^x dq^u \wedge dq^v, \end{aligned} \tag{15.14}$$

called hyper-Kähler form, which is a generalization of the Kähler form $K = g_{i\bar{j}}dz^i \wedge dz^{\bar{j}}$ (here J is used to split the form into z and $\bar{z}$), and is covariantly constant ($\nabla K^x = dK^x + \epsilon^{xyz}\omega^y \wedge K^z = 0$, where ω is the – independent in the first-order formalism – $(SU(2)$ part of the) spin connection).

The Kähler spaces have symmetries, described by Killing vectors k_I^i (and k_I^u), that is, we have symmetries under the transformations

$$z^i \to z^i + \epsilon^I k_I^i, \tag{15.15}$$

and similarly for q^u and k_I^u.

These (k_I^i, k_I^u) are the objects appearing in the Lagrangian mentioned earlier. These Killing vectors are holomorphic, that is,

$$\partial_{\bar{j}} k_I^i = 0. \tag{15.16}$$

In the case of the hyper-Kähler geometry, the Killing vectors k_I^u are tri-holomorphic (holomorphic with respect to each complex structure). Note that in both cases, the symmetries are associated with the gauge fields A_μ^I.

Note that the Killing vector condition in complex coordinates is

$$\nabla_i k_j + \nabla_j k_i = 0; \quad \nabla_{\bar{i}} k_j + \nabla_j k_{\bar{i}} = 0, \tag{15.17}$$

(the complex conjugate of the first condition gives the condition for $(\bar{i}\bar{j})$, but gives nothing new), but we define

$$k_j = g_{i\bar{j}} k^{\bar{j}}; \quad k_{\bar{j}} = g_{\bar{j}i} k^i. \tag{15.18}$$

The Killing vectors $k_I = k_I^i \partial_i$, as usual, satisfy a Lie algebra

$$[k_I, k_J] = f_{IJ}{}^K k_K. \tag{15.19}$$

Finally, the object $\mathcal{P}_I$ is called the momentum map, and it satisfies

$$k_I^i = i g^{i\bar{j}} \partial_{\bar{j}} \mathcal{P}_I. \tag{15.20}$$

This does not fix completely $\mathcal{P}_I$, but we can impose a condition that is equivalent to

$$\frac{i}{2} g_{i\bar{j}} (k_I^i k_J^{\bar{j}} - k_J^i k_I^{\bar{j}}) = \frac{1}{2} f_{IJ}{}^K \mathcal{P}_K. \tag{15.21}$$

Moreover, if the Kähler potential is exactly invariant under the transformations of the isometry group G and not only up to Kähler transformations ($K' = K + \text{Re} f(z)$, which don't change $g_{i\bar{j}} = \partial_i \partial_{\bar{j}} K$), that is, if

$$k_I^i \partial_i K + k_I^{\bar{i}} \partial_{\bar{i}} K = 0, \tag{15.22}$$

then we can write also

$$i \mathcal{P}_I = k_I^i \partial_i K = -k_I^{\bar{i}} \partial_{\bar{i}} K. \tag{15.23}$$

On hyper-Kähler manifolds, we can define a tri-holomorphic momentum map $\mathcal{P}_I^x$.

The Killing vectors k_I^i, with label that belongs to the isometry group G of the manifold, have been written with symplectic indices (like X^I and F_I), since they are embedded in the symplectic group, specifically by the relation*

$$(k_K^i \partial_i + k_K^{\bar{i}} \partial_{\bar{i}}) \begin{pmatrix} X^I \\ F_I \end{pmatrix} = T_K \begin{pmatrix} X^I \\ F_I \end{pmatrix}, \tag{15.24}$$

where T_K are matrices in the symplectic group, chosen to be block-diagonal, with the blocks being (for the purposes of gauging the symmetries G) the adjoint representation of the group G, that is, (f_{KJ}^I= structure constants of the group G, embedded in $Sp(2n + 2; \mathbb{R})$)

$$T_K = \begin{pmatrix} f_{KJ}^I & 0 \\ 0 & -f_{KJ}^I \end{pmatrix} \in Sp(2n + 2, R). \tag{15.25}$$

---

* Note that, in general, on the right-hand side of (15.24), there can be a compensating $U(1)$ Kähler transformation acting on (X^I, F_I), which thus changes the Kähler potential as $K \to K + f(X) + \bar{f}(\bar{X})$.

We can now finally define the notion of special Kähler geometry. Note that we have n coordinates z^i on this space, but $n + 1$ X^I's. We can define

$$F_I = \frac{\partial F}{\partial X^I}, \qquad (15.26)$$

in terms of a prepotential F like in the rigid case, though we can in fact define special geometry without a reference to an F.

We need to impose the constraint

$$i(\bar{X}^I F_I - \bar{F}_I X^I) = 1 \qquad (15.27)$$

for two reasons.

The first is that if not, the left-hand side of the above will appear in front of the Einstein action, and the second is that in any case, we must impose some constraints, since we have $n + 1$ X^I's, but only n z^i's.

The coordinates X^I are also *covariantly holomorphic*, that is, we have

$$\nabla_{\bar{i}} X^I \equiv \left(\partial_{\bar{i}} - \frac{1}{2} \partial_{\bar{i}} K \right) X^I = 0, \qquad (15.28)$$

where K is the Kähler potential.

Also, because of dimensionality reasons (X has dimension 1, whereas $F(X)$ needs to have dimension 2, since F_{AB} acts as gauge coupling function), $F(X)$ needs to be a homogenous function of degree 2 (so that dimensions work out for $F(X)$) in the X's, that is, under $X^I \rightarrow \lambda X^I$, we should have $F(X) \rightarrow \lambda^2 F(X)$. That in turn means that $F_I = \partial_I F$ is homogenous of degree 1, so scales the same as X^I. We thus need to redefine as

$$X^I = e^{K/2} Z^I(z) \Rightarrow F_I = e^{K/2} F_I(Z(z)), \qquad (15.29)$$

where $F^I(Z(z))$ is obtained by replacing $X \rightarrow Z$ in F_I, or rather in F and then doing $\partial F(Z)/\partial Z^I$. Here $K = K(z, \bar{z})$ is the Kähler potential. After this transformation, we have

$$e^{-K(z,\bar{z})} = i[\bar{Z}^I(z) F_I(Z(z)) - Z^I(z) \bar{F}_I(\bar{Z}(\bar{z}))]. \qquad (15.30)$$

Note that now the coordinates Z^I are (regular) *holomorphic*, that is, $\partial_{\bar{i}} Z = 0$. We see that the rescaling was done such as to cancel the extra term in the covariant holomorphic derivative, leaving just the usual derivative in the holomorphicity condition.

If then, moreover, the Riemann tensor for the space takes the form

$$R^i_{\ jk}{}^l = \delta^i_j \delta^l_k + \delta^i_k \delta^l_j - e^{2K} \mathcal{W}_{jkm} \bar{\mathcal{W}}^{mil}, \qquad (15.31)$$

where

$$\mathcal{W}_{ijk} = i F_{IJK}(Z(z)) \frac{\partial Z^I}{\partial z^i} \frac{\partial Z^J}{\partial z^j} \frac{\partial Z^K}{\partial z^k},$$

$$F_{IJK} = \frac{\partial}{\partial Z^I} \frac{\partial}{\partial Z^J} \frac{\partial}{\partial Z^K} F(Z), \qquad (15.32)$$

we call the space special Kähler.

We should note however that the derivatives of F are not so well-defined in some sense, since the X^I's satisfy the constraint (15.27), though we take derivatives as if the X's are independent.

Finally, in the gauge field kinetic terms in our general bosonic Lagrangian, the matrix $\mathcal{N}_{IJ}$ has the form

$$\mathcal{N}_{IJ} = \bar{F}_{IJ} + 2i\frac{\text{Im}(F_{IK})\text{Im}(F_{JL})X^K X^L}{\text{Im}(F_{K'L'})X^{K'}X^{L'}}. \tag{15.33}$$

We then easily find that it satisfies

$$\mathcal{N}_{IJ}X^J = F_{IJ}X^J. \tag{15.34}$$

Using the constraint (15.27), we can prove that in fact we have

$$F_I = \mathcal{N}_{IJ}X^J, \tag{15.35}$$

and moreover

$$\partial_i \bar{F}_I = \mathcal{N}_{IJ}\partial_i \bar{X}^J. \tag{15.36}$$

These two conditions together define the matrix $\mathcal{N}_{IJ}$, called the *period matrix*, in the general case, even when F_I is defined without a prepotential F.

We can choose a set of coordinates z^i on the special Kähler manifold called special coordinates by

$$z^A = \frac{X^A}{X^0}; \quad A = 1, ..., n, \tag{15.37}$$

that is, $Z^0(z) = 1, Z^A(z) = z^A$.

15.1.3 Very special geometry and duality symmetries

We can classify the special Kähler manifolds according to the form of the prepotential F. For example, in the case of

$$F(X) = \frac{d_{ABC}X^A X^B X^C}{X^0}, \tag{15.38}$$

we call it *very special geometry*.

The kinetic term for the gauge fields contains the period matrix as a coupling function, that is, generalizing the coupling constants to scalar field-dependent objects. It is written as

$$\mathcal{L}_1 = \frac{1}{4}(\text{Im}\mathcal{N}_{IJ})\mathcal{F}^I_{\mu\nu}\mathcal{F}^{\mu\nu J} - \frac{i}{8}(\text{Re}\mathcal{N}_{IJ})\epsilon^{\mu\nu\rho\sigma}\mathcal{F}^I_{\mu\nu}\mathcal{F}^J_{\rho\sigma} = \frac{1}{2}\text{Im}(\mathcal{N}_{IJ}\mathcal{F}^{+I}_{\mu\nu}\mathcal{F}^{+\mu\nu J}), \tag{15.39}$$

where $\mathcal{F}^{\pm}_{\mu\nu}$ are the self-dual and anti-self-dual parts, defined by

$$\mathcal{F}^{\pm}_{\mu\nu} = \frac{1}{2}(\mathcal{F}_{\mu\nu} \pm \frac{1}{2}\epsilon_{\mu\nu\rho\sigma}\mathcal{F}^{\rho\sigma}), \tag{15.40}$$

and where $\epsilon^{0123} = i$.

We define the objects (standard in nonlinear electromagnetic theories)

$$G^{\mu\nu}_{+I} \equiv 2i\frac{\partial\mathcal{L}}{\partial\mathcal{F}^{+I}_{\mu\nu}} = \mathcal{N}_{IJ}\mathcal{F}^{+J\mu\nu},$$

$$G^{\mu\nu}_{-I} \equiv -2i\frac{\partial\mathcal{L}}{\partial\mathcal{F}^{-I}_{\mu\nu}} = \bar{\mathcal{N}}_{IJ}\mathcal{F}^{-J\mu\nu}. \tag{15.41}$$

Now we can form the objects

$$\begin{pmatrix} \mathcal{F}^+ \\ G^+ \end{pmatrix}; \quad \begin{pmatrix} X^I \\ F_I \end{pmatrix}, \tag{15.42}$$

and then on them we have a set of *duality symmetries*.

The simplest case of such symmetries is for the Maxwell equations in the vacuum,

$$dF = 0; \quad d * F = 0, \tag{15.43}$$

which are symmetric under electric-magnetic duality, the exchange of $F \to *F$, $*F \to -F$ (note that $*^2 = -1$), or electric field with magnetic field. We need to exchange as well electric and magnetic charges, in particular, the units e with g. This is not a symmetry like gauge invariance, which doesn't change anything in the physics; in this case, the form of physical processes will be different in general, in particular, due to the fact that charges are modified (if physics is the same, we say we have a self-duality).

In the case at hand, we have a group of duality symmetries $Sp(2n+2;\mathbb{R})$, the symplectic group of $(2n+2) \times (2n+2)$ matrices with real coefficients, defined as the matrices M satisfying

$$M^T \Omega M = \Omega, \tag{15.44}$$

where the constant $(2n+2) \times (2n+2)$ matrix Ω is given by

$$\Omega = \begin{pmatrix} \mathbf{0} & \mathbf{1} \\ -\mathbf{1} & \mathbf{0} \end{pmatrix}. \tag{15.45}$$

In order to be an invariance of the allowed charges at the quantum level (the "charge lattice"; for electromagnetism, we have the Dirac quantization condition $q_e q_m = 2\pi n$), the group is restricted to integer coefficients, that is, $Sp(2n+2;\mathbb{Z})$.

The group acts on the above-defined vectors, that is,

$$\begin{pmatrix} \mathcal{F}^+ \\ G^+ \end{pmatrix} \to \begin{pmatrix} \tilde{\mathcal{F}}^+ \\ \tilde{G}^+ \end{pmatrix} = M \begin{pmatrix} \mathcal{F}^+ \\ G^+ \end{pmatrix},$$

$$\begin{pmatrix} X^I \\ F_I \end{pmatrix} \to \begin{pmatrix} \tilde{X}^I \\ \tilde{F}_I \end{pmatrix} = M \begin{pmatrix} X^I \\ F_I \end{pmatrix}, \tag{15.46}$$

where now, in general, we have

$$\tilde{F}_I = \frac{\partial \tilde{F}}{\partial \tilde{X}^I}, \tag{15.47}$$

where $\tilde{F}$ is another prepotential.

Note, however, that there are counterexamples to this: one can have a situation that after this duality transformation (so in another "symplectic frame"), there actually is no corresponding prepotential $\tilde{F}$, only $\tilde{F}_I, \tilde{X}^I$.

15.2 Other supergravity theories

If we minimally couple the gravitinos in the four-dimensional $\mathcal{N} = 2$ multiplet to an abelian gauge field, we obtain gauged supergravity. In fact, as we said, the gauged supergravity is only a deformation by the coupling constant g of the ungauged model, so the abelian gauge field is in fact the one in the $\mathcal{N} = 2$ supergravity multiplet. The new gravitino transformation law is

$$\delta\psi_\mu^i = D_\mu(\omega(e,\psi))\epsilon^i + g\gamma_\mu\epsilon^i + gA_\mu\epsilon^i. \qquad (15.48)$$

Thus we have a constant term $(g\gamma_\mu\epsilon^i)$ in this transformation law, so it is natural to find that we must add a constant term in the action as well, namely a cosmological constant term, $\int e\Lambda$. This cosmological constant is negative, leading to the fact that the simplest background of gauged supergravity is anti-de Sitter (AdS). Unlike the ungauged supergravity, it does not admit a Minkowski background. Thus, in fact, gauged supergravity is AdS supergravity.

The next possible generalization in four dimensions is the $\mathcal{N} = 3$ supergravity multiplet, composed of the supergravity multiplet $(2, 3/2)$, 2 gravitino multiplets $(3/2, 1)$, and a vector multiplet $(1, 1/2)$. Together, they correspond to the fields $\{e_\mu^a, \psi_\mu^i, A_\mu^i, \lambda\}$, for $i = 1, 2, 3$.

We can also minimally couple the $\mathcal{N} = 3$ multiplet with gauge fields, and as before, the gauge fields have to be the same three gauge fields in the ungauged multiplet. We also find that we must add a negative cosmological constant, and find again that gauged supergravity is AdS supergravity. The difference is that now, under the gauge coupling deformation, the gauge fields become nonabelian (the ungauged model had abelian vector fields).

The next possibility is the $\mathcal{N} = 4$ supergravity multiplet, which is the first to also contain scalars. It is composed of the $\mathcal{N} = 1$ multiplets $(2, 3/2)$, $3 \times (3/2, 1)$, $3 \times (1, 1/2)$, $(1, 0)$, together making $\{e_\mu^a, \psi_\mu^i, A_\mu^k, B_\mu^k, \lambda^i, \phi\}$, where $i = 1, ..., 4$, $k = 1, 2, 3$, A_μ^k are vectors, B_μ^k are axial vectors, and ϕ is scalar. The model can be obtained as a KK dimensional reduction of $\mathcal{N} = 1$ supergravity in 10 dimensions, on a torus T^6. The same comments as above apply for the gauging of this model. But in general, we can gauge a subset of the vectors, so there are various gaugings possible.

The $\mathcal{N} = 5$ supergravity multiplet is composed of the $\mathcal{N} = 1$ multiplets $(2, 3/2)$, $4 \times (3/2, 1)$, $6 \times (1, 1/2)$, $5 \times (1/2, 0)$, together making the graviton, 5 gravitini, 10 vectors, 11 spin 1/2 fermions, and 10 real scalars. The $\mathcal{N} = 6$ supergravity multiplet is composed of the $\mathcal{N} = 1$ multiplets $(2, 3/2)$, $5 \times (3/2, 1)$, $11 \times (1, 1/2)$, $15 \times (1/2, 0)$, together making the graviton, 6 gravitini, 16 vectors, 26 spin 1/2 fermions, and 30 real scalars.

We could imagine that we could have $\mathcal{N} = 7$ supergravity, but if we impose this susy, we obtain $\mathcal{N} = 8$ as well, so the next model is in fact $\mathcal{N} = 8$ supergravity. It is the maximal possible model in four dimensions. The reason is that we want multiplets with at most spin 2, since models with higher spin have no consistent interactions. But when filling a multiplet, we have a finite number of helicities possible, and in case of maximum spin 2, these helicities are filled by the $\mathcal{N} = 8$ model.

The $\mathcal{N} = 8$ supergravity multiplet can be obtained by KK dimensional reduction on T^7 of $\mathcal{N} = 1$ supergravity in 11 dimensions. In fact, 11 dimensions is the maximal dimension from which we can reduce to obtain $\mathcal{N} = 8$ in four dimensions, since the eight four-dimensional gravitini make up a single gravitino in 11 dimensions, but would make less than one gravitino in higher dimensions. The field content of $\mathcal{N} = 8$ supergravity is given in Exercise 2 of Chapter 5.

We can have other supergravity theories in all dimensions and with various super-symmetries (such that when reducing to four dimensions, we would get at most eight supersymmetries). All of the ungauged models can be obtained from the $\mathcal{N} = 1$ 11-dimensional model by torus reductions and truncations. A torus reduction of an ungauged model will always give an ungauged model. There are various gaugings possible (not clear if all have been found). Reducing a model on a nontrivial space (with nonabelian symmetries) leads to a gauged model, but it is not clear if we can obtain all possible gauged models from some nontrivial reduction.

In Chapter 16, we will give more precisely the Lagrangian and susy transformation rules of the maximal supergravity in $d = 4, 5$, both the ungauged and the gauged models, and the 10-dimensional type IIB (ungauged) model. Together with the following $\mathcal{N} = 1$ supergravity in 11 dimensions, this is enough to obtain many other models by reduction and/or truncation.

15.3 $\mathcal{N} = 1$ supergravity in 11 dimensions

The model was found by Cremmer, Julia, and Sherk [26]. Due to its uniqueness, it plays a special role. The field content is: e^a_μ, $\psi_{\mu\alpha}$, and $A_{\mu\nu\rho}$. We know that $\mathcal{N} = 1$ always means we have e^m_μ and $\psi_{\mu\alpha}$, but in 11 dimensions, we see we also need the antisymmetric 3-form $A_{\mu\nu\rho}$. We can check that the number of on-shell degrees of freedom matches. e^m_μ has $9 \times 10/2 - 1 = 44$ degrees of freedom (symmetric traceless transverse tensor), and the 3-form has $9 \times 8 \times 7/(1 \times 2 \times 3) = 84$, for a total of 128 bosonic degrees of freedom. The gravitini have $8 \times 32/2 = 128$ degrees of freedom, so we indeed have matching on-shell.

The kinetic terms in the Lagrangian are

$$\mathcal{L} = -\frac{e}{2}R(e, \omega) - \frac{e}{2}\bar{\psi}_\mu \Gamma^{\mu\nu\rho} D_\nu(\omega)\psi_\rho - \frac{e}{48}F^2_{\mu\nu\rho\sigma}, \tag{15.49}$$

where we have defined

$$F_{\mu\nu\rho\sigma} = 24\partial_{[\mu}A_{\nu\rho\sigma]} \equiv \partial_\mu A_{\nu\rho\sigma} + 23 \text{ terms} \tag{15.50}$$

(the antisymmetrization is with strength one).

In 11 dimensions, the C matrix is antisymmetric, $C^T = -C$, and satisfies

$$C\gamma_\mu C^{-1} = -\gamma^T_\mu, \tag{15.51}$$

so that we have the Majorana spinor relations

$$\bar{\lambda}\Gamma^{A_1}...\Gamma^{A_n}\chi = (-)^n \bar{\chi}\Gamma^{A_n}...\Gamma^{A_1}\lambda. \tag{15.52}$$

We expect the supersymmetry transformation laws to be

$$\delta e^m_\mu = \frac{1}{2}\bar{\epsilon}\gamma^m\psi_\mu,$$

$$\delta\psi_\mu = D_\mu(\omega)\epsilon + \text{more}, \tag{15.53}$$

and some susy law for $A_{\mu\nu\rho}$.

We define a supercovariant extension of $\omega(e)$ in the same way as in $d = 4$, by

$$\hat{\omega}_{\mu mn} = \omega_{\mu mn}(e) + \frac{1}{4}(\bar{\psi}_\mu\gamma_m\psi_n - \bar{\psi}_\mu\gamma_n\psi_m + \bar{\psi}_m\gamma_\mu\psi_n), \tag{15.54}$$

and also a supercovariant extension of $F_{\alpha\beta\gamma\delta}$ by

$$\hat{F}_{\alpha\beta\gamma\delta} = 24\left[\partial_{[\alpha}A_{\beta\gamma\delta]} + \frac{1}{16\sqrt{2}}\bar{\psi}_{[\alpha}\Gamma_{\beta\gamma}\psi_{\delta]}\right]. \tag{15.55}$$

Then the Lagrangian is

$$\mathcal{L} = -\frac{e}{2k^2}R(e,\omega) - \frac{e}{2}\bar{\psi}_\mu\Gamma^{\mu\nu\rho}D_\rho\left(\frac{\omega+\hat{\omega}}{2}\right) - \frac{e}{48}F^2_{\mu\nu\rho\sigma}$$

$$- \frac{3D}{4}k\left[\bar{\psi}_\mu\Gamma^{\mu\alpha\beta\gamma\delta}{}_\nu\psi^\nu + 12\bar{\psi}^\alpha\gamma^{\beta\gamma}\psi^\delta\right](F_{\alpha\beta\gamma\delta} + \hat{F}_{\alpha\beta\gamma\delta})$$

$$+ Ck\epsilon^{\mu_1\cdots\mu_{11}}F_{\mu_1\dots\mu_4}F_{\mu_5\dots\mu_8}A_{\mu_9\mu_{10}\mu_{11}}, \tag{15.56}$$

and the susy laws are

$$\delta e^m_\mu = \frac{k}{2}\bar{\epsilon}\gamma^m\psi_\mu,$$

$$\delta\psi = \frac{1}{k}D_\mu(\hat{\omega})\epsilon + D(\Gamma^{\alpha\beta\gamma\delta}{}_\mu - 8\delta^\alpha_\mu\Gamma^{\beta\gamma\delta})\epsilon\hat{F}_{\alpha\beta\gamma\delta},$$

$$\delta A_{\mu\nu\rho} = E\bar{\epsilon}\Gamma_{[\mu\nu}\psi_{\rho]}. \tag{15.57}$$

Imposing susy invariance of the above action, we fix the constants C, D, E to

$$C = -\frac{\sqrt{2}}{6\cdot(24)^2}; \quad D = \frac{\sqrt{2}}{6\cdot48}; \quad E = -\frac{\sqrt{2}}{8}. \tag{15.58}$$

Note that here ω satisfied its own equation of motion, $\delta I/\delta\omega = 0$, that is, we have a 1.5 order formalism, and is found to be

$$\omega_{\mu mn} = \hat{\omega}_{\mu mn} - \frac{1}{8}(\bar{\psi}^\alpha\Gamma_{\alpha\mu mn\beta}\psi^\beta). \tag{15.59}$$

This is unlike four dimensions, where $\omega = \hat{\omega}$.

We can find the susy algebra (gauge algebra) by demanding closure on the vielbein. Indeed, we know that in general, the susy algebra closes on the vielbein e^m_μ even on-shell. We find

$$[\delta_Q(\epsilon_1), \delta_Q(\epsilon_2)] = \delta_E(\xi^\nu) + \delta_Q(-\xi^\nu\psi_\nu) + \delta_{l.L.}(\lambda_{mn}) + \delta_{Maxwell}(\Lambda_{\mu\nu}),$$

$$\lambda_{mn} = \xi^\nu\hat{\omega}^{mn}_\nu + \bar{\epsilon}_2(\gamma^{mn\alpha\beta\gamma\delta} - 24e^{m\alpha}e^{n\beta}\gamma^{\gamma\delta})\epsilon_1\hat{F}_{\alpha\beta\gamma\delta},$$

$$\Lambda_{\mu\nu} = -\frac{1}{2}\bar{\epsilon}_2\Gamma_{\mu\nu}\epsilon_1 - \xi^\sigma A_{\sigma\mu\nu}, \tag{15.60}$$

and $\xi^\mu = \frac{1}{2}\bar{\epsilon}_2\gamma^\mu\epsilon_1$ as before.

We see that unlike in four dimensions, now we have an extra symmetry on the right-hand side of the susy commutator, namely the Maxwell symmetry (gauge invariance) $\delta_{\text{Maxwell}} A_{\mu\nu\rho} = \partial_{[\mu} \Lambda_{\nu\rho]}$, since in general we can have any of the symmetries of the theory on the right-hand side of the commutator. Also the parameters of the various transformations are different than in three dimensions and four dimensions.

15.4 Off-shell and superspace

In four dimensions, only in the $\mathcal{N} = 1$ and $\mathcal{N} = 2$ are the auxiliary fields known, and considering the other dimensions, also a few other cases are known. But in general, not even auxiliary fields are known, let alone a full superspace formulation like we had for $\mathcal{N} = 1$ in three dimensions and four dimensions.

But we do know a partial superfield formulation in a few cases, that gives, imposing constraints and Bianchi identities, the *on-shell* supergravity, namely its equations of motion.

For example, in the case of $\mathcal{N} = 1$ in 11 dimensions, this is due to Brink and Howe, and Cremmer and Ferrara.

The superfield formulation is in the super-geometric approach, with E_Λ^M and Ω_Λ^{MN}, written in terms of independent Ω_Λ^{mn}, as we saw before. The new feature about 11 dimensions is that now we need to add also a superfield $A_{\Lambda\Pi\Sigma}$, that is, a super-3-form on superspace. In general (for other supergravities), we need some other superfields than E and Ω, but there is no general prescription for what kind of superfields.

From $A_{\Lambda\Pi\Sigma}$, we can define

$$A = A_{\Lambda\Pi\Sigma} dz^\Lambda \wedge dz^\Pi \wedge dz^\Sigma \tag{15.61}$$

and its field strength, $H = dA$, and then we can flatten the indices, getting

$$H = E^M E^N E^P E^Q H_{MNPQ}, \tag{15.62}$$

where $E^M = E_\Lambda^M dz^\Lambda$ and $dz^\Lambda = (dx^M, d\theta^\alpha)$.

We can then also define super-torsions and super-curvatures in the usual way. Note that H_{MNPQ} is also a generalized super-curvature (just like the YM curvature F_{MN}^I was before). The Bianchi identities and constraints are then written in terms of the torsions and curvatures and of H_{MNPQ}, and we obtain the 11-dimensional supergravity equations of motion. More details can be found for instance in [23] and references therein.

Important concepts to remember

- In rigid $\mathcal{N} = 2$ susy, the scalars in the vector multiplet are in a Kähler manifold.
- When coupling $\mathcal{N} = 2$ sugra with vector multiplets and hypermultiplets, the scalars in the vector multiplets live in a special Kähler manifold, and the scalars in the hypermultiplets in a hyper-Kähler or quaternionic manifold, together forming special geometry.

- A Kähler manifold is a complex manifold that has $g_{i\bar{j}} = \partial_i \partial_{\bar{j}} K$, and a hyper-Kähler manifold has three complex structures satisfying the quaternionic algebra.
- There are holomorphic Killing vectors k_I^i, related to the momentum map $\mathcal{P}_I$ by $k_I^i = ig^{i\bar{j}} \partial_{\bar{j}} \mathcal{P}_I$.
- In special Kähler geometry, we have the constraint $i(F_I \bar{X}^I - \bar{F}_I X^I) = 1$, everything is most of the times written in terms of a prepotential F, and the Riemann tensor satisfies a constraint.
- The period matrix satisfies $\mathcal{N}_{IJ} X^J = F_I$ and $\partial_{\bar{i}} \bar{F}_I = \mathcal{N}_{IJ} \partial_{\bar{i}} \bar{F}^J$.
- The vectors (X^I, F_I) and $(\mathcal{F}^+, G^+)$ are acted upon by symplectic transformations in $Sp(2n + 2; \mathbb{Z})$, that are duality symmetries.
- Gauged supergravity is AdS supergravity and is an extension by a gauge coupling parameter of the ungauged models.
- All ungauged models can be obtained from torus reductions and truncations from $\mathcal{N} = 1$ supergravity in 11 dimensions.
- Gauged supergravities are obtained from reduction on nontrivial spaces of ungauged models.
- In $\mathcal{N} = 1$ supergravity in 11 dimensions, the fields are e_μ^a, $\psi_{\mu\alpha}$, and $A_{\mu\nu\rho}$.
- Like in four dimensions, ω satisfies its own equation of motion, but unlike in four dimensions, it is different from its supercovariant extension.
- The gauge algebra has a Maxwell transformation also.
- There is a superspace formulation for $\mathcal{N} = 1$ 11-dimensional supergravity in the supergeometric approach, with constraints and Bianchis in terms of torsions and curvatures and $H = dA$ with flat indices on superspace, in which the on-shell supergravity is obtained, that is, the equations of motion.

References and further reading

For more about $\mathcal{N} = 2$ supergravity and special geometry, see [24] and [25]. For $\mathcal{N} = 1$ supergravity, see [5] and the original paper [26].

Exercises

(1) Consider the prepotential $F = (X^1)^3 / X^0$ and the symplectic transformation

$$S = \begin{pmatrix} A & B \\ C & D \end{pmatrix} = \begin{pmatrix} 1 & 0 & 0 & 0 \\ 0 & 0 & 0 & 1/3 \\ 0 & 0 & 1 & 0 \\ 0 & -3 & 0 & 0 \end{pmatrix}. \tag{15.63}$$

Calculate the transformed $(\tilde{X}^I, \tilde{F}_I)$ and from them the new $\tilde{F}(\tilde{X})$.

(2) Check that $\hat{F}_{\alpha\beta\gamma\delta}$ is supercovariant.

(3) Prove that for

$$e^{-k(z,\bar{z})} = i[\bar{Z}^I(z) F_I(Z(z)) - Z^I(z) \bar{F}_I(\bar{Z}(z))], \tag{15.64}$$

we obtain

$$R^i{}_{jk}{}^l = \delta^i_j \delta^l_k + \delta^i_k \delta^l_j - e^{2k} W_{jkm} W^{mil},$$

$$W_{ijk} = i F_{IJK}(Z(z)) \frac{\partial Z^I}{\partial z^i} \frac{\partial Z^J}{\partial z^j} \frac{\partial Z^K}{\partial z^k}. \tag{15.65}$$

(4) Prove that the $\sim FF\psi$ type terms in the susy variation of the 11-dimensional sugra, $\delta_{\text{susy}} S$ vanish. You need to use the 11-dimensional Majorana spinor relation

$$\bar{\psi} \Gamma^{A_1} ... \Gamma^{A_n} \chi = (-)^n \bar{\chi} \Gamma^{A_n} ... \Gamma^{A_1} \psi, \tag{15.66}$$

and gamma matrix identities which you should prove.

PART II

APPLICATIONS

16

$AdS_7 \times S^4$ nonlinear KK compactification of 11-dimensional supergravity and related notions

In this chapter, we will study the nonlinear KK compactification (and reduction) of 11-dimensional supergravity on an $AdS_7 \times S^4$ background. The reasons for focusing on this one are the uniqueness of 11-dimensional supergravity, the maximal supersymmetry of the background, and the resulting 7-dimensional gauged supergravity (since, as we said, AdS supergravity is gauged supergravity), and the fact that it is the only one of the maximal supersymmetry examples that is completely worked out. Other nonlinear KK compactifications are simpler and/or can be obtained from this one.

We also describe here, for completeness (and for the diligent reader) the gauged supergravities and maximal supergravities in various dimensions, as well as how to gauge subgroups of the general supergravity group. Finally, we consider the modified supergravities in 10 and 11 dimensions with modified "local Lorentz" covariance and their "exceptional field theory" generalization, as well as the geometric approach to supergravity of d'Auria and Fré.

16.1 $\mathcal{N} = 1$ 11-dimensional supergravity and compactifications

As we already said, the $\mathcal{N} = 1$ supergravity in 11 dimensions is unique, in that it is the maximal dimension in which we can lift ("oxidize") the $\mathcal{N} = 8$ supergravity in four dimensions, which is the maximal supergravity with spins ≤ 2 (for higher spin, there is no known way to have interactions with a finite number of fields). In higher dimensions, the eight gravitini of four dimensions will form only part of a gravitino. In 10 dimensions, there are 2 possible maximal supergravities, that is, with $\mathcal{N} = 2$, IIA with 2 gravitini of different chiralities and IIB with 2 gravitini of the same chirality. They correspond to low energy limits of the IIA and IIB string theories. The IIA supergravity is obtained by the circle reduction of the $\mathcal{N} = 1$ 11-dimensional supergravity (the 11-dimensional gravitino splits into two 10-dimensional gravitini of different chiralities), but the IIB is not obtained by any dimensional reduction (though in the full string theory it is nonperturbatively related to the 11-dimensional supergravity). Therefore the 11-dimensional supergravity and the IIB 10-dimensional supergravity are the important cases of supergravities, from which we can obtain the rest.

The maximally supersymmetric backgrounds of these two theories are for 11-dimensional supergravity, Minkowski, $AdS_4 \times S^7$, $AdS_7 \times S^4$, and the pp (parallel plane) waves obtained as a Penrose limit of the $AdS_4 \times S^7$ and $AdS_7 \times S^4$. For IIB supergravity, we have Minkowski, $AdS_5 \times S^5$, and the pp wave obtained as a Penrose limit of $AdS_5 \times S^5$.

Therefore the nontrivial cases of relevance for compactification (except the pp waves that are just limits) are $AdS_4 \times S^7$, $AdS_7 \times S^4$, and $AdS_5 \times S^5$.

The first case to be studied was the full nonlinear KK reduction of 11-dimensional supergravity on $AdS_4 \times S^7$, by de Wit and Nicolai [27, 28], where the ansatz and proof are not fully complete, though it is almost so (it turns out to be very difficult to complete). Here the initial hope was to obtain a nontrivial theory (with nonabelian gauge fields) in four dimensions, hopefully, relevant to phenomenology. However, in the $AdS_4 \times S^7$ solution, the radius of S^7 is equal (up to a factor of 2) to the scale of AdS_4, therefore, by making S^7 small enough so that it is unobservable, we are also making AdS_4 very small, which certainly contradicts experiments. It was found to be impossible to decouple the scale of AdS from the scale of S, so the phenomenological avenue does not work.

Instead, since 1997, AdS/CFT was found to be another application. In AdS/CFT, a string theory (or its supergravity limit) in an $AdS_{d+1} \times X$ background is related to a gauge theory on the d-dimensional boundary of AdS_{d+1}. The compactification in the $AdS_p \times S^q$ cases leads to *gauged supergravities in the lower dimension*. The most important cases are the maximally supersymmetric cases $AdS_5 \times S^5$ (dual to four-dimensional $\mathcal{N} = 4$ SYM, the most interesting case), $AdS_4 \times S^7$, and $AdS_7 \times S^4$, dual to theories of M2-branes and M5-branes, respectively, which are however less understood. For the $AdS_5 \times S^5$ case (and for the reduction of the *original* 10-dimensional type IIB supergravity, see later on in this chapter), we only have complete results for subsets of fields (further consistent truncations of the maximal supergravity), but no complete result for the full ansatz. For the $AdS_4 \times S^7$ case, we have an almost complete result due to de Wit and Nicolai. Therefore the only known full result is for $AdS_7 \times S^4$ [20]. From it, we can derive other results, by further consistent truncations of the maximal seven-dimensional gauged supergravity. We can also consider further KK reductions of the maximal gauged supergravity (even though we derive the seven-dimensional gauged supergravity as arising on an AdS_7 background, once we obtain the gauged supergravity, we can consider a compactification ansatz of seven dimensions instead of the AdS_7 background).

Before we turn to the analysis of the $AdS_7 \times S^4$ compactification, we mention a potential problem. We want to obtain the known maximal gauged supergravity action in seven dimensions, from the compactification of 11-dimensional supergravity on $AdS_7 \times S^4$. In 11-dimensions, we have a gauge field 3-form $A_{\Lambda\Pi\Sigma}$ with a kinetic term with two derivatives, $\sim F^2 + \epsilon FFA$, but in seven dimensions, we have a gauge field 3-form $S_{\alpha\beta\gamma,A}$ with a kinetic term with one derivative, $\sim m^2 S^2 + m\epsilon S\partial S$. Certainly a simple linear KK compactification of the type $A_{\alpha\beta\gamma} \propto S_{\alpha\beta\gamma,A}$ will not work, since it will give by reduction of an action with two derivatives. It follows that we must write an action with a single derivative, specifically a first-order action. Indeed, we know how that works for instance for the Maxwell action, $-\int(\partial_{[\mu}A_{\nu]})^2$, which has two derivatives, but can be rewritten by the introduction of an auxiliary field as $\int[(F_{\mu\nu})^2 - 2F^{\mu\nu}\partial_{[\mu}A_{\nu]}]$, that is, with a single derivative (here $F_{\mu\nu}$ is independent, with field equation $F_{\mu\nu} = \partial_{[\mu}A_{\nu]}$).

Therefore we must write a first-order form for the 11-dimensional supergravity. In principle we have two options, we can write a first-order action for the spin connection $\omega_{\Lambda MN}$ or the 3-form $A_{\Lambda\Pi\Sigma}$. The first option was found not to work, so we need to use the second.

16.2 First-order formulation of 11-dimensional supergravity

Such a formulation was not available before, so we need to define it. The Lagrangian is

$$
\mathcal{L} = -\frac{E}{2k^2} R(E, \Omega) - \frac{E}{2} \bar{\Psi}_\Lambda \Gamma^{\Lambda\Pi\Sigma} D_\Lambda \left(\frac{\Omega + \hat{\Omega}}{2} \right) \Psi_\Sigma
$$

$$
+ \frac{E}{48} (\mathcal{F}_{\Lambda\Pi\Sigma\Omega} \mathcal{F}^{\Lambda\Pi\Sigma\Omega} - 48 \mathcal{F}^{\Lambda\Pi\Sigma\Omega} \partial_\Lambda A_{\Pi\Sigma\Omega})
$$

$$
- \frac{k\sqrt{2}}{6} \epsilon^{\Lambda_0 \dots \Lambda_{10}} \partial_{\Lambda_0} A_{\Lambda_1 \Lambda_2 \Lambda_3} \partial_{\Lambda_4} A_{\Lambda_5 \Lambda_6 \Lambda_7} A_{\Lambda_8 \Lambda_9 \Lambda_{10}}
$$

$$
- \frac{\sqrt{2}k}{8} E [\bar{\Psi}_\Pi \Gamma^{\Pi\Lambda_1 \dots \Lambda_4 \Sigma} \Psi_\Sigma + 12 \bar{\Psi}^{\Lambda_1} \Gamma^{\Lambda_2 \Lambda_3} \Psi^{\Lambda_4}] \frac{1}{24} \left(\frac{F + \hat{F}}{2} \right)_{\Lambda_1 \dots \Lambda_4}, \tag{16.1}
$$

where the field strength of A is

$$
F_{\Lambda\Pi\Sigma\Omega} \equiv \partial_\Lambda A_{\Pi\Sigma\Omega} + 23 \text{ terms} = 24 \partial_{[\Lambda} A_{\Pi\Sigma\Omega]}, \tag{16.2}
$$

and the equation of motion of $\mathcal{F}_{\Lambda\Pi\Sigma\Omega}$ is

$$
\mathcal{F}_{\Lambda\Pi\Sigma\Omega} = F_{\Lambda\Pi\Sigma\Omega}. \tag{16.3}
$$

We see that the only part added to the Lagrangian is in the $\mathcal{F}^2 - 48\mathcal{F}\partial A$ term. We then redefine the difference between $\mathcal{F}$ and F as a new field $\mathcal{B}$, but with flat indices, multiplied by vielbeins,

$$
\mathcal{F}_{\Lambda\Pi\Sigma\Omega} = \partial_\Lambda A_{\Pi\Sigma\Omega} + 23 \text{terms} + \frac{\mathcal{B}_{MNPQ} E_\Lambda^M \dots E_\Omega^Q}{\sqrt{E}}. \tag{16.4}
$$

Then we can write the susy transformation rules as

$$
\delta E_\Lambda^M = \frac{k}{2} \bar{\epsilon} \Gamma^M \Psi_\Lambda
$$

$$
\delta \Psi_\Lambda = \frac{D_\Lambda(\hat{\Omega})\epsilon}{k} + \frac{\sqrt{2}}{12} (\Gamma^{\Lambda_1 \dots \Lambda_4}{}_\Lambda - 8\delta_\Lambda^{\Lambda_1} \Gamma^{\Lambda_2 \Lambda_3 \Lambda_4}) \epsilon \frac{\hat{F}_{\Lambda_1 \dots \Lambda_4}}{24}
$$

$$
+ \frac{1}{24} \left(b \Gamma_\Lambda^{\Lambda_1 \dots \Lambda_4} \frac{\mathcal{B}_{\Lambda_1 \dots \Lambda_4}}{\sqrt{E}} - a \Gamma^{\Lambda_1 \Lambda_2 \Lambda_3} \frac{\mathcal{B}_{\Lambda\Lambda_1 \Lambda_2 \Lambda_3}}{\sqrt{E}} \right) \epsilon
$$

$$
\delta A_{\Lambda_1 \Lambda_2 \Lambda_3} = -\frac{\sqrt{2}}{8} \bar{\epsilon} \Gamma_{[\Lambda_1 \Lambda_2} \Psi_{\Lambda_3]}
$$

$$
\delta \mathcal{B}_{MNPQ} = \sqrt{E} \bar{\epsilon} [a \Gamma_{MNP} E_Q^\Lambda R_\Lambda(\Psi) + b \Gamma_{MNPQ\Lambda} R^\Lambda(\Psi)], \tag{16.5}
$$

where R_Λ is the gravitino field equation,

$$
R^\Lambda(\Psi) = \frac{1}{E} \frac{\delta \mathcal{L}}{\delta \bar{\Psi}} = -\Gamma^{\Lambda\Pi\Sigma} D_\Pi \Psi_\Sigma - \frac{\sqrt{2}}{4} k \left(\frac{\hat{F}_{\Lambda_1 \dots \Lambda_4}}{24} \right) \Gamma^{\Lambda\Lambda_1 \dots \Lambda_5} \Psi_{\Lambda_5} - 3\sqrt{2}k \frac{\hat{F}^{\Lambda\Pi\Sigma\Omega}}{24} \Gamma_{\Pi\Sigma} \Psi_\Omega.
$$

$$\tag{16.6}$$

We note that $\mathcal{B} = 0$ is a field equation, so its susy variation had to be proportional to the gravitino field equation.

Here a, b are free constants, which perhaps could be fixed by the closure of an algebra. However, in this context, they are fixed by requiring to obtain the maximal seven-dimensional gauged supergravity by compactification.

The action admits a background of $AdS_7 \times S^4$ type, with

$$F_{\mu\nu\rho\sigma} = \frac{3}{\sqrt{2}} m (\det e_\mu^{(0)m}(x)) \epsilon_{\mu\nu\rho\sigma}, \tag{16.7}$$

where $\mu = 1, ..., 4$ are indices on S^4 and $m = 1/R_{AdS_7}$. The Einstein equations of motion in the background are

$$R_{\mu\nu} - \frac{1}{2} g_{\mu\nu}^{(0)} R = \frac{1}{6} (F_{\mu\wedge\Pi\Sigma} F_\nu^{\wedge\Pi\Sigma} - \frac{1}{8} g_{\mu\nu}^{(0)} F^2) = -\frac{9}{4} g_{\mu\nu}^{(0)} m^2$$

$$R_{\alpha\beta} - \frac{1}{2} g_{\alpha\beta}^{(0)} R = \frac{1}{48} g_{\alpha\beta}^{(0)} F^2 = \frac{9}{4} g_{\alpha\beta}^{(0)} m^2. \tag{16.8}$$

The solution involves a constant Riemann tensor, namely

$$R_{\mu\nu}^{mn}(e^{(0)4}) = m^2 (e_\mu^{(0)m}(x) e_\nu^{(0)n}(x) - e_\nu^{(0)m}(x) e_\mu^{(0)n}(x))$$

$$R_{\alpha\beta}^{ab}(e^{(0)4}) = -\frac{1}{4} m^2 (e_\alpha^{(0)a}(x) e_\beta^{(0)b}(x) - e_\beta^{(0)a}(x) e_\alpha^{(0)a}(x)). \tag{16.9}$$

Note that for a space of constant curvature, the Riemann tensor can only be constructed out of vielbein, with the unique possible structure allowed by symmetries being the one in the brackets. The prefactor is positive for S^4 (space of positive curvature) and negative for AdS_7 (space of negative curvature). Moreover, the coefficients on the right-hand sides of the Einstein tensors above are equal and of opposite sign, a reflection of the fact that in 11 dimensions, there is no cosmological constant (but then we can have a positive one on S^4 and a negative one on AdS_7).

The ansatz in (16.7) is called Freund–Rubin or spontaneous KK compactification. It was first written for compactification to four dimensions, namely for the $AdS_4 \times S^7$ background, as a way to justify the fact that we live in only four noncompact dimensions. Namely, if we have an antisymmetric tensor field strength, a natural thing is to choose a constant value for it. In the case of 11 dimensions, we can choose $F_{\alpha\beta\gamma\delta} \propto \epsilon_{\alpha\beta\gamma\delta}$ (with noncompact indices), in which case we obtain an $AdS_4 \times S^7$ background, or $F_{\mu\nu\rho\sigma} \propto \epsilon_{\mu\nu\rho\sigma}$ like here, in which case we obtain an $AdS_7 \times S^4$ background. In general, if we have an antisymmetric tensor field, we have sphere compactifications. For instance, in $\mathcal{N} = 1$ 10-dimensional supergravity, we have a field $H_{\mu\nu\rho}$ ($H = dB$, where B is the antisymmetric tensor field that couples to the string), which means that the value $H_{\mu\nu\rho} \propto \epsilon_{\mu\nu\rho}$ will give a Freund–Rubin (spontaneous) compactification on S^3.

16.3 Consistent truncations and nonlinear ansätze

In order to proceed, we must understand the notion of nonlinear ansatz and its relation with consistent truncations.

As we mentioned, the KK expansion is always valid, since it is just a generalized Fourier theorem. But the KK reduction ansatz is not, except in the case of the torus T^n, when it is always valid. In general the KK reduction ansatz is not consistent (i.e., valid), except *at the linearized level*, that is, for terms quadratic in the action.

Indeed, making a truncation to just the lowest mode ϕ_0, and putting the rest to zero ($\phi_q = 0$) is in general not a solution of the higher-dimensional (D-dimensional) equations of motion. If it is a solution to the higher-dimensional equations of motion, we say we have a *consistent truncation*.

What can go wrong? We can have terms in the equation of motion of the fields ϕ_n that we want to put to zero, which are *sources* for them, depending only on ϕ_0, that is,

$$(\Box - m^2 + ...)\phi_n = f(\phi_0). \tag{16.10}$$

This corresponds in the action to terms that are linear in ϕ_n, for $n > 0$, $\phi_n f(\phi_0)$. In such a case, the truncation to ϕ_0 is inconsistent, since putting ϕ_n for $n > 0$ to zero contradicts the equations of motion.

But in the quadratic action, namely at a linearized level (for a quadratic reduction), we can always linearly redefine the fields to get rid of $\phi_n\phi_0$ terms and obtain consistency.

To better understand the issues, consider a ϕ^3 interaction in the higher dimension, $\int d^D x\sqrt{-g^{(D)}}\lambda\phi^3$, and substitute the general KK expansion $\phi(x,y) = \sum_{q,I_q}\phi_q^{I_q}Y_q^{Y_q}$ in it, and focus on a single such term, namely on

$$\lambda \int d^d\vec{x}\sqrt{\det g_{\mu\nu}^{(0)}}\phi_q^{I_q}(\vec{x})\phi_0^{I_0}(\vec{x})\phi_0^{J_0}(\vec{x}) \times \int d^n\vec{y}\sqrt{\det g_{mn}^{(0)}}Y_q^{I_q}(\vec{y})Y_0^{I_0}(\vec{y})Y_0^{J_0}(\vec{y}). \tag{16.11}$$

Then in general, if such term is nonzero, from the equations of motion of ϕ_n, we will get

$$(\Box - m_q^2)\phi_q^{I_q}(\vec{x}) = \lambda C\phi_0^{I_0}(\vec{x})\phi_0^{J_0}(\vec{x}), \tag{16.12}$$

where

$$C = \int d^n\vec{y}\sqrt{\det g_{mn}^{(0)}}Y_q^{I_q}(\vec{y})Y_0^{I_0}(\vec{y})Y_0^{J_0}(\vec{y}). \tag{16.13}$$

In this general case, the truncation is inconsistent, since $\phi_n = 0$ is not a solution to the equations of motion.

But if $C = 0$, then the truncation is consistent. This is indeed what happens for the torus, since there $Y_0^{I_0}(\vec{y}) = 1$, so we obtain for S^1

$$C_1 = \int dy Y^{I_n} = \int dy e^{\frac{2\pi i n y}{R}} = 0 \tag{16.14}$$

for $n \neq 0$, and for T^n, $C_{T^n} = \prod_n C_1 = 0$.

So the torus truncation is always consistent *if we keep all the $n = 0$ fields*. For instance, in the original KK reduction (five dimensions on a circle, down to four dimensions), if we keep all the fields, we have a consistent truncation. But if we put $\varphi = 1$, amounting to putting one ϕ_0 to zero, we have an inconsistent truncation.

More generally, if we have some global symmetry G for the fields in the KK expansion, and under the dimensional reduction ansatz, we keep ALL the singlets of G (fields that do not transform under G), then we obtain the same result. Indeed, if $Y_0^{I_0}$ and $Y_0^{J_0}$ are singlets, then $Y_0^{I_0}Y_0^{J_0}$ is also a singlet, whereas $Y_q^{I_q}$ is not, since we assumed we keep all the singlets.

Then by spherical harmonic orthogonality, or rather by the need of G-group invariance, we have

$$\int Y_q^{I_q}(Y_0^{I_0} Y_0^{J_0}) = 0. \tag{16.15}$$

The simplest example of nonlinear KK ansatz is the one needed to get the correct d-dimensional Einstein–Hilbert action (in Einstein frame) from the D-dimensional Einstein–Hilbert action. Namely we need to write

$$g_{\mu\nu}(\vec{x}, \vec{y}) = g_{\mu\nu}(\vec{x}) \left[\frac{\det g_{mn}(\vec{x}, \vec{y})}{\det g_{mn}^{(0)}(\vec{y})} \right]^{-\frac{1}{d-2}}, \tag{16.16}$$

as we already saw before.

16.4   Linearized ansatz for reduction on S^4

We will denote by y the noncompact coordinates and by x the compact coordinates.

- **Metric ansatz** At the linearized level, the metric splits into a background and a fluctuation,

$$g_{\Lambda\Pi} = E_\Lambda^M E_\Pi^M = g_{\Lambda\Pi}^{(0)} + k h_{\Lambda\Pi}. \tag{16.17}$$

The ansatz for the fluctuation with seven-dimensional indices is

$$h_{\alpha\beta}(y, x) = h_{\alpha\beta}(y) - \frac{g_{\alpha\beta}^{(0)}(y)}{5} (h_{\mu\nu}(y, x) g^{(0)\mu\nu}(x)), \tag{16.18}$$

where $h_{\alpha\beta}(y)$ is the seven-dimensional graviton fluctuation, and the second term is needed in order to diagonalize the kinetic term. As we saw, at the nonlinear level, we need to make a rescaling between the Jordan frame and the Einstein frame. The extra term is the linearization of that rescaling.

The ansatz for the fluctuation with mixed indices is the gauge fields times the corresponding Killing vectors,

$$h_{\mu\alpha}(y, x) = B_{\alpha,IJ}(y) V_\mu^{IJ}(x). \tag{16.19}$$

Here the indices $I, J = 1, ..., 4$ are in a spinor representation of $SO(5) = USp(4)$, the invariance group of the four-sphere, or equivalently the fundamental representation of $USp(4)$. Corresponding to each invariance, we have a gauge field. The representation is antisymmetric, so we have $5 \times 4/2 = 10$ Killing vectors and 10 corresponding gauge fields.

The ansatz for the fluctuation with compact indices is

$$h_{\mu\nu}(y, x) = S_{IJKL}(y) \eta_{\mu\nu}^{IJKL}(x), \tag{16.20}$$

where the representation $IJKL$ is a 14 representation of $USp(4)$ with Young tableau in the shape of a Box, that is, antisymmetric in IJ and KL and symmetric in IK and JL. The $\eta_{\mu\nu}^{IJKL}$ is the corresponding spherical harmonic.

- **Gravitino ansatz**

  The ansatz for the gravitino with compact index is

  $$\Psi_\mu(y,x) = \lambda_{J,KL}(y)\gamma_5^{1/2}\eta_{\mu\nu}^{JKL}(x), \tag{16.21}$$

  where we have defined (there is an ambiguity in general in taking the square root of γ_5)

  $$\sqrt{\gamma_5} \equiv \frac{i-1}{2}(1+i\gamma_5), \tag{16.22}$$

  and the JKL is in a 16 representation of $SO(5) = USp(4)$, with Young tableau in the shape of a gun, that is, antisymmetric in KL and symmetric in JK.

  The ansatz for the gravitino with noncompact index is

  $$\Psi_\alpha(y,x) = \psi_{\alpha I}(y)\gamma_5^{\pm 1/2}\eta^I(x) - \frac{1}{5}\tau_\alpha\gamma_5\gamma^\mu\Psi_\mu(y,x). \tag{16.23}$$

  Here $\eta^I(x)$ is a Killing spinor, and the term subtracted, with the gamma trace of Ψ_μ, is again needed in order to diagonalize the kinetic term of the gravitino.

- **Antisymmetric tensor ansatz**

  The antisymmetric tensor with only compact indices is written only in terms of the trace of the graviton,

  $$A_{\mu\nu\rho}(y,x) = \frac{\sqrt{2}}{40}\sqrt{g^{(0)}}\epsilon_{\mu\nu\rho\sigma}D^\sigma h_\lambda^\lambda, \tag{16.24}$$

  meaning that these components mix with the trace of the graviton under the dimensional reduction, an example of a more general phenomenon.

  The antisymmetric tensor with only one noncompact index is written as

  $$A_{\alpha\mu\nu}(y,x) = \frac{i}{12\sqrt{2}}B_{\alpha,IJ}(y)\bar\eta^I(x)\gamma_{\mu\nu}\gamma_5\eta^J(x). \tag{16.25}$$

  Note that naively, we would have said that the $A_{\alpha\mu\nu}$ are vectors labeled by $\mu\nu$, that is, $4 \times 3/2 = 6$ of them, whereas in the off-diagonal metric $h_{\alpha\mu}$, we would have said there are vectors labeled by μ, that is, four of them. However, an important lesson is that in writing a KK ansatz, fields of the same spin are always grouped together, transforming in a representation of a symmetry group. So we cannot just write 4 of the vectors in $h_{\mu\alpha}$ and the other 6 in $A_{\alpha\mu\nu}$, we must write all 10 of them in both. *The counting of degrees of freedom still has to match though, and in fact that is a very important and nontrivial constraint on the symmetry groups that appear after KK compactification: the total number of fields of a given spin, obtained by naively counting, like four in $h_{\mu\alpha}$ and six in $A_{\alpha\mu\nu}$ above, must fill up some representation of the symmetry group.*

  There is no independent field with only one compact index, that is, at the linearized level,

  $$A_{\alpha\beta\mu} = 0. \tag{16.26}$$

  Finally, the antisymmetric tensor with only noncompact indices is

  $$A_{\alpha\beta\gamma}(y,x) = \frac{1}{6}A_{\alpha\beta\gamma,IJ}(y)\phi_5^{IJ}(x), \tag{16.27}$$

  where IJ are antisymmetric and Ω-traceless, that is, $5 \times 4/2 - 1 =$ five-dimensional representation of $USp(4)$.

- **Auxiliary field**

  Like we mentioned, in order to get from the action with two derivatives for $A_{\alpha\beta\gamma}$ in 11 dimensions to the action with one derivative for $A_{\alpha\beta\gamma,IJ}$ in seven dimensions, we need to add an auxiliary field. In the nonlinear case, we can do it in 11 dimensions, but at the linearized level, we can just add by hand an auxiliary field $B_{\alpha\beta\gamma,IJ}$ in seven dimensions. The point is to rotate (rewrite) the action $\sim (\partial A)^2 + m^2 A^2 + B^2$ into (as) two actions with one derivative $m\epsilon S\partial S + m^2 S^2$ and $m\epsilon G\partial G - m^2 G^2$, and then we drop the G, thus obtaining a constraint. The procedure in effect decomposes $\Box - m^2$ into $\epsilon\partial + m$ and $\epsilon\partial - m$. Thus, we have (the constraint expresses both B and A in terms of the same S)

$$B_{\alpha\beta\gamma,IJ} = \frac{1}{5}\left(S_{\alpha\beta\gamma,IJ} + \frac{1}{6}\epsilon_{\alpha\beta\gamma}{}^{\delta\epsilon\eta\zeta} D_\delta S_{\epsilon\eta\zeta,IJ}\right). \tag{16.28}$$

  We note that in fact, the right-hand side is exactly the equation of motion for $S_{\alpha\beta\gamma,IJ}$, as it should be, since $B = 0$ is supposed to be an equation of motion, since B is an auxiliary field.

16.5 Review of spherical harmonics on S^4

We remember a few facts about the spherical harmonics on S^4, already described previously.

The spherical harmonics are constructed from the Killing spinors, which on S^4 are defined by the condition

$$D^{(0)}_\mu \eta^I = \frac{i}{2}\gamma_\mu \eta^I. \tag{16.29}$$

They also satisfy orthonormality,

$$\bar{\eta}^I \eta^J = \Omega^{IJ}, \tag{16.30}$$

and completeness

$$\eta^\alpha_J \bar{\eta}^I_\beta = -\delta^\alpha_\beta, \tag{16.31}$$

where

$$\eta^\alpha_J = \eta^{\alpha I}\Omega_{IJ}. \tag{16.32}$$

As usual, the Killing spinor is a commuting spinor, in order to satisfy the spin-statistics theorem, as we reduce an 11-dimensional anticommuting spinor to a seven-dimensional anticommuting spinor times a Killing spinor.

The scalar field harmonic ϕ_5^{IJ} is written in terms of η^I as

$$\phi_5^{IJ} = \bar{\eta}^I \gamma_5 \eta^J. \tag{16.33}$$

Since $C\gamma_5$ is antisymmetric in four Euclidean dimensions and η^I is commuting, IJ is an antisymmetric representation. It is also Ω-traceless, thus ϕ_5^{IJ} is in a $\underline{5}$ representation of $USp(4)$. We can thus multiply with a constant matrix (Clebsch–Gordan coefficient) taking us from the IJ representation of $USp(4)$ to the vector representation of $SO(5)$. There is only one possible coefficient, namely $(\gamma^A)_{IJ}$, so that we build the normalized object

$$Y^A = \frac{1}{4}(\gamma^A)_{IJ}\phi_5^{IJ}. \tag{16.34}$$

These scalar spherical harmonics act as five-dimensional Euclidean embedding coordinates for the four-sphere, satisfying $Y^A Y^A = 1$.

The Killing vectors are written in terms of the Killing spinors using the general formula,

$$V_\mu^{IJ} = \bar{\eta}^I \gamma_\mu \eta^J, \tag{16.35}$$

and they satisfy the Killing vector equation,

$$D_{(\mu}^{(0)} V_{\nu)} = 0. \tag{16.36}$$

Since $C\gamma_\mu$ is symmetric and the Killing spinors are commuting, the representation is symmetric in IJ, that is, $4 \times 5/2 = 10$-dimensional. This is the same as the antisymmetric representation of $SO(5)$, therefore, we can write an object with A, B $SO(5)$ indices by multiplying with the unique Clebsch–Gordan coefficient for this transformation, $(\gamma^{AB})_{IJ}$, that is, the normalized object is

$$V_\mu^{AB} = -\frac{i}{8}(\gamma^{AB})_{IJ} V_\mu^{IJ}. \tag{16.37}$$

This object can be written in terms of Y^A as

$$V_\mu^{AB} = Y^{[A} D_\mu^{(0)} Y^{B]}. \tag{16.38}$$

We can also define the *conformal Killing vectors* C_μ^{IJ}, satisfying the conformal Killing vector equation,

$$D_{(\mu}^{(0)} C_{\nu)} = \frac{1}{4} g_{\mu\nu}^{(0)} (D^{(0)\rho} C_\rho). \tag{16.39}$$

In terms of the Killing spinor, they are

$$C_\mu^{IJ} = \bar{\eta}^I \gamma_\mu \gamma_5 \eta^J. \tag{16.40}$$

Since $C\gamma_\mu \gamma_5$ is antisymmetric, the representation is antisymmetric, and moreover Ω-traceless, that is, again five-dimensional, so again we can multiply with the Clebsch–Gordan coefficient $(\gamma^A)_{IJ}$, defining the normalized object

$$C_\mu^A = \frac{i}{4} C_\mu^{IJ} (\gamma^A)_{IJ}. \tag{16.41}$$

We can again write it in terms of Y^A, as

$$C_\mu^A = D_\mu^{(0)} Y^A. \tag{16.42}$$

The spherical harmonic for $h_{\mu\nu}$ is actually the sum of two spherical harmonics with the same symmetry, but different eigenvalues of $\Box$,

$$\eta_{\mu\nu}^{IJKL} = \eta_{\mu\nu}^{IJKL}(-2) - \frac{1}{3}\eta_{\mu\nu}^{IJKL}(-10). \tag{16.43}$$

Finally, the gravitino Ψ_μ spherical harmonic is again a sum of two independent spherical harmonics with different eigenvalues of the kinetic operator,

$$\eta_\mu^{JKL} = \eta_\mu^{JKL}(-2) + \eta_\mu^{JKL}(-6). \tag{16.44}$$

16.6 Nonlinear ansatz

We now turn to the nonlinear version of the ansatz.

The ansatz for E_α^a and E_α^m is standard, as we explained before, namely E_α^a gives the seven-dimensional vielbein, rescaled in order to get to Einstein frame,

$$E_\alpha^a(y,x) = e_\alpha^a(y)\Delta^{-1/5}(y,x)$$
$$\Delta(y,x) = \frac{\det E_\mu^m}{\det e_\mu^{(0)m}}, \tag{16.45}$$

and E_α^m with flattened indices is equal to the gauge fields times the Killing vectors,

$$E_\alpha^m(y,x) = B_\alpha^\mu(y,x)E_\mu^m$$
$$B_\alpha^\mu(y,x) = -2B_\alpha^{AB}V^{\mu,AB}, \tag{16.46}$$

where B_α^{AB} is the $SO(5)$ gauge field, and V_μ^{AB} is the corresponding Killing vector.

The gravitini and the susy parameter ϵ need to be rotated as before and also rescaled by powers of the same Δ, for the same reason: to get the standard kinetic term. We have

$$\Psi_a = \Delta^{1/10}(\gamma_5)^{-p}\psi_a - \frac{A}{5}\tau_a\gamma_5\gamma^m\Delta^{1/10}(\gamma_5)^q\psi_m$$
$$\Psi_m = \Delta^{1/10}(\gamma_5)^q\psi_m$$
$$\epsilon(y,x) = \Delta^{-1/10}(\gamma_5)^{-p}\varepsilon(y,x). \tag{16.47}$$

Then the ansatz for the new objects $\psi_\alpha, \psi_m, \varepsilon$ is written in terms of physical spinors and Killing spinors, but with a matrix that relates the two types of indices, I in the gauge group $SO(5)_g$ and I' in the local composite symmetry group $SO(5)_c$:

$$\psi_\alpha(y,x) = \psi_{\alpha I'}(y)U^{I'}{}_I(y,x)\eta^I(x)$$
$$\psi_m(y,x) = \lambda_{J'K'L'}(y)U^{J'}{}_J(y,x)U^{K'}{}_K(y,x)U^{L'}{}_L(y,x)\eta_m^{JKL}(x)$$
$$\varepsilon(y,x) = \varepsilon_{I'}(y)U^{I'}{}_I(y,x)\eta^I(x). \tag{16.48}$$

Here $A = \pm 1, p = \pm 1/2, q = \pm 1/2$ and $U^{I'}{}_I$ is a complicated $USp(4)$ matrix that satisfies the relation

$$(\tilde{\Omega} \cdot U^T \cdot \Omega)^I{}_{I'} = -(U^{-1})^I{}_{I'}. \tag{16.49}$$

The local composite symmetry is a symmetry defined by a gauge field that is not independent, but rather it is a composite made up of scalars. It will be explained better later.

We can now write the ansatz for E_μ^m,

$$E_\mu^m = \frac{1}{4}\Delta^{2/5}\Pi_A^i C_\mu^A C^{mB}\text{Tr}(U^{-1}\gamma^i U\gamma_B). \tag{16.50}$$

While the ansatz for the vielbein looks complicated, the metric element looks simple:

$$ds_{11}^2 = \Delta^{-2/5}g_{\alpha\beta}dy^\alpha dy^\beta + \Delta^{4/5}T_{AB}^{-1}(dY^A + 2B^{AC}Y^C)(dY^B + 2B^{BD}Y^D). \tag{16.51}$$

The ansatz for the 4-form field strength is more complicated,

$$
\frac{\sqrt{2}}{3} F_{(4)} = \epsilon_{ABCDE} \left(-\frac{1}{3} DY^A \wedge DY^B \wedge DY^C \wedge DY^D \frac{(T \cdot Y)^E}{Y \cdot T \cdot Y} \right.
$$
$$
+ \frac{4}{3} DY^A \wedge DY^B \wedge DY^C \wedge D \left[\frac{(T \cdot Y)^D}{Y \cdot T \cdot Y} \right] Y^E
$$
$$
\left. + 2 F_{(2)}^{AB} \wedge DY^C \wedge DY^D \frac{(T \cdot Y)^E}{Y \cdot T \cdot Y} + F_{(2)}^{AB} \wedge F_{(2)}^{CD} Y^E \right) + d(\mathcal{A}). \quad (16.52)
$$

Here we used the notation

$$
F_{(2)}^{AB} = 2(dB^{AB} + 2(B \cdot B)^{AB})
$$
$$
DY^A = dY^A + 2(B \cdot Y)^A, \quad (16.53)
$$

for the field strength of the gauge field and the covariant derivative of the scalar harmonic. We also used the notation

$$
T^{AB} = (\Pi^{-1})_i^A (\Pi^{-1})_j^B \delta^{ij}, \quad (16.54)
$$

for the object called *T-tensor*.

Here Π_i^A are the scalar fields in seven dimensions, living in a coset that is $Sl(5, \mathbb{R})/SO(5)$ in the ungauged case, with i an index in the $SO(5)_c$ composite symmetry, and A an index in the $SO(5)_g$ gauge symmetry (in the ungauged case, it is in $Sl(5, \mathbb{R})$).

The ansatz for $\mathcal{A}$ is the same as in the linearized case,

$$
\mathcal{A}_{\alpha\beta\gamma} = \frac{8i}{\sqrt{3}} S_{\alpha\beta\gamma,B} Y^B. \quad (16.55)
$$

The ansatz for the auxiliary antisymmetric tensor field is again the equation of motion of $S_{\alpha\beta\gamma,A}$, just that this time it is a nonlinear equation,

$$
\frac{\mathcal{B}_{\alpha\beta\gamma\delta}}{\sqrt{E}} = \frac{i}{2\sqrt{3}} \epsilon_{\alpha\beta\gamma\delta\epsilon\eta\zeta} \frac{\delta S^{(7)}}{\delta S_{\epsilon\eta\zeta,A}} Y^A
$$
$$
= -24\sqrt{3} i \nabla_\alpha S_{\beta\gamma\delta,A} Y^A + \sqrt{3} i \epsilon_{\alpha\beta\gamma\delta}{}^{\epsilon\eta\zeta} T^{AB} S_{\epsilon\eta\zeta,B} Y^A
$$
$$
+ g \epsilon_{ABCDE} F_{[\alpha\beta}^{BC} F_{\gamma\delta]}^{DE} Y^A + 2 - \text{fermi}. \quad (16.56)
$$

After this nonlinear KK ansatz, the full supergravity action and transformation rules for $\mathcal{N} = 4$ (maximal) seven-dimensional gauged supergravity are found.

16.7 * Gauged supergravities, maximal supergravities in various dimensions, and gaugings

We have described the nonlinear compactification on S^4, but we now want to understand a bit about the gauged supergravity that is the endpoint of the nonlinear compactification. In order to do that, since the seven-dimensional gauged supergravity is a bit of a particular case, we will analyze all the relevant maximal gauged supergravities in various dimensions.

Compared to the ungauged supergravities, in the gauged supergravities the fields and symmetries get rearranged.

16.7.1 $d = 4\,\mathcal{N} = 8$ (maximal) gauged supergravity

Ungauged

The fields are the graviton e_μ^m, eight gravitini ψ_μ^i, with $i = 1, ..., 8$, fermions χ_{ijk}, vectors A_μ^{IJ}, and 70 scalars that form a matrix $\mathcal{V}$ in the coset $E_7/SU(8)$.

Symmetries:

- $SO(8)$ global invariance with indices $I, J = 1, ..., 8$, which organizes the vector fields A_μ^{IJ}, which are however still abelian.
- $SU(8)$ local composite symmetry with fundamental indices $i, j = 1, ..., 8$. The gravitini are fundamental under it. This is a composite local symmetry, in that there is no independent gauge field, but rather the gauge field $\mathcal{B}_\mu{}^i{}_j$ is made up of the fields in $\mathcal{V}$.
- global E_7 symmetry, acting on $\mathcal{V}(x)$. The transformation of $\mathcal{V}(x)$ is $\mathcal{V}(x) \to U(x)\mathcal{V}(x)E^{-1}$ where $U(x) \in SU(8)$ and $E \in E_7$.

The scalar fields are decomposed under the $SO(8)$ and $SU(8)$ groups as

$$\mathcal{V} = \begin{pmatrix} u_{ij}{}^{IJ} & v_{ijKL} \\ v^{klIJ} & u^{kl}{}_{IJ} \end{pmatrix} \Rightarrow \mathcal{V}^{-1} = \begin{pmatrix} u^{ij}{}_{IJ} & -v_{klIJ} \\ -v^{ijKL} & u_{kl}{}^{KL} \end{pmatrix}, \tag{16.57}$$

where the u and v matrices satisfy

$$u^{ij}{}_{IJ}u_{kl}{}^{IJ} - v^{ijIJ}v_{klIJ} = \delta_{kl}^{ij}, \quad u^{ij}{}_{IJ}v^{klIJ} - v^{ijIJ}u^{kl}{}_{IJ} = 0$$

$$u^{ij}{}_{IJ}u_{ij}{}^{KL} - v_{ijIJ}v^{ijKL} = \delta_{IJ}^{KL}, \quad u^{ij}{}_{IJ}v_{ijKL} - v_{ijIJ}u^{ij}{}_{KL} = 0$$

$$\left(u^{ij}{}_{KI}u_{kl}{}^{JK} - v^{ijKI}v_{klJK}\right)_{[IJ]} = \frac{2}{3}\delta_{[k}^{[i}\left(u^{j]m}{}_{KI}u_{l]m}{}^{JK} - v^{j]mKI}v_{l]mJK}\right)_{[IJ]}$$

$$\left(u^{ij}{}_{KI}v^{klJK} - v^{ijKI}u^{kl}{}_{JK}\right)_{[IJ]} = \frac{\eta}{24}\epsilon^{ijklmnpq}\left(u_{mn}{}^{KI}v_{pqJK} - v_{mnKI}u_{pq}{}^{JK}\right)_{[IJ]}, \tag{16.58}$$

and where $[IJ]$ refers to antisymmetrization in IJ.

Then the composite gauge field is written as

$$\mathcal{B}_\mu{}^i{}_j = \frac{2}{3}\left(u^{ik}{}_{IJ}\partial_\mu u_{jk}{}^{IJ} - v^{ikIJ}\partial_\mu v_{jkIJ}\right), \tag{16.59}$$

and the physical scalars are found by computing

$$D_\mu\mathcal{V}\cdot\mathcal{V}^{-1} = -\frac{1}{4}\sqrt{2}\begin{pmatrix} 0 & \mathcal{A}_\mu^{ijkl} \\ \mathcal{A}_{\mu mnpq} & 0 \end{pmatrix}. \tag{16.60}$$

In the above, the physical scalars are

$$\mathcal{A}_\mu^{ijkl} = -2\sqrt{2}\left(u^{ij}{}_{IJ}\partial_\mu v^{klIJ} - v^{ijIJ}\partial_\mu u^{kl}{}_{IJ}\right), \tag{16.61}$$

whereas $\mathcal{A}_{mnpq}$ are related to them via

$$\mathcal{A}_\mu^{ijkl} = \frac{1}{24}\eta\epsilon^{ijklmnpq}\mathcal{A}_{\mu mnpq}. \tag{16.62}$$

Covariant derivatives are with respect to both the local Lorentz and $SU(8)$ local composite symmetry, that is,

$$D_\mu\epsilon^i = \partial_\mu\epsilon^i - \frac{1}{2}\omega_{\mu ab}\sigma^{ab}\epsilon^i + \frac{1}{2}\mathcal{B}_\mu{}^i{}_j\epsilon^j. \tag{16.63}$$

The Lagrangian is (for $\kappa_N = 1$)

$$
\begin{aligned}
e^{-1}\mathcal{L} = {} & -\frac{1}{2}R(e,\omega) - e^{-1}\frac{1}{2}\epsilon^{\mu\nu\rho\sigma}(\bar\psi_\mu^i\gamma_\nu D_\rho\psi_{\sigma i} - \bar\psi_\mu^i\overleftarrow{D}_\rho\,\gamma_\nu\psi_{\sigma i}) \\
& - \frac{1}{12}(\bar\chi^{ijk}\gamma^\mu D_\mu\chi_{ijk} - \bar\chi^{ijk}\overleftarrow{D}_\mu\,\gamma^\mu\chi_{ijk}) - \frac{1}{96}\mathcal{A}_\mu^{ijkl}\mathcal{A}_{ijkl}^\mu \\
& - \frac{1}{8}\left[F_{\mu\nu IJ}^+(2S^{IJ,KL} - \delta_{KL}^{IJ})F^{+\mu\nu}{}_{KL} + h.c.\right] \\
& - \frac{1}{2}\left[F_{\mu\nu IJ}^+ S^{IJ,KL}O^{+\mu\nu KL} + h.c.\right] \\
& - \frac{1}{4}\left[O_{\mu\nu}^{+\,IJ}(S^{IJ,KL} + u^{ij}{}_{IJ}v_{ijKL})O^{+\mu\nu KL} + h.c.\right] \\
& - \frac{1}{24}\left[\bar\chi_{ijk}\gamma^\nu\gamma^\mu\psi_{\nu l}(\hat{\mathcal{A}}_\mu^{ijkl} + \mathcal{A}_\mu^{ijkl}) + h.c.\right] \\
& - \frac{1}{2}\bar\psi_\mu^{[i}\psi_\nu^{j]}\bar\psi_i^\mu\psi_j^\nu + \frac{1}{4}\sqrt{2}\left[\bar\psi_\lambda^i\sigma^{\mu\nu}\gamma^\lambda\chi_{ijk}\bar\psi_\mu^j\psi_\nu^k + h.c.\right] \\
& + \left[\frac{1}{144}\eta\epsilon_{ijklmnpq}\bar\chi^{ijk}\sigma^{\mu\nu}\chi^{lmn}\bar\psi_\mu^p\psi_\nu^q + \frac{1}{8}\bar\psi_\lambda^i\sigma^{\mu\nu}\gamma^\lambda\chi_{ikl}\bar\psi_{\mu j}\gamma_\nu\chi^{jkl} + h.c.\right] \\
& + \frac{1}{864}\sqrt{2}\eta\left[\epsilon^{ijklmnpq}\bar\chi_{ijk}\sigma^{\mu\nu}\chi_{lmn}\bar\psi_\mu^r\gamma_\nu\chi_{pqr} + h.c.\right] \\
& + \frac{1}{32}\bar\chi^{ikl}\gamma^\mu\chi_{jkl}\bar\chi^{jmn}\gamma_\mu\chi_{imn} - \frac{1}{96}(\bar\chi^{ijk}\gamma^\mu\chi_{ijk})^2,
\end{aligned} \tag{16.64}
$$

where

$$
\begin{aligned}
O_{\mu\nu}^{+\,ij} = {} & -\frac{1}{144}\sqrt{2}\eta\epsilon^{ijklmnpq}\bar\chi_{klm}\sigma_{\mu\nu}\chi_{npq} \\
& - \frac{1}{2}\bar\psi_{\lambda k}\sigma_{\mu\nu}\gamma^\lambda\chi^{ijk} + \frac{1}{2}\sqrt{2}\bar\psi_\rho^i\gamma^{[\rho}\sigma_{\mu\nu}\gamma^{\sigma]}\psi_\sigma^i \\
O_{\mu\nu}^{+\,ij} \equiv {} & u^{ij}{}_{IJ}O_{\mu\nu}^{+\,IJ}
\end{aligned} \tag{16.65}
$$

are bilinear operators and the scalar field matrix $S^{IJ,KL}$ is fixed by the equation

$$
(u^{ij}{}_{IJ} + v^{ijIJ})S^{IJ,KL} = u^{ij}{}_{KL}. \tag{16.66}
$$

Moreover, the field strengths $F_{\mu\nu IJ}^+$, $F_{\mu\nu}^{-\ IJ}$ and the object $\bar F_{\mu\nu ij}^+$ are constructed out of the abelian gauge fields A_μ^{IJ}, the scalars in u, v, and S and the fermions in O^+ as follows:

$$
\begin{aligned}
\bar F_{\mu\nu ij}^+ &\equiv u_{ij}{}^{IJ}F_{1\mu\nu IJ}^+ + v_{ijIJ}F_{2\mu\nu IJ}^+ \\
F_{1\mu\nu IJ}^+ &\equiv \frac{1}{2}(G_{\mu\nu IJ}^+ + F_{\mu\nu IJ}) \\
F_{2\mu\nu IJ}^+ &\equiv \frac{1}{2}(G_{\mu\nu IJ}^+ - F_{\mu\nu IJ}) \\
G^{+\mu\nu}{}_{IJ} &\equiv 2S^{IJ,KL}(F_{\mu\nu,KL}^+ + O_{\mu\nu}^{+\ KL}) - F_{\mu\nu IJ}^+ \\
F_{\mu\nu}^{IJ} &= \partial_\mu A_\nu^{IJ} - \partial_\nu A_\mu^{IJ} = F_{\mu\nu IJ}^+ + F_{\mu\nu}^{-IJ}.
\end{aligned} \tag{16.67}
$$

Finally, the susy transformation rules are

$$\delta \mathcal{V} \cdot \mathcal{V}^{-1} = -2\sqrt{2} \begin{pmatrix} 0 & \bar{\epsilon}_{[i}\chi_{jkl]} \\ \bar{\epsilon}^{[m}\chi^{npq]} + \frac{1}{24}\eta\epsilon^{mnpqrstu}\bar{\epsilon}_r\chi_{stu} & 0 \end{pmatrix}$$

$$\delta\chi^{ijk} = -\hat{A}_\mu^{ijkl}\gamma^\mu \epsilon_l + 2\sigma^{\mu\nu}\hat{\bar{F}}_{\mu\nu}^{-[ij}\chi^{k]} - \frac{1}{24}\sqrt{2}(\eta\epsilon^{ijklmnpq}\bar{\chi}_{lmn}\chi_{pqr})\epsilon^r$$

$$\delta A_\mu^{IJ} = -(u_{ij}{}^{IJ} + v_{ijIJ})(\bar{\epsilon}_k\gamma_\mu\chi^{ijk} + 2\sqrt{2}\bar{\epsilon}^i\psi_\mu^j) + h.c.$$

$$\delta\psi_\mu^i = 2D_\mu\epsilon^i + \frac{1}{2}\sqrt{2}\hat{\bar{F}}_{\rho\sigma}^{-ij}\sigma^{\rho\sigma}\gamma_\mu\epsilon_j$$

$$\qquad - \left(\frac{1}{144}\eta\epsilon^{ijklmnpq}\bar{\chi}_{klm}\sigma^{\rho\sigma}\chi_{npq}\right)\gamma_\mu\sigma_{\rho\sigma}\epsilon_j$$

$$\qquad + \frac{1}{4}\left(\bar{\chi}^{ijk}\gamma^\nu\chi_{jkl}\right)\gamma_\nu\gamma_\mu\epsilon^l + \frac{1}{2}\sqrt{2}\left(\bar{\psi}_{\mu k}\gamma^\nu\chi^{ijk}\right)\gamma_\nu\epsilon_j$$

$$\delta e_\mu^a = \bar{\epsilon}^i\gamma^a\psi_{\mu i} + h.c. \tag{16.68}$$

Gauged

We gauge the global $SO(8)$ symmetry. This means that now the vectors A_μ^{IJ} are non-abelian, and the IJ indices are not just labels anymore but rather gauge group indices. Then nonabelian field strengths are

$$F_{\mu\nu}^{IJ} = 2\partial_{[\mu}A_{\nu]}^{IJ} - 2gA_{[\mu}^{IK}A_{\nu]}^{KJ} \equiv F_{\mu\nu IJ}^+ + F_{\mu\nu IJ}^-. \tag{16.69}$$

We make derivatives on $SO(8)$ tensors covariant by introducing (besides the $\mathcal{B}_\mu{}^i{}_j$ $SU(8)_c$ composite connection term) the order g A_μ^{IJ} connection term, for example, for the scalars $u_{ij}{}^{IJ}$,

$$D_\mu u_{ij}{}^{IJ} = \partial_\mu u_{ij}{}^{IJ} + \mathcal{B}_\mu{}^k{}_{[i}u_{j]k}{}^{IJ} - 2gA_\mu^{K[I}u_{ij}{}^{J]K}. \tag{16.70}$$

Both $\mathcal{A}_\mu$ and $\mathcal{B}_\mu$ have order g^2 terms in them now, since the covariant derivative in the definition of $\mathcal{A}_\mu$, (16.60) is also $SO(8)$-covariant, and ∂_μ in the definition of $\mathcal{B}_\mu$, (16.59), is replaced by the $SO(8)$-covariant derivative.

Besides the covariantization of derivatives, the only difference in the Lagrangian are the introduction of the terms of order g and g^2,

$$\mathcal{L}_g + \mathcal{L}_{g^2} = ge\left[\sqrt{2}A_{1ij}\bar{\psi}_\mu^i\sigma^{\mu\nu}\psi_\nu^j + \frac{1}{6}A_{2i}{}^{jkl}\bar{\psi}_\mu^i\gamma^\mu\chi_{jkl}\right.$$

$$\left.+ \frac{1}{144}\sqrt{2}\eta\epsilon^{ijkpqrlm}A_2{}^n{}_{pqr}\bar{\chi}_{ijk}\chi_{lmn} + h.c.\right]$$

$$+ g^2 e\left[\frac{3}{4}|A_1^{ij}|^2 - \frac{1}{24}|A_{2jkl}^i|^2\right]. \tag{16.71}$$

In the susy rules, the only change (besides the covariantization of derivatives with respect to $SO(8)$) is the addition of the

$$\delta_g\psi_\mu^i = -\sqrt{2}g\bar{\epsilon}_j\gamma_\mu A_1^{ji}$$

$$\delta_g\bar{\chi}^{jkl} = -2g\bar{\epsilon}^i A_{2i}{}^{jkl}, \tag{16.72}$$

where

$$A_1^{ij} \equiv \frac{4}{21}T_k{}^{ikj}, \quad A_{2i}{}^{jkl} \equiv -\frac{4}{3}T_i{}^{[jkl]}, \tag{16.73}$$

and the T-tensor is defined as

$$T_i^{jkl} \equiv (u^{kl}{}_{IJ} + v^{klIJ})(u_{im}{}^{JK} u^{jm}{}_{KI} - v_{imJK} v^{jmKI}).$$ (16.74)

In conclusion, besides the covariantization of derivatives, the only changes are encapsulated in the T-tensor above, and its resulting A_1 and A_2 combinations.

16.7.2 $d = 5 \, \mathcal{N} = 8$ (maximal) gauged supergravity

Ungauged

The fields are the graviton e_μ^m, the eight gravitini ψ_μ^a, $a = 1, ..., 8$, the vectors $A_\mu^{\alpha\beta}$, the spinors λ_{abc} and the scalars $\Pi_{\alpha\beta}{}^{ab}$.

Symmetries:

– global $SO(6)$ invariance. The indices $\alpha, \beta = 1, ..., 8$ are spinors of $SO(6)$ invariance. The representation $[\alpha\beta]$ is antisymmetric Ω-traceless, that is $7 \times 8/2 - 1 = 27$ representation for the vectors. But this representation is reducible and can be decomposed into irreducible representations using projectors (whose explicit form we will not need)

$$A_\mu^{\alpha\beta} = \mathbf{P}(15)^{\alpha\beta}{}_{\gamma\delta} A_\mu^{\gamma\delta} + \mathbf{P}^1(6)^{\alpha\beta}{}_{\gamma\delta} A_\mu^{\gamma\delta} + \mathbf{P}^2(6)^{\alpha\beta}{}_{\gamma\delta} A_\mu^{\gamma\delta}.$$ (16.75)

The first term, in the 15 representation, is called $B_\mu^{\alpha\beta}$, with field strengths $G_{\mu\nu}^{\alpha\beta}$, the other terms, in the two 6 representations, have field strengths that are called $S_{\mu\nu}^{(1)\alpha\beta}$ and $S_{\mu\nu}^{(2)\alpha\beta}$.

– local composite symmetry $USp(8)_c$ invariance, with fundamental indices $a, b = 1, ..., 8$ and composite gauge field $Q_{\mu a}{}^b$.

– global E_6 symmetry. The scalars are in the coset $E_6/USp(8)_c$, with vielbein $\Pi_{\alpha\beta}{}^{ab}$ (which is an E_6 matrix), that is, with indices in $USp(8)$ and $SO(6)$.

The Lagrangian is (for $\kappa_N^2 = 2$)

$$\begin{aligned}
e^{-1}\mathcal{L} = &-\frac{1}{4}R - \frac{1}{8}g^{\mu\sigma}g^{\nu\lambda}\Pi_{\alpha\beta}{}^{ab}\Pi_{\gamma\delta,ab}F_{\mu\nu}^{\alpha\beta}F_{\sigma\lambda}^{\gamma\delta} + \frac{1}{24}g^{\mu\nu}P_\mu^{abcd}P_{\nu,abcd} \\
&-\frac{i}{2}\bar{\psi}_\mu^a\gamma^{\mu\nu\rho}D_\nu\psi_{\rho a} + \frac{i}{12}\bar{\lambda}^{abc}\gamma^\mu D_\mu\lambda_{abc} \\
&+ i\Pi_{\alpha\beta,ab}F_{\mu\nu}^{\alpha\beta}\left(\frac{1}{4}\bar{\psi}_{[\rho}^a\gamma^\rho\gamma^{\mu\nu}\gamma^{\sigma]}\psi_\sigma^b + \frac{1}{4\sqrt{2}}\bar{\psi}_\rho^c\gamma^{\mu\nu}\gamma^\rho\lambda^{ab}{}_c\right. \\
&\left.+\frac{1}{8}\bar{\lambda}^{acd}\gamma^{\mu\nu}\lambda^b{}_{cd}\right) + \frac{i\sqrt{2}}{6}P_\rho^{abcd}\bar{\psi}_{\mu a}\gamma^\rho\gamma^\mu\lambda_{bcd} \\
&+\frac{e^{-1}}{12}\epsilon^{\mu\nu\rho\sigma\tau}F_{\mu\nu}^{\alpha\beta}\omega_{\beta\gamma}F_{\rho\sigma}^{\gamma\delta}\omega_{\delta\epsilon}A_\tau^{\epsilon\zeta}\omega_{\zeta\alpha}.
\end{aligned}$$ (16.76)

Here raising and lowering of flat symplectic indices $a, b, ...$ is done with a $USp(8)$ symplectic metric Ω_{ab}. The spinor representation of $SO(6)$ has two invariant tensors, $\omega_{\alpha\beta}$ and $C_{\alpha\beta}^{-1}$. The $A_\mu^{\alpha\beta}$ fields are symplectic-traceless, that is, $A_\mu^{\alpha\beta}\omega_{\alpha\beta} = 0$.

The susy transformation laws will be written below, in the gauged case (so for the above they are obtained in the $g = 0$ limit).

Gauged

We gauge the $SO(6)$ global group, making nonabelian the gauge field $B_\mu^{\alpha\beta}$, which is in the 15 representation, which is the antisymmetric ($6 \times 5/2 = 15$, adjoint) representation, as it should. We should mention, however, that in general, it is possible to gauge only a subgroup of the global symmetry group, and a subset of the vector fields. This is the reason that there are many gauged supergravities available. We write $SO(6)$ covariant derivatives on tensors, for instance in the definition

$$(\Pi^{-1})_{ab}{}^{\alpha\beta}(\delta_\alpha^\gamma \delta_\beta^\delta \partial_\mu + 2g B_{\mu\alpha}{}^\gamma \delta_\beta^\delta)\Pi_{\gamma\delta}{}^{cd} = 2Q_{\mu[a}{}^{[c}\delta_{b]}^{d]} + 2P_{\mu ab}{}^{cd}, \qquad (16.77)$$

where the antisymmetric part is the $USp(8)_c$ composite connection Q, and the symmetric part is called P. Thus this is a decomposition of an E_6 matrix into an $USp(8)$ part and a $E_6/USp(8)$ coset part.

Like in the previous case, the gauged Lagrangian is obtained by introducing terms with $gS_{\alpha\beta}^{(i)}$, making derivatives covariant, and introducing scalar terms involving T-tensors. Now we also have the change of the ungauged CS term $\sim \epsilon FFA$ to the gauged one,

$$\Omega_5[B] = \epsilon^{\mu\nu\rho\sigma\tau} \text{Tr}\left[\Gamma_7\left(G_{\mu\nu}G_{\rho\sigma}B_\tau - gG_{\mu\nu}B_\rho B_\sigma B_\tau + \frac{2}{5}g^2 B_\mu B_\nu B_\rho B_\sigma B_\tau\right)\right]. \quad (16.78)$$

The gauged Lagrangian is then

$$\begin{aligned}
e^{-1}\mathcal{L}_g = &-\frac{1}{4}R - \left(\frac{2}{3}A_{ab}^{(1)}A^{(1)ab} - \frac{3}{4}A_{abc,d}^{(2)}A^{(2)abc,d}\right) \\
&- \frac{1}{8}g^{\mu\sigma}g^{\nu\lambda}\Pi_{\alpha\beta}{}^{ab}\Pi_{\gamma\delta,ab}\left(G_{\mu\nu}^{\alpha\beta} + gS_{\mu\nu}^{(1)\alpha\beta} + gS_{\mu\nu}^{(2)\alpha\beta}\right) \times \\
&\times \left(G_{\sigma\lambda}^{\gamma\delta} + gS_{\sigma\lambda}^{(1)\gamma\delta} + gS_{\sigma\lambda}^{(2)\gamma\delta}\right) + \frac{1}{24}g^{\mu\nu}P_\mu^{abcd}P_{\nu abcd} \\
&- \frac{1}{4}g\,e^{-1}\epsilon^{\mu\nu\rho\sigma\lambda}\omega_{\alpha\beta}C_{\gamma\delta}^{-1}S_{\mu\nu}^{(1)\alpha\gamma}\nabla_\rho S_{\sigma\lambda}^{(2)\beta\delta} \\
&- \frac{i}{2}\bar\psi_\mu^a \gamma^{\mu\nu\rho}\nabla_\nu \psi_{\rho a} + \frac{i}{12}\bar\lambda^{abc}\gamma^\mu \nabla_\mu \lambda_{abc} \\
&+ A_{ab}^{(1)}\bar\psi_\mu^a \gamma^{\mu\nu}\psi_\nu^b + A_{abc,d}^{(2)}\bar\lambda^{abc}\gamma^\mu\psi_\mu^d + A_{abc,def}^{(3)}\bar\lambda^{abc}\lambda^{def} \\
&+ i\Pi_{\alpha\beta,ab}\left(G_{\mu\nu}^{\alpha\beta} + gS_{\mu\nu}^{(1)\alpha\beta} + gS_{\mu\nu}^{(2)\alpha\beta}\right)\left(\frac{1}{4}\bar\psi_{[\rho}^a\gamma^\rho\gamma^{\mu\nu}\gamma^\sigma\psi_{\sigma]}^b\right. \\
&\left.+ \frac{1}{4\sqrt{2}}\bar\psi_\rho^c\gamma^{\mu\nu}\gamma^\rho\lambda^{ab}{}_c + \frac{1}{8}\bar\lambda^{acd}\gamma^{\mu\nu}\lambda^b{}_{cd}\right) \\
&- \frac{1}{12}e^{-1}\Omega_5[B] + \frac{i\sqrt{2}}{6}P_\rho^{abcd}\bar\psi_{\mu a}\gamma^\rho\gamma^\mu\lambda_{bcd}, \qquad (16.79)
\end{aligned}$$

where the scalar functions $A_{ab}^{(1)}, A_{abc,d}^{(2)}$ are of order g, and defined by

$$A_{ab}^{(1)} = -\frac{i}{4}gT_{ac,bd}^{(2)}\Omega^{cd} = -\frac{2i}{15}gT_{ca,db}^{(1)}\Omega^{cd}$$

$$A_{ab[c,d]}^{(2)} = \frac{i\sqrt{2}}{12}g\tilde{T}_{ab,cd}^{(2)}$$

$$A_{ab(c,d)}^{(2)} = -\frac{i\sqrt{2}}{9}g\tilde{T}_{ab,cd}^{(1)}, \qquad (16.80)$$

where tilde on the T's means the symplectic traceless part, and $A^{(3)}_{abc,def}$ is defined implicitly, via

$$\nabla_\mu A^{(2)}_{abc,g} - 2\sqrt{2} A^{(3)}_{abc,def} P_\mu{}^{def}{}_g - \frac{\sqrt{2}}{3} A^{(1)}_{dg} P_\mu{}^d{}_{abc} = 0, \tag{16.81}$$

and finally the T-tensors are defined by

$$T^{(1)}_{ab,cd} = \Pi^{-1}{}_{ab}{}^{\epsilon\zeta} \Pi^{-1}{}_{cf}{}^{\alpha\beta} (\Lambda_{\epsilon\zeta})_{\alpha\beta}{}^{\gamma\delta} \Pi_{\gamma\delta}{}^{fg} \Omega_{gd}$$

$$T^{(2)}_{ab,cd} = \Pi_{\alpha\beta,ab} \omega^{\alpha\gamma} C^{\beta\delta} \Pi_{\gamma\delta,cd}$$

$$T^{(3)}_{ab,cdef} = \Pi^{-1}{}_{ab}{}^{\epsilon\zeta} \Pi^{-1}{}_{[cd}{}^{\alpha\beta} (\Lambda_{\epsilon\zeta})_{\alpha\beta}{}^{\gamma\delta} \Pi_{\gamma\delta,ef]}$$

$$(\Lambda_{\epsilon\zeta})_{\alpha\beta}{}^{\gamma\delta} = P(\mathbf{15})^{\eta[\gamma}{}_{\epsilon\zeta} C^{-1}_{\eta[\alpha} \delta_{\beta]}{}^{\delta]}, \tag{16.82}$$

and they satisfy

$$T^{(1)}_{a[b,c]|d} \Omega^{ad} = 0, \quad T^{(1)}_{[ab,cd]\mathbf{594}} = 0$$

$$T^{(3)}_{ab,cdef} \Omega^{ac} \Omega^{bd} = 0, \quad T^{(3)}_{[ab,cdef]\mathbf{792}} = 0. \tag{16.83}$$

The susy transformation rules are

$$\delta e^m_\mu = -i\bar{\epsilon}^a \gamma^m \psi_{\mu a}$$

$$\Pi^{-1}{}_{ab}{}^{\alpha\beta} \delta \Pi_{\alpha\beta,cd} = -2i\sqrt{2} \bar{\epsilon}_{[a} \lambda_{bcd]} - \frac{3i\sqrt{2}}{2} \Omega_{[ab} \bar{\epsilon}^e \lambda_{cd]e}$$

$$\delta B^{\alpha\beta}_\mu = i P(\mathbf{15})^{\alpha\beta}{}_{\gamma\delta} \Pi^{-1}{}_{ab}{}^{\gamma\delta} \left(2\bar{\epsilon}^a \psi^b_\mu + \frac{\sqrt{2}}{2} \bar{\epsilon}_c \gamma_\mu \lambda^{abc} \right)$$

$$\delta S^{(i)\alpha\beta}_{\mu\nu} = \frac{2i}{g} P^i(\mathbf{6})^{\alpha\beta}{}_{\gamma\delta} \nabla_{[\mu} \left[\Pi^{-1}{}_{ab}{}^{\gamma\delta} \left(2\bar{\epsilon}^a \psi^b_{\nu]} + \frac{\sqrt{2}}{2} \bar{\epsilon}_c \gamma_{\nu]} \lambda^{abc} \right) \right]$$

$$\quad + i\omega^{\alpha\gamma} C^{-1}_{\gamma\delta} P^i(\mathbf{6})^{\delta\beta}{}_{\epsilon\zeta} C^{\epsilon\eta} C^{\zeta\tau} \Pi_{\eta\tau}{}^{ab} \left(2\bar{\epsilon}^a \gamma_{[\mu} \psi_{\nu]b} + \frac{1}{2\sqrt{2}} \bar{\epsilon}_c \gamma_\mu \lambda^{abc} \right)$$

$$\delta \psi_{\mu a} = D_\mu \epsilon_a - \frac{2i}{3} A^{(1)}_{ab} \gamma_\mu \epsilon^b$$

$$\quad - \frac{1}{6} \Pi_{\alpha\beta,ab} \left(G^{\alpha\beta}_{\rho\sigma} + g S^{(1)\alpha\beta}_{\rho\sigma} + g S^{(2)\alpha\beta}_{\rho\sigma} \right) \left(\gamma^{\rho\sigma} \gamma_\mu + 2\gamma^\rho \delta^\sigma_\mu \right) \epsilon^b$$

$$\delta \lambda_{abc} = 6i A^{(2)}_{abc,d} \epsilon^d + \sqrt{2} \gamma^\mu P_{\mu abcd} \epsilon^d$$

$$\quad - \frac{3}{2\sqrt{2}} \pi_{\alpha\beta,[ab} \left(G^{\alpha\beta}_{\rho\sigma} + g S^{(1)\alpha\beta}_{\rho\sigma} + g S^{(2)\alpha\beta}_{\rho\sigma} \right) \gamma^{\rho\sigma} \epsilon_{c]}. \tag{16.84}$$

16.7.3 $d = 7 \, \mathcal{N} = 4$ (maximal) gauged supergravity

Ungauged

The fields are the graviton e^a_α, the four gravitini $\psi^{I'}_\alpha$, the vectors B^{AB}_α, the scalars Π^i_A, the spinors $\lambda^{I'}_i$ and the 3-form $S_{\alpha\beta\gamma,A}$, with field strength

$$F_{\alpha\beta\gamma\delta,A} = 4\nabla_{[\alpha} S_{\beta\gamma\delta],A}. \tag{16.85}$$

The symmetries are

- global $SO(5)$ with fundamental index A and vectors B_α^{AB} with abelian field strengths

$$F_{\alpha\beta}^{AB} = \partial_\alpha B_\beta^{AB} - \partial_\beta A_\alpha^{AB} \qquad (16.86)$$

  labeled by it, and spinor index $I = 1, ..., 4$.
- local composite $SO(5)_c$ symmetry with spinor indices $I' = 1, ..., 4$ and fundamental (vector) indices $i = 1, ..., 5$. For instance, the spinors $\lambda_i^{I'}$ are vector-spinors of $SO(5)_c$, which are gamma-traceless, $\gamma^i \lambda_i^{I'} = 0$.
- global $Sl(5;\mathbb{R})$ invariance. The scalars are in the coset $Sl(5;\mathbb{R})/SO(5)_c$, with vielbein Π_A^i.

We will write the action and susy rules in the gauged case, and the ungauged case is obtained by putting $g = 0$.

Gauged

The global $SO(5)$ is gauged to $SO(5)_g$, and the scalars Π_A^i have indices in $SO(5)_c$ and $SO(5)_g$. The $SO(5)_g$ covariant derivatives now have the gauge fields, which become nonabelian. For instance,

$$(\Pi^{-1})_i{}^A (\delta_A^B \partial_\alpha + g B_{\alpha A}{}^B) \Pi_B{}^k \delta_{kj} = Q_{\alpha ij} + P_{\alpha ij}, \qquad (16.87)$$

where $Q_{\alpha ij}$ is the antisymmetric part, giving the $SO(5)_c$ composite connection appearing in $\nabla_\alpha = \partial_\alpha + Q_\alpha$, and $P_{\alpha ij}$ is the symmetric part, that also appears in $D_\alpha = \partial_\alpha + Q_\alpha + P_\alpha$.

The Lagrangian is (for $\kappa_N = 1$)

$$
\begin{aligned}
e^{-1}\mathcal{L} = &-\frac{1}{2}R + \frac{m^2}{4}(T^2 - 2T_{ij}T^{ij}) - \frac{1}{2}P_{\alpha ij}P^{\alpha ij} - \frac{1}{4}\left(\Pi_A{}^i \Pi_B{}^j F_{\alpha\beta}^{AB}\right)^2 \\
&+ \frac{1}{2}\left(\Pi^{-1}{}_i{}^A S_{\alpha\beta\gamma,A}\right)^2 + \frac{m}{48}e^{-1}\epsilon^{\alpha\beta\gamma\delta\epsilon\eta\zeta}\delta^{AB} S_{\alpha\beta\gamma,A} F_{\delta\epsilon\zeta,B} - \frac{1}{2}\bar\psi_\alpha \tau^{\alpha\beta\gamma}\nabla_\beta\psi_\gamma \\
&- \frac{1}{2}\bar\lambda^i \tau^\alpha \nabla_\alpha \lambda_i - \frac{m}{8}(8T^{ij} - T\delta^{ij})\bar\lambda_i\lambda_j + \frac{m}{2}T^{ij}\bar\lambda_i\gamma_j\tau^\alpha\psi_\alpha + \frac{1}{2}\bar\psi_\alpha\tau^\beta\tau^\alpha\gamma^i\lambda^j P_{\beta ij} \\
&+ \frac{m}{8}T\bar\psi_\alpha\tau^{\alpha\beta}\psi_\beta + \frac{1}{16}\bar\psi_\alpha\left(\tau^{\alpha\beta\gamma\delta} - 2g^{\alpha\beta}g^{\gamma\delta}\right)\gamma_{ij}\psi_\delta \Pi_A{}^i \Pi_B{}^j F_{\beta\gamma}^{AB} \\
&+ \frac{1}{4}\bar\psi_\alpha\tau^{\beta\gamma}\tau^\alpha\gamma_i\lambda_j\Pi_A{}^i\Pi_B{}^j F_{\beta\gamma}^{AB} + \frac{1}{32}\bar\lambda_i\gamma^j\gamma_{kl}\gamma^i\tau^{\alpha\beta}\lambda_j\Pi_A{}^k\Pi_B{}^l F_{\alpha\beta}^{AB} \\
&+ \frac{im}{8\sqrt{3}}\bar\psi_\alpha\left(\tau^{\alpha\beta\gamma\delta\epsilon} + 6g^{\alpha\beta}\tau^\gamma g^{\delta\epsilon}\right)\gamma^i\psi_\epsilon\,\Pi^{-1}{}_i{}^A S_{\beta\gamma\delta,A} \\
&- \frac{im}{4\sqrt{3}}\bar\psi_\alpha\left(\tau^{\alpha\beta\gamma} - 3g^{\alpha\beta}\tau^\gamma\delta\right)\lambda^i\Pi^{-1}{}_i{}^A S_{\beta\gamma\delta,A} - \frac{im}{8\sqrt{3}}\bar\lambda^i\tau^{\alpha\beta\gamma}\gamma^j\lambda_i\Pi^{-1}{}_j{}^A S_{\beta\gamma\delta,A} \\
&+ \frac{ie^{-1}}{16\sqrt{3}}\epsilon^{\alpha\beta\gamma\delta\epsilon\eta\zeta}\epsilon_{ABCDE}\delta^{AG}S_{\alpha\beta\gamma,G}F_{\delta\epsilon}^{BC}F_{\eta\zeta}^{DE} + \frac{m^{-1}}{8}e^{-1}\Omega_5[B] - \frac{m^{-1}}{16}e^{-1}\Omega_3[B],
\end{aligned}
$$
$$\qquad (16.88)$$

where $m = g/2$, $\Omega_3[B]$ and $\Omega_5[B]$ are the CS forms for B_α^{AB}, normalized to

$$d\Omega_3[B] = (\mathrm{Tr}\,F^2)^2, \quad d\Omega_5[B] = \mathrm{Tr}\,F^4, \qquad (16.89)$$

and the T-tensors are

$$T_{ij} = \Pi^{-1}{}_i{}^A \Pi^{-1}{}_j{}^B \delta_{AB}. \qquad (16.90)$$

As before, we see that the gauging procedure just introduces covariant derivatives with respect to B_μ^{AB}, nonabelianizes the field strength and the CS terms, and then one adds some terms with $gS_{\alpha\beta\gamma,A}$ and with the T-tensors.

The susy transformation rules are

$$\delta e_\alpha^a = \frac{1}{2}\bar\epsilon\tau^a\psi_\alpha$$

$$\Pi_A{}^i\Pi_B{}^j\delta B_\alpha^{AB} = \frac{1}{4}\gamma^{ij}\psi_\alpha + \frac{1}{8}\bar\epsilon\tau_\alpha\gamma^k\gamma^{ij}\lambda_k$$

$$\delta S_{\alpha\beta\gamma,A} = -\frac{i\sqrt{3}}{8m}\Pi_A{}^i\left(2\bar\epsilon\gamma_{ijk}\psi_{[\alpha} + \bar\epsilon\tau_{[\alpha}\gamma^l\gamma_{ijk}\lambda_l\right)\Pi_B{}^j\Pi_C{}^k F_{\beta\gamma]}^{BC}$$
$$\qquad - \frac{i\sqrt{3}}{4m}\delta_{ij}\Pi_A{}^j D_{[\alpha}\left(2\bar\epsilon\tau_\beta\gamma^i\psi_{\gamma]} + \bar\epsilon\tau_{\beta\gamma]}\lambda^i\right)$$
$$\qquad + \frac{i\sqrt{3}}{12}\delta_{AB}\Pi^{-1}{}_i{}^B\left(3\bar\epsilon\tau_{\alpha\beta}\gamma^i\psi_{\gamma]} - \bar\epsilon\tau_{\alpha\beta\gamma}\lambda^i\right)$$

$$\Pi^{-1}{}_i{}^A\delta\Pi_A{}^j = \frac{1}{4}\left(\bar\epsilon\gamma_i\lambda^j + \bar\epsilon\gamma^j\lambda_i\right)$$

$$\delta\psi_\alpha = \nabla_\alpha\epsilon - \frac{m}{20}T\tau_\alpha\epsilon - \frac{1}{40}\left(\tau_\alpha{}^{\beta\gamma} - 8\delta_\alpha^\beta\tau^\gamma\right)\gamma_{ij}\epsilon\Pi_A{}^i\Pi_B{}^j F_{\beta\gamma}^{AB}$$
$$\qquad + \frac{im}{10\sqrt{3}}\left(\tau_\alpha{}^{\beta\gamma\delta} - \frac{9}{2}\delta_\alpha^\beta\tau^{\gamma\delta}\right)\gamma^i\epsilon\Pi^{-1}{}_i{}^A S_{\beta\gamma\delta,A}$$

$$\delta\lambda_i = \frac{1}{16}\tau^{\alpha\beta}\left(\gamma_{kl}\gamma_i - \frac{1}{5}\gamma_{kli}\right)\epsilon\Pi_A{}^k\Pi_B{}^l F_{\alpha\beta}^{AB}$$
$$\qquad + \frac{im}{20\sqrt{3}}\tau^{\alpha\beta\gamma}\left(\gamma_i^j - 4\delta_i^j\right)\epsilon\Pi^{-1}{}_j{}^A S_{\alpha\beta\gamma,A}$$
$$\qquad + \frac{m}{2}\left(T_{ij} - \frac{1}{5}T\delta_{ij}\right)\gamma^j\epsilon + \frac{1}{2}\tau^\alpha\gamma^j\epsilon P_{\alpha ij}. \qquad (16.91)$$

16.7.4 Massive type IIA supergravity in $d = 10$

Type IIA supergravity in $d = 10$ is obtained by dimensional reduction on a circle of 11-dimensional supergravity, and is the low-energy limit of type IIA string theory at low coupling. At strong coupling, string theory becomes 11-dimensional M-theory (with the string coupling g_s acting as the radius R_{11}), whose low-energy limit is 11-dimensional supergravity.

The bosonic fields are the "NS-NS sector": metric $g_{\mu\nu}$, antisymmetric tensor $B_{\mu\nu}$ (B_2 as a form) and dilaton ϕ; the "RR sector": $U(1)$ gauge field A_μ (A_1 as a form) and $A_{\mu\nu\rho}$ (A_3 as a form). The field strengths are $H_3 = dB_2$, $F_2 = dA_1$, and $F_4 = dA_3$.

The fermionic fields are two Majorana–Weyl fermions (dilatinos) λ^i and two Majorana–Weyl gravitinos ψ_μ^i, both of opposite chiralities, all obtained from the one 11-dimensional gravitino $\Psi_M = (\psi_\mu^i, \psi_{11}^i)$.

Its bosonic action is written in a simple form in the "string frame," relevant for string theory, in which the Einstein action has an $e^{-2\phi}$ prefactor, where ϕ is called the dilaton, and its VEV gives the string coupling through $g_s = e^{-\langle\phi\rangle}$. The action is

$$S = \frac{1}{2\kappa^2} \int d^{10}x\sqrt{-g} \left\{ e^{-2\phi} \left[R + 4\partial_\mu \phi \partial^\mu \phi - \frac{1}{2}|H_3|^2 \right] - \frac{1}{2}(|F_2|^2 + |\tilde{F}_4|^2) \right\}$$
$$- \frac{1}{4\kappa^2} \int B_2 \wedge F_4 \wedge F_4, \qquad (16.92)$$

where we have defined

$$|F_n|^2 \equiv \frac{e}{n!} F_{M_1 \ldots M_n} F^{M_1 \ldots M_n},$$
$$\tilde{F}_4 = F_4 - A_1 \wedge H_3. \qquad (16.93)$$

and where the (Einstein frame) Newton's constant in 10 dimensions, κ_{10}, is given by

$$\kappa_{10}^2 = \kappa^2 g_s^2, \quad \kappa = \frac{1}{2\pi}(2\pi l_s)^8, \quad l_s = \sqrt{2\alpha'}, \qquad (16.94)$$

in terms of the string theory quantities string length l_s and string coupling g_s.

The susy laws will be given and used in this form in Chapter 19.

However, Romans found that there is an extension of this type IIA supergravity with a mass parameter, hence named "massive type IIA."

It is found by introducing a A_9 9-form field, with a 10-form field strength $F_{10} = dA_9$, and field equation $d * F_{10} = 0$ (since the kinetic term would be $\propto F_{10} * F_{10}$), solved by F_{10}=constant ($*F_{10}$ is a scalar), or rather one can consider it to be *piece-wise constant*, with the possibility to have domain walls where the value changes, called D8-branes in string theory.

The bosonic action is found by the replacement

$$F_2 \to F_2' = F_2 + MB_2, \quad F_4 \to F_4' = F_4 + \frac{M}{2}B_2 \wedge B_2, \Rightarrow \tilde{F}_4 \to \tilde{F}_4 + \frac{M}{2}B_2 \wedge B_2. \quad (16.95)$$

The full Lagrangian (except for 4-fermi terms, which are usually neglected) *in the usual, Einstein frame*, as found by Romans, is (for $\kappa_N^2 = 2$)

$$e^{-1}\mathcal{L} = -\frac{R}{4} + \frac{1}{2}\bar{\psi}_m \Gamma^{mnp} D_n \psi_p + \frac{1}{2}\bar{\lambda}\Gamma^m D_m \lambda + \frac{1}{2}(D^m \phi)(D_m \phi)$$
$$- \frac{1}{\sqrt{2}}(\partial_n \phi)\bar{\lambda}\Gamma^m \Gamma^n \psi_m + \frac{1}{48}\xi^{-2}\tilde{F}^{mnpq}\tilde{F}_{mnpq}$$
$$+ \frac{1}{12}\xi^4 H^{mnp} H_{mnp} + \frac{M^2}{4}\xi^{-6} B^{mn} B_{mn} - \frac{1}{24 \cdot 48}(\epsilon FFB)$$
$$+ \frac{1}{96}\xi^{-1}\tilde{F}_{abcd}\left[\bar{\psi}^m \Gamma_{[m}\Gamma^{abcd}\Gamma_{n]}\psi^n + \frac{1}{\sqrt{2}}\bar{\lambda}\Gamma^m \Gamma^{abcd}\psi_m + \frac{3}{4}\bar{\lambda}\Gamma^{abcd}\lambda \right]$$
$$- \frac{1}{24}\xi^2 H_{abc}\left(\bar{\psi}^m \Gamma_{[m}\Gamma^{abc}\Gamma_{n]}\Gamma_{11}\psi^n + \sqrt{2}\bar{\lambda}\Gamma^m \Gamma^{abc}\Gamma_{11}\psi_m \right)$$
$$- \frac{1}{8}M\xi^{-3}B_{ab}\left[\bar{\psi}^m \Gamma_{[m}\Gamma^{ab}\Gamma_{n]}\Gamma_{11}\psi^n + \frac{3}{\sqrt{2}}\bar{\lambda}\Gamma^m \Gamma^{ab}\Gamma_{11}\psi_m + \frac{5}{4}\bar{\lambda}\Gamma^{ab}\Gamma_{11}\lambda \right]$$
$$+ \frac{1}{8}M\xi^{-5}\bar{\psi}_m \Gamma^{mn}\psi_n + \frac{5}{8\sqrt{2}}M\xi^{-5}\bar{\lambda}\Gamma^m \psi_m - \frac{21}{32}M\xi^{-5}\bar{\lambda}\lambda + \frac{M^2}{8}\xi^{-10}, \quad (16.96)$$

where the field A_m is a Stueckelberg field for B_{mn}, so it was gauged away by a gauge transformation of B_{mn}, leaving B_{mn} massive and not gauge invariant (Higgs-like mechanism).

Here we have

$$\xi \equiv e^{\phi/2}$$

$$H_{mnp} \equiv 3\partial_{[m}B_{np]}, \quad \tilde{F}_{mnpq} = F_{mnpq} + 6MB_{[mn}B_{pq]} = 4\partial_{[m}A_{npq]} + 6MB_{[mn}B_{pq]}$$

$$(\epsilon FFB) = e^{-1}\epsilon^{mnpqrstuvw}\left(F_{mnpq}F_{rstu}B_{vw} + 4MF_{mnpq}B_{rs}B_{tu}B_{vw}\right.$$

$$\left. + \frac{36}{5}M^2 B_{mn}B_{pq}B_{rs}B_{tu}B_{vw}\right). \tag{16.97}$$

The susy variations of the fields are

$$\delta e_m^a = \bar{\epsilon}\Gamma^a\psi_m$$

$$\delta B_{mn} = \theta_{mn} + 2M^{-1}D_{[m}\theta_{n]}$$

$$\delta A_{mnp} = \theta_{mnp} - 6B_{[mn}\theta_{p]}$$

$$\delta\phi = \frac{1}{\sqrt{2}}\bar{\lambda}\epsilon$$

$$\delta\psi_m = D_m\epsilon - \frac{M}{32}\xi^{-5}\Gamma_m\epsilon - \frac{M}{32}\xi^{-3}B_{np}\left(\Gamma_m{}^{np} - 14\delta_m{}^n\Gamma^p\right)\Gamma_{11}\epsilon$$

$$+ \frac{1}{48}\xi^2 H_{npq}\left(\Gamma_m{}^{npq} - 9\delta_m{}^n\Gamma^{pq}\right)\Gamma_{11}\epsilon$$

$$+ \frac{1}{128}\xi^{-1}\tilde{F}_{npqr}\left(\Gamma_m{}^{npqr} - \frac{20}{3}\delta_m{}^n\Gamma^{pqr}\right)\epsilon$$

$$\delta\lambda = \frac{1}{\sqrt{2}}(\partial_m\phi)\Gamma^m\epsilon - \frac{5}{8\sqrt{2}}M\xi^{-5}\epsilon + \frac{3}{8\sqrt{2}}M\xi^{-3}B_{mn}\Gamma^{mn}\Gamma_{11}\epsilon$$

$$+ \frac{1}{12\sqrt{2}}\xi^2 H_{mnp}\Gamma^{mnp}\Gamma_{11}\epsilon - \frac{1}{96\sqrt{2}}\xi^{-1}\tilde{F}_{mnpq}\Gamma^{mnpq}\epsilon, \tag{16.98}$$

where

$$\theta_m \equiv \xi^3\left[-\frac{1}{2}\bar{\psi}_m\Gamma_{11}\epsilon + \frac{3}{4\sqrt{2}}\bar{\lambda}\Gamma_m\Gamma_{11}\epsilon\right]$$

$$\theta_{mn} \equiv \xi^{-2}\left[\bar{\psi}_{[m}\Gamma_{n]}\Gamma_{11}\epsilon + \frac{1}{2\sqrt{2}}\bar{\lambda}\Gamma_{mn}\Gamma_{11}\epsilon\right]$$

$$\theta_{mnp} \equiv \xi\left[\frac{3}{2}\bar{\psi}_{[m}\Gamma_{np]}\epsilon - \frac{1}{4\sqrt{2}}\bar{\lambda}\Gamma_{mnp}\epsilon\right]. \tag{16.99}$$

16.7.5 Type IIB supergravity in $d = 10$

Type IIB supergravity in $d = 10$ contains two chiral fermions and two gravitini *of the same chirality*, as opposed to type IIA above, that contains fermions of opposite chiralities (and all come from the dimensional reduction of a gravitino in 11 dimensions). It, however, also contains a self-dual 5-form field strength, which means that there is no Lorentz invariant formulation for the action, without auxiliary fields.

The fields are the same bosonic "NS-NS sector," $(g_{\mu\nu}, B_{\mu\nu}, \phi)$. In the "RR sector," we have antisymmetric tensors $A_{\mu\nu}$ (A_2 as a form), with field strength $F_3 = dA_2$ and $A_{\mu\nu\rho\sigma}$ (A_4^+ as a form) with self-dual field strength F_5, and a scalar a (or A_0 as a form).

The fermions are again two Majorana–Weyl dilatinos λ^i and two Majorana–Weyl gravitino ψ_μ^i, but now with the same chirality.

The action can be written, but without incorporating the self-duality condition, which must be added separately. Then the action *in string frame* is

$$S = \frac{1}{2\kappa^2} \int d^{10}x\sqrt{-g} \left\{ e^{-2\phi} \left[R + 4\partial_\mu\phi\partial^\mu\phi - \frac{1}{2}|H_3|^2 \right] - \frac{1}{2}\left(|F_1|^2 + |\tilde{F}_3|^2 + \frac{1}{2}|\tilde{F}_5|^2 \right) \right.$$
$$\left. - \frac{1}{4\kappa^2} \int A_4 \wedge H_3 \wedge F_3 \right\}, \tag{16.100}$$

where

$$\tilde{F}_3 \equiv F_3 - A_0 \wedge H_3$$
$$\tilde{F}_5 \equiv F_5 - \frac{1}{2}A_2 \wedge H_3 + \frac{1}{2}B_2 \wedge F_3. \tag{16.101}$$

The field equations are consistent with the self-duality condition

$$* \tilde{F}_5 = \tilde{F}_5, \tag{16.102}$$

since the equations of motion and Bianchi identity for $\tilde{F}_5$ are

$$d * \tilde{F}_5 = d\tilde{F}_5 = H_3 \wedge F_3. \tag{16.103}$$

The action can be put into an $SL(2, \mathbb{R})$-invariant form *in the Einstein frame*,

$$g^E_{\mu\nu} = e^{-\phi/2} g_{\mu\nu}, \tag{16.104}$$

by defining the covariant objects

$$\tau \equiv a + ie^{-\phi}$$
$$\mathcal{M}_{ij} \equiv \frac{1}{\text{Im } \tau} \begin{pmatrix} |\tau|^2 & -\text{Re } \tau \\ -\text{Re } \tau & 1 \end{pmatrix}$$
$$F_3^i = \begin{pmatrix} H_3 \\ F_3 \end{pmatrix} \equiv F_3$$
$$\tilde{F}_5 = F_5 + \frac{1}{2}\epsilon_{ij}A_2^i \wedge F_3^j, \tag{16.105}$$

and from them constructing also

$$\frac{1}{4}\text{Tr}\left[\partial^\mu \mathcal{M} \partial_\mu \mathcal{M}^{-1} \right] = -\frac{1}{2}\frac{\partial_\mu\bar{\tau}\partial^\mu\tau}{(\text{Im } \tau)^2} - \frac{1}{2}|F_1|^2, \tag{16.106}$$

such that the transformation rules under a matrix

$$\Lambda = \begin{pmatrix} a & b \\ c & d \end{pmatrix} \in SL(2, \mathbb{R}) \tag{16.107}$$

are

$$F_3 \to \Lambda F_3, \quad \mathcal{M} \to (\Lambda^{-1})^T \mathcal{M} \Lambda, \quad \tilde{F}_5 \to \tilde{F}_5, \quad \tau' = \frac{a\tau + b}{c\tau + d}. \tag{16.108}$$

Then the invariant action is

$$S_{\text{IIB}} = \frac{1}{2\kappa^2} \int d^{10}x\sqrt{-g_E} \left[R(g_E) + \frac{1}{4}\text{Tr}\left(\partial_\mu \mathcal{M} \partial^\mu \mathcal{M}^{-1} \right) - \frac{1}{2}F_3^i \mathcal{M}_{ij} F_3^j - \frac{1}{2}|\tilde{F}_5|^2 \right]$$
$$- \frac{\epsilon^{ij}}{8\kappa^2} \int C_4 \wedge F_3^i \wedge F_3^j. \tag{16.109}$$

16.7.6 Other gaugings: *ISO*(7) example

In supergravity, one can gauge subgroups of the full global symmetry group, and even gauge non-compact groups G or even non-semisimple, and often times these can be interpreted as nonlinear ansatze for consistent truncations for compactifications on nontrivial spaces.

ISO(7) example

In particular, it was found that there is an $ISO(7) = SO(7) \ltimes \mathbb{R}^7$ (so, non-compact and non-semisimple) gauging of the $N = 8$ $d = 4$ supergravity, which arises under consistent truncation (nonlinear reduction ansatz) of *massive* type IIA supergravity on $AdS_4 \times S^6$. In fact, there is only an usual, "purely electric," gauging, with gauge coupling g, and a "dyonic" (so, "electric + magnetic") gauging, which also has a magnetic coupling $m = gc$, with $c \neq 0$.

The fields of the gauging split, into $SL(7)$ representations ($SO(7) \subset SL(7)$, with indices $I = 1, ..., 7$), as:

- metric $g_{\mu\nu}$;
- scalars $\mathcal{V}^{IJij}, \mathcal{V}^{I8ij}, \tilde{\mathcal{V}}_{IJ}^{ij}, \tilde{\mathcal{V}}_{I8}^{ij}$, with $i, j = 1, ..., 8$, filling up all the scalars in the ungauged coset.
- vector $A_\mu^{IJ}, A_\mu^I, \tilde{A}_{\mu IJ}, \tilde{A}_{\mu I}$, again filling the ungauged $SO(8)$-invariant gauge fields, now with their magnetic counterparts with tildes. Note that we can say that A_μ^{IJ} *gauge $SO(7)$, while A_μ^I gauge $\mathbb{R}^7$, together gauging $ISO(7)$*. The partners with tildes are magnetic counterparts.
- 2-forms $A_{\mu\nu I}{}^J, A_{\mu\nu}^I$
- 3-forms $A_{\mu\nu\rho}^{IJ}$.

The gauge fields have field strengths

$$F_{(2)}^{IJ} = dA_{(1)}^{IJ} - g\delta_{KL}A_{(1)}^{IK} \wedge A_{(1)}^{LJ}$$

$$F_{(2)}^I = dA_{(1)}^I - g\delta_{JK}A_{(1)}^{IJ} \wedge A_{(1)}^K + \frac{m}{2}A_{(1)}^{IJ} \wedge \tilde{A}_{(1)J} + mA_{(2)}^I$$

$$\tilde{F}_{(2)IJ} = d\tilde{A}_{(1)IJ} + g\delta_{K[I}A_{(1)}^{KL} \wedge \tilde{A}_{(1)J]L} + g\delta_{K[I}A_{(1)}^K \wedge \tilde{A}_{(1)J]} - m\tilde{A}_{(1)I} \wedge \tilde{A}_{(1)J} + 2g\delta_{K[I}A_{(2)J]}{}^K$$

$$\tilde{F}_{(2)I} = d\tilde{A}_{(1)I} - \frac{g}{2}\delta_{IJ}A_{(1)}^{JK} \wedge \tilde{A}_{(1)K} + g\delta_{IJ}A_{(2)}^J, \tag{16.110}$$

and the 2-forms and 3-forms also have (even more) complicated fields strengths.

The action is even more complicated than the usual gauging, and has a scalar potential of order g^2, as well as terms of order g and mg. We will not reproduce it here.

16.7.7 Some general properties of gaugings

As we saw, gaugings can be done for subgroups G of the full global symmetry group. In general, we consider some *embedding tensor* $\Theta_A{}^a$, defining the subgroup of the total G_{tot} that is gauged, a linear combination, roughly

$$X_A = \Theta_A{}^a t_a, \tag{16.111}$$

of the generators $t_a \in G_{\text{tot}}$, where A is in the *fundamental* representation of the global symmetry group G, since $A = 1, ..., n_V$, with n_V the number of (abelian) gauge fields in the ungauged theory (and $a = 1, ..., dim(G)$ is in the adjoint representation).

The gauging is done via the Noether procedure, of adding terms with gauge fields times the Noether current in the Lagrangian, and then order by order in the action and transformation rules. As we said in Chapter 15, for gauged supergravity, we have schematically

$$\delta \psi_\mu^i = D_\mu(\omega(e, \psi))\epsilon^i + g\gamma_\mu\epsilon^i + gA_\mu\epsilon^i + ..., \tag{16.112}$$

where $D_\mu + gA_\mu \sim \mathcal{D}_\mu$ is the gauge – and gravitationally – covariant derivative, so we have order g terms in the susy law of the gravitinos, one constant, related to the existence of the AdS background (the susy variation of the cosmological constant is related to this extra term in the susy variation of the gravitinos), the other proportional to the gauge fields.

But, more precisely, one considers the *fermion shift tensors* A^{ij}, B^{Mi} in the susy laws of the gravitini ψ_μ^i and spin 1/2 (dilatini) χ^M,

$$\delta \psi_i^\mu = \mathcal{D}_\mu(\omega(e, \psi))\epsilon^i - gA^{ij}\gamma_\mu\epsilon^j + ... = \delta_{(0),\text{cov}}\psi_\mu^i - gA^{ij}\gamma_\mu\epsilon^j$$
$$\delta \chi^M = \delta_{(0),\text{cov}}\chi^M - gB^{Mi}\epsilon_i, \tag{16.113}$$

(note that $D_\mu \to \mathcal{D}_\mu$ in the covariantized $\delta_{(0),\text{cov}}$), together with a matrix C^{MN} related to (made up of) A^{ij}, B^{Mi} by susy, together forming the fermion mass matrix

$$e^{-1}\Delta_{(1)}\mathcal{L} = g\left[2\bar{\psi}_\mu^i\gamma^{\mu\nu}\psi_\nu^j A_{ij} + \bar{\chi}^M B_{Mi}\gamma^\mu\psi_\mu^i + \bar{\chi}^M\chi^N C_{MN}\right] + h.c. \tag{16.114}$$

Indeed, in this way the $D_\mu\epsilon^i$ contributions coming from the susy variation $\delta_{(0)}$ of the fermion mass terms are canceled by variation of the kinetic non-mass terms in the Lagrangian.

As in the maximally supersymmetric examples from this subsection, the matrices A_{ij}, B_{Mi}, C_{MN} are made up from the *T-tensor*, which in general is a contraction of two scalar group matrices (the matrix, invariant under the full symmetry group G, that contains the scalars) with the embedding tensor, one matrix in the fundamental representation, with indices $A, \tilde{A}$, and another in the adjoint representation, with indices $a, \tilde{a}$, so

$$T_{\tilde{A}}^{\tilde{a}} = \Theta_A{}^a \mathcal{V}_{\tilde{A}}^A \mathcal{V}_a{}^{\tilde{a}}. \tag{16.115}$$

Then this generic T-tensor decomposes in irreducible representations, and one usually describes the T-tensors as these irreducible representations, as we saw in the maximally supersymmetric cases in $D = 4, 5, 7$. For instance, in $D = 4$ we had $T_i{}^{jkl}$, decomposed in $A_1^{ij} \propto T_k{}^{ikj}$ and $A_{2i}{}^{jkl} \propto T_i{}^{[jkl]}$, with $A^{ij} \to A_1^{ij}, B^{Mi} \to A_{2i}{}^{jkl}$ and $C^{MN} \to \epsilon^{ijkpqrlm}A_2{}^n{}_{pqr}$ (so $M \to jkl$), while in $D = 5$ we had $T_{ab,cd}^{(1)}, T_{ab,cd}^{(2)}, T_{ab,cdef}^{(3)}$ and $A^{ij} \to A_{ab}^{(1)} \propto T_{ca,db}^{(1)}\Omega^{cd}, A^{Mi} \to A_{abc,d}^{(2)}$ (so $i \to a, M \to abc$), with $A_{ab[c,d]}^{(2)} \propto T_{ab,cd}^{(2)}, A_{ab(c,d)}^{(2)} \propto T_{ab,cd}^{(1)}$ and $C^{MN} \to A_{abc,def}^{(3)}$ defined implicitly, by (16.81), in terms of $A_{ab,cd}^{(1)}$ and $A_{ab,cd}^{(2)}$. Finally, in $D = 7$, we only had T_{ij}, and $A^{ij} \to T\delta^{I'J'}, B^{Mi} \to T^{ij}(\gamma_j)_{I'}{}^{J'}$ and $C^{MN} \to (T^{ij} - \frac{T}{8}\delta^{ij})\delta^{I'J'}$ (so $i \to I', M \to iI'$).

At order g^2, one needs a contribution to the potential, in order to cancel the $\delta_{(1)}$ of the fermion mass terms, and this is

$$V = \frac{g^2}{\mathcal{N}}\left(B^{Mi}B_{Mi} - 12A^{ij}A_{ij}\right), \tag{16.116}$$

where $\mathcal{N} = \delta_i^i$ is the number of supercharges. This is necessary in order for the condition

$$\delta_i^j V = g^2 \left(B^{Mj} B_{Mi} - 12 A^{jk} A_{ik} \right) \tag{16.117}$$

to hold, which is called the *potential Ward identity*, a type of supersymmetric Ward identity. The resulting scalar potential is quadratic in the embedding tensor $\Theta_A{}^a$.

The potential Ward identity is a generalization of the condition mentioned before, that the $g\gamma_\mu \epsilon^i$ term in $\delta \psi_\mu^i$ was necessary in order to cancel the variation of the cosmological constant term. Now, the variation of the $\det e$ multiplying the full scalar potential V (not just the cosmological constant),

$$(\delta \det e)V = e(\bar{\epsilon}^i \gamma^\mu \psi_{\mu i} + \bar{\epsilon}_i \gamma^\mu \psi_\mu^i)V, \tag{16.118}$$

is cancelled by the extra variation of the fermion mass terms $\bar{\psi}_\mu^i A_{ij} \gamma^{\mu\nu} \delta_{(1)} \psi_\nu^j$.

16.8 * Modified supergravities in 10 and 11 dimensions: modified "local Lorentz" covariance, "exceptional field theory," and the geometric approach to supergravity

For various reasons, it is easier to modify the higher-dimensional theory than to work with the standard theories. Of course, in this chapter, we have presented a minimal modification, by introducing an auxiliary antisymmetric tensor field. But there are other, more radical and important, modifications.

In particular, in order to obtain a simple nonlinear dimensional reduction to a gauged supergravity, we can modify the higher-dimensional theory in order to already have the gauge invariance needed before compactification. That is the case of the original 11-dimensional theory of de Wit and Nicolai with modified "local Lorentz" invariance $SO(3,1) \times SU(8)$, which was generalized then to other cases, in particular the ones that go generically under the name of "exceptional field theory."

Another important modification was to use a "geometric" approach to supergravity, in terms of forms, and (almost) without the need of a metric to write the action. This was pioneered by d'Auria and Fré, who rewrote 11-dimensional supergravity in this way.

16.8.1 Eleven-dimensional supergravity with $SO(3,1) \times SU(8)$ ("local Lorentz") covariance

The goal, as done by de Wit and Nicolai, is to rewrite the 11-dimensional supergravity in terms of quantities defined for the compactification on $AdS_4 \times S^7$, just that without any restriction on the fields, that is, still full dependence on both x (for AdS_4) and y (for S^7).

Therefore, we start with the usual 11-dimensional vielbein $E_M{}^A$, put into a gauge (for the local Lorentz symmetry) that breaks $SO(11)$ down to $SO(1,3) \times SO(7)$, as

$$E_M{}^A = \begin{pmatrix} e_\mu{}^\alpha & B_\mu{}^m e_m{}^a \\ 0 & e_m{}^a \end{pmatrix} \Rightarrow E_A{}^M = \begin{pmatrix} e_\alpha{}^\mu & -e_\alpha{}^\mu B_\mu{}^m \\ 0 & e_a{}^m \end{pmatrix}. \tag{16.119}$$

Further, we redefine

$$e_\mu{}^\alpha \equiv \Delta^{-1/2} e'_\mu{}^\alpha$$
$$\Delta \equiv \det e_m{}^a, \qquad (16.120)$$

and now the new *11-dimensional fields* are $e'_\mu{}^a, e_m{}^a, B_\mu{}^m$. But we want to have $SU(8)$ invariance, so in fact we need to change $e_m{}^a$ to

$$e_{AB}^m = i\Delta^{-1/2} e_a{}^m \Gamma_{CD}^a U^C{}_A U^D{}_B, \qquad (16.121)$$

where the matrix $U^A{}_B$ is subject to a $SU(8)$ gauge transformation,

$$U^A{}_B \rightarrow V^A{}_C U^C{}_B, \quad V^A{}_C \in SU(8). \qquad (16.122)$$

Note that now one has the covariant derivative

$$D_\mu = e_\mu{}^\alpha \partial_\alpha = e_\mu{}^\alpha E_\alpha{}^M \partial_M = \partial_\mu - B_\mu{}^m \partial_m. \qquad (16.123)$$

There are some new spin connection variables, $\omega_M{}^{AB}$, that are derived from the vielbein ones, in the usual way (though now they are defined in the gauge (16.119)).

The fermions also split in the usual way, relevant for compactification,

$$\Psi_M \rightarrow (\Psi_\mu, \Psi_m), \qquad (16.124)$$

but now we also make the further redefinitions, which were needed for the nonlinear $AdS_4 \times S^7$ compactification (similar to the ones we wrote for the $AdS_7 \times S^4$ case),

$$\psi'_\mu \equiv e'_\mu{}^\alpha \Delta^{-1/4}(i\gamma_5)^{-1/2}\left(\Psi_\alpha - \frac{1}{2}\gamma_5\gamma_\alpha\Gamma^a\Psi_a\right)$$
$$\psi'_a \equiv \Delta^{-1/4}(i\gamma_5)^{-1/2}\Psi_a, \qquad (16.125)$$

and further, after the chiral projections (in terms of eight-dimensional $SU(8)$ indices, spinorial indices $A = 1, ..., 8$)

$$\psi_\mu{}^A \equiv \frac{1}{2}(1+\gamma_5)\psi'_{\mu A}, \quad \psi_{\mu A} \equiv \frac{1}{2}(1-\gamma_5)\psi'_{\mu A}, \qquad (16.126)$$

and rewriting in terms of three-index spinors, we get the $SU(8)$ covariant spinors

$$\chi^{ABC} \equiv (1+\gamma_5)\frac{3i\sqrt{2}}{4}\Gamma^a_{[AB}\psi'_{aC]}$$
$$\chi_{ABC} \equiv (1-\gamma_5)\frac{3i\sqrt{2}}{4}\Gamma^a_{[AB}\psi'_{aC]}. \qquad (16.127)$$

As in the case discussed in most of this chapter, the redefinition of the 4-form field strength is the most complicated. The fermionic Lagrangian also becomes very long and cumbersome under the above transformations.

One observation is that, in order to stay in the gauge (16.119), we must do compensating transformations for the various transformation laws. Thus, for instance, for the susy transformation, we must perform a compensating (local Lorentz) rotation with parameter $\Omega_A{}^\alpha$, such that

$$\delta_{\text{susy}} E_m{}^\alpha + E_m{}^A \Omega_A{}^\alpha = 0 \Rightarrow \Omega_{a\alpha} = -\frac{1}{2}\bar\epsilon\tilde\Gamma_\alpha\Psi_a. \qquad (16.128)$$

But for the redefined fields that becomes nicer,

$$\delta e'^{\,\alpha}_{\mu} = \frac{1}{2}\bar{\epsilon}'\gamma^\alpha\psi'_\mu = \frac{1}{2}\bar{\epsilon}^A\gamma^\alpha\psi_{\mu A} + h.c.$$

$$\delta B^m_\mu = \frac{\sqrt{2}}{8}e^m_{AB}\left(2\sqrt{2}\bar{\epsilon}^A\psi_\mu{}^B + \bar{\epsilon}^C\gamma'_\mu\chi^{ABC}\right) + h.c., \qquad (16.129)$$

and also we have for the susy parameter

$$\epsilon' = \Delta^{1/4}(i\gamma_5)^{-1/2}\epsilon, \quad \epsilon^A \equiv \frac{1}{2}(1+\gamma_5)\epsilon'_A, \quad \epsilon_A = \frac{1}{2}(1-\gamma_5)\epsilon'_A. \qquad (16.130)$$

The variation of e^m_{AB} becomes

$$\delta e^m_{AB} = -\sqrt{2}\Sigma_{ABCD}e^{mCD},$$

$$\Sigma_{ABCD} = \bar{\epsilon}_{[A}\chi_{BCD]} + \frac{1}{24}\epsilon_{ABCDEFGH}\bar{\epsilon}^E\chi^{FGH}. \qquad (16.131)$$

Other susy variations are more complicated, and other transformation laws (coordinate, local Lorentz) are done similarly, with compensating transformations, in order to stay in the gauge.

The redefinition of A_{MNP} components is more complicated, but roughly speaking, as in the usual case:

- $A_{\mu mn}$ gives rise to gauge fields A^{mn}_μ that, together with the $B^m_\mu \equiv A^{m8}_\mu$ above, give rise to the required $A^{\tilde{A}\tilde{B}}_\mu$ in the adjoint of $SO(8)$ (the same split, called A^{IJ}_μ and A^I_μ in the case of the $ISO(7)$ gauging above).
- $A_{\mu\nu m}$ is dual to scalars A^m that, together with the scalars A_{mnp} and the scalars e^m_{AB}, will fill representations of $SU(8)$.
- $A_{\mu\nu\rho}$ gives no new states (its field strength is constant)

As we see, the logic was: we have to gauge fix the local Lorentz invariance from $SO(10,1)$ to $SO(3,1) \times SO(7)$, and then extended from $SO(7)$ to $SU(8)$, through the fact that the spinor indices of $SO(7)$, $A, B = 1, ..., 8$, are changed to transform in $SU(8)$.

But the $d = 4$ $\mathcal{N} = 8$ supergravity was invariant under E_7, or more precisely $E_{7(7)}$, whose maximal compact subgroup is $SU(8)$ (and the scalars live in the coset $E_{7(7)}/SU(8)$). Similarly, in $d = 5$, the maximal supergravity is invariant under $E_{6(6)}$, whose maximal compact subgroup is $USp(8)$ (and the scalars live in the coset $E_{6(6)}/USp(8)$), and in $d = 3$, the maximal supergravity is invariant under $E_{8(8)}$, whose maximal compact subgroup is $SO(16)$ (and the scalars live in $E_{8(8)}/SO(16)$). We will see later on in the book that one can generalize this to $E_{d'(d')}$ for 11-dimensional supergravity compactified on $T^{d'}$ down to $11 - d'$ dimensions. $E_{d'(d')}$ is a noncompact form of $E_{d'}$.

So it would be important to have a generalization of 11-dimensional supergravity invariant not only under the maximal compact subgroup, $SU(8)$ for $d = 4$, but rather under $E_{d'(d')}$.

16.8.2 Exceptional field theory

This generalization was done in several papers and goes under the name of "Exceptional Field Theory," EFT (since the most relevant cases are $E_{6(6)}, E_{7(7)}$, and $E_{8(8)}$, noncompact forms of the exceptional groups E_6, E_7, and E_8). Also since there are some ideas borrowed from the T-duality invariant formulation of string theory in the supergravity limit in which the number of coordinates is doubled, but subject to a constraint (which, when solved explicitly, gives back the usual formulation), known as "Double Field Theory" (DFT).

$E_{6(6)}$ **EFT**

In the case of $E_{6(6)}$, relevant to $d - 5$ maximal supergravity, the construction is as follows. Again, one considers the split into $d = 5$ coordinates x^μ and the rest, now six coordinates, but extended to 27 coordinates Y^M, with conjugate derivatives ∂_M, and transforming under the **27** representation of $E_{6(6)}$, yet subject to the $E_{6(6)}$-invariant constraint

$$d^{MNK}\partial_N\partial_K A = 0, \quad d^{MNK}\partial_N A\partial_K B = 0, \tag{16.132}$$

where A, B are arbitrary fields and gauge parameters, and d^{MNK} is one of the two $E_{6(6)}$-invariant tensors, d^{MNK} and d_{MNK}, with $d_{MPQ}d^{NPQ} = \delta_M^N$, $M, N = 1, ..., 27$. Like in the case of DFT, solving explicitly these constraints gives back the six original coordinates.

The gauge fields are still in the **27** representation of $E_{6(6)}$, A_μ^M, even though the adjoint is 78-dimensional, with generators $\alpha = 1, ..., 78$. The gauge transformations are

$$\delta A_\mu^M = \mathcal{D}_\mu \Lambda^M, \tag{16.133}$$

where, in general, we have the covariant derivatives, for the five-dimensional coordinates, on a V^M,

$$\mathcal{D}_\mu V^M = \partial_\mu V^M - A_\mu^K\partial_K V^M + \frac{1-3\lambda}{3}\partial_K A_\mu^K V^M \\ + V^K\partial_K A_\mu^M - 10d^{MNP}d_{PKL}\partial_N A_\mu^K V^L, \tag{16.134}$$

and the "weight" λ of Λ^M is 1/3.

The field strength of A_μ^M is

$$\mathcal{F}_{\mu\nu}^M = 2\partial_{[\mu}A_{\nu]}^M - [A_\mu, A_\nu]_E^M + 10d^{MNK}\partial_K B_{\mu\nu N}, \tag{16.135}$$

where the "E-bracket" is

$$[\Lambda_2, \Lambda_1]_E^M \equiv 2\Lambda_{[2}^K\partial_K\Lambda_{1]}^M - 10d^{MNP}d_{KLP}\Lambda_{[2}^K\partial_N\Lambda_{1]}^L, \tag{16.136}$$

and the $B_{\mu\nu}^M$ 2-forms are also in the bosonic spectrum of the theory, together with the vielbein e_μ^a, and a scalar symmetric matrix $\mathcal{M}_{MN}$, transforming as a symmetric tensor.

So the full bosonic spectrum is

$$\{e_\mu^a, \mathcal{M}_{MN}, A_\mu^M, B_{\mu\nu M}\}. \tag{16.137}$$

The bosonic action is

$$S = \int d^5x d^{27}Y\, e\left[\hat{R} + \frac{1}{24}g^{\mu\nu}\mathcal{D}_\mu\mathcal{M}^{MN}\mathcal{D}_\nu\mathcal{M}_{MN} - \frac{1}{4}\mathcal{M}_{MN}\mathcal{F}^{\mu\nu M}\mathcal{F}_{\mu\nu}^N \right. \\ \left. + V(\mathcal{M}, e)\right] + S_{\text{top}}, \tag{16.138}$$

where the topological action is written in six physical dimensions as

$$S_{\text{top}} = \kappa\int d^{27}Y\int_{\mathcal{M}_6}\left(d_{MNK}\mathcal{F}^M\wedge\mathcal{F}^N\wedge\mathcal{F}^K - 14d^{MNK}\mathcal{H}_M\wedge\partial_N\mathcal{H}_K\right), \tag{16.139}$$

and becomes a five-dimensional boundary term, $\mathcal{H}_M$ is the field strength of $\mathcal{B}_M$, the scalar potential is

$$V = \frac{1}{24}\mathcal{M}^{MN}\partial_M\mathcal{M}^{KL}(12\partial_L\mathcal{M}_{NK} - \partial_N\mathcal{M}_{KL}) - e^{-1}(\partial_M e)\partial_N\mathcal{M}^{MN}$$
$$- \mathcal{M}^{MN}e^{-1}(\partial_M e)e^{-1}(\partial_N e) - \frac{1}{4}\mathcal{M}^{MN}(\partial_M g^{\mu\nu})(d_N g_{\mu\nu}), \qquad (16.140)$$

and the "improved Riemann tensor" is

$$\hat{R}_{\mu\nu}^{ab} = R_{\mu\nu}{}^{ab} + \mathcal{F}_{\mu\nu}^M e^{a\rho}\partial_M e_\rho^b. \qquad (16.141)$$

The constant κ is free at this point, but is fixed to $\kappa^2 = 5/18$ by the requirement of gauge invariance ("generalized diffeomorphisms") under transformations with $\xi^\mu(x, Y)$, as

$$\delta e_\mu^a = \xi^\nu \mathcal{D}_\nu e_\mu^a + \mathcal{D}_\mu \xi^\nu e_\nu^a$$
$$\delta\mathcal{M}_{MN} = \xi^\mu \mathcal{D}_\mu \mathcal{M}_{MN}$$
$$\delta A_\mu{}^M = \xi^\nu \mathcal{F}_{\nu\mu}{}^M + \mathcal{M}^{MN}g_{\mu\nu}\partial_N\xi^\nu$$
$$\delta B_{\mu\nu M} = \frac{1}{16\kappa}\xi^\rho e\epsilon_{\mu\nu\rho\sigma\tau}\mathcal{M}_{MN}\mathcal{F}^{\rho\sigma N} - d_{MKL}A_{[\mu}^K \delta A_{\nu]}^L. \qquad (16.142)$$

The formulations with $E_{7(7)}$ and $E_{8(8)}$ invariance are similarly constructed.

Also a formulation of $d = 10$ type IIB supergravity with $USp(8)$ invariance (the maximal compact subgroup of $E_{6(6)}$) was written, and shown that it gives a full nonlinear ansatz for compactification on $AdS_5 \times S^5$ to the maximal gauged supergravity in $d = 5$.

16.8.3 The geometric approach to supergravity: d'Auria and Fré's 11-dimensional supergravity with $OSp(1|32)$ invariance

Already Cremmer, Julia, and Sherk, in their original paper on 11-dimensional supergravity, noted that the super-group $OSp(1|32)$ is the minimal grading of the (usual, bosonic) group $Sp(32)$, which itself is the maximal bosonic group preserving the Majorana property of an $SO(10, 1)$ spinor, and so $OSp(1|32)$ is therefore a natural candidate to be a generalized invariance group of 11-dimensional supergravity.

Under $SO(10, 1) \subset OSp(1|32)$, the generators of $OSp(1|32)$ split into

$$(P^A, J^{AB}, Z_5^{A_1...A_5}, Q_\alpha). \qquad (16.143)$$

So it is natural to ask whether 11-dimensional supergravity can be written as a gauging of $OSp(1|32)$, which was investigated by d'Auria and Fré. P^A, J^{AB}, and Q_α naturally couple to the (gauge, i.e., one-forms) fields e_A, ω_{AB}, and ψ^α (vielbein, spin connection and gravitino, respectively), but we are left with $Z_5^{A_1...A_5}$, which would seem to couple to a gauge field (one-form) $B_{A_1...A_5}$. At first, it would seem like

$$F_7 = dB_{A_1...A_5} \wedge e^{A_1} \wedge ... \wedge e^{A_5} \qquad (16.144)$$

is dual to the $F_4 = dA_3$ present in the usual form of 11-dimensional supergravity, $F_7 = *F_4$. However, then the $OSp(1|32)$ algebra would imply a curvature (through the usual gauge theory relations $F^a = dA^a + f^a{}_{bc} A^b \wedge A^c$, corresponding to the algebra $[T_a, T_b] = f_{ab}{}^c T_c$) for $B^{A_1 \ldots A_5}$ of

$$R^{A_1 \ldots A_5} = \mathcal{D}B^{A_1 \ldots A_5} - \frac{i}{2}\bar{\psi}\Gamma^{A_1 \ldots A_5} \wedge \psi + \alpha_5 \epsilon^{A_1 \ldots A_5 BCDEFG} B_{BCDHI} B_{EFG}{}^{HI}, \qquad (16.145)$$

that is nonabelian, so violates the Coleman–Mandula theorem (since B corresponds to a bosonic nonabelian internal symmetry of the theory with nontrivial commutation relations with the Poincaré one, given by P^A and J^{AB}), and the possible Wigner–Inönü contraction (rescaling some of the generators to zero or infinity) of the theory that makes the curvature abelian doesn't reproduce 11-dimensional supergravity.

The basic problem is then that $A_{\mu\nu\rho}$ of 11-dimensional supergravity is a 3-form, not a one-form. In order to have a good YM-type, *geometric (meaning, written only in terms of forms, without the need to introduce a metric or the Hodge star operation on the space) theory*, we can either: (a) write A_3 as a *composite* of one-forms or (b) write an n-form generalization of YM theory, or more precisely of the theory on super-Lie groups. The latter was explored by d'Auria and Freé, and called "Cartan Integrable Systems" (CIS). In CIS, the gauge potential p-forms $\Pi^{M(p)}$ have curvatures

$$R^{M(p+1)} = d\Pi^{M(p)} + \sum_{n=1}^{N} \frac{1}{n} C^{M(p)}_{N_1(p_1) \ldots N_n(p_n)} \Pi^{N_1(p_1)} \ldots \Pi^{N_n(p_n)}, \qquad (16.146)$$

where $C^{M(p)}_{N_1(p_1) \ldots N_n(p_n)}$ are constants, that appear in some *corresponding* Generalized Maurer-Cartan Equations (GMCES), $d\Theta^{A(p)} + \sum C\Theta \wedge \ldots \wedge \Theta = 0$ (replacing Π's with Θ's, all the curvature vanish), $M, N_1, \ldots, N_n$ are indices in the algebra, and $p_1, \ldots, p_n$ are degrees of the forms. The integrability conditions of the above relations become Bianchi identities, $\nabla R^{M(p)} = 0$.

The geometric Lagrangian in this case is written as a d-form (in d-dimensional space), made by combining (with $\wedge$) curvatures $R^{M(p+1)}$ with (polynomials in) $\Pi^{N(p)}$'s.

d'Auria and Fré found a CIS with 1-forms (ω^{AB}, e^A, ψ), corresponding respectively to the generators (J_{AB}, P_A, Q), together with the 3-form A and a 0-form $F_{A_1 \ldots A_4}$, where the 0-form is needed in order to give a *first-order formulation* for the kinetic action involving the curvature of A, $\int (dA_3) * (dA_3)$. The curvatures of these forms are

$$R^{AB} = d\omega^{AB} - \omega^{AC} \wedge \omega_C{}^B$$

$$R^A = \mathcal{D}e^A - \frac{i}{2}\bar{\psi} \wedge \Gamma^A \psi$$

$$\rho = \mathcal{D}\psi$$

$$R^{\square} = dA - \frac{1}{2}\bar{\psi} \wedge \Gamma^{AB}\psi \wedge e_A \wedge e_B, \qquad (16.147)$$

and where $\mathcal{D}$ is the covariant derivative involving the spin connection ω.

However, because of the presence of the 0-form, the Lagrangian is not completely geometric. Indeed, one can only find the action of the schematic type

$$S \sim \int_{M_{11}} (R^{AB} \wedge \ldots + R^A \wedge \ldots + \bar{\rho} \wedge \ldots + R^{\square} \wedge \ldots) + \int R^{\square} \wedge R^{\square} \wedge A$$

$$+ \int \left[-F_{A_1 \ldots A_4} F^{A_1 \ldots A_4} (e^{\wedge})^{11} + F_{A_1 \ldots A_4} R^{\square} \wedge (e^{\wedge})^7 \right], \qquad (16.148)$$

and the terms on the second line are not geometric, since one cannot find a way to write $\int (dA_3) * (dA_3)$ as a geometric object.

More precisely, the action is

$$
\begin{aligned}
S = \int_{M_{11}} \Bigg[& -\frac{1}{9}\epsilon_{A_1...A_{11}} R^{A_1 A_2} \wedge e^{A_3} \wedge ... \wedge e^{A_{11}} - 840 R^{\square} \wedge R^{\square} \wedge A \\
& +\frac{7i}{30}\epsilon_{B_1...B_{11}} R^A \wedge e_A \wedge \bar{\psi} \wedge \Gamma^{B_1...B_5}\psi \wedge e^{B_6} \wedge ... \wedge e^{B_{11}} \\
& +2\bar{\rho} \wedge \Gamma_{C_1...C_8}\psi \wedge e^{C_1} \wedge ... \wedge e^{C_8} \\
& -84 R^{\square} \wedge \left(i\bar{\psi} \wedge \Gamma_{A_1...A_5}\psi \wedge e^{A_1} \wedge ... \wedge e^{A_5} - 10 A \wedge \bar{\psi} \wedge \Gamma_{AB}\psi \wedge e^A \wedge e^B \right) \\
& +\frac{1}{4}\bar{\psi} \wedge \Gamma^{A_1 A_2}\psi \wedge \bar{\psi} \wedge \Gamma^{A_3 A_4}\psi \wedge e^{A_5} \wedge .... \wedge e^{A_{11}}\epsilon_{A_1...A_{11}} \\
& -210\bar{\psi} \wedge \Gamma^{A_1 A_2}\psi \wedge \bar{\psi} \wedge \Gamma^{A_3 A_4}\psi \wedge e^{A_1} \wedge .... \wedge e_{A_4} \wedge A \\
& -\frac{1}{330}F_{A_1...A_4}F^{A_1...A_4}\epsilon_{C_1...C_{11}}e^{C_1} \wedge .... \wedge e^{C_{11}} + 2\epsilon^{A_1...A_{11}}F_{A_1...A_4}R^{\square} \wedge e_{A_5} \wedge ... \wedge e_{A_{11}} \Bigg].
\end{aligned}
$$
$$(16.149)$$

One notes also that the CIS above is equivalent with a usual supergroup formulation, if one exchanges the 3-form A for the one-forms B^{AB}, $B^{A_1...A_5}$, and η (an extra 32-component spinor). The decomposition of A into one-forms is lengthy, but is of the type

$$A = B_{AB} \wedge e^A \wedge e^B + \text{nonlinear}, \tag{16.150}$$

where the lengthy part is nonlinear in fields other than the vielbein e^A.

The gauge field on the supergroup (usually written as $A = A^{\tilde{A}}T_{\tilde{A}}$, where $T_{\tilde{A}}$ are generators and $A^{\tilde{A}}$ are fields) is now

$$\omega^{AB}J_{AB} + e^A P_A + \psi Q + \eta Q' + B^{AB}Z_{AB} + B^{A_1...A_5}Z_{A_1...A_5}, \tag{16.151}$$

so the supergroup has extra generators Z_{AB}, $Z_{A_1...A_5}$, and Q'. But the remarkable fact is that the resulting superalgebra is actually the M theory algebra (explained later on in the book), with Z_{AB} and $Z_{A_1...A_5}$ identified with the 2-form and 5-form central charges of M theory, but now together with the extra spinor charge Q', whose presence modifies the algebra as follows (the two values in brackets correspond to two possible supergroups)

$$[Q, P_A] = i \begin{pmatrix} 1 \\ 0 \end{pmatrix} \Gamma_A Q'$$

$$[Q, Z_{A_1 A_2}] = \begin{pmatrix} \frac{1}{5} \\ -\frac{1}{2} \end{pmatrix} \Gamma_{A_1 A_2} Q'$$

$$[Q, Z_{A_1...A_5}] = \begin{pmatrix} \frac{1}{240} \\ -\frac{1}{144} \end{pmatrix} \Gamma_{A_1...A_4} Q', \tag{16.152}$$

and everything else stays the same (and Q' commutes with everything). This means that the algebra becomes just the M theory algebra if Q' is Wigner–Inönü contracted away, by $Q' \to aQ' \to 0$. That sounds great, except it is still not a completely geometric formulation, since one still has the 0-form $F_{A_1...A_4}$; only A_3 has been replaced by one-forms.

Finally, we note that the geometric approach described here has been extended, and used in the case of other supergravities.

Important concepts to remember

- The nontrivial backgrounds for supergravities with maximal susy are $AdS_4 \times S^7$, $AdS_7 \times S^4$, and $AdS_5 \times S^5$.
- In seven dimensions, we have self-duality in odd dimensions for $S_{\alpha\beta\gamma,A}$. In order to obtain it from KK reduction, we need a first-order formulation in 11 dimensions.
- The Freund–Rubin ansatz for spontaneous compactification is a constant antisymmetric tensor field strength. It gives spaces of $AdS \times S$ type.
- The truncation to the zero modes (KK reduction) is a priori inconsistent at the nonlinear level, that is, it could not satisfy the D-dimensional equations of motion.
- On a torus, or if we have some global symmetry group G, and keep ALL the singlets under the symmetry, the linear KK reduction is consistent.
- Sometimes, a nonlinear redefinition of fields, or equivalently a nonlinear KK reduction ansatz from the beginning, will turn make a reduction ansatz consistent.
- All fields of the same spin are grouped together under KK reduction. The counting should work, but all the fields of same spin appear in all the components giving such fields.
- In seven dimensions, we rotate the action with two derivatives plus an auxiliary field action (no derivatives), into two actions with one derivative, decomposing $\Box - m^2$ into $\epsilon\partial + m$ and $\epsilon\partial - m$.
- All the spherical harmonics on S^4 are built from the Killing spinors η^I.
- In the nonlinear ansatz for the fermions, we have a matrix $U^{I'}{}_I$ that relates $SO(5)_c$ to $SO(5)_g$ indices and appears also in the ansatz for the compact vielbein E^m_μ.
- The ungauged supergravities have a global symmetry, a local composite symmetry, and vectors in a global symmetry that is a subgroup of the larger global symmetry.
- Gauging corresponds to making local the global symmetry of the vectors, and these vectors becoming nonabelian.
- Massive-type IIA supergravity is a deformation with a mass parameter M of the 10-dimensional theory obtained by circle reduction of the 11-dimensional supergravity.
- Type IIB supergravity is a 10-dimensional theory with two chiral fermions of the same chirality and with an $SL(2, \mathbb{R})$ duality symmetry.
- Gauging can be for any global symmetry group, a subgroup of the total, a non-compact one, or even non-semisimple.
- One can modify the 11-dimensional and 10-dimensional supergravity theories (or, in general, in any dimension), by changing the "local Lorentz" symmetry to a generalized one. If one has an exceptional group (and also in general, by extension), one says one has an "exceptional field theory."
- One can write a modification of supergravities, in particular of 11-dimensional one, in the geometric approach, that is written in terms of forms (so without need for a metric), except for the presence of a 0-form $F_{A_1...A_4}$ (for a first-order formulation).

References and further reading

For more details on the nonlinear dimensional reduction on $AdS_7 \times S^4$, see [20]. For the (almost complete) nonlinear dimensional reduction on $AdS_4 \times S^7$, see [27] and [28]. For the maximal $d = 5$ gauged supergravity, see [29]. For the $\mathcal{N} = 8$ supergravity, see [30]. For the reduction of the $\mathcal{N} = 1$ supergravity in 11 dimensions with $SO(1,3) \times SU(8)$ "local Lorentz" invariance defined in [31] down to $\mathcal{N} = 8$ gauged supergravity in four dimensions, see [32]. For the generalization of this construction to "exceptional field theories" with E_6, E_7, E_8 covariance (of "local Lorentz" type), see [33, 34, 35, 36] and for the use of this construction in 10-dimensional type IIB supergravity for the nonlinear reduction on $AdS_5 \times S^5$, see [37, 38]. The $ISO(7)$ $\mathcal{N} = 8$ supergravity in $d = 4$ gauging, its nonlinear ansatz for consistent truncation of massive type IIA in $d = 10$ on $AdS_4 \times S^6$ (as well as gravity dual pair solutions) were found in [39], following [40]. Reviews on gauged supergravities are found in [41, 42]. Massive type II A supergravity in 10 dimensions was found by Romans in [43], where one can find more details about the theory. The original geometric 11-dimensional supergravity is in [44], and a review of the geometric approach is in [45, 46, 47].

Exercises

(1) Fierzing $\eta^{[K}\bar{\eta}^{J]}\eta^I$, prove that

$$\gamma_5 \eta^I \phi_5^{JK} - \gamma_\mu \gamma_5 \eta^I C_\mu^{JK} = 4\eta^{[K}\Omega^{J]I} - \eta^I \Omega^{JK}, \tag{16.153}$$

where $\phi_5^{JK} = \bar{\eta}^J \gamma_5 \eta^K$, $C_\mu^{JK} = \bar{\eta}^J \gamma_\mu \gamma_5 \eta^K$ and using that in four Euclidean dimensions, C is antisymmetric, $C\gamma_\mu$ is symmetric, and $C\gamma_5$ is antisymmetric.

(2) Show that in three dimensions, the $U(1)$ invariant actions

$$\mathcal{L} = -\frac{1}{2}m^2 A_\mu A^\mu + \frac{1}{2}m\epsilon^{\mu\nu\rho}A_\mu \partial_\nu A_\rho \tag{16.154}$$

("self-dual in odd dimensions") and ($F_{\mu\nu} = \partial_{[\mu}A_{\nu]}$)

$$\mathcal{L} = -\frac{1}{4}F_{\mu\nu}F^{\mu\nu} - \frac{1}{2}m\epsilon^{\mu\nu\rho}A_\mu \partial_\nu A_\rho \tag{16.155}$$

("topologically massive") are equivalent, by writing a "master action," by defining $F^\mu = \epsilon^{\mu\nu\rho}\partial_\nu A_\rho$ and writing a first-order action for an independent field f^μ with equation of motion $f^\mu = F^\mu$.

(3) Prove that the terms proportional to B_{MNPQ} in the susy variation of the 11-dimensional first-order supergravity action, δS, cancel.

(4) At $B_\alpha^{AB} = 0$, $\phi = 0$, $\lambda = 0$, $\Pi_i^A = \delta_i^A$, the $A_{\Lambda\Pi\Sigma}$ ansatz reduces to

$$A_{\mu\nu\rho} = -\frac{1}{2\sqrt{2}}\frac{D_\sigma^{(0)}}{\Box^{(0)}}(\epsilon_{\mu\nu\rho\sigma}\sqrt{g^{(0)}}) \tag{16.156}$$

(background) and

$$A_{\alpha\beta\gamma} = -\frac{i\sqrt{6}}{6}S_{\alpha\beta\gamma,A}Y^A, \tag{16.157}$$

and the ansatz for B is

$$\frac{B_{\alpha\beta\gamma\delta}}{\sqrt{E}} = -24\sqrt{3}i\nabla_{[\alpha}S_{\beta\gamma\delta],A}Y^A. \tag{16.158}$$

Substitute in the 11-dimensional action to find the 7-dimensional quadratic action for S,

$$e^{-1}\mathcal{L} = \frac{1}{2}S_{\alpha\beta\gamma,A}S^{\alpha\beta\gamma}{}_{,B}\delta^{AB} + \frac{1}{48}me^{-1}\epsilon^{\alpha\beta\gamma\delta\epsilon\eta\zeta}\delta^{AB}S_{\alpha\beta\gamma,A}F_{\delta\epsilon\eta\zeta,B}. \tag{16.159}$$

17

(Abelian and nonabelian) T-dualities and other solution-generating techniques: TsT, $O(d, d)$, and null Melvin twist

In this chapter, we will study T-duality, of the Abelian and nonabelian kind, as well as other solution-generating techniques based on it: the TsT and $O(d, d)$ transformations that are also T-duality generalizations, and the null Melvin twist that is an algorithmic procedure based on Abelian T-dualities.

17.1 Abelian T-duality

Abelian T-duality is a perturbative duality symmetry of string theory: it was proven to be valid at all orders in perturbation theory (though not proven non-perturbatively). Moreover, it is a symmetry defined on the worldsheet of the string, acting on the worldsheet fields. The spacetime fields, the metric $G_{\mu\nu}(X)$, the Kalb–Ramond B-field $B_{\mu\nu}(X)$, and the dilaton $\phi(X)$ are functions of $X(\sigma, \tau)$, the worldsheet scalars defining the position of the string. Then there is also a dependence on the Ramond–Ramond (RR) antisymmetric tensor ($n = p+1$-form) fields $C^{(n)}$, though they are harder to describe from the point of view of the string worldsheet. All in all then, the T-duality symmetry acts on the fields of 10-dimensional supergravity, which is the low energy ($\alpha' \to 0$) limit of string theory. Indeed, by self-consistency, strings must move in backgrounds of low energy strings, that is, supergravity backgrounds.

Therefore T-duality is a solution-generating technique for supergravity: it takes us between a supergravity solution and another.

In its simplest form, T-duality inverts the radius of a compact dimension, $R \leftrightarrow \alpha'/R$, which can be understood (since the metric in the compact dimension 0 is $g_{00} = R^2/\alpha'$) as inverting the metric, $g_{00} \leftrightarrow 1/g_{00}$, as well as exchanging the quantized string momenta p in units of the radius with the string winding w around the compact circle, in units of the radius, that is,

$$p \cdot R = n \leftrightarrow m = \frac{w}{R}. \tag{17.1}$$

In general, it is an action on the supergravity fields found by Buscher, so known as the "Buscher rules."

The string action in Polyakov, or first-order, form is, in Euclidean space (an Euclidean signature on the worldsheet)

$$S_P = \frac{1}{4\pi\alpha'} \int d^2\sigma \left[\sqrt{\gamma}\gamma^{ab} g_{\mu\nu}(X)\partial_a X^\mu \partial_b X^\nu + \epsilon^{ab} B_{\mu\nu}(X)\partial_a X^\mu \partial_b X^\nu + \alpha'\sqrt{\gamma}\mathcal{R}^{(2)}\Phi(x) \right], \tag{17.2}$$

where γ^{ab} is the intrinsic (independent) metric on the worldsheet and $\mathcal{R}^{(2)}$ is the two-dimensional Ricci scalar. The dilaton term S_ϕ, in the case of constant dilaton $\Phi = \Phi_0$, becomes an integer, since in two dimensions we have the relation

$$\frac{1}{4\pi} \int \sqrt{\gamma} \mathcal{R}^{(2)} = \chi = 2(1 - g), \qquad (17.3)$$

where χ is a topological number called the Euler character, and g is the genus of the two-dimensional surface ($g = 0$ for a sphere, $g = 1$ for a torus, etc.), so

$$S_{\Phi_0} = 2\Phi_0(1 - g). \qquad (17.4)$$

To understand T-duality, we start with a simpler case (as an exercise), with $B_{\mu\nu} = 0$ and $\Phi = \Phi_0$, and consider an isometric coordinate denoted by X^0. We want to isolate the terms in the action written in terms of it and do T-duality on them. Then,

$$S_P = +\frac{1}{4\pi\alpha'} \int d^2\sigma \sqrt{\gamma} \gamma^{ab} \left[g_{00} \partial_a X^0 \partial_b X^0 + 2g_{0i} \partial_a X^0 \partial_b X^i + g_{ij} \partial_a X^i \partial_b X^j \right]. \qquad (17.5)$$

A general duality transformation procedure in the path integral is defined as follows. Find a master path integral, corresponding to a master action, in terms of some variables plus a Lagrange multiplier, such that, if we eliminate the Lagrange multiplier, we go back to the original action. But if, instead, we eliminate the original variable, we obtain a dual action, in terms of the Lagrange multiplier, now taking the role of dynamical variable.

In our case, we find the master integral

$$S_{\text{master}} = \frac{1}{4\pi\alpha'} \int d^2\sigma \left\{ \sqrt{\gamma} \gamma^{ab} \left[g_{00} V_a V_b + 2g_{0i} V_a \partial_b X^i + g_{ij} \partial_a X^i \partial_b X^j \right] + 2\epsilon^{ab} \hat{X}^0 \partial_a V_b \right\}. \qquad (17.6)$$

Here V_a is a new variable, and we have imposed on it the constraint $\epsilon^{ab} \partial_a V_b = 0$, with the Lagrange multiplier $\hat{X}^0$. If we thus solve for $\hat{X}^0$, the constraint means that $V_a = \partial_a X^0$, and we are back to the previous S_P.

But if instead, we solve for V_a, obtaining

$$V_a = \frac{\hat{g}_{00}}{\sqrt{\gamma}} \epsilon^{bc} \gamma_{ca} \partial_b \hat{X}^0 - g_{0i} \partial_a \hat{X}^i, \qquad (17.7)$$

where $\hat{X}^i = X^i$ and $\hat{g}_{00} = 1/g_{00}$, and replace V_a in the master action, we obtain the dual action,

$$S_{\text{dual}} = \frac{1}{4\pi\alpha'} \int d^2\sigma \left[\sqrt{\gamma} \gamma^{ab} \hat{g}_{\mu\nu} \partial_a \hat{X}^\mu \partial_b \hat{X}^\nu + \epsilon^{ab} \hat{B}_{\mu\nu} \partial_a \hat{X}^\mu \partial_b \hat{X}^\nu \right], \qquad (17.8)$$

where $\hat{g}_{\mu\nu}, \hat{B}_{\mu\nu}$ are the dual metric and B-field,

$$\hat{g}_{ij} = g_{ij} - \frac{g_{0i} g_{0j}}{g_{00}}, \hat{B}_{0i} = \frac{g_{0i}}{g_{00}}. \qquad (17.9)$$

The relation between X^0 and $\hat{X}^0$ is obtained by equating the two formulas for V_a,

$$\partial_a X^0 = \frac{\hat{g}_{00}}{\sqrt{\gamma}} \epsilon^{bc} \gamma_{ca} \partial_b \hat{X}^0 - g_{0i} \partial_a \hat{X}^i$$

$$= \frac{\hat{g}_{00}}{\sqrt{\gamma}} \epsilon_{ba} \partial^b \hat{X}^0 - g_{0i} \partial_a \hat{X}^i. \qquad (17.10)$$

This is nothing but a generalization of Poincaré (or Hodge, in mathematical language) duality, or also Maxwell duality in two dimensions, $\tilde{F} = *F$, namely of

$$\partial_a X^0 = \frac{1}{\sqrt{\gamma}} \epsilon_{ba} \partial^b \hat{X}^0. \tag{17.11}$$

For a general background (with nontrivial B field and dilaton as well), the Buscher T-duality rules become

$$\hat{g}_{00} = \frac{1}{g_{00}}, \hat{g}_{0i} = \frac{B_{0i}}{g_{00}}, \quad \hat{g}_{ij} = g_{ij} - \frac{g_{0i}g_{0j} - B_{0i}B_{0j}}{g_{00}}$$

$$\hat{B}_{0i} = \frac{g_{0i}}{g_{00}}, \quad \hat{B}_{ij} = B_{ij} + \frac{g_{0i}B_{0j} - B_{0i}g_{0j}}{g_{00}}. \tag{17.12}$$

To these transformations, one must add a transformation of the dilaton, found by Buscher, that comes from a one-loop quantum calculation on the worldsheet, giving the one-loop determinant rewritten as a shift of the dilaton,

$$\hat{\Phi} = \Phi - \frac{1}{2} \log g_{00}. \tag{17.13}$$

These rules can be extended also to the RR antisymmetric tensor fields. Note that T-duality takes us between type IIA supergravity (or string theory) and type IIB one, and vice versa. Type IIA has odd form fields, and type IIB has even form fields.

Then the rules to get from type IIB to type IIA are as follows. The type IIA form fields are

$$C^{(2n+1)}_{\mu_1...\mu_{2n+1}} = C^{(2n+2)}_{\mu_1...\mu_{2n+1}0} + (2n+1)B_{[\mu_1|0|}C^{(2n)}_{\mu_2...\mu_{2n+1}]}$$

$$+ 2n(n+1)\frac{1}{g_{00}}B_{[\mu_1|0|}g_{\mu_2|0|}C^{(2n)}_{\mu_3...\mu_{2n+1}]0}$$

$$C^{(2n+1)}_{\mu_1...\mu_{2n}0} = C^{(2n)}_{\mu_1...\mu_{2n}} - 2n\frac{1}{g_{00}}g_{[\mu_1|0|}C^{(2n)}_{\mu_2...\mu_{2n}]0}. \tag{17.14}$$

17.2 Nonabelian T-duality

Nonabelian T-duality is a transformation on compact space, invariant under the action of a general nonabelian group G. In some sense, we perform simultaneously a T-duality in all the directions of the group. In general, this transformation is not proven to be a duality symmetry of string theory: there can, in principle, be nontrivial topological issues on the string worldsheet (for the topology of the group G versus the one of the worldsheet). Nevertheless, it is also used as a solution-generating technique. Indeed, after we find the solution, it could be easier (for instance, on a computer) to check whether it is a solution of supergravity.

One thing to stress is that, unlike the Abelian case, nonabelian T-duality is uni-directional, so it is not reversible.

So the idea is defined for any group G, but in practice, we only use it for $G = SU(2)$. The reason is that $SU(2)$ has three generators, which corresponds to three compact space directions. Higher nonabelian groups would be as follows. In the SU series, we would have $SU(3)$, with 8 generators, corresponding to 8 compact coordinates, which would be only relevant for 11-dimensional supergravity compactified to 3 dimensions, a rather special case. Since $SU(2) \simeq SO(3)$, next in the SO series, we have $SO(4)$, but we have that $SO(4) \simeq SO(3) \times SO(3)$, so this case can be understood by applying the $SU(2)$ case twice. Next is $SO(5)$, which has $5 \times 4/2 = 10$ generators, so it is not relevant.

Consider the left-invariant one-forms L^i on the group manifold, satisfying

$$dL^i = \frac{1}{2}f^i{}_{jk}L^j \wedge L^k. \tag{17.15}$$

In the case of $G = SU(2)$, the left-invariant forms correspond to Eq. (17.15) for the case $f_{ijk} = \epsilon_{ijk}$.

In terms of these one-forms, invariance of the metric and B-field means that we can always write them as

$$ds^2 = G_{\mu\nu}dX^\mu dX^\nu + 2G_{\mu i}dX^\mu L^i + g_{ij}L^iL^j,$$

$$B = B_{\mu\nu}dX^\mu \wedge dX^\nu + B_{\mu i}dX^\mu \wedge L^i + \frac{1}{2}b_{ij}L^i \wedge L^j, \tag{17.16}$$

and we also have $\Phi = \Phi(X)$, where $i, j = 1, 2, 3$ for the three $SU(2)$ invariant coordinates and $\mu, \nu = 1, ..., 7$ for the transverse ones.

In terms of the usual Euler angles for $SU(2)$ θ, ψ, ϕ, the L^i's are given by

$$L_1 = \frac{1}{\sqrt{2}}(-\sin\psi d\theta + \cos\psi \sin\theta d\phi),$$

$$L_2 = \frac{1}{\sqrt{2}}(\cos\psi d\theta + \sin\psi \sin\theta d\phi),$$

$$L_3 = \frac{1}{\sqrt{2}}(d\psi + \cos\theta d\phi), \tag{17.17}$$

where $\theta \in [0, \pi], \phi \in [0, 2\pi], \psi \in [0, 4\pi]$.

Consider normalized $SU(2)$ generators t_i, $\text{Tr}[t_it_j] = \delta_{ij}$, so

$$t_i = \frac{\tau_i}{\sqrt{2}} \tag{17.18}$$

in terms of the Pauli matrices τ_i, and group elements g,

$$g = e^{i\phi\tau_3/2}e^{i\theta\tau_2/2}e^{i\psi\tau_3/2}. \tag{17.19}$$

Further, consider light-cone coordinates on the worldsheet, $x^{\pm} = (\sigma \pm \tau)/\sqrt{2}$ and, since the differentials $(d\theta, d\psi, d\phi)$ are on the worldsheet, d refers to $\partial_{\pm}$. Then, the left-invariant form components are

$$L^i_{\pm} = -i\,\text{Tr}[t^ig^{-1}\partial_{\pm}g] \in \mathbb{R}, \tag{17.20}$$

since by $-i\,\text{Tr}[t^i...]$ we select the ith element of the Lie algebra, which is obtained from $g^{-1}dg$.

Define also the objects (natural, since $G_{\mu\nu}$ and $B_{\mu\nu}$ are obtained in string theory as the symmetric traceless and the antisymmetric parts of two string oscillators acting on a vacuum, $a^{\dagger}_{\mu}a^{\dagger}_{\nu}|0\rangle$)

$$Q_{\mu\nu} \equiv G_{\mu\nu} + B_{\mu\nu}, Q_{\mu i} \equiv G_{\mu i} + B_{\mu i},$$

$$Q_{i\mu} = G_{i\mu} + B_{i\mu}, E_{ij} = g_{ij} + b_{ij}. \tag{17.21}$$

Then, the string Polyakov action in the global $SU(2)$-invariant form is

$$S_P = \frac{1}{4\pi\alpha'} \int d^2\sigma \left[Q_{\mu\nu}\partial_+X^\mu \partial_-X^\nu + Q_{\mu i}\partial_+X^\mu L_-^i + Q_{i\mu}L_+^i \partial_-X^\mu + E_{ij}L_+^i L_-^j \right]. \quad (17.22)$$

To obtain the nonabelian T-duality transformation on L^i, we first gauge the global G invariance to a local one, by introducing a gauge field A. This means that in the action, we replace derivatives with covariant derivatives, so

$$\partial_\pm g \rightarrow D_\pm g = \partial_\pm g - A_\pm g \Rightarrow$$
$$L_i \rightarrow \tilde{L}_i = -i\,\mathrm{Tr}\left[t^i g^{-1} D_\pm g \right]. \quad (17.23)$$

But then, in order not to change the number of degrees of freedom, we must impose triviality of the gauge field, by imposing $\Gamma_{ab} = 0$ as a constraint, with a Lagrange multiplier v, so Lagrange multiplier term

$$-i\,\mathrm{Tr}[vF_{+-}] = -i\epsilon^{ab}\,\mathrm{Tr}[vF_{ab}], \quad (17.24)$$

so a natural generalization of the Lagrange multiplier term in the Abelian case, $\hat{X}^0 \epsilon^{ab} \partial_a V_b$. Here v is in the adjoint of G and, as usual,

$$F_{+-} = \partial_+A_- - \partial_-A_+ - [A_+, A_-]. \quad (17.25)$$

If we integrate out (solve for) the Lagrange multiplier first, $A_\pm$ is trivial, so is (gauge) equivalent to zero, meaning we can put $A_\pm = 0$, and we obtain the original action.

If instead we integrate out $A_\pm$ first, we must obtain the dual action. However, we have an issue of number of degrees of freedom: we seem to have too many, since we are left both with g and with the Lagrange multipliers v^i. The solution is that we need to fix a gauge for the local symmetry, thus getting rid of the degrees of freedom in g.

The simplest gauge is to set $g = 1$, in which case we obtain

$$L_\pm^i = i\,\mathrm{Tr}[t^i A_\pm] = iA_\pm^i. \quad (17.26)$$

We partially integrate the Lagrange multiplier term to

$$-i \int d^2\sigma\,\mathrm{Tr}[vF_{+-}] = \int d^2\sigma \left\{ \mathrm{Tr}\left[+i(\partial_+v)A_- - i(\partial_-v)A_+ \right] - iA_+fA_- \right\}, \quad (17.27)$$

where

$$A_+fA_- \equiv A_+^i f_{ij} A_-^j, \quad f_{ij} \equiv f_{ij}{}^k v_k. \quad (17.28)$$

Then, with the previously discussed Lagrange multiplier term, and replacing $L^i = iA^i$ in the action, we obtain

$$S = \frac{1}{4\pi\alpha'} \int d^2\sigma \left[Q_{\mu\nu}\partial_+X^\mu \partial_-X^\nu + Q_{\mu i}\partial_+X^\mu(+iA_-^i) + Q_{i\mu}\partial_-X^\mu(+iA_+^i) \right.$$
$$\left. +E_{ij}(iA_+^i)(iA_-^j) + i\partial_+v^i A_-^i - i\partial_-v^i A_+^i - A_+^i f_{ij} A_-^j \right]. \quad (17.29)$$

Varying it with respect to A_+^i and A_-^j, we obtain

$$f_{ij}A_-^j = -i\partial_-v_i - E_{ij}A_-^j + iQ_{\mu i}\partial_-X^\mu,$$
$$A_-^i f_{ij} = +i\partial_+v_j - E_{ij}A_+^i + iQ_{\mu j}\partial_+X^\mu, \quad (17.30)$$

respectively, and defining

$$M_{ij} \equiv E_{ij} + f_{ij}, \tag{17.31}$$

we solve for A_+^i, A_-^i as

$$A_-^i = -iM_{ij}^{-1}(\partial_- v_j - Q_{j\mu}\partial_- X^\mu),$$
$$A_+^i = +iM_{ij}^{-1}(\partial_+ v_j + Q_{\mu j}\partial_+ X^\mu). \tag{17.32}$$

Replacing in the action, we find the dual action

$$S_{\text{dual}} = \frac{1}{4\pi\alpha'} \int d^2\sigma \left[Q_{\mu\nu}\partial_+ X^\mu \partial_- X^\nu + (\partial_+ v_i + Q_{\mu i}\partial_+ X^\mu)M_{ij}^{-1}(\partial_- v_j - Q_{j\mu}\partial_- X^\mu) \right]. \tag{17.33}$$

Also, in the same way as in the Abelian case, one can find the one-loop determinant on the worldsheet that modifies the dilaton as

$$\Phi(x, v) = \Phi(x) - \frac{1}{2}\log(\det M). \tag{17.34}$$

Note that for the $g = 1$ gauge fixing, used in Eqs. (17.26) to (17.29) (corresponding to $\theta = \psi = \phi = 0$), we have

$$(M^{(-1)})^{ij} = \frac{1}{\det M}[(\det g)g^{ij} + y^i y^j - \epsilon^{ijk}g_{kl}y^l], \tag{17.35}$$

where

$$b_{ij} \equiv \epsilon_{ijk}b_k, \quad y_i = b_i + v_i. \tag{17.36}$$

In other gauges, we define

$$D^{ij} = \frac{1}{2}\text{Tr}[\tau^i g \tau^j g^{-1}], \quad \hat{v}^i = D_{ji}v^j, \tag{17.37}$$

and then replace v_i with $\hat{v}^i$ everywhere.

Finally, we must define the action on RR fields. For that, we must first find what happens to the vielbeins, not just the metric. Consider the vielbeins

$$e^A = e_\mu^A dx^\mu, \quad A = 1, ..., 7,$$
$$e^a = e_j^a L^j + e_\mu^a dx^\mu, \quad i = 1, 2, 3, \tag{17.38}$$

in terms of which we have the metric

$$ds^2 = \eta_{AB}e^A e^B + e^a e^a, \tag{17.39}$$

or, equivalently

$$G_{\mu\nu} = \eta_{AB}e^A e^B + K_{\mu\nu}, K_{\mu\nu} = e_\mu^a e_\nu^b \delta_{ab},$$
$$g_{ij} = e_i^a e_j^a, G_{\mu i} = e_i^a e_\mu^a. \tag{17.40}$$

Then, the nonabelian T-dualization acts differently on the left $(-)$ and right $(+)$ on the worldsheet, which means that the left- and right-vielbeins are related by a fixed Lorentz transformation (this is fine, since the metric is invariant under the local Lorentz transformation of the vielbeins),

$$\hat{e}_+^a = \Lambda^a{}_b \hat{e}_-^b. \tag{17.41}$$

Note that $\pm$ refers now to the components of the differential d for the one-forms.

This Lorentz transformation in the vector representation has a corresponding matrix Ω in the spinor representation, meaning that when acting with $\Lambda^a{}_b$ on gamma matrices, that is equivalent to an action with Ω on their spinor space,

$$\Omega^T \Gamma^a \Omega = \Lambda^a{}_b \Gamma^b. \tag{17.42}$$

Define now the RR sector (antisymmetric fields) of supergravity in the "democratic formalism," meaning we not only consider the standard $C^{(n)}$'s but also their Hodge duals. Then, define the polyforms, separately for the type IIA and type IIB cases,

$$IIB: \quad P = \frac{e^\phi}{2} \sum_{n=0}^{4} \slashed{F}_{2n+1}, \quad IIA: \quad \hat{P} = \frac{e^{\hat{\phi}}}{2} \sum_{n=0}^{5} \slashed{\hat{F}}_{2n}, \tag{17.43}$$

where, as usual, slash refers to contraction with gamma matrices, so

$$\slashed{F}_n \equiv F_{\mu_1 \dots \mu_n} \Gamma^{\mu_1} \dots \Gamma^{\mu_n}. \tag{17.44}$$

Then, the transformation of the RR fields is defined by the action of Ω on the polyforms,

$$\hat{P} = P \cdot \Omega^{-1}. \tag{17.45}$$

17.3 TsT transformation

Yet another solution-generating technique that is based on a generalization of T-duality is the TsT transformation. The name stands for T-duality (T), shift (s), T-duality (T).

For the case of a *simple TsT transformation*, we consider an isometry direction ϕ_1 and another direction ϕ_2. We do first a T-duality on ϕ_1, then a shift on ϕ_2 proportional to ϕ_1, so $\phi_2 \to \phi_2 + \gamma\phi_1$, then a T-duality back on ϕ_1.

After the T-duality, the shift takes us from

$$ds^2 = \hat{g}_{11}d\phi_1^2 + \hat{g}_{22}d\phi_2^2 + 2\hat{g}_{12}d\phi_1 d\phi_2 + 2\hat{g}_{23}d\phi_2 d\phi_3 + 2\hat{g}_{24}d\phi_2 d\phi_4 + \dots \tag{17.46}$$

to

$$ds^2 = \hat{g}_{11}d\phi_1^2 + \hat{g}_{22}(d\phi_2 + \gamma d\phi_1)^2 + 2\hat{g}_{12}d\phi_1(d\phi_2 + \gamma d\phi_1)$$
$$+ 2\hat{g}_{23}(d\phi_2 + \gamma d\phi_1)d\phi_3 + \dots, \tag{17.47}$$

that is, we have the transformation for the metric

$$\hat{G}_{11} = \hat{g}_{11} + \gamma^2 \hat{g}_{22} + 2\gamma \hat{g}_{12},$$
$$\hat{G}_{1i} = \hat{g}_{1i} + \gamma \hat{g}_{2i}, \; \forall i \neq 1, \tag{17.48}$$

and similar relations for other fields (if there are any).

A *general TsT transformation* is found by first defining, as before $E_{ij} = g_{ij} + B_{ij}$, and then transforming the metric and B-field via the matrix relation

$$(E_{ij}^{-1} + \Theta_{ij})^{-1} = E'_{ij}, \tag{17.49}$$

or, explicitly,

$$((g+B)^{-1} + \Theta)^{-1} = g' + B', \tag{17.50}$$

together with the usual T-dual invariant dilaton relation,

$$e^{-2\Phi}\sqrt{-g} = e^{-2\Phi'}\sqrt{-g'}. \tag{17.51}$$

Here Θ_{ij} is a matrix deformation parameter. If

$$\Theta^{mn} = -2\eta r^{ij} v_i^m v_j^n, \tag{17.52}$$

where η is a scalar deformation parameter, v_i^m are Killing vectors, and $r^{ij} = -r^{ji}$ are constants, then the supergravity equations of motion are special, and they are expected to reduce to the so-called "Classical Yang–Baxter Equations," CYBE, that will be discussed in Chapter 22.

If moreover r_{ij} has a *single* nonzero component and v_i^m are commuting Killing vectors, so we have an Abelian r-matrix, then the TsT transformation goes back to being the simple TsT transformation from before.

As an aside, note that at $B = 0$, the relation

$$[g^{-1\,ij} + \Theta^{ij}]^{-1} = g'_{ij} + B'_{ij} \tag{17.53}$$

reduces to the relation between open and closed string variables in the case of noncommutative geometry studied by Seiberg and Witten.

17.3.1 Example

Consider, as an example of the simple TsT transformation, the case of the $AdS_3 \times S^3 \times T^4$ solution with B field,

$$ds^2 = e^{2\rho}(-dt^2 + dx^2) + d\rho^2 + \frac{1}{4}(\sigma_1^2 + \sigma_2^2 + \sigma_3^2) + ds^2(T^4),$$

$$H = -2e^{-2\rho}dt \wedge dx \wedge d\rho + \frac{1}{4}\sigma_1 \wedge \sigma_2 \wedge \sigma_3, \tag{17.54}$$

and a constant dilaton ϕ_0, so that $B_{01}(\rho) = -2e^{-2\rho}$ in the simplest gauge ($B_{\rho 0} = B_{\rho 1} = 0$).

After a T-duality in time t, we get

$$ds^2 = -e^{-2\rho}dt^2 + 2dtdx + d\rho^2 + \frac{1}{4}(\sigma_1^2 + \sigma_2^2 + \sigma_3^2) + ds^2(T^4),$$

$$\Phi = \Phi_0 - \frac{1}{2}\log e^{2\rho},$$

$$B_{01} = 0. \tag{17.55}$$

After the shift $x \to x + \gamma t$, we find

$$ds^2 = dt^2(-e^{-2\rho} + 2\gamma) + 2dtdx + d\rho^2 + \frac{1}{4}(\sigma_1^2 + \sigma_2^2 + \sigma_3^2) + ds^2(T^4),$$

$$\Phi = \Phi_0 - \frac{1}{2}\log e^{2\rho},$$

$$B_{01} = 0. \tag{17.56}$$

And after the T-duality back on t, we find

$$ds^2 = \frac{e^{2\rho}(-dt^2 + dx^2)}{1 + 2\gamma e^{2\rho}} + d\rho^2 + \frac{1}{4}(\sigma_1^2 + \sigma_2^2 + \sigma_3^2) + ds^2(T^4),$$

$$\Phi - \Phi_0 - \frac{1}{2}\log e^{2\rho} - \frac{1}{2}\log(e^{-2\rho} - 2\gamma) \Rightarrow e^{2\Phi} = e^{2\Phi_0}\frac{e^{2-\rho}}{e^{-2\rho} - 2\gamma},$$

$$B_{01} = -\frac{e^{2\rho}}{1 + 2\gamma e^{2\rho}} \Rightarrow H_{01\rho} = -\frac{2e^{2\rho}}{(1 + 2\gamma e^{2\rho})^2}. \tag{17.57}$$

17.4 $O(d, d)$ transformation

These are generalizations of the TsT transformations so that TsT transformations are a subset of them.

$O(d, d)$ transformations are transformations on a *doubled* set of variables, that is, a $2d \times 2d$ matrix. Consider a supergravity background with Killing vectors

$$v_i = v_i^m \partial_m, \tag{17.58}$$

and g_{mn}, B_{mn}, and Φ such that the Lie derivatives along v_i vanish,

$$\mathcal{L}_{v_i} g_{mn} = \mathcal{L}_{v_i} B_{mn} = \mathcal{L}_{v_i} \Phi = 0. \tag{17.59}$$

But, since B_{mn} has a gauge invariance such that $H = dB$ is invariant, we can actually relax the condition $\mathcal{L}_{v_i} B_{mn} = 0$ and only require that

$$\mathcal{L}_{v_i} B = -d\tilde{v}_i, \tag{17.60}$$

(so that $\mathcal{L}_{v_i} H = 0$), which defines the one-forms $\tilde{v}_i$.

Then, define the generalized ($2d$-dimensional) Killing vectors

$$V_i^M = (v_i^m, \tilde{v}_{im}), \tag{17.61}$$

such that they form an algebra, just like v_i themselves, now through the Courant bracket,

$$[V_i, V_j]_C = f_{ij}{}^k V_K. \tag{17.62}$$

Define also the $O(d, d)$ invariant matrix

$$\eta_{MN} = \begin{pmatrix} 0 & \delta_m^n \\ \delta_n^m & 0 \end{pmatrix}, \tag{17.63}$$

and through it, the inner products of generalized Killing vectors,

$$\mathcal{G}_{ij} \equiv \eta_{MN} V_i^M V_j^N = v_i \cdot \tilde{v}_j + v_j \cdot \tilde{v}_i. \tag{17.64}$$

The generalized ($2d$-dimensional) metric is defined as

$$\mathcal{H}_{MN} \equiv \begin{pmatrix} g_{mn} - B_{mp}g^{pq}B_{qn} & B_{mp}g^{pn} \\ -g^{mp}B_{pn} & g^{mn} \end{pmatrix}, \tag{17.65}$$

and the usual T-duality invariant dilaton is

$$e^{-2d} \equiv e^{-2\Phi}\sqrt{-g}. \tag{17.66}$$

Further define the object

$$(T_{ij})_M{}^N \equiv V_{iM}V_j{}^N - V_{jM}V_i{}^N = -(T_{ji})_M{}^N \tag{17.67}$$

that satisfies the $O(d,d)$ algebra property

$$(T_{ij})_M{}^P \eta_{PN} + \eta_{MP}(T_{ij}^T)^P{}_N = 0, \tag{17.68}$$

meaning that the T_{ij} form a subalgebra of $O(d,d)$. From them, define the $O(d,d)$ matrix

$$h = e^{-\eta r^{ij}T_{ij}}, \tag{17.69}$$

where $r^{ij} = -r^{ji}$ are constants.

Then, $O(d,d)$ acts with it on $\mathcal{H}_{MN}$, through the transformation

$$\mathcal{H}'_{MN} = h_M{}^P \mathcal{H}_{PQ}(h^T)^Q{}_N, \quad e^{-2d'} = e^{-2d}. \tag{17.70}$$

This transformation is known to be a symmetry of string theory for constant T_{ij}. If V_i^M depend on coordinates, T_{ij} are also not constants, and then the $O(d,d)$ is a solution-generating technique that, nevertheless, like in the nonabelian T-duality case, is not guaranteed to give a solution of supergravity (though, of course, it is much easier to test whether something is a solution, for instance, on a computer, than to find a new solution to the equations of motion).

There is a subgroup of $O(d,d)$ transformations, called beta transformations, that are equivalent to Yang–Baxter deformations, in which case the r^{ij} solve CYBE. But we will talk more about this in Chapter 22.

In the special case that $\tilde{v}_i = 0$, so

$$h = \begin{pmatrix} \delta_m^n & 0 \\ \Theta^{mn} & \delta_n^m \end{pmatrix}, \quad \Theta^{mn} = -2\eta r^{ij}v_i^m v_j^n, \tag{17.71}$$

we obtain the previous case of the general TsT transformation. If moreover $[v_i, v_j] = 0$, we have the case of a simple TsT transformation.

17.5 Null Melvin twist

This is yet another solution-generating technique, that can be applied to a solution, and is composed of several steps that include Abelian T-dualities.

The basic observation for the construction of the null Melvin twist is that a "twisted rotation," to be defined shortly, is still an invariance of supergravity. Consider an S^7 at $r = \infty$, for a solution of 10-dimensional supergravity. Then the transverse coordinates are t and y.

The twist of the rotation of S^7 along y is defined as follows. Consider x_i as Cartesian coordinates for the S^7, so (r, S^7) are described by x_i, with $x_i x_i = 1$, $i = 1, ..., 8$.

The twist is defined by a y-dependent phase in each of the four independent planes in the eight Cartesian coordinates,

$$
\begin{aligned}
x_1 + ix_2 &\rightarrow e^{i\alpha y}(x_1 + ix_2), \\
x_3 + ix_4 &\rightarrow e^{i\alpha y}(x_3 + ix_4), \\
x_5 + ix_6 &\rightarrow e^{i\alpha y}(x_5 + ix_6), \\
x_7 + ix_8 &\rightarrow e^{i\alpha y}(x_7 + ix_8).
\end{aligned}
\tag{17.72}
$$

This implies that

$$
d\Omega_7^2 \rightarrow d\Omega_7^2 + \alpha\sigma\, dy + \alpha^2 dy^2,
\tag{17.73}
$$

where

$$
\frac{r^2\sigma}{2} \equiv x_1 dx_2 - x_2 dx_1 + x_3 dx_4 - x_4 dx_3 + x_5 dx_6 - x_6 dx_5 + x_7 dx_8 - x_8 dx_7.
\tag{17.74}
$$

Then the null Melvin twist is obtained by combining it with T-dualities and boosts in the following procedure:

1. Boost along y with parameter γ.
2. T-dualize along y (so y must be an isometry of the solution).
3. Twist the rotation of S^7 along y with α.
4. T-dualize back on y.
5. Boost back on y with $-\gamma$.
6. Do a double-scaling limit, taking $\alpha \rightarrow 0$, $\gamma \rightarrow \infty$, keeping

$$
\beta \equiv \frac{1}{2}\alpha e^{\gamma} = \text{fixed}.
\tag{17.75}
$$

Note that this is a *null* twist because we take the boost $\gamma \rightarrow \infty$ in the limit, leading to a null direction.

Example: Apply the procedure to flat space, with metric

$$
ds^2 = -dt^2 + dy^2 + dr^2 + r^2 d\Omega_7^2.
\tag{17.76}
$$

Then 1, 2 are trivial now.

3. After the twist, we have

$$
ds^2 = -dt^2 + dy^2(1 + \alpha^2 r^2) + \sum_{i=1}^{8} dx_i^2 + 2\alpha \left(\frac{\alpha r^2}{2}\right) dy, \quad B = 0.
\tag{17.77}
$$

4. After T-duality on y, we have

$$
\begin{aligned}
g_{yy} &= \frac{1}{1 + \alpha^2 r^2}, \quad B = \frac{\sigma \wedge dy}{1 + \alpha^2 r^2}\alpha r^2 \\
g_{11} &= 1 - \frac{\alpha^2 x_1^2}{1 + \alpha^2 r^2}, \quad g_{12} = +\frac{\alpha^2 x_1 x_2}{1 + \alpha^2 r^2}, \quad \text{and so on,}
\end{aligned}
\tag{17.78}
$$

so the solution is

$$ds^2 = -dt^2 + \frac{dy^2}{1 + \alpha^2 r^2} + \sum_{i=1}^{8} dx_i^2 - \frac{\alpha^2 (x_1 dx_1 - x_2 dx_2)^2 + \dots}{1 + \alpha^2 r^2},$$

$$B = \frac{\sigma \wedge dy}{1 + \alpha^2 r^2} \alpha r^2. \tag{17.79}$$

5. For the boost on y, we have

$$dy \to \cosh \gamma \, dy + \sinh \gamma \, dt, \quad dt \to \cosh \gamma \, dt + \sinh \gamma \, dy. \tag{17.80}$$

6. Then, taking $\alpha \to 0, \gamma \to \infty$, we write first

$$\frac{1}{1 + \alpha^2 r^2} \simeq 1 - \alpha^2 r^2, \quad dy \simeq \frac{e^\gamma}{2} (dy' + dt'). \tag{17.81}$$

Then, keeping $\beta = \frac{\alpha e^\gamma}{2}$ fixed in the limit, finally we obtain the solution

$$ds^2 = -dt'^2 + dy'^2 - \beta^2 r^2 (dt' + dy')^2 + \sum_{i=1}^{8} dx_i^2,$$

$$B = \beta r^2 \sigma \wedge (dy' + dt'), \tag{17.82}$$

which is a supersymmetric pp wave sourced by B.

Important concepts to remember

- T-duality is a solution-generating technique, taking us from a supergravity solution to another.
- Abelian T-duality exchanges $R \leftrightarrow \alpha'/R$ and $p \cdot R = n \leftrightarrow m = w/R$, and acts on the supergravity fields through the Buscher rules.
- Abelian T-duality is an example of duality in path integral, coming from a master action in which we can solve either for one field or another, and it is a generalization of Poincaré (or Hodge) duality in two dimensions (Maxwell duality in two dimensions).
- Nonabelian T-duality is a transformation for a solution invariant under a group G, written in terms of the left-invariant one-forms L^i, usually applied to $SU(2)$, and generalizing (to several directions at once) Abelian T-duality.
- Nonabelian T-duality is unidirectional (not reversible) and is defined by gauging the G ($SU(2)$) global symmetry to a local one, with a gauge field A, then imposing triviality of A via a Lagrange multiplier term for $F = 0$. For the dual theory, we solve for A and fix a gauge for the group element G, usually $g = 1$.
- For the RR field transformation under nonabelian T-duality, the polyforms in democratic formalism on the two sides are related by the spinor representation version of the local Lorentz transformation relating the left- and right-vielbeins, e_+ and e_-.
- A simple TsT transformation is a T-duality on ϕ_1, a shift of ϕ_2 by ϕ_1, then T-duality back. A general TsT transformation is defined by $(E_{ij}^{-1} + \Theta_{ij})^{-1} = E'_{ij}$, with $E_{ij} = g_{ij} + B_{ij}$, and Θ_{ij} the deformation parameter.
- For $\Theta^{mn} = -2\eta r^{ij} v_i^m v_j^n$, with v_i^m Killing vectors and r^{ij} constants, the supergravity equation reduce to CYBE. If r_{ij} has a single nonzero component and the Killing vectors commute, we are back to a simple TsT.

- An $O(d, d)$ transformation generalizes further the TsT one, by acting on a doubled space, with $\mathcal{L}_{v_i} B = -d\tilde{v}_i$ and $V_i^M = (v_i^m, \tilde{v}_{im})$, defined via the action of a certain $O(d, d)$ transformation on the metric in the $2d$-dimensional space.
- If V_i^M are constants, the supergravity equations are satisfied, and if $\tilde{v}_i = 0$, we go back to TsT transformations.
- A null Melvin twist is a combination of two T-dualities, two boosts, a twisted rotation, and a double scaling limit leading to a null direction. It is also a solution-generating technique.

References and further reading

For a quick review of T-duality, see, for instance, [48], and for a quick review of nonabelian T-duality, see, for instance, [49]. The Buscher rules were derived in [50, 51]. The transformation of RR fields under T-duality was done in [52]. See also [53]. For a quick review of TsT and $O(d, d)$ transformations, see, for instance, [54]. For the null Melvin twist, see, for instance, [55].

Exercises

(1) Calculate the Abelian T-dual of the solution with metric

$$H = H(r_3^2 dx_3^2 + r_2^2 dx_2^2) + \frac{r_1^2}{H}(dx_1 + mx_3 dx_2)^2, \qquad (17.83)$$

and trivial B-field and dilaton, first on x_1, and then on x_2.

(2) Calculate the TsT-transformed solution for the $AdS_5 \times S^5$ solution, with metric

$$R^{-2} ds^2 = z^2 \left(-dt^2 + \sum_{i=1}^{3} dx_i^2 \right) + dz^2 + ds^2(\Omega_5), \qquad (17.84)$$

in the directions t, x_1, in the same way as it was done for $AdS_3 \times S^3 \times T^4$ in the text.

(3) Calculate the nonabelian T-dual of the $AdS_3 \times S^3 \times T^4$ solution in the S^3 directions.

(4) Calculate the null Melvin twist of flat space plus a constant B field $B_{ty} = \mu$.

18

Extremal and black *p*-brane solutions of supergravity; Tseytlin's harmonic function rule

After having described solution-generating techniques for supergravity, in this chapter, we will describe how to find the most important solutions of supergravity, the extremal and black *p*-brane solutions, after which we will describe Tseytlin's harmonic function rule about how to combine them into a new solution.

18.1 Introduction

The solutions of supergravity we will consider are some extended objects in *p* spatial dimensions, "*p*-branes" and, since supergravity is the low energy ($\alpha' \to 0$) limit of string theory, they will correspond to (extended) objects in string theory. In a famous paper of Joe Polchinski in 1995, he identified the supergravity solutions with extended objects in string theory defined abstractly, called D-branes. These solutions are either electrically charged or magnetically charged.

To understand them, we start with a more profound understanding of the objects they generalize, the electrically charged solution (electron) and magnetically charged one (monopole) of electromagnetism.

The Bianchi identity,

$$dF = 0 \ \text{ or } \ \partial_{[M}F_{NP]} = 0 \ \text{ or } \ \partial^N * F_{NP} = 0, \tag{18.1}$$

implies that *F* can be written in terms of a gauge field *A*,

$$F = dA \ \text{ or } \ F_{\mu\nu} = \partial_\mu A_\nu - \partial_\nu A_\mu. \tag{18.2}$$

The equations of motion for (electrically charged) particles, with a corresponding source *J* are

$$d * F = *J \ \text{ or } \ \partial_{[M} * F_{NP]} = *J_{MNP} \ \text{ or } \ \partial^N F_{NP} = J_P, \tag{18.3}$$

where the source is

$$J^M = qu^M \delta^{d-1}(y^M - y^M(\tau)) \to \text{static case: } J^0 = q\delta^{d-1}(\vec{y}), \ (*J)_{123} = q\delta^3(\vec{y}). \tag{18.4}$$

However, because of Maxwell duality (duality present in the absence of sources, and extended to the presence of sources, if we include magnetic sources), we can consider monopoles (magnetically charged particles) with source *X*, in which case we have

$$dF = X \ \text{ or } \ \partial_{[M}F_{NP]} = X_{MNP} \ \text{ or } \ \partial^M F_{MN} = *X_N, \tag{18.5}$$

which implies that we can only write F as coming from an A with the addition of a (singular) part, that cannot be written globally like that, but only on patches:

$$X = d\omega \Rightarrow F = dA + \omega. \tag{18.6}$$

A Dirac monopole, that is, a singular solution, like the electron, has a source that is, in the static case,

$$X_{123} = g\delta^3(\vec{y}) = *X_N. \tag{18.7}$$

Using the equations of motion $dF = X$ and $d * F = *J$ and the Stokes theorem, we can calculate the electric and magnetic charges as

$$q = \int_{S^2 = \partial M_3} *F = \int_{M_3} *J$$
$$g = \int_{S^2 = \partial M_3} F = \int_{M_3} X. \tag{18.8}$$

We can also have non-singular monopoles, the 't Hooft monopoles, which are solutions to the nonabelian theory, but *without source*. At large distances, the monopoles look abelian and like Dirac's solution, but near the core they are nonabelian and nonsingular.

We want to generalize these ideas to the supergravity case. We consider D-p-brane solutions, extended in p space directions.

We will start by considering "extremal" solutions, the equivalent of solutions with mass equal charge, $M = Q$ for the four-dimensional Reissner–Nordstrom solutions. In this case, we rather have mass per unit volume (tension) equals charge per unit volume, $T_p = Q_p$.

We consider solutions in flat space background (asymptotically flat). In principle, one can consider also AdS backgrounds, though this is quite different, as we will see in a later chapter. The extremal solutions considered are supersymmetric, and we will analyze the supersymmetry better in the next chapter.

We will consider the $p = d - 1$ electric branes and the corresponding magnetically charged branes, or "generalized monopoles," both arising from an action for a A_d d-form $(d = p + 1)$ with $(d + 1)$-form field strength F_{d+1} and action

$$-\int \sqrt{-g} \frac{F_{d+1}^2}{2(d+1)}. \tag{18.9}$$

The generalization of the equations of motion are

$$d(*F)_{D-d-1} = (*J)_{D-d}, \tag{18.10}$$

where the electrically charged current for the p-brane is

$$(*J)_{1...D-d} = e_d \delta^{D-d}(\vec{y}), \tag{18.11}$$

while the Bianchi identity modified by the generalized monopole source is

$$dF_{d+1} = X_{d+2} = d\omega_{d+1}, \tag{18.12}$$

which means that we have

$$F_{d+1} = dA_d + \omega_{d+1}, \tag{18.13}$$

and the magnetically charged current is

$$X_{1\ldots d+2} = g_{D-d-2}\delta^{d+2}(\vec{y}). \tag{18.14}$$

The electric and magnetic charges are found by integration, again using the equations of motion $dF = X$ and $d * F = *J$ and the Stokes theorem, as

$$e_d = \int_{S^{D-d-1}=\partial M^{D-d}} (*F)_{D-d-1} = \int_{M^{D-d}} *J_{D-d}$$

$$g_{D-d-2} = \int_{S^{d+1}=\partial M^{d+2}} F_{d+1} = \int_{M^{d+2}} X_{d+2}, \tag{18.15}$$

and they satisfy Dirac's, or the Dirac–Schwinger–Zwanziger quantization relation

$$e_d g_{D-d-2} = 4\pi \frac{n}{2} \quad n \in \mathbb{Z}. \tag{18.16}$$

18.2 Actions and equations of motion

The action that we will solve will include a source term, which is the p-brane action, understood as a generalization of the particle action.

The action for a particle of mass m moving on a worldline is just m times the worldline length, but written in such a way as to be a function of $x^\mu(\tau)$, the worldline variables:

$$S = -m(c^2) \int d\tau = -m \int d\tau \sqrt{-\dot{X}^\mu \dot{X}_\mu}, \tag{18.17}$$

where $\dot{X}^\mu = dX^\mu/d\tau$.

This is a second-order form of the action, being nonlinear in the field variable $X^\mu(\tau)$ and not having auxiliary fields. We can go to a first-order form by introducing an independent worldline vielbein (einbein, in this case) $e(\tau)$ (thus equal to $\sqrt{\gamma(\tau)}$, with $\gamma(\tau)$ the one-dimensional metric on the worldline).

The second-order form of the action is then

$$S_P = \frac{1}{2} \int d\tau \left(e^{-1}(\tau) \frac{dX^\mu}{d\tau} \frac{dX_\mu}{d\tau} - e(\tau) m^2 \right), \tag{18.18}$$

and we note that we could have guessed the action, up to numbers, on general principles. Since $e^{-1}(\tau) = \sqrt{\gamma(\tau)} \gamma^{11}(\tau)$ and $e(\tau) = \sqrt{\gamma}$, the two terms are the actions for a scalar field and a cosmological constant in one dimension, respectively.

We next consider the gauge for worldline diffeomorphisms (general coordinate transformations) $e(\tau) = 1$. Then we couple the particle to background fields. The easiest is to couple to electromagnetism, through the standard source coupling, for the source $j^\mu(\vec{x}) = q\frac{dX^\mu}{d\tau}\delta^3(\vec{x} - \vec{X}(\tau))$, giving

$$\int d^4x A_\mu(\vec{x}) j^\mu(\vec{x}) = \int d\tau A_\mu(\vec{X}(\tau)) q \frac{dX^\mu(\tau)}{d\tau}. \tag{18.19}$$

Incidentally, note that, in order to obtain the Lorentz force from the above source term, by varying with respect to $X^\mu(\tau)$, we must vary both the explicit term in $\frac{dX^\mu}{d\tau}$ and the implicit one in $A_\mu(\vec{X}(\tau))$, obtaining the two terms in $qF_{\mu\nu}u^\nu = q(\partial_\mu A_\nu - \partial_\nu A_\mu)\frac{dX^\nu}{d\tau}$.

We now generalize this to the case of a p-brane source. The result is the p-brane action in first-order formulation,

$$S_d = T_d \int d^d \xi \left[-\frac{1}{2} \sqrt{-\gamma} \gamma^{ij} E_i^M E_j^N g_{MN}(\vec{X}(\xi)) e^{\frac{a(d)\varphi}{d}} - \frac{d-2}{2} \sqrt{-\gamma} \right.$$
$$\left. -\frac{1}{d!} \epsilon^{i_1 \dots i_d} E_{i_1}^{M_1} \dots E_{i_d}^{M_d} A_{M_1 \dots M_d} \right], \tag{18.20}$$

where T_d is the brane tension, that is, energy per unit p-volume, γ_{ij} is an independent worldvolume metric, the spacetime fields are the metric, RR $d = p + 1$-form field and the dilaton, $\{g_{MN}, A_{M_1 \dots A_d}, \phi\}$, $a(d)$ are some fixed, dimension d-dependent numbers, and E_i^M are the pullbacks from spacetime to the worldvolume,

$$E_i^M = \begin{cases} \partial_i X^M, & \text{bosonic case} \\ \partial_i X^M - \bar{\theta} \Gamma^M \partial_i \theta, & \text{supersymmetric case.} \end{cases} \tag{18.21}$$

Here we will study the bosonic case only, but in a later chapter we will also consider the supersymmetric case.

The supergravity action, for the spacetime fields $\{g_{MN}, A_{M_1 \dots A_d}, \phi\}$ is standard, except the d-form field has an action that has a dilaton factor, as it happens to be the case in supergravity,

$$I_D(d) = \frac{1}{2\kappa_N^2} \int d^D x \sqrt{-g} \left[R - \frac{1}{2} (\partial_M \varphi)^2 - \frac{1}{2(d+1)} e^{-ad(d)\varphi} F_{d+1}^2 \right]. \tag{18.22}$$

We look for two types of solutions:

- elementary, that is, electrically charged, solutions to $I_D(d) + S_d$ (supergravity plus source), analogous to the electron solution being a solution of $-\int d^4 x \frac{F_{\mu\nu}^2}{4} + \int j^\mu A_\mu$.
- solitonic, that is, magnetically charged, solutions of $I_D(d)$ only, analogous to the 't Hooft monopole (in that it needs no source). Note however that it is still for an abelian d-form field: there are no self-interaction (nonabelian) form fields, analogous to Yang-Mills. The nontrivial core is due to the interaction of gravity with the form field now, unlike the 't Hooft monopole case.

We will also consider *black p-brane* solutions, which are generalizations of the above. The extremal member of the family is the fundamental solution and is the only supersymmetric and BPS saturated one.

We start with the elementary solution, and first write the equations of motion.

The equation of motion for A_d is (considering that its kinetic term in $I_D(d)$ has also an $e^{-a(d)\varphi}$ dilaton factor in front, so this appears inside $d*$ in the equation of motion)

$$d * (e^{-a(d)\varphi} F) = 2\kappa_N^2 (-)^{d^2} * J. \tag{18.23}$$

The field A_d, in the case of the elementary solution, still satisfies the Bianchi identity $dF = 0$, so $F = dA$.

The equations of motion of the spacetime fields, from $I_D(d) + S_d$, are:

1. The Einstein's equations, $2\kappa_N^2 \frac{\delta}{\delta g_{MN}}(I_D(d) + S_d)$, giving

$$\sqrt{-g}\left\{R^{MN} - \frac{1}{2}g^{MN}R - \frac{1}{2}\left[\partial^M\varphi\partial^N\varphi - \frac{1}{2}(\partial\varphi)^2\right]\right.$$
$$\left. -\frac{1}{2}\left(F^M{}_{M_1...M_d}F^{NM_1...M_d} - \frac{1}{2(d+1)}g^{MN}F^2\right)e^{-a(d)\varphi}\right\}$$
$$= -2\kappa_N^2\frac{\delta S_d}{\delta g_{MN}} \equiv \kappa_N^2\sqrt{-g}T^{MN}(d-1-\text{brane}), \tag{18.24}$$

where

$$T_{MN} = -T_d\int d^d\xi\sqrt{-\gamma}\,\gamma^{ij}\partial_iX^M\partial_jX^N e^{\frac{a(d)\varphi}{d}}\frac{\delta^D(x-X(\xi))}{\sqrt{-g}}. \tag{18.25}$$

2. The A_d equation, $2\kappa_N^2\frac{\delta}{\delta A_{M_1...M_d}}(S_d + I_D)$, giving in components

$$\partial_N(\sqrt{-g}e^{-a(d)\varphi}F^{NM_1...M_d}) = 2\kappa_N^2 T_d\int d^d\xi\,\epsilon^{i_1...i_d}\partial_{i_1}X^{M_1}...\partial_{i_d}X^{M_d}\delta^D(x-X(\xi))$$
$$\equiv 2\kappa_N^2\sqrt{-g}J^{M_1...M_d}. \tag{18.26}$$

3. The dilaton φ equation, $2\kappa_N^2\frac{\delta}{\delta\varphi}(S_d + I_D(d))$, giving

$$\partial_M(\sqrt{-g}g^{MN}\partial_N\varphi) + \frac{a(d)}{2(d+1)}\sqrt{-g}e^{-a(d)\varphi}F_{d+1}^2$$
$$= \frac{a(d)}{d}\kappa_N^2 T_d\int d^d\xi\sqrt{-g}\,\gamma^{ij}\partial_iX^M\partial_jX^N g_{MN}(X(\xi))e^{\frac{a(d)}{d}\varphi}\delta^D(x-X(\xi)). \tag{18.27}$$

Then we have the $d-1$-brane equations, coming only from S_d, by varying with respect to its worldvolume fields, $\gamma^{ij}(\xi)$ and $X^M(\xi)$.

4. The brane equation, $\frac{\delta}{\delta X^M}S_d$, gives

$$\partial_i\left(\sqrt{-\gamma}\,\gamma^{ij}\partial_jX^N g_{MN}e^{\frac{a(d)}{d}\varphi}\right) - \frac{1}{d!}\partial_{i_1}X^{M_1}...\partial_{i_d}X^{M_d}F_{MM_1...M_d}$$
$$-\frac{1}{2}\sqrt{-\gamma}\,\gamma^{ij}\partial_iX^N\partial_jX^P\partial_M\left(g_{NP}e^{\frac{a(d)}{d}\varphi}\right) = 0. \tag{18.28}$$

5. The independent brane metric equation, or embedding equation for the induced metric, $-\frac{2}{\sqrt{-g}}\frac{\delta}{\delta\gamma^{ij}}S_d$, gives

$$\gamma_{ij} = \partial_iX^M\partial_jX^N g_{MN}e^{\frac{a(d)}{d}\varphi}. \tag{18.29}$$

For solitonic, or magnetic, brane solutions, one can consider the above equations, just that we put $T_{MN}(d-1-\text{brane}) = 0$, $J^{M_1...M_d} = 0$, the right-hand side of the φ equation to zero, and the brane equations are put to zero.

18.3 The electric *p*-brane solution

To write an ansatz for the solution, split the coordinates as $X^M = (x^\mu, y^m)$, and require Poincaré invariance ($ISO(1, d-1)$) in the d coordinates parallel to the brane, and rotational invariance $SO(D-d)$ in the transverse coordinates.

That fixes the ansatz for the metric and the antisymmetric field to be

$$ds^2 = e^{2A}\eta_{\mu\nu}dx^\mu dx^\nu + e^{2B}\delta_{mn}dy^m dy^n$$

$$A_{\mu_1\ldots\mu_d} = -\frac{1}{d_g}\epsilon_{\mu_1\ldots\mu_d}e^C, \tag{18.30}$$

where the epsilon tensor is defined by $\epsilon^{01\ldots d-1} = +1$ and

$$\epsilon_{\mu_1\ldots\mu_d} = g_{\mu_1\nu_1}\ldots g_{\mu_d\nu_d}\epsilon^{\nu_1\ldots\nu_d}, \tag{18.31}$$

so

$$A_{01\ldots d-1} = -e^C. \tag{18.32}$$

Moreover, $ISO(1, d-1)$ and $SO(d)$ invariance means that all scalar functions, A, B, C and φ must be functions of $y = \sqrt{\delta_{mn}y^m y^n}$ only (we could in principle also consider functions of $x = \sqrt{\eta_{\mu\nu}x^\mu x^\nu}$, but it is not clear what that could mean physically). Then we also have the field strength ansatz

$$F_{m\mu_1\ldots\mu_d} = -\frac{1}{d_g}\epsilon_{\mu_1\ldots\mu_d}\partial_m e^C. \tag{18.33}$$

We consider also the static gauge choice for the worldvolume diffeomorphisms (general coordinate transformations),

$$X^\mu = \xi^\mu, \quad y^m = \text{const.} \tag{18.34}$$

Define $\tilde{d} \equiv D - d - 2$. Then, after a long algebra, the Einstein's equations become:
1a'. The $\{\mu\nu\}$ components give

$$e^{(d-2)A+\tilde{d}B}\delta^{mn}\left[(d-1)\partial_m\partial_n A + \frac{d(d-1)}{2}\partial_m A\partial_n A + (\tilde{d}+1)\partial_m\partial_n B\right.$$
$$\left. +\frac{\tilde{d}(\tilde{d}+1)}{2}\partial_m B\partial_n B + \tilde{d}(\tilde{d}-1)\partial_m A\partial_n B + \frac{1}{4}e^{-2dA+2C-a(d)\varphi}\partial_m C\partial_n C + \frac{1}{4}\partial_m\varphi\partial_n\varphi\right]$$
$$= -\kappa_N^2 T_d e^{(d-2)A+\frac{a(d)}{2}\varphi}\delta^{D-d}(y). \tag{18.35}$$

1b'. The mn components give

$$e^{dA+(\tilde{d}-2)B}\left[-\tilde{d}\partial^m\partial^n B + \tilde{d}\delta^{mn}\partial^2 B + \tilde{d}\partial^m\partial^n B + \frac{\tilde{d}(\tilde{d}-1)}{2}(\partial B)^2\delta^{mn}\right.$$
$$-d\partial^m\partial^n A + d\delta^{mn}(\partial^2 A) - d\partial^m A\partial^n A + \frac{d(d+1)}{2}\delta^{mn}(\partial A)^2$$
$$+d\left(\partial^m A\partial^n B + \partial^m B\partial^n A + (\tilde{d}-1)\delta^{mn}\frac{\partial A\cdot\partial B}{2}\right) - \frac{1}{2}\partial^m\varphi\partial^n\varphi + \frac{1}{4}\delta^{mn}(\partial\varphi)^2\right]$$
$$-\frac{1}{2}e^{dA+(\tilde{d}-2)B+2C-a(d)\varphi}\left[-\partial^m C\partial^n C + \frac{1}{2}\delta^{mn}(\partial C)^2\right] = 0. \tag{18.36}$$

Observations:

– The $\mu\nu$ A_d energy–momentum tensor is

$$\frac{1}{2d!}\left[F^\mu_{\ldots}F^{\nu\ldots} - \frac{1}{2(d+1)}g^{\mu\nu}F^2\right] = \frac{1}{4}\eta^{\mu\nu}(\partial_m C)^2 e^{-2(d+1)A-2B+2C}. \tag{18.37}$$

– The mn A_d energy–momentum tensor is

$$\frac{1}{2d!}\left[F^m_{\ ...}F^{n...} - \frac{1}{2(d+1)}g^{mn}F^2\right] = \frac{1}{2}e^{-2dA-4B+2C}\left[\partial^m C \partial^n C - \frac{1}{2}g^{mn}F^2\right]. \quad (18.38)$$

The rest of the equations are as follows.

2'. The A_d equation of motion becomes

$$\partial_m\left(e^{-dA+\tilde{d}B-a(d)\varphi}\partial^m e^C\right) = 2\kappa_N^2 T_d \delta^{D-d}(\vec{y}). \quad (18.39)$$

3'. The dilaton φ equation of motion becomes

$$\delta^{mn}\partial_m(e^{dA+\tilde{d}B}\partial_n\varphi) - \frac{a(d)}{2}e^{-dA+\tilde{d}B-a(d)\varphi}(\partial_m e^C)^2 = a(d)\kappa_N^2 T_d e^{dA+\frac{a(d)}{2}\varphi}\delta^{D-d}(\vec{y}). \quad (18.40)$$

4'. The brane equation of motion becomes

$$\partial_m\left(e^{dA+\frac{a(d)}{2}\varphi} - e^C\right) = 0. \quad (18.41)$$

The solution is obtained as follows.
From 4', we find

$$e^C = K + e^{dA+\frac{a(d)}{2}\varphi}. \quad (18.42)$$

If $K = 0$, we find

$$\frac{a(d)}{2}\varphi = C - dA. \quad (18.43)$$

From a linear combination of the $\{mn\}$ components of the Einstein equation (1b'), defining

$$X \equiv dA + B\tilde{d}, \quad (18.44)$$

we find the equation

$$X'' + \frac{2\tilde{d}+1}{4}X' + X'^2 = 0, \quad (18.45)$$

solved by

$$X = C + \ln\left[1 + \frac{1}{2K^{\tilde{d}}y^{2\tilde{d}}}\right] \quad \text{OR}$$
$$X = C. \quad (18.46)$$

If we consider the particular solution $X = C = 0$, then we have

$$Ad + B\tilde{d} = 0. \quad (18.47)$$

Then, solving the A equation 2', we find

$$\delta^{mn}\partial_m\partial_n e^{-C} = -2\kappa_N^2 T_d \delta^{D-d}(\vec{y}). \quad (18.48)$$

We define harmonic function $f(y)$ by

$$\delta^{mn}\partial_m\partial_n f(y) = \delta^{D-d}(y), \quad (18.49)$$

and thus reduce the previous equation to

$$\delta^{mn}\partial_m\partial_n e^{-C} = -2\kappa_N^2 T_d \delta^{mn}\partial_m\partial_n f(y),$$
(18.50)

so that

$$e^{-C} = e^{-C_0} - 2\kappa_N^2 T_d f(y).$$
(18.51)

Solving for the harmonic function $f(y)$, we finally find

$$e^{-C} = \begin{cases} e^{-C_0} - \kappa_N^2 \frac{T_d}{\pi} \log y, & \text{if } D - d = 2 \\ e^{-C_0} + 2\kappa_N^2 \frac{T_d}{\tilde{d}\Omega_{\tilde{d}+1}} \frac{1}{y^{\tilde{d}}}, & \text{otherwise} \end{cases}.$$
(18.52)

Then, 3' is solved by

$$dA = (C - C_0)\left(1 - \frac{a^2(d)}{4}\right).$$
(18.53)

Also, from 1b', we obtain the consistency condition for the constants $a(d)$,

$$a^2(d) = 4 - \frac{2d\tilde{d}}{d + \tilde{d}}.$$
(18.54)

Then the rest of the equations are identically satisfied (0=0).
Finally then, the solution we want is

$$ds^2 = e^{2A}\eta_{\mu\nu}dx^\mu dx^\nu + e^{2B}\delta_{mn}dy^m dy^n$$
$$A_{\mu_1\dots\mu_d} = -\frac{1}{d_g}\epsilon_{\mu_1\dots\mu_d}e^C,$$
(18.55)

where

$$A = \frac{\tilde{d}}{2(d+\tilde{d})}(C - C_0),$$
$$B = -\frac{d}{2(d+\tilde{d})}(C - C_0),$$
$$\frac{a(d)}{2}\varphi = C - dA$$
$$a^2(d) = 4 - 2\frac{d\tilde{d}}{d+\tilde{d}}$$
$$e^{-C} = e^{-C_0} - 2\kappa_N^2 T_d f(y),$$
(18.56)

where

$$f(y) = \begin{cases} -\frac{1}{d\Omega_{\tilde{d}+1}} \frac{1}{y^{\tilde{d}}} & \tilde{d} > 0 \\ \frac{1}{2\pi}\log y, & \tilde{d} = 0 \end{cases}.$$
(18.57)

Observations:

1. A very important observation is that, if we look at the *coefficients* of the delta functions on the right-hand sides of the equations, *on the solution* (not in general!), they are =0 for the Einstein equation and the dilaton equation, and they are only nonzero for the antisymmetric tensor equation.

2. We note first that $B - A = -\frac{1}{2}(C - C_0)$, so we find that

$$e^{2B-2A} = \frac{e^{-C}}{e^{-C_0}} \equiv H \tag{18.58}$$

is a harmonic function, just like $f(y)$. In terms of it, the solution looks nicer,

$$ds^2 = e^{2A}d\vec{x}^2 + e^{2B}d\vec{y}^2 = e^{2B}(e^{2A-2B}d\vec{x}^2 + d\vec{y}^2)$$
$$= H^{\frac{d}{d+\tilde{d}}}\left(H^{-1}d\vec{x}^2 + d\vec{y}^2\right)$$
$$e^{2\varphi} = H^{-a(d)}. \tag{18.59}$$

In the string frame, where the gravity action is $\int d^D x \sqrt{-g} e^{-2\varphi} R + ...$, related to the Einstein frame by the relation $g_{\mu\nu}^{(s)} = e^{\varphi/2} g_{\mu\nu}^{(E)}$, we have the metric

$$ds^2 = H^{\frac{d}{d+\tilde{d}} - \frac{a(d)}{4}}\left(H^{-1}d\vec{x}^2 + d\vec{y}^2\right). \tag{18.60}$$

18.4 Magnetic p-branes and duality

To write down the equations that the magnetic solutions need to satisfy, we put some terms to zero, and we have no brane equations. Then, the ansatz is the same, but now with $\mu = 0, ..., \tilde{d} - 1$ (taking $\tilde{d}$ values instead of d values) and $m = \tilde{d}, ..., D - 1$ (taking $D - \tilde{d} = d + 2$ values instead of $D - d$ values). Moreover, instead of an ansatz for the form field $A_{\tilde{d}}$, we must write an ansatz for the field strength $F_{\tilde{d}+1}$, since in the presence of a magnetic charge, $A_{\tilde{d}}$ is only defined on patches.

To define the ansatz, we first consider the electric ansatz and rewrite it a bit.

The electric solution, as we will shortly prove, has

$$\frac{1}{\sqrt{2}\kappa_N} e^{-a(d)\varphi} * F_{\tilde{d}+1} = e_d \frac{\epsilon_{\tilde{d}+1}}{\Omega_{\tilde{d}+1}}, \tag{18.61}$$

where $\epsilon_{\tilde{d}+1}$ is the constant epsilon tensor form, and $\Omega_{\tilde{d}+1}$ is the volume of the unit sphere, satisfying the relation

$$\int_{S^D} \epsilon_D = \Omega_D. \tag{18.62}$$

Moreover, the form itself is

$$\epsilon_D = \epsilon^{m_1...m_{D+1}} \frac{y^{m_{D+1}}}{|y|^{D+1}} dx_1 \wedge ... \wedge dx_D, \tag{18.63}$$

where $x_1, ..., x_D$ are coordinates on the sphere, and together with y^{D+1}, they make coordinates in the $(D + 1)$-dimensional Euclidean space, and $|y|$ is the radius in this latter space.

Note that in the case of electromagnetism in four dimensions, we have

$$\epsilon^{ijk} \frac{y^i}{y^3} dx^j \wedge dx^k = \frac{y^i}{|y|^3} d^2 S_i, \tag{18.64}$$

relating the form integration with usual integration.

Taking into account the fact that $(e^C)' = -(e^{-C})'e^{2C}$, and the fact that the electric charge is written in terms of the tension as

$$e_d = \sqrt{2\kappa_N}T_d(-)^{(D-d)(d+1)} \tag{18.65}$$

for the "extremal" solution that we consider ("mass equal charge"), we can write the components of the field strength of the electric solution as (substituting the form of e^{-C} and the ansatz for $A_{01...d-1}$)

$$F_{y01...d-1} = -(e^C)' = -2\kappa_N^2 \frac{T_d}{\Omega_{\tilde{d}+1}} \frac{1}{|y|^{\tilde{d}+1}} e^{2C}, \tag{18.66}$$

which is indeed consistent with (18.61), since it implies that

$$(*F_{\tilde{d}+1})_{m_1...m_{\tilde{d}+1}} = \epsilon^{m_1...m_{\tilde{d}+2}} \frac{y^{m_{\tilde{d}+2}}}{|y|}(e^C)'. \tag{18.67}$$

Then, the magnetically charged solution has an ansatz analog to the Dirac monopole, in that it is singular at the origin (and $dF \neq 0, d*F = 0$), and is the dual of the electron ansatz. Yet, the magnetic solution is also a solution of the supergravity action only, without the source, so in that respect it is like the 't Hooft monopole. Unlike the 't Hooft monopole, the core is not nonabelian, since there is no nonabelian p-form antisymmetric tensor theory (with self-interactions); the smoothness that means there is no source is due to the nontrivial interaction of the p-form with gravity.

The ansatz is then

$$\frac{1}{\sqrt{2\kappa_N}}F_{d+1} = g_{\tilde{d}}\frac{\epsilon_{d+1}}{\Omega_{d+1}}, \tag{18.68}$$

which implies

$$F_{m_1...m_{d+1}} = \frac{1}{dg}\epsilon_{m_1...m_{d+2}} \frac{y^{m_{d+2}}}{|y|^{d+2}} \frac{g_{\tilde{d}}\sqrt{2\kappa_N}}{\Omega_{\tilde{d}+1}}. \tag{18.69}$$

Then, define the harmonic function

$$e^{-C} = 1 + \frac{K_{\tilde{d}}}{|y|^{\tilde{d}}} \equiv H, \tag{18.70}$$

where the constant is

$$K_{\tilde{d}} \equiv \frac{\kappa_N^2 T_d}{\tilde{d}\Omega_{\tilde{d}+1}}. \tag{18.71}$$

Then, the $(d+1)$-form field strength becomes simply

$$F_{m_1...m_{d+1}} = \epsilon_{m_1...m_{d+1}m_{d+2}} \partial^{m_{d+2}} H. \tag{18.72}$$

Moreover, then we obtain the solution with

$$e^{2A} = \left(1 + \frac{K_{\tilde{d}}}{|y|^d}\right)^{-\frac{d}{d+\tilde{d}}}$$

$$e^{2B} = \left(1 + \frac{K_{\tilde{d}}}{|y|^d}\right)^{\frac{\tilde{d}}{d+\tilde{d}}}$$

$$e^{2\varphi} = \left(1 + \frac{K_{\tilde{d}}}{|y|^d}\right)^{a(d)}$$

$$a(d) = 2 - \frac{d\tilde{d}}{d+\tilde{d}}. \tag{18.73}$$

Since then we can add for free a *delta function source with zero coefficient* and formally have the same equations as in the electric case, except for the antisymmetric tensor equation and for replacing d with $\tilde{d}$ and then using $a(\tilde{d}) = -a(d)$, it means that we have an electric/magnetic duality.

18.4.1 Duality

$\tilde{I}_D(\tilde{d})$ coupled to a $\tilde{d} - 1$-brane with source $\tilde{S}_{\tilde{d}}$ has *elementary* $\tilde{d} - 1$-brane solutions and *solitonic* $d - 1$-brane solutions and the canonical (Einstein frame) g_{MN} metric is the same in both cases, whereas $\tilde{F}_{\tilde{d}+1}$ is dual to F_{d+1},

$$\tilde{F}_{\tilde{d}+1} = e^{-a(d)\varphi} * F_{d+1}. \tag{18.74}$$

We can define the sigma model metrics

$$g_{MN}(d) = e^{\frac{a(d)}{d}\varphi} g_{MN,\text{can}}$$

$$g_{MN}(\tilde{d}) = e^{\frac{a(\tilde{d})}{\tilde{d}}\varphi} g_{MN,\text{can}}. \tag{18.75}$$

Then, in terms of them, we have the supergravity action in the respective frames, written as

$$I_D(d) = \frac{1}{2\kappa_N^2} \int d^D x \sqrt{-g} e^{-(D-2)\frac{a(d)}{2d}\varphi} \left[R - \frac{1}{2}\left(1 - \frac{a^2}{2d^2}(D-1)(D-2)\right)(\partial\varphi)^2 \right.$$

$$\left. - \frac{1}{2(d+1)}F_{d+1}^2 \right]$$

$$\tilde{I}_D(\tilde{d}) = \frac{1}{2\kappa_N^2} \int d^D x \sqrt{-g} e^{-(D-2)\frac{a(\tilde{d})}{2\tilde{d}}\varphi} \left[R - \frac{1}{2}\left(1 - \frac{\tilde{a}^2}{2\tilde{d}^2}(D-1)(D-2)\right)(\partial\varphi)^2 \right.$$

$$\left. - \frac{1}{2(\tilde{d}+1)}F_{\tilde{d}+1}^2 \right]. \tag{18.76}$$

We define the couplings, as usual, by the (VEV of the) dilaton factor in front of the actions, so

$$e^{-(D-2)\frac{a(d)}{2d}\varphi} \equiv \frac{1}{g^2}, \quad e^{-(D-2)\frac{a(\tilde{d})}{2\tilde{d}}\varphi} \equiv \frac{1}{\tilde{g}^2}. \tag{18.77}$$

Then we note that we have the relation

$$[g(d)]^d = \frac{1}{[g(\tilde{d})]^{\tilde{d}}}, \tag{18.78}$$

which means a strong/weak duality, consistent with the fact that the magnetic solution is a fundamental solution to the dual theory and vice versa.

18.5 Generalizations: black p-branes

In four dimensions, the Reissner–Nordstrom black holes have $Q \leq M$, and the extremal case is $Q = M$, while the non-extremal, or black, case is $Q < M$. We have the same situation in the case of p-brane solutions: we can write non-extremal, or black, solutions as well.

These solutions will have horizons at $r > 0$ and are the black strings and p-branes. Like in the Reissner–Nordstrom case, the black solutions have two horizons, at $r = r_+$ and $r = r_-$.

The $\tilde{d} - 1$ (note that we consider d exchanged with $\tilde{d}$ with respect to the previous case, or otherwise think of the magnetic brane case!) black brane solutions are (for the metric in string frame, $g^{(S)}_{\mu\nu} = e^{\varphi/2} g_{\mu\nu}$):

$$F = Q\epsilon_{d+1}$$

$$ds^2_{\text{string}} = -\left[1 - \left(\frac{r_+}{r}\right)^d\right]\left[1 - \left(\frac{r_-}{r}\right)^d\right]^{\gamma_x - 1} dt^2 + -\left[1 - \left(\frac{r_+}{r}\right)^d\right]^{-1}\left[1 - \left(\frac{r_-}{r}\right)^d\right]^{\gamma_r} dr^2$$

$$+ r^2\left[1 - \left(\frac{r_-}{r}\right)^d\right]^{\gamma_r + 1} d\Omega^2_{\tilde{d}+1} + \left[1 - \left(\frac{r_-}{r}\right)^d\right]^{\gamma_x} dx^i dx^i$$

$$e^{-2\varphi} = \left[1 - \left(\frac{r_-}{r}\right)^d\right]^{\gamma_\varphi}, \tag{18.79}$$

where

$$\gamma_r = \delta\left(\frac{-a-1}{2}\right) + \frac{d-2}{2}$$

$$\gamma_x = \delta\left(\frac{-a+1}{2}\right)$$

$$\gamma_\varphi = -\delta(-2a_4 - d)$$

$$\delta = \left(\frac{a^2}{2} + (4-d)d + 2\right)^{-1}$$

$$Q = \left[\delta d^2 \frac{(r_+ r_-)^d}{2}\right]^{1/2}. \tag{18.80}$$

The fundamental solution is extremal, so the two horizons coincide, $r_+ = r_- \equiv r_0$, so we obtain

$$ds^2_{\text{string}} = \left[1 - \left(\frac{r_0}{r}\right)^d\right]^{\gamma_d}(-dt^2 + dx^i dx^i) + \left[1 - \left(\frac{r_0}{r}\right)^d\right]^{\gamma_r - 1} dr^2$$

$$+ \left[1 - \left(\frac{r_0}{r}\right)^d\right]^{\gamma_r + 1} r^2 d\Omega^2_{\tilde{d}+1} Q = \left[\delta d^2 \frac{r_0^{2d}}{2}\right]^{1/2}, \tag{18.81}$$

where $\gamma_d \equiv \gamma_x$.

18.5.1 D-p-branes in 10 dimensions

However, for D-p-branes in 10 dimensions, it is better to express the solutions differently. The extremal solution can be written as

$$ds^2_{\text{string}} = H_p^{-1/2}(-d\bar{r}^2 + d\bar{x}_p^2) + H_p^{1/2}(d\bar{r}^2 + \bar{r}^2 d\Omega^2_{8-p})$$

$$e^{-2\varphi} = H_p^{\frac{p-3}{2}}$$

$$A_{01...p} = -\frac{1}{2}(H_p^{-1} - 1), \tag{18.82}$$

where the harmonic function H_p satisfies

$$\Delta_{(9-p)} H_p = -[(7-p)\Omega_{8-p} 2C_p] Q_p \delta^{(9-p)}(x^i), \tag{18.83}$$

so it is solved by

$$H_p = 1 + \frac{2C_p Q_p}{\bar{r}^{7-p}}, \tag{18.84}$$

and where the coordinate $\bar{r}$ is related to the previous r by

$$r^{7-p} = \bar{r}^{7-p} + r_0^{7-p} = \bar{r}^{7-p} + 2C_p Q_p, \tag{18.85}$$

and

$$C_p = \frac{1}{(D-p-3)\Omega_{D-p-2}}. \tag{18.86}$$

The black p-brane solutions are found then by adding a "blackening factor" $f(\bar{r})$:

$$ds^2_{\text{string}} = H_p^{-1/2}(-f(\bar{r})dt^2 + d\bar{x}_p^2) + H_p^{1/2}\left(\frac{dr^2}{f(\bar{r})} + \bar{r}^2 d\Omega^2_{8-p}\right), \tag{18.87}$$

where

$$f(\bar{r}) = 1 - \frac{\tilde{C}_p \mu_p}{\bar{r}^{7-p}}, \tag{18.88}$$

and

$$\mu_p \equiv T_p - C_p Q_p > 0 \tag{18.89}$$

is the excess mass with respect to extremality.

18.5.2 Fundamental string and magnetic (NS)5-brane solutions

These are other important cases of p-branes in 10 dimensions, besides the D-p-branes discussed above. They are charged under the Kalb–Ramond NS–NS field $B_{\mu\nu}$, with field strength $H_{\mu\nu\rho} = \partial_{[\mu} B_{\nu\rho]}$.

The 5-brane solution corresponds to the case $a = 2, \gamma_r = -1, \gamma_x = 0, \gamma_\varphi = 1, d = 2$ (so that $p = \tilde{d} - 1 = 8 - d - 1 = 5$) in the previous (eq. (18.79)), giving

$$ds^2_{\text{string}} = -dt^2 + \sum_{i-1}^{5} dx_i dx_i + \left(1 + \frac{r_0^2}{y^2}\right)(dy^2 + y^2 d\Omega_3^2)$$

$$e^{2\varphi} = 1 + \frac{r_0^2}{y^2} \equiv H_5$$

$$H = Q\epsilon_3. \tag{18.90}$$

Here we have used the notation with $\bar{r} \to y$, so with respect to (18.79), we have $r^2 = y^2 + r_0^2$.

After a conformal transformation with

$$e^{2A} = \left(1 + \frac{K_6}{y^2}\right)^{-1/4} = e^{-\varphi/2}, \tag{18.91}$$

where $K_6 = r_0^2$, leading to the Einstein frame, we find

$$ds^2_{\text{Einstein}} = \left(1 + \frac{K_6}{y^2}\right)^{-1/4}(-dt^2 + dx_i dx_i) + \left(1 + \frac{K_6}{y^2}\right)^{3/4}(dy^2 + y^2 d\Omega_3^2). \tag{18.92}$$

To obtain the fundamental string solution, we dualize to electric charge, by replacing a with $-a$ and d with $\tilde{d}$.

Then the black string solution, corresponding in (18.79) to $a' = -2, \gamma_r = -2/3, \gamma_x = 1, \gamma_\varphi = -1$, and written in terms of y ($\bar{r}$) defined by $y^6 = r^6 - r_-^6$, is

$$ds^2_{\text{string}} = -\frac{1 - C/y^6}{1 + r_-^6/y^6}dt^2 + \frac{dx^2}{1 + r_-^6/y^6} + \frac{dy^2}{1 - C/y^6} + y^2 d\Omega_7^2$$

$$e^{-2\varphi} = 1 + \frac{r_-^6}{y^6}$$

$$H = Qe^{2\varphi} * \epsilon_7$$

$$C = r_+^6 - r_-^6. \tag{18.93}$$

If $C = 0$, we obtain the extremal solution. Using a conformal transformation with

$$e^{2B} = \left(1 + \frac{r_0^6}{y^6}\right)^{1/4}, \tag{18.94}$$

we obtain the Einstein frame metric

$$ds^2_{\text{Einstein}} = \left(1 + \frac{r_0^6}{y^6}\right)^{-3/4}(-dt^2 + d\vec{x}^2) + \left(1 + \frac{r_0^6}{y^6}\right)^{1/4}(dy^2 + y^2 d\Omega_7^2). \tag{18.95}$$

18.6 Tseytlin's harmonic function rule for intersecting branes

Tseytlin found a rule, based on supersymmetry, that allows one to write down a solution representing intersecting branes.

The first observation is that one can write several parallel branes of the same type, just by choosing a more general harmonic function H_p,

$$H_p = 1 + \sum_I \frac{2C_p Q_{p,I}}{r_I^{7-p}}, \quad r_I \equiv |\vec{x} - \vec{x}_I|. \tag{18.96}$$

That can be understood mathematically as just the fact that harmonic functions sum up, so if we have a source for the harmonic functions given by delta functions based at $\vec{x}_I$, the total harmonic function is the sum of the individual ones.

But we can also understand this physically, since for a supersymmetric theory (such that the equations of motion will reduce to the single equation for one harmonic function, as we did in the first part of the chapter) we have $\{Q, Q\} \sim H$ or, in the massive and charged case relevant here, $\{Q, Q\} \sim M \pm Z$, where Z is a central charge, represented by the electric or topological (magnetic) charges. This means that on the vacuum, or BPS background with $M = |Z|$, there are vanishing $a^\dagger$'s, corresponding to supersymmetric directions, or moduli, that don't change the energy of the system. In particular, we can move the brane in transverse directions without energy change. When introducing a second parallel brane, that becomes nontrivial, and says that the relative distance between the branes is arbitrary, or that the relative motion of the branes experiences no force (the sum of all the forces, gravity, scalar, antisymmetric tensors, and fermions, exerted on the branes, vanishes).

This physical understanding can be extended to the case of intersecting branes preserving some supersymmetry, in which case we can write independently the harmonic functions of each brane in the solution, as if the other brane doesn't exist.

Therefore, Tseytlin's harmonic function rule says that for the extremal (BPS, supersymmetric) intersecting brane solution, we can consider independent harmonic factors for each brane, and then the total factor for each coordinate is the product of the factors of the various branes.

Formally then,

$$ds^2 = (\text{overall conf. factor}) \left[\sum_k \left(\prod_{j=1}^{n_k} H_j^{-1} \right) dy_k^2 - dt^2 \prod_{j=1}^{n} H_k^{-1} + \sum_{i=1}^{T} dx_i dx_i \right], \tag{18.97}$$

and the overall conformal factor is the product of the conformal factors of each brane. For n branes, the parallel coordinates are y_k for each, having n_k of them. Here H_k are the harmonic functions, and T is the number of coordinates transverse to all the branes (overall transverse). Moreover, the field strengths of the antisymmetric tensors add up.

The harmonic functions H_k usually depend only on the overall transverse coordinates (not on the coordinates transverse to the corresponding brane), so the solutions are "delocalized" over the relative transverse coordinates (the ones not overall transverse). This means that the solution above is not the most general one, but is one that can be written algorithmically.

Note that

$$\delta^{mn} \partial_m \partial_n H = 0, \tag{18.98}$$

or rather, there are sums of delta functions with coefficients, but on the solution, the coefficients vanish, as we already said, so we can put them to zero. Then we have

$$H_k = C + \sum_I \frac{C_I}{(r_{k,I})^n}, \quad r_{k,I} = |\vec{x} - \vec{x}_{k,I}|, \tag{18.99}$$

and $n = T_k - 2$, with T_k the number of overall transverse coordinates.

Important concepts to remember

- p-brane solutions are elementary, or electric, the analog of the electron in electromagnetism, and solitonic, or magnetic, the analog of monopoles in electromagnetism.
- Extremal solutions, with "$M = Q$" are the fundamental objects of string theory, while black solutions (non-extremal) have horizons and temperature, so they decay.
- The elementary electric solutions are solutions of the supergravity action $I_D(d)$ and the source action S_d, while the solitonic magnetic solutions are solutions only of $I_D(d)$, without source, just like the 't Hooft monopole, though they are abelian and singular at 0, just like the Dirac monopole.
- The electric and magnetic solutions are dual, so we can exchange the action with a dual action, and then the role of elementary and solitonic are reversed. Moreover, the duality is a strong/weak duality, as expected.
- D-p-branes in 10 dimensions have string metric $ds^2 = H_p^{-1/2}(-dt^2 + d\vec{x}_p^2) + H_p^{1/2}d\vec{x}_\perp^2$, dilaton $e^\varphi = H_p^{(3-p)/4}$ and antisymmetric tensor $A_{01...p} = -\frac{1}{2}(H_p^{-1} - 1)$, in terms of the harmonic function H_p.
- Black, non-extremal solutions are found by multiplying dt^2 by the blackening factor $f(r)$ and dividing dr^2 by the same.
- The NS5-brane solution has string metric $ds^2 = -dt^2 + d\vec{x}_5^2 + H_5 d\vec{x}_\perp^2$, and the fundamental string F1 solution has $ds^2 = H_1^{-1}(-dt^2 + dx^2) + d\vec{x}_\perp^2$, and both are charged with respect to the NS–NS B-field $B_{\mu\nu}$.
- We can construct a solution with identical parallel branes by adding their harmonic functions.
- Tseytlin's harmonic function rule says that for extremal supersymmetric intersecting brane solutions we can consider independent harmonic factors for each brane, and the total factor for each coordinate is the product of factors of the various branes.

References and further reading

For a review on extremal and black brane solitons, see [56]. For Tseytlin's harmonic function rule, see the original papers [57, 58].

Exercises

(1) Prove that the $\mu\nu$ components of the Einstein equation are given by the formula in the text,

$$
e^{(d-2)A+\tilde{d}B}\delta^{mn}\left[(d - 1)\partial_m\partial_n A + \frac{d(d-1)}{2}\partial_m A\partial_n A + (\tilde{d} + 1)\partial_m\partial_n B \right.
$$

$$
\left. + \frac{\tilde{d}(\tilde{d}+1)}{2}\partial_m B\partial_n B + \tilde{d}(\tilde{d}-1)\partial_m A\partial_n B + \frac{1}{4}e^{-2dA+2C-a(d)\varphi}\partial_m C\partial_n C + \frac{1}{4}\partial_m\varphi\partial_n\varphi \right]
$$

$$
= -\kappa_N^2 T_d e^{(d-2)A+\frac{a(d)}{2}\varphi}\delta^{D-d}(y). \tag{18.100}
$$

(2) Why can we say that the fundamental D2-brane solution is fundamental (with source), whereas the magnetically charged D4-brane is solitonic (without source), given how we derived the solutions?

(3) Write down the black (magnetically charged with respect to $B_{\mu\nu}$) NS5-brane solution.

(4) Use Tseytlin's harmonic function rule to write down the solution for a (D3-brane intersecting with a fundamental string (F1) over a point (a worldline in time) in type IIB supergravity.

19 Supersymmetry of solutions, classification via susy algebra, intersecting brane solutions

In this chapter we continue the analysis of solutions of supergravity, with their supersymmetry, their classification according to the susy algebra, including solutions not discussed in Chapter 18, and the general properties of intersecting brane solutions.

19.1 Algebra vs. susy background

The solutions from Chapter 18 are BPS saturated (i.e., they saturate the Bogomolnyi–Prasad–Sommerfeld bound). This is a bound coming from supersymmetry: we have shown in Chapter 4 that the $\mathcal{N} = 2$ algebra in four dimensions (easily generalizable to higher $\mathcal{N}$ and dimension) reduces to

$$\{a_\alpha, a_\beta^\dagger\} = 2(M - Z)\delta_{\alpha\beta}, \quad \{b_\alpha, b_\beta^\dagger\} = 2(M + Z)\delta_{\alpha\beta}, \tag{19.1}$$

meaning that we have the bound

$$M \geq |Z|, \tag{19.2}$$

where Z is a central charge in the algebra (an operator), realized on the fields as either an electric charge or a magnetic (topological) charge for a supergravity solution.

The BPS-saturated solution has $M = Z$ and thus preserves 1/2 of supersymmetry,

$$\{a_\alpha, a_\beta^\dagger\} = 0, \quad a_\alpha = \frac{1}{\sqrt{2}}\left[Q_\alpha^1 + \epsilon_{\alpha\dot\beta}\bar{Q}_{2\dot\beta}\right]. \tag{19.3}$$

Acting on the vacuum state $|0\rangle$, we obtain the condition of the form

$$(\text{lin.comb.of})Q_\alpha|0\rangle = 0, \tag{19.4}$$

saying that the BPS ground state must be supersymmetric.

But then, in terms of the action on the fields of supergravity, this means that, *on the background of the solution*, the susy variation of the fermions, which is bosonic, so nontrivial in VEV, must vanish (the variation of the bosons is fermionic, so on the VEV vanishes),

$$\delta_S \psi's = 0. \tag{19.5}$$

19.2 Mass of solutions

In order to decide whether a certain solution is BPS, we must understand what is "M" (nontrivial in a theory of gravity) and "Z," in particular, also how to calculate them. Z is a central charge carried by a p-brane or similar object (more on that later) and M must be thought of as tension $T_p = \text{mass}/p\text{-volume } V_p$.

In order to understand how to calculate the mass, we remember that, in asymptotically flat geometries like the solutions from Chapter 18, we have the Arnowitt–Deser–Misner (ADM) formalism for calculation of masses and momenta in general relativity.

One first must put the canonical (Einstein frame) metric in the form, valid at least at infinity in r, which for $d = 4$ is

$$ds^2 = -N^2 dt^2 + \sum_{i,j=1}^{3} h_{ij}(N^i dt + dx^i)(N^j dt + dx^j), \tag{19.6}$$

and which is always possible, with a simple generalization for arbitrary dimension.

Then the ADM mass, or energy, is given by

$$(E =)M = \frac{1}{16\pi} \lim_{r\to\infty} \sum_{i,j=1}^{3}(\partial_i h_{ij} - \partial_j h_{ii})N^i dA, \tag{19.7}$$

and the ADM momentum is given by

$$P_j = \frac{1}{8\pi} \lim_{r\to\infty} \sum_{i=1}^{3} \int (K_{ij}N^i - K^i{}_i N_j)dA, \tag{19.8}$$

together forming a Lorentz vector.

Here $r = \sqrt{x^2 + y^2 + z^2}$ and the extrinsic three-curvature tensor is

$$K_{ij} = \frac{1}{2N}[N_{i|j} + N_{j|i} - \partial_t g_{ij}]. \tag{19.9}$$

We see that, in order to be able to use this ADM formalism, we need to have $N^i \neq 0$, so a $dx^i dt$ term in the metric (off-diagonal), which is not the case for our solutions, that are all diagonal in the metric. But that is easily fixed by a general coordinate transformation, or change of gauge, leading to $N_i \neq 0$.

Instead of doing a change of gauge, which is more complicated, for the mass, there is the alternative formulation as the integral of the 00 component of the energy–momentum tensor in the first approximation, $\theta_{00}^{(1)}$. Then

$$\frac{M}{V_p} = \int d^{D-p-1}x \theta_{00}^{(1)} = \frac{1}{\kappa_N^2} \int d^{D-p-1}x \left[R_{00}^{(1)} - \frac{1}{2}\eta_{00} R^{(1)\lambda}{}_\lambda \right]. \tag{19.10}$$

For a static metric, $\partial_0 = 0$, and a metric depending only on the transverse coordinates, so $\partial_\mu = 0$ (parallel to the brane), we obtain

$$\frac{M}{V_p} = \int d^{D-p-1}x u^m \left(r^{D-p-2} d\Omega_{D-p-2} \right) (\partial_m h_{qq} + \partial_m h_{\mu\mu} - \partial_q h_{mq}). \tag{19.11}$$

As an example, we will consider the fundamental string ($p = 1$) case. In Einstein frame, we have

$$e^{2A} = e^{\frac{2(D-4)}{D-2}E}, \quad \varphi = aE$$

$$e^{2B} = e^{-\frac{4}{D-2}E}, \quad B_{01} = -e^{E}$$

$$e^{-E} = \begin{cases} 1 + \frac{M}{r^{D-4}}, & D > 4 \\ 1 - 8G\mu \ln r, & D = 4. \end{cases} \tag{19.12}$$

We see that in the $D = 4$ case, there is no Minkowski asymptotics, so the ADM formalism cannot even be used.

Then, for $D = 4$, using the θ_{00} formula, we find

$$\frac{M}{V} = \frac{M}{L} = -\frac{1}{2\kappa_N^2} \int_{\partial M} d\Sigma^m \left(\partial_m e^A + (D-3) \partial_m e^B \right). \tag{19.13}$$

But, since $A + (D-3)B = -E$, we find that

$$\frac{M}{L} = \mu, \tag{19.14}$$

so the mass parameter in the e^{-E} harmonic function was indeed the mass per unit length.

19.3 Supersymmetry of solutions

As we saw, the supersymmetry condition is that, in the background, we must have $\delta_S \psi = 0$ for all fermions.

Consider first, as a simple example, 10-dimensional strings in NS-NS background only, that is, fields g_{MN}, B_{MN}, φ. Then the supersymmetry variations of the gravitino ψ_M and dilatino (spin 1/2 spinor) λ in Einstein frame give

$$\delta \psi_M = D_M \epsilon + \frac{1}{96} e^{-\varphi/2} \left(\Gamma_M{}^{NPQ} - 9\delta_M^N \Gamma^{PQ} \right) H_{NPQ} \epsilon = 0,$$

$$\delta \lambda = -\frac{1}{2\sqrt{2}} \Gamma^M \partial_M \varphi \epsilon + \frac{1}{24\sqrt{2}} e^{-\varphi/2} \Gamma^{MNP} H_{MNP} \epsilon = 0. \tag{19.15}$$

These two conditions each break up into a scalar condition on the x dependence of ϵ and a condition on the constant spinor it multiplies. It is a nontrivial fact that both fermionic equations (for ψ_M and for λ) are solved by the same conditions.

We use first, as an example, the fundamental string (F1) solution. Since in the solution, the parallel directions $0, 1$ and the transverse directions $2, ..., 9$ are different, we use the gamma matrix split

$$\Gamma_A = \{\gamma_\alpha \otimes \mathbb{1}, \gamma_3 \otimes \Sigma_a\}, \tag{19.16}$$

where $\alpha = 0, 1$, $a = 2, ..., 9$, and

$$\gamma_3 = \gamma_0 \gamma_1, \quad \Gamma_9 = \Sigma_2 ... \Sigma_9, \quad \Gamma_{11} = \Gamma_0 ... \Gamma_9. \tag{19.17}$$

In 10 dimensions, we use Majorana–Weyl fermions, so in particular we have

$$\Gamma_{11}\epsilon = \epsilon, \quad \Gamma_{11} = \gamma_3 \otimes \Gamma_9. \tag{19.18}$$

We use the ansatz for the susy parameter on the F1 solution

$$\epsilon(x, y) = \epsilon(x) \otimes \eta(y), \tag{19.19}$$

and substituting, together with the F1 solution, in the susy conditions $\delta\psi_M = 0$, $\delta\lambda = 0$, we obtain the solution

$$\epsilon(x, y) = e^{3\varphi(y)/8}\epsilon_0 \otimes \eta_0, \tag{19.20}$$

where

$$(\mathbb{1} - \gamma_3)\epsilon_0 = 0, \quad (\mathbb{1} - \Gamma_9)\eta_0 = 0. \tag{19.21}$$

Now, in 11 dimensions, there was a single chirality condition (that related half the components with the other half), while now we have two, for 2 dimensions and for 8 dimensions (each relating half the components with the other half), so we have broken 1/2 susy.

The way this comes about is that the nontrivial components of H_{MNP} are $H_{y01} = \partial_y B_{01}$, so for instance the λ equation becomes

$$-\frac{1}{2\sqrt{2}}\Gamma^y \left(\partial_y\varphi - \frac{1}{12}e^{-\varphi/2}\Gamma^{01}\partial_y B_{01}\right)\epsilon = 0, \tag{19.22}$$

which has a solution only if ϵ satisfies

$$\Gamma_0\Gamma_1\epsilon \equiv \gamma_3 \otimes \mathbb{1}\epsilon = \epsilon \Rightarrow \gamma_3\epsilon_0 = \epsilon_0, \tag{19.23}$$

which means that the solution breaks 1/2 susy, and if moreover

$$\partial_y\varphi = \frac{1}{12}e^{-\varphi/2}\partial_y B_{01}, \tag{19.24}$$

which is a condition defining the F1 solution (not the susy parameter).

Then the ψ_M equation gives the y dependence of ϵ, since it becomes, for $M = y$,

$$\partial_y\epsilon + \frac{1}{4}\omega_y^{01}\epsilon - \frac{3}{32}e^{-\varphi/2}\partial_y B_{01}\epsilon = 0. \tag{19.25}$$

The type IIA supergravity action in string frame is

$$S_{\text{string}} = \frac{1}{2\kappa_N^2}\int d^{10}x\sqrt{-g}\left\{e^{-2\phi}\left[R + 4\partial_\mu\phi\partial^\mu\phi - \frac{1}{2}|H_3|^2\right] - \frac{1}{2}(|F_2|^2 + |\tilde{F}_4|^2)\right\}$$
$$-\frac{1}{4\kappa_N^2}\int B_2 \wedge F_4 \wedge F_4, \tag{19.26}$$

where

$$|F_n|^2 \equiv \frac{e}{n!}F_{M_1...M_n}F^{M_1...M_n}. \tag{19.27}$$

The most general fermionic supersymmetry transformation rules for the various string (and corresponding supergravity) theories in 10 dimensions *in string frame* are in the conventions of Becker–Becker–Schwarz.

For the type IIA supergravity (note that, with the definition $F^{(n)} \equiv \frac{1}{n!} F_{M_1 \ldots M_n} \Gamma^{M_1 \ldots M_n}$, the reduction from 11 dimensions gives $F^{(4)} = e^{4\phi/3} \tilde{F}^{(4)} + e^{\phi/3} H^{(3)} \Gamma_{11}$),

$$\delta\psi_\mu = \left[D_\mu - \frac{1}{4}\frac{1}{2!}\Gamma^{\nu\rho}H_{\mu\nu\rho} - \frac{1}{8}e^\phi F^{(2)}_{\nu\rho}\Gamma_\mu{}^{\nu\rho}\Gamma_{11} + \frac{e^\phi}{8}\frac{1}{4!}\Gamma^{\nu\rho\sigma\tau}F^{(4)}_{\nu\rho\sigma\tau}\Gamma_\mu \right]\epsilon,$$

$$\delta\lambda = \left[-\frac{1}{3}\Gamma^\mu\partial_\mu\phi\Gamma_{11} + \frac{1}{6\cdot 3!}H_{\mu\nu\rho}\Gamma^{\mu\nu\rho} - \frac{e^\phi}{4}\frac{1}{2}F^{(2)}_{\mu\nu}\Gamma^{\mu\nu} + \frac{e^\phi}{12}\frac{1}{4!}\Gamma^{\mu\nu\rho\sigma}\tilde{F}^{(4)}_{\mu\nu\rho\sigma} \right]\epsilon, \tag{19.28}$$

where

$$\tilde{F}_4 = dA_3 + A_1 \wedge H_3. \tag{19.29}$$

For the type IIB supergravity,

$$\delta\psi_\mu = \left(D_\mu + \frac{i}{8}e^\phi\Gamma^\nu F^{(1)}_\nu\Gamma_\mu + \frac{i}{16}e^\phi\frac{1}{5!}\tilde{F}_{\mu_1 \ldots \mu_5}\Gamma^{\mu_1 \ldots \mu_5}\Gamma_\mu \right)\epsilon$$

$$- \frac{1}{8}\left(\frac{2}{2!}\Gamma^{\nu\rho}H_{\mu\nu\rho} + ie^\phi\Gamma^{\nu\rho\sigma}\tilde{F}^{(3)}_{\nu\rho\sigma}\Gamma_\mu \right)\epsilon^*$$

$$\delta\lambda = \frac{1}{2}(\partial_\mu\phi - ie^\phi\partial_\mu C_0)\Gamma^\mu\epsilon + \frac{1}{4}\left(ie^\phi\Gamma^{\mu\nu\rho}\tilde{F}^{(3)}_{\mu\nu\rho} - \frac{1}{3}\Gamma^{\mu\nu\rho}H_{\mu\nu\rho} \right)\epsilon^*, \tag{19.30}$$

where

$$\tilde{F}_3 = F_3 - C_0 H_3, \quad \tilde{F}_5 = F_5 - \frac{1}{2}C_2 \wedge H_3 + \frac{1}{2}B_2 \wedge F_3. \tag{19.31}$$

For 11-dimensional supergravity (low energy limit of M theory),

$$\delta\psi_M = D_M\epsilon + \frac{1}{12}\left(\Gamma_M\frac{1}{3!}\Gamma^{NPQR}F_{NPQR} - \frac{3}{3!}\Gamma^{NPQ}F_{MNPQ} \right)\epsilon. \tag{19.32}$$

For the heterotic supergravity (with $SO(32)$ or $E_8 \times E_8$ Yang–Mills fields F_2, YM adjoint gauginos χ and dilatino partners of supergravity λ),

$$\delta\psi_\mu = D_\mu\epsilon - \frac{1}{4}\frac{1}{3!}\Gamma^{\nu\rho}\tilde{H}^{(3)}_{\mu\nu\rho}\epsilon,$$

$$\delta\lambda = -\frac{1}{2}\Gamma^\mu\partial_\mu\phi\epsilon + \frac{1}{4}\Gamma^{\mu\nu\rho}\tilde{H}^{(3)}_{\nu\rho\sigma}\epsilon,$$

$$\delta\chi = -\frac{1}{2\cdot 2!}\Gamma^{\mu\nu}F^{(2)}_{\mu\nu}\epsilon, \tag{19.33}$$

where

$$\tilde{H}_3 = dB_2 + \frac{l_s^2}{4}\omega_3 \tag{19.34}$$

satisfies (Tr is the trace in the adjoint representation of the gauge group)

$$d\tilde{H}_3 = \frac{l_s^2}{4}\left(\operatorname{tr} R \wedge R - \frac{1}{30}\operatorname{Tr} F \wedge F \right). \tag{19.35}$$

For the type I supergravity (with $SO(32)$ Yang–Mills field F_2, YM adjoint gauginos χ and dilatino λ),

$$\delta\psi_\mu = D_\mu\epsilon - \frac{1}{8}e^\phi \tilde{F}^{(3)}_{\nu\rho\sigma}\Gamma^{\nu\rho\sigma}\Gamma_\mu\epsilon,$$

$$\delta\lambda = \frac{1}{2}\slashed{\partial}\phi\epsilon + \frac{1}{4}e^\phi\Gamma^{\nu\rho\sigma}\tilde{F}^{(3)}_{\nu\rho\sigma}\epsilon,$$

$$\delta\chi = -\frac{1}{2}F^{(2)}_{\mu\nu}\Gamma^{\mu\nu}\epsilon, \qquad (19.36)$$

where

$$\tilde{F}_3 = dC_2 + \frac{l_s^2}{4}\omega_3,$$

$$\omega_3 = \omega_{\mathrm{L}} - \omega_{\mathrm{YM}}, \qquad (19.37)$$

and the Lorentz and YM CS 3-forms are given by

$$\omega_{\mathrm{L}} = \mathrm{tr}\left(\omega\wedge d\omega + \frac{2}{3}\omega\wedge\omega\wedge\omega\right),$$

$$\omega_{\mathrm{YM}} = \mathrm{tr}\left(A\wedge dA + \frac{2}{3}A\wedge A\wedge A\right). \qquad (19.38)$$

We note that in all these cases, the same rule applies for brane solutions like in the case of the F1 solution: With the ansatz $\epsilon(x,y) = \epsilon(x)\otimes\eta(y)$, we only have a solution if we have the condition on the susy parameter ϵ,

$$\Gamma^{\mu_1\ldots\mu_{p+1}}F_{y\mu_1\ldots\mu_{p+1}}\epsilon = \epsilon \Rightarrow \Gamma^{01\ldots p}\epsilon_0 = \epsilon_0, \qquad (19.39)$$

where we have used the fact that $F_{y\mu_1\ldots\mu_{p+1}} = \partial_y C_{\mu_1\ldots\mu_{p+1}} \propto \epsilon_{\mu_1\ldots\mu_{p+1}}$.

We note that ϵ_0 is a special (constant) solution, and then the x or y dependence is fixed by one of the spinor variation equations, usually the gravitino one, since it has $\partial_\mu\epsilon +$ more $= 0$. The rest of the spinor equations are equations for the fields of the background, which must be satisfied if the solution is supersymmetric.

In conclusion, the supersymmetry of the p-brane solution is defined generically by the condition

$$\Gamma^{01\ldots p}\epsilon_0 = \epsilon_0, \qquad (19.40)$$

which breaks half susy, since it relates half of the components of ϵ_0 to the other half, through the matrix $\Gamma^{01\ldots p}$.

19.4 Supergravity solutions corresponding to fundamental objects = states in string theory

In Chapter 18, we have shown D-brane, fundamental string (F1), and NS5-brane solutions in string theory. But those are not the only brane-like, fundamental object solutions of supergravity. There are also the M2-brane and M5-brane solutions of 11-dimensional supergravity, that will be described in a moment, which correspond to the fundamental object of M-theory (the string theory at strong coupling, which in the low energy limit becomes 11-dimensional supergravity), the M2-brane or fundamental membrane, and its magnetic dual, the solitonic M5-brane.

But also, in both 10 and 11 dimensions, as well as in all the lower dimensions until 4 (by compactification), there are also solutions that involve only the metric. These are as follows:

1. **The pp wave (or momentum)**, with solution

$$ds^2 = -dt^2 + dx^2 + (H_0 - 1)(dx^+)^2 + d\vec{y}^2_{D-2} = 2dx^+dx^-$$
$$+ (H_0 - 1)(dx^+)^2 + d\vec{y}^2_{D-2}, \tag{19.41}$$

where $x^\pm = (x \pm t)/\sqrt{2}$, $r = \sqrt{\vec{y}^2}$, and H_0 is a harmonic function, defined with a 1 in front, just like the harmonic functions for the other branes ($H_p = 1 + K/r^{D-p-3}$, with K related to the charge, and for $p = 0$ when dimensionally reduced from 11 to 10 dimensions), though of course flat space corresponds now to no $(dx^+)^2$ term, hence the $H_0 - 1$, so

$$H_0(r) = 1 + \frac{K}{r^{D-4}}. \tag{19.42}$$

The harmonic function satisfies the Poisson equation

$$\Delta_{D-2}H_0 = -K(D-4)\Omega_{D-3}\delta^{(D-2)}(\vec{y}). \tag{19.43}$$

Note that the pp waves are the only solutions of gravity for which gravity linearizes, namely the linearized approximation to the (usually highly nonlinear) Einstein equations is actually exact, namely the Poisson equation discussed previously. We call the above "the" pp wave only in the context of fundamental objects of supergravity, otherwise a general pp-wave solution has the form

$$ds^2 = dx^+dx^- + H(x^+, x^i)(dx^+)^2 + d\vec{y}^2_\perp, \tag{19.44}$$

and other important special cases are the following: the shockwave, with source, and hence also H, proportional to $\delta(x^+)$; the sourceless wave, for which $\Delta H = 0$; and the maximally supersymmetric pp wave, a Penrose limit of a maximally supersymmetric supergravity solution of the type $AdS \times S$, for which $H = H(x^i)$ only, but ΔH is a constant.

Of course, the most relevant pp wave is the one for 11-dimensional supergravity, which corresponds to a state of a momentum propagating in the x direction. This is a basic state of the 11-dimensional supergravity and also of the corresponding M theory.

This pp wave, or momentum wave, reduces to D0-branes in 10-dimensional type IIA supergravity when KK reducing from 11 dimensions.

2. **The KK monopole**, which is a generalization of the KK monopole of the original five-dimensional KK theory, by adding some more flat spatial directions. Again, the most relevant case is $D = 11$, giving solitonic states of 11-dimensional supergravity and its embedding M theory. Then, we denote the solution by KK6, since it is like a $p = 6$ brane solution, in that it has a flat (6+1)-dimensional space.

The solution is

$$ds_{11}^2 = d\vec{x}_{(1,6)}^2 + ds_{TN}^2(y)$$

$$ds_{TN}^2(y) = H_6(y) \sum_{i=1,2,3} dy^i dy^i + H_6^{-1}(y) \left(d\psi_{TN} + \sum_i V_i(y) dy^i \right)^2, \tag{19.45}$$

where V_i is related to the harmonic function H_6 (again, in the general form of H_p, for $p = 6$, when reduced from 11 to 10 dimensions), by

$$\epsilon^{ijk} \partial_j V_k = \partial^i H_6$$

$$H_6 = 1 + \frac{K}{|y|}. \tag{19.46}$$

TN stands for Taub–NUT solution (found in two independent papers by Taub, and by Newman, Tamburini, and Unti (NUT)), and ψ_{TN} is the KK direction (on which we are compactifying, in this particular case from 11 dimensions to 10 dimensions).

The compactification of 11-dimensional supergravity down to 10-dimensional type IIA supergravity is done through the ansatz

$$ds_{11}^2 = e^{-\frac{2}{3}\phi} G_{\mu\nu}^{S(10)} dx^\mu dx^\nu + e^{\frac{4}{3}\phi}(dx^{10} + A_\mu dx^\mu)^2$$

$$A_{\mu\nu 11} = B_{\mu\nu}^{IIA}$$

$$A_{\mu\nu\rho} = A_{\mu\nu\rho}^{IIA}, \tag{19.47}$$

where $G_{\mu\nu}^{S(10)}$ is the 10-dimensional metric in string frame.

Type II B supergravity is then T-dual to type IIA supergravity.

Finally, we write the 11-dimensional supergravity solutions of the p-brane type:

– The M2-brane solution, electrically charged under $A_{\mu\nu\rho}$, is

$$ds_{11}^2 = H_2^{-2/3}(r)(-dt^2 + d\vec{x}_{(2)}^2) + H_2^{1/3}(r)(dr^2 + r^2 d\Omega_7^2)$$

$$F_{(4)} = d(H_2^{-1}(r)) \wedge dt \wedge dx_1 \wedge dx_2$$

$$H_2(r) = 1 + \frac{32\pi^2 N l_P^6}{r^6} = 1 + \frac{\kappa_{N,11}^2 T_2}{3\Omega_7} \frac{1}{r^6}$$

$$T_2 = \frac{1}{l_P^6}, \tag{19.48}$$

where T_2 is the M2-brane tension.

– The M5-brane solution, magnetically charged under $A_{\mu\nu\rho}$, is

$$ds_{11}^2 = H_5^{-1/3}(r)(-dt^2 + d\vec{x}_{(5)}^2) + H_5^{+2/3}(r)(dr^2 + r^2 d\Omega_4^2)$$

$$F_{(4)} = *(dt \wedge dx^1 \wedge ... \wedge dx^5 \wedge d(H_5^{-1}(r)))$$

$$H_5(r) = 1 + \frac{\pi N l_P^3}{r^3} = 1 + \frac{\kappa_{N,11} T_5}{3\Omega_4} \frac{1}{r^3}$$

$$T_5 = \frac{1}{l_P^6}, \tag{19.49}$$

where T_5 is the M5-brane tension.

19.5 Classification of solutions by the susy algebra

As we saw, the N-extended susy algebra in four dimensions is

$$\{Q_\alpha^i, Q_\beta^j\} = \delta^{ij}(C\gamma^\mu)_{\alpha\beta}P_\mu + U^{ij}C_{\alpha\beta} + V^{ij}(C\gamma_5)_{\alpha\beta}, \tag{19.50}$$

where U^{ij} and V^{ij} are antisymmetric central charges, so operators, carried (on the fields) by solitonic objects, just like P_μ (the momentum) can be thought of as "charges" carried by elementary particles.

We can extend these notions to the susy algebras in higher dimensions.

The $\mathcal{N} = 1$, 10-dimensional susy algebra has now central charges that have Lorentz indices, so will correspond to extended objects. It is

$$\{Q_\alpha, Q_\beta\} = (C\Gamma^M P)_{\alpha\beta}P_M + (C\Gamma^{MNPQR}P)_{\alpha\beta}Z_{MNPQR}^+, \tag{19.51}$$

where P is the chiral projector, and Z_{MNPQR}^+ is a self-dual antisymmetric tensor in 10 dimensions.

More important to us will be the $\mathcal{N} = 1$ *susy algebra in 11 dimensions*,

$$\{Q_\alpha, Q_\beta\} = (C\Gamma^M)_{\alpha\beta}P_M + (C\Gamma^{MN})_{\alpha\beta}Z_{MN} + (C\Gamma^{MNPQR})_{\alpha\beta}Z_{MNPQR}, \tag{19.52}$$

where Z_{MN} and Z_{MNPQR} are antisymmetric tensor central charges, and Q_α is a 32-component spinor.

Since the left-hand side is a symmetric matrix, it has $32 \times 33/2 = 528$ components (thus degrees of freedom). We now show that the right-hand side has the same number of degrees of freedom, in the central charges, thus we have included all possible central charges to describe the system.

P_M has 11 components, Z_{MN} has $11 \times 10/2 = 55$, and Z_{MNPQR} has $11 \times 10 \times 9 \times 8 \times 7/(1 \times 2 \times 3 \times 4 \times 5) = 462$. Then we have $11 + 55 + 462 = 528$ components, equaling the number of components on the left-hand side of the susy algebra.

But now we want to understand what are the states that carry these various types of charges.

In 11-dimensional supergravity (and the corresponding M theory), there are M2-brane and M5-brane solutions. They should carry some of the above charges. Indeed, consider the case where $M, N, ...$ are spatial, $i, j, ...$, then Z_{ij} are charges carried by M2-branes extended in the corresponding spatial directions i, j, and Z_{ijklm} are charges carried by M5-branes extended in directions i, j, k, l, m.

Further, P_i are "charges" (momenta) carried by the pp waves moving in direction i, and P_0 is the mass M of a fundamental particle. The charges Z_{0ijkl} can be dualized to 6-brane charges, via

$$Z_{0ijkl} = \epsilon_{0ijkli_1...i_6}Z^{i_1...i_6}, \tag{19.53}$$

and we remember that we do have a "6-brane" object, the KK6 monopole, which therefore will carry these charges.

Finally, we are left with the Z_{0i} charges, which can be dualized to 9-brane charges, as $\epsilon_{0ii_1...i_9}Z^{i_1...i_9}$, so there should be some 9-brane object that carries them. In 11 dimensions, this should be a "domain wall," but such object doesn't exist in supergravity, and it only makes sense in the embedding M theory, where a spatial boundary of the space can take the form of a 9-brane, on which M2-branes can end.

Next, we consider the $\mathcal{N}=IIA$ *susy algebra in 10 dimensions*. It involves two super-charges of different chiralities, which will be separated by their eigenvalue of Γ_{11}. Then the susy algebra is

$$\{Q_\alpha, Q_\beta\} = (C\Gamma^M)_{\alpha\beta} P_M + (C\Gamma_{11})_{\alpha\beta} Z + (C\Gamma^M \Gamma_{11})_{\alpha\beta} Z_M + (C\Gamma^{MN})_{\alpha\beta} Z_{MN}$$
$$+ (C\Gamma^{MNPQ}\Gamma_{11})_{\alpha\beta} Z_{MNPQ} + (C\Gamma^{MNPQR})_{\alpha\beta} Z_{MNPQR}. \qquad (19.54)$$

Again, the left-hand side has 528 degrees of freedom. On the right-hand side, we have P_M with 10 components, Z with one, Z_M with 10, Z_{MN} with $10 \times 9/2 = 45$, Z_{MNPQ} with $10 \times 9 \times 8 \times 7/(1 \times 2 \times 3 \times 4) = 210$ and Z_{MNPQR} with $10 \times 9 \times 8 \times 7/(1 \times 2 \times 3 \times 4 \times 5) = 252$, for a total of $10 + 1 + 10 + 45 + 210 + 252 = 528$, so again we have matching. That was not surprising, since from the reduction from 11 dimensions, Z_{MN} in 11 dimensions gives Z_{MN} and $Z_{11M} \rightarrow Z_M$, and Z_{MNPQR} in 11 dimensions gives Z_{MNPQR} and $Z_{11MNPQ} \rightarrow Z_{MNPQ}$.

The charges are carried by supergravity states as follows. Again, P_i is carried by the pp wave or momentum, and P_0 by the mass of a fundamental particle. Z is carried by a D0-brane (D-particle) solution, which is charged under the gauge field A_μ. Z_i is carried by a fundamental string in direction i, which is (electrically) charged under the gauge field $B_{\mu\nu}$. Z_{ij} is carried by a D2-brane in directions ij, which is (electrically) charged under $A_{\mu\nu\rho}$. Z_{0i} is dualized to a $Z_{i_1 \ldots i_8}$, carried by a D8-brane, or domain wall, which is a solution that appears in *massive* supergravity, and actually corresponds to just a different mass on the sides of the domain wall. Z_{ijkl} is carried by a D4-brane, which is (magnetically) charged with respect to $A_{\mu\nu\rho}$. Z_{0ijk} is carried by a D6-brane, which is (magnetically) charged with respect to A_μ. Z_{ijklm} is carried by a NS5-brane, which is (magnetically) charged with respect to $B_{\mu\nu}$. Z_{0ijkl} is dualized to a $\tilde{Z}_{ijklm}$, which is carried by a KK5 monopole, which is (magnetically) charged with respect to the off-diagonal metric.

Finally, we are left with Z_0, which is dualized to a $Z_{i_1 \ldots i_9}$, carried by a "9-brane," that is now space-filling, so it doesn't actually correspond to anything in supergravity. But in string theory, it corresponds to a so-called "Chan–Paton factor," which acts as an endpoint for open strings. So this is the only object that needs to be described in string theory.

Finally, consider the $\mathcal{N} = IIB$ *susy algebra in 10 dimensions*. Now the supercharges Q_α^i, $i = 1, 2$ have the same chirality, and the susy algebra is

$$\{Q_\alpha^i, Q_\beta^j\} = \delta^{ij}(C\Gamma^M P)_{\alpha\beta} P_M + (C\Gamma^M P)\tilde{Z}_M^{ij} + \epsilon^{ij}(C\Gamma^{MNP}P)_{\alpha\beta} Z_{MNP}$$
$$+ \delta^{ij}(C\Gamma^{MNPQR}P)_{\alpha\beta} Z_{MNPQR}^+ + (C\Gamma^{MNPQR}P)_{\alpha\beta}(\tilde{Z}^+)_{MNPQR}^{ij}, \qquad (19.55)$$

where P is the chiral projector and tilde on Z^{ij} refers to the symmetric traceless tensor of $SO(2)$, or equivalently a $U(1)$ doublet, so gives two antisymmetric tensor charges, and Z_{MNPQR}^+ refers to a self-dual antisymmetric tensor in 10 dimensions.

Again the left-hand side has 528 degrees of freedom. On the right-hand side, we have 10 components for P_M, 2 times 10 components for $\tilde{Z}_M^{ij}$, $10 \times 9 \times 8/(1 \times 2 \times 3) = 120$ for Z_{MNP}, $10 \times 9 \times 8 \times 7 \times 6/(1 \times 2 \times 3 \times 4 \times 5)/2 = 126$ for Z_{MNPQR}^+ and 2×126 for $\tilde{Z}_{MNPQR}^{ij}$, for a total of $10 + 2 \times 10 + 120 + 126 + 2 \times 126 = 528$, so again we have matching.

The charges carried by the supergravity fields are as follows. Again P_i is the momentum, carried by the pp wave in direction i, and $P_0 = M$ is the mass of an elementary particle. $\tilde{Z}^{ij}_m$ are carried by the F1- and D1-branes extended in spatial direction m, (electrically) charged under $B_{\mu\nu}$ and $C_{\mu\nu}$, respectively. Z_{ijk} are carried by D3-branes, charged under $C^+_{\mu\nu\rho\sigma}$. Z_{0ij} are carried by D7-branes, which are (magnetically) charged under C. Z^+_{ijklm} are carried by KK5-branes (and Z_{0ijkl} are equal to them by self-duality), magnetically charged under the off-diagonal metric. $\tilde{Z}^{ij}_{mnpqr}$ are carried by NS5-branes and D5-branes (and $\tilde{Z}^{ij}_{0mnpq}$ are equal to them by self-duality), which are (magnetically) charged under $B_{\mu\nu}$ and $C_{\mu\nu}$, respectively.

Finally, we are left with $\tilde{Z}^{ij}_0$, which are dualized to $\tilde{Z}^{ij}_{m_1...m_9}$, which are carried by D9-branes, which again are space-filling objects that are not supergravity solutions, but are objects defined in string theory, as Chan–Paton factors.

19.6   Intersecting brane solutions

We have seen that the condition for supersymmetry of a brane solution is $\Gamma^{01...p}\epsilon = \epsilon$. Then in the presence of two intersecting branes, we have two such conditions,

$$\Gamma_{\mu_1...\mu_{p+1}}\eta = 0, \;\; (1) \;\; \Gamma_{\nu_1...\nu_{p'+1}}\eta = 0. \;\; (2) \tag{19.56}$$

In this case, if the branes are different, we obtain at most 1/4 susy, since each condition halves the number of components.

As we saw in Chapter 18, we can use Tseytlin's harmonic function rule if we have an intersection, obtaining in this way a delocalized solution.

The *composition rules* for brane intersections are as follows:

1) p-branes of the same type (self-)intersect only over a $(p-2)$-brane.
2) $p \perp q$ branes intersect only over a $(\min_{\{p,q\}} - 1)$-brane.
3) n *orthogonal* intersecting branes preserve at least $1/2^n$ susy.

The harmonic function rule, for $1/2^n$ susy, means that we superpose BPS states, so also gives a harmonic function.

The large separation limit and the coincident branes limit together with exchange symmetry and charge conservation should determine the background uniquely. The question is how to solve the supergravity equations in general (not for delocalized branes).

The basic intersections in $D = 11$ dimensions are as follows:

$M2 \perp M2(0)$, meaning a self-intersection of membranes over a point, $M5 \perp M5(3)$, $M2 \perp M5(1)$, as well as $M5 \perp M5(1)$, which is an overlap, rather than intersection.

Using these basic intersections, and combining them, one can use T- and S-dualities, as solution-generating techniques, and we should obtain all the relevant intersections of branes.

The basic intersections in $D = 10$ dimensions are the following:

– Self-intersections $Dp \perp Dp(p-2)$.
– Branes ending on branes $Dp \perp Dq(p-1)$ for $p < q$.
– Branes within branes $Dp \perp Dq(p)$, for $p < q$.

– One can also add pp waves and KK monopoles as well.

For dimensional reduction, or reversely, for "dimensional oxidation," we can use the BPS condition, stating that we can put separated, parallel branes of the same type without energy cost, to construct a periodic array.

For instance, for the original KK case, for the five-dimensional black holes solution, with harmonic function

$$H = 1 + \frac{\tilde{Q}}{r^2},$$ (19.57)

we can create a periodic array, thus equivalently compactifying the corresponding dimension (with the period of the array being 2π times the radius of the compactified direction), obtaining

$$H = 1 + \sum_{k=-\infty}^{+\infty} \frac{\tilde{Q}}{r^2 + (z + 2\pi k R)^2}$$
$$= 1 + \frac{Q \sinh r/R}{2r(\cosh r/R - \cos z/R)}.$$ (19.58)

For the brane within brane solution, if we consider a p-brane inside a $(p + 4)$-brane, that preserves 1/2 of supersymmetry still. The general solution is

$$ds_{10}^2 = (H_p H_{p+4})^{-1/2}(-dt^2 + d\vec{x}_p^2) + H_p^{1/2}H_{p+4}^{-1/2}(dy_1^2 + \dots + dy_4^2)$$
$$+ (H_p H_{p+4})^{1/2}d\vec{z}^2$$ (19.59)

where

$$H_p = 1 + \sum_i \frac{Q_i}{\left[|\vec{x} - \vec{x}_{0i}|^2 + \frac{4Q}{(p-1)^2}|\vec{z} - \vec{z}_{0i}|^{p-1}\right]^{\frac{p+1}{p-1}}}$$
$$H_{p+4} = \frac{Q}{|\vec{z} - \vec{z}_0|^{3-p}}$$ (19.60)

satisfy

$$\partial_z^2 H_p + H_{p+4}\partial_{\vec{x}}^2 H_p = 0,$$
$$\partial_z^2 H_{p+4} = 0,$$ (19.61)

and Q_i is proportional to $N_p N_{p+4}$, and Q is proportional to N_{p+4} (the number of $(p + 4)$-branes).

The most general scenario would have

$$ds^2 = (H_1 H_2)^{-1/2}(-dt^2 + d\vec{w}^2) + H_1^{1/2}H_2^{-1/2}d\vec{x}^2$$
$$+ H_2^{+1/2}H_1^{-1/2}d\vec{y}^2 + (H_1 H_2)^{1/2},$$ (19.62)

and would satisfy

$$\partial_z^2 H_1 + H_2\partial_{\vec{x}}^2 H_1 = 0$$
$$\partial_z^2 H_2 + H_1\partial_{\vec{x}}^2 H_2 = 0$$ (19.63)

and the constraint

$$\partial_{\vec{x}}H_1 \partial_{\vec{y}}H_2 = 0.$$ (19.64)

Important concepts to remember

- The condition for susy of the BPS solutions is that the supersymmetric variation of the fermions on the background should vanish, $\delta_s \psi = 0$.
- The mass (and momenta) of the BPS solutions is calculated usually not through the ADM parametrization (which would require a change of gauge for the BPS gravitational solution), but rather as $\int d^{D-p-1} x \theta_{00}^{(1)}$.
- The condition for the susy of the Dp-brane solutions is found to be $\Gamma^{01\cdots p} \epsilon_0 = \epsilon_0$, which breaks 1/2 susy. $\epsilon(x)$, related to ϵ_0 by a scalar function, is found usually from $\delta \psi_\mu = 0$, and the rest of the susy conditions are conditions on the BPS solution.
- In 11-dimensional M theory, we have the M2-brane, the M5-brane, the KK momentum or pp wave, and the KK monopole based on the Taub–NUT solution, while in string theory, we have the Dp-branes, fundamental string F1, and NS5-brane, as well as KK momentum or pp wave and KK monopole.
- All of the states correspond to all of the (central) charges in the susy algebra, realized on fields as electric or magnetic-type charges. All of the previously mentioned supergravity states, plus the 9-branes of M theory (end of the world branes) and 9-branes of string theory (Chan–Paton factors, world filling branes), make up all of the susy algebra charges.
- Intersecting brane solutions have composition rules: Dp self-intersects over $p - 2$, Dp and Dq over D-min$_{p,q} - 1$, and branes within branes are Dp inside Dq; one can add pp waves and KK monopoles to all.
- Compactification of a BPS solution can be realized by considering a periodic array of these BPS branes (which still preserves the BPS condition).
- For a susy solution, the brane within brane must be Dp inside D$(p+4)$: give the harmonic function H_{p+4} and find H_p from it.

References and further reading

For the supersymmetry of brane solutions, see, for instance, the review [56]. For the definition of M theory, see Witten's original paper, [59]. For the classification by supersymmetry algebra, see [60], and also [61]. For general brane intersections, see [62], and for partially localized brane intersection solutions, see [63].

Exercises

(1) Write down the solution for a pp wave inside an M5-brane in 11-dimensional supergravity. Is it supersymmetric, and why?

(2) Show that the type IIA and type IIB susy algebras are compatible with T-duality, meaning that the algebras obtained by reduction to nine dimensions are equivalent.

(3) Write down the type IIB solution for a D1-brane ending on a D5-brane.

(4) Write down explicitly the type IIA solution for D2-branes inside D6-branes.

U-duality group acting on supergravity theories and on solutions, M theory unification

In this chapter, we define the full duality group of supergravities in various dimensions, called U-duality, and their string theory counterparts, and define M theory unification, which is related to it.

20.1 Generalities

For a (linearized) compactification of 10-dimensional supergravity on a torus T^d, we have the KK reduction metric

$$ds^2_{10} = g_{ij}(dx^i + A^i_\mu dx^\mu)(dx^j + A^j_\nu dx^\nu) + g_{\mu\nu}dx^\mu dx^\nu, \qquad (20.1)$$

where $i, j = 1, ..., d$ are indices on the torus and μ, ν on the reduced space.

As we saw in previous chapters, we can define the T-duality invariant dilaton (actually, we had defined it first for a single compactified direction, but the generalization is obvious)

$$e^{-2\phi_d} = \sqrt{\det g_{mn}} e^{-2\phi}, \qquad (20.2)$$

and we can also define an $O(d, d)$ matrix, thus $2d \times 2d$, on a doubled space,

$$M = \begin{pmatrix} g^{-1} & g^{-1}B \\ -Bg^{-1} & g - Bg^{-1}B \end{pmatrix}, \qquad (20.3)$$

obeying the $O(d, d)$ relation

$$M^T \eta M = \eta, \quad \eta = \begin{pmatrix} 0 & \mathbb{1}_d \\ \mathbb{1}_d & 0 \end{pmatrix}. \qquad (20.4)$$

Then g_{mn} and B_{mn}, via M, parametrize the coset

$$M \in G/H = \frac{SO(d, d; \mathbb{R})}{SO(d) \times SO(d)}, \qquad (20.5)$$

where, as usual, $G = SO(d, d; \mathbb{R})$ is the isometry group and $H = SO(d) \times SO(d)$ is the local (Lorentz) invariance group.

However, while G is a symmetry group for supergravity, In string theory, that is, considering the quantum mechanics associated with it, $SO(d, d; \mathbb{R})$ is not a symmetry anymore.

What this means is that string states are quantized, and the symmetry must also be a symmetry of this quantized spectrum. But we don't need to know much about string theory, we can treat this spectrum also in supergravity, through the equivalence of supergravity states with string states, plus the notion of quantization of charges.

So, for instance, strings have integer momenta m_i on a compact direction i (as any particle does), but also integer windings w^i, in fact belonging to an integer, self-dual lattice Γ_p, but we will not use that notion here. Consider then the $2d$-dimensional "charge vector" $m \equiv (m_i, w^i)$.

Then we have the mass formula for states, which must be supplemented by the null matching condition for the vector,

$$\text{mass}^2 = m^T M m = (m_i + B_{ij} w^j) g^{ik} (m_k + B_{kl} w^l) + w^i g_{ij} w^j$$
$$||m||^2 = 0, \tag{20.6}$$

where m is the charge vector and M is the coset element.

This mass formula and constraint must be invariant under the symmetry group. But for a general $SO(d, d; \mathbb{R})$ transformation on M and m, the formula and constraint are not invariant. They are only invariant if the transformation has integer coefficients, so if we have $SO(d, d; \mathbb{Z})$ instead.

In fact, the same condition on integer coefficients is found from the conditions for D-brane states, which must also be imposed, as also part of the spectrum.

If we consider the compactified theory, as we said in the previous chapter, type IIA and type IIB supergravities are T-dual, so it doesn't matter which of the two we consider (as they are equivalent in the lower dimension). For this compactified type II supergravity, we also have an $Sl(2; \mathbb{R})$ symmetry (called S-duality), found as a symmetry of the equations of motion, under which the type IIB dilaton and axion, $e^{-\phi}$ and a, transform as a complex scalar, and the NS–NS and R–R fields $B_{\mu\nu}$ and $C_{\mu\nu}$ form a doublet. Again, quantum mechanics (the need for invariance of the quantized spectrum) breaks this symmetry to $Sl(2; \mathbb{Z})$.

Then the product of these two groups is part of a larger, U-duality group,

$$SO(d, d; \mathbb{Z}) \ltimes Sl(2; \mathbb{Z}) \subset \text{U} - \text{duality}, \tag{20.7}$$

where the symbol $\ltimes$ signifies that the two groups actually don't commute in general, so the group generated by their commutation relation is the one above.

To define U-duality, we must turn to M theory. But before that, we must learn better what M theory is.

20.2 M theory unification

In string theory, there is a unique nonperturbative definition of the theory. If type IIA string theory becomes in the low energy limit $\alpha' \to 0$ the type IIA supergravity, at strong coupling g_s (in the nonperturbative regime), it becomes this M theory, which itself at low energy $l_P \to 0$ becomes the 11-dimensional supergravity.

The KK reduction of 11-dimensional supergravity was defined in the previous chapter. Moreover, string theory doesn't have any free parameters, so the string coupling is actually the VEV of the exponential of the dilaton field,

$$\langle e^{\phi} \rangle = g_s. \tag{20.8}$$

Then the KK reduction metric from 11 to 10 dimensions,

$$ds_{11}^2 = e^{-\frac{2}{3}\phi} G_{\mu\nu} dx^\mu dx^\nu + e^{\frac{4}{3}\phi} (dx^{10} + A_\mu dx^\mu)^2, \tag{20.9}$$

means that the compact metric $g_{10,10}$, which must be interpreted as R^2 in 11-dimensional units, meaning in l_P units (the only dimensional object in the 11-dimensional theory), must equal the VEV of the dilaton factor in the above metric,

$$\left(\frac{R}{l_P}\right)^2 = e^{\frac{4}{3}\langle\phi\rangle} = g_s^{4/3} \Rightarrow R = l_P g_s^{2/3}, \tag{20.10}$$

or that

$$g_s = (R/l_P)^{3/2}. \tag{20.11}$$

Under the KK reduction, the 11-dimensional Einstein–Hilbert action must give the 10-dimensional Einstein–Hilbert action in string frame, as we can check (since under the KK reduction ansatz $\sqrt{-g^{(11)}} g_{(11)}^{\mu\nu} = \sqrt{-g^{s,(10)}} g_{s,(10)}^{\mu\nu} e^{-2\phi}$), so

$$\begin{aligned}
\frac{1}{2\kappa_{N,11}^2} \int d^{11}x \sqrt{-g} R^{(11)} &= \frac{2\pi R}{2\kappa_{N,11}^2} \int d^{10}x \sqrt{-g^{s,(10)}} e^{-2\phi} R^{(10)} + \dots \\
&= \frac{2\pi R}{(2\pi)^8 l_P^9} \int d^{10}x \sqrt{-g^{s,(10)}} e^{-2\phi} R^{(10)} + \dots \\
&= \frac{1}{g_s^2 (2\pi)^7 (\sqrt{\alpha'})^8} \int d^{10}x \sqrt{-g^{s,(10)}} e^{-2(\phi-\langle\phi\rangle)} R^{(10)} + \dots \\
&= \frac{1}{2\kappa_{N,10}^2} \int d^{10}x \sqrt{-g^{s,(10)}} e^{-2(\phi-\langle\phi\rangle)} R^{(10)} + \dots, \tag{20.12}
\end{aligned}$$

from which we define that

$$l_P^9 = R g_s^2 (\sqrt{\alpha'})^8, \tag{20.13}$$

which, together with the previously derived $R = l_P g_s^{2/3}$, means that

$$l_P = g_s^{1/3} \sqrt{\alpha'} \Rightarrow R = g_s \sqrt{\alpha'}. \tag{20.14}$$

Thus, as expected, the 11-dimensional radius is just the string coupling in string units.

Moreover, the M theory KK modes (momentum modes) on the 11th dimension give masses of 10-dimensional BPS states, "extremal black hole solutions,"

$$M_n = c \frac{n}{g_s}, \tag{20.15}$$

where c is the quantum of electric charge corresponding to them and n is an integer. These are in fact the masses of the D0-branes from the previous chapter.

One thing to note is that, while generically a fundamental particle has a mass independent of the coupling, $M_{\text{fundam}} \sim \mathcal{O}(1)$, and a soliton has a mass that is $M \sim 1/g^2$ (think for instance of the instanton, whose YM action has $1/g^2$ in front, and then a g-independent on-shell action, thus of order 1; the same is true for a monopole, and so on, for every standard field theory soliton), here we have $M \sim 1/g_s$. So these D0-branes (and actually all Dp-branes, as we will see) are something new, in between the fundamental particles and solitons.

We can in fact match the whole spectrum of M theory compactified on S^1 with the spectrum of 10-dimensional type IIA supergravity (or string theory). Defining a wrapped brane as a brane longitudinal to the compact direction (the direction is parallel to the brane) and an unwrapped brane as a brane transverse to the compact direction (the direction is transverse to the brane), we have the equivalence:

- a wrapped M2-brane with reduced (10-dimensional) tension (the tension in 11-dimensions is $1/l_P^3$)

$$T_1 = \frac{R_{10}}{l_p^3} = \frac{1}{l_s^3} \tag{20.16}$$

is reinterpreted as a fundamental string F1 in the IIA side.

- an unwrapped M2-brane with reduced (10-dimensional) tension

$$T_2 = \frac{1}{l_P^3} = \frac{1}{g_s l_s^3} \tag{20.17}$$

is reinterpreted as D2-brane in the IIA side.

- a wrapped M5-brane with reduced (10-dimensional) tension (the tension in 11 dimensions is $1/l_P^6$)

$$T_4 = \frac{R_{10}}{l_P^6} = \frac{1}{g_s l_s^5} \tag{20.18}$$

is reinterpreted as a D4-brane in the IIA side.

- an unwrapped M5-brane with reduced (10-dimensional) tension

$$T_5 = \frac{1}{l_P^6} = \frac{1}{g_s^2 l_s^6} \tag{20.19}$$

is reinterpreted as a NS5-brane in the IIA side.

- a wrapped KK (momentum) mode with reduced (10-dimensional) tension

$$T_0 = \frac{1}{R_{10}} = \frac{1}{g_s l_s} \tag{20.20}$$

is reinterpreted as a D0-brane in the IIA side.

- an unwrapped KK (momentum) mode with reduced (10-dimensional) tension

$$T_0 = \frac{1}{R_i} = \frac{1}{R_i} \tag{20.21}$$

is still a KK (momentum) mode on x^i in the IIA side.

- a wrapped KK6 monopole with reduced (10-dimensional) tension

$$T_5 = \frac{R_{10} R_{TN}^2}{l_P^9} = \frac{R_{TN}^2}{g_s^2 l_s^8} \tag{20.22}$$

is reinterpreted as a KK5 monopole in the IIA side.

- an unwrapped KK6 monopole with $R_{TN} = R_{10}$ with reduced (10-dimensional) tension

$$T_6 = \frac{R_{10}^2}{l_P^9} = \frac{1}{g_s l_s^7} \tag{20.23}$$

is reinterpreted as a D6-brane on the type IIA side.

Thus indeed, the M theory spectrum matches the type IIA spectrum. Moreover, we see that in general, the Dp-brane tension is

$$T_p = \frac{1}{g_s l_s^{p+1}}.$$

(20.24)

20.3 String duality web

Moreover, all string theories are obtained from M theory.

Type IIB theory is T-dual to type IIA (under circle compactification with radius R_9), which was derived from M theory as above. Then, more directly, M theory on a torus T^2 of vanishing area, with modulus $\tau(x, y)$ (the modulus of a 2-torus is a vector on the complex plane; the 2-torus is obtained by periodic identifications of the complex plane with τ and with the unit in the real direction) is related to type IIB theory on S^1, where now

$$\tau(x, y) = a + i e^{-\phi},$$

(20.25)

where a is the axion and ϕ the dilaton. The other type IIB parameters, in terms of M theory, are

$$g_s = \frac{R_{10}}{R_9}, \quad l_s^2 = \frac{l_P^3}{R_{10}}, \quad R_B = \frac{l_P^3}{R_9 R_{10}},$$

(20.26)

with R_B the T-dual compactification radius in type IIB.

The T-duality *between type IIA and type IIB* exchanges the corresponding spectra as follows:

- KK (momentum) modes with masses $M_n = n/R_i$ are exchanged with winding modes with

$$M_n = n \frac{R_i}{l_s^2}$$

(20.27)

and vice versa.
- Wrapped Dp-branes with reduced (nine-dimensional) tension

$$T_{p-1} = \frac{R_i}{g_s l_s^{p+1}}$$

(20.28)

are reinterpreted as unwrapped D$(p-1)$-branes with reduced (nine-dimensional) tension

$$T_{p-1} = \frac{1}{g_s l_s^{p}}.$$

(20.29)

- Wrapped NS5-branes with reduced (nine-dimensional) tension

$$T_4 = \frac{R_i}{g_s^2 l_s^6}$$

(20.30)

are interpreted as the same wrapped NS5-branes on the other side.

- Unwrapped NS5-branes with reduced (nine-dimensional) tension

$$T_5 = \frac{1}{g_s^2 l_s^6} \tag{20.31}$$

are reinterpreted as unwrapped KK5 monopoles with reduced (nine-dimensional) tension

$$T_5 = \frac{R_i^2}{g_s^2 l_s^6} \tag{20.32}$$

on the other side.

We can combine the M theory to IIA match with the IIA to IIB match to obtain a direct match between *M theory and type IIB* as:

- An M2-brane on x^{10}, with reduced (nine-dimensional) tension

$$T_1 = \frac{R_{10}}{l_P^3} = \frac{1}{l_s^2} \tag{20.33}$$

is reinterpreted as a fundamental string F1 in IIB.
- An M2-brane on x^9, with reduced (nine-dimensional) tension

$$T_1 = \frac{R_9}{l_P^3} = \frac{1}{g_s l_s^2} \tag{20.34}$$

is reinterpreted as a D1-brane in IIB.
- An M5-brane on x^9, x^{10}, with reduced (nine-dimensional) tension

$$T_5 = \frac{R_9 R_{10}}{l_P^6} = \frac{1}{g_s l_s^4} \tag{20.35}$$

is reinterpreted as a D3-brane in IIB.
- A KK6 monopole on x^9, with reduced (nine-dimensional) tension

$$T_5 = \frac{R_9 R_{10}^2}{l_P^9} = \frac{1}{g_s l_s^6} \tag{20.36}$$

is reinterpreted as a D5-brane in IIB.
- A KK6 monopole on x^{10}, with reduced (nine-dimensional) tension

$$T_5 = \frac{R_9^2 R_{10}}{l_P^9} = \frac{1}{g_s^2 l_s^6} \tag{20.37}$$

is reinterpreted as an NS5-brane in IIB.
- An unwrapped M2-brane, with reduced (nine-dimensional) tension

$$T_2 = \frac{1}{l_P^3} = \frac{R_B}{g_s l_s^4} \tag{20.38}$$

is reinterpreted as a wrapped D3-brane in IIB.
- An unwrapped M5-brane, with reduced (nine-dimensional) tension

$$T_5 = \frac{1}{l_P^6} = \frac{R_B^2}{g_s^2 l_s^8} \tag{20.39}$$

is reinterpreted as a wrapped KK5 monopole with $R_{TN} = R_{10}$ in IIB.

- An M5-brane on x^{10}, with reduced (nine-dimensional) tension

$$T_4 = \frac{R_{10}}{l_P^6} = \frac{R_B}{g_s l_s^5} \qquad (20.40)$$

is reinterpreted as a wrapped D5-brane in IIB.
- An M5-brane on x^9, with reduced (nine-dimensional) tension

$$T_4 = \frac{R_9}{l_P^6} = \frac{R_B}{g_s^2 l_s^6} \qquad (20.41)$$

is reinterpreted as a wrapped NS5-brane in IIB.
- An unwrapped KK6 monopole, with $R_{TN} = R_{10}$, with reduced (nine-dimensional) tension

$$T_6 = \frac{R_{10}^2}{l_P^9} = \frac{R_B}{g_s l_s^8} \qquad (20.42)$$

is reinterpreted as a wrapped D7-brane.

Thus we see that we can obtain directly the type IIB spectrum from the M theory spectrum.

Besides the type IIA and type IIB string theories and the corresponding supergravities, there are also three other string theories with only $\mathcal{N} = 1$ and YM fields (so an $\mathcal{N} = 1$ YM multiplet coupled to the $\mathcal{N} = 1$ supergravity multiplet), the heterotic $E_8 \times E_8$, heterotic $SO(32)$ and type I $SO(32)$.

In fact, these unique cases are obtained as the only solutions to the Green–Schwarz anomaly cancellation mechanism: in 10-dimensional theories with chiral fermions, we need to cancel the local anomalies, which is a highly nontrivial constraint (for instance, in the Standard Model, this anomaly cancellation condition is satisfied). In 10 dimensions, this amounts to the cancellation of hexagon diagram one-loop anomalies with fermions running in the loop, and it was shown to be possible only in the cases of IIA (fermions of both chiralities), IIB, gauge groups $E_8 \times E_8$, $SO(32)$, and $SO(16) \times SO(16)$ (while the latter doesn't correspond to a nice theory).

So, besides the IIA and IIB theories, we also have the $\mathcal{N} = 1$ theories: a heterotic $E_8 \times E_8$ theory, a heterotic $SO(32)$ theory and a type I $SO(32)$ theory, all of which have a corresponding supergravity theory in the low energy $\alpha' \to 0$ limit.

The two heterotic theories are T-dual to each other, and the two $SO(32)$ theories (heterotic and type I) are S-dual to each other (a strong/weak duality).

So all we need to finish the identification of the string/supergravity theories from M theory is a way to get one of the three extra theories: that is obtained by putting M theory on an interval $S^1/\mathbb{Z}_2$, and considering two M9 end-of-the-world branes at the ends of the interval. The length of the interval is πR_{10}, so in the weak coupling limit of string theory it vanishes (since $R_{10} = g_s l_s$). In this way we obtain the heterotic $E_8 \times E_8$ theory at strong coupling. Note that M2-branes can end on these M9-branes: M9-branes are boundary conditions for membranes, just like D-branes are boundary conditions for strings. In fact, by dimensional reduction, this corresponds to strings F1 ending on the world-filling 9-branes of string theory. The two E_8 gauge groups come from strings with both ends on 9-branes, giving as massless states YM gauge fields, lifted to 11 dimensions to M2-branes with both ends *on the same M9-brane.*

In conclusion, we see that all the possible string theories are obtained from M theory, as advertised, and the spectra of the string theories are also obtained from the spectra of M theory.

In the original paper of Witten on M theory, U-duality was used to define M theory better (since M theory and U-duality were defined more or less at the same time), but we can, reversely, use M theory to define U-duality better, which is what we will do next.

20.4 U-duality from M theory

We saw that, in type II supergravity (note that by compactification, type IIB is T-dual to type IIA, so they are equivalent, and we therefore need not specify which of the two type II supergravities we are talking about), the U-duality group must contain at least $SO(d, d;\mathbb{R}) \times Sl(2;\mathbb{R})$, but in quantum mechanics, the quantization of BPS states means that we must restrict to just integer coefficients, $SO(d, d;\mathbb{Z}) \ltimes Sl(2;\mathbb{Z})$.

To derive the possible symmetry groups in the lower dimensions, we KK reduce the 11-dimensional susy algebra (since, as we saw, the susy algebra gives also the full BPS states by dimensional reduction).

- Compactifying the 11-dimensional susy algebra on T^5 (so $d = 5$), down to $D = 6$, we obtain

$$\{Q_\alpha^a, Q_\beta^b\} = \Omega^{ab}(\gamma^\mu)_{\alpha\beta}P_\mu$$
$$\{Q_\alpha^a, \bar{Q}_{\bar\beta}^b\} = \delta_{\alpha\bar\beta}Z^{ab}, \tag{20.43}$$

where $a, b = 1, .., 4$, meaning the supercharges transform in $USp(4) = SO(5)$, with Ω^{ab} their invariant matrix. But we see that we have independent transformations for Q and $\bar{Q}$, so the total R-symmetry group is $SO(5) \times SO(5)$. Note that R-symmetry acts on the fermions, in particular on the supercharges, as we have here.
- Compactifying the 11-dimensional susy algebra on T^6 ($d = 6$), down to $D = 5$, we obtain

$$\{Q_{\alpha A}, Q_{\beta B}\} = (C\gamma^\mu)_{\alpha\beta}\Omega_{AB}P_\mu + C_{\alpha\beta}Z_{AB}, \tag{20.44}$$

where $A, B = 1, ..., 8$, which means that the supercharges transform in $USp(8)$, with Ω_{AB} the invariant symplectic form. Therefore the R-symmetry is now $USp(8)$.
- Compactifying the 11-dimensional supergravity on T^7 ($d = 7$), down to $D = 4$, we obtain

$$\{Q_{\alpha A}, Q_{\beta B}\} = \epsilon_{\alpha\beta}Z_{AB}$$
$$\{Q_{\dot\alpha\bar A}, Q_{\dot\beta,\bar B}\} = \epsilon_{\dot\alpha\dot\beta}Z_{\bar A\bar B}^*, \tag{20.45}$$

where $A, \bar A = 1, ..., 8$, so the supercharges transform in $SU(8)$, meaning the R-symmetry group is $SU(8)$.

The other cases, similarly found, are as follows:

- For compactification on S^1, there is nothing.
- For compactitication on T^2, there is an $SO(2)$ R-symmetry.
- For compactification on T^3, there is an $SO(2) \times U(1)$ R-symmetry.
- For compactification on T^4, there is an $SO(5)$ R-symmetry.
- For compactification on T^8, down to $D = 3$, there is an $SO(16)$ R-symmetry.

But considering 11-dimensional supergravity on $T^{d'}$, where now we denote $d' = d + 1$ (so that d is the dimension of the torus for string theory compactification), there is also an extra Lorentz symmetry for $T^{d'}$, equal to $Sl(d';\mathbb{R})$ (just like for T^2, we had $Sl(2;\mathbb{R})$).

This means that the continuous symmetry group of compactified supergravity, $G_{d'}$, contains at least

$$G_{d'} \supset SO(d' - 1, d' - 1;\mathbb{R}) \ltimes Sl(d';\mathbb{R}). \tag{20.46}$$

But already Cremmer and Julia showed that for 11-dimensional supergravity on $T^{d'}$, the symmetry group is $G_{d'} = E_{d'(d')}$, which is a noncompact form of $E_{d'}$, understood as the exceptional groups E_6, E_7, E_8, and their usual extensions (using Dynkin diagrams), using identifications to classic groups. One then obtains the following table:

D	d'	$G_{d'} = E_{d'(d')}$	$H_{d'}$=maximal compact subgroup
10	1	$\mathbb{R}^+$	1
9	2	$Sl(2;\mathbb{R}) \times \mathbb{R}^+$	$U(1)$
8	3	$Sl(3;\mathbb{R}) \times Sl(2;\mathbb{R})$	$SO(2) \times U(1)$
7	4	$Sl(5;\mathbb{R})$	$SO(5)$
6	5	$SO(5,5;\mathbb{R})$	$SO(5) \times SO(5)$
5	6	$E_{6(6)}$	$USp(8)$
4	7	$E_{7(7)}$	$SU(8)$
3	8	$E_{8(8)}$	$SO(16)$

The scalars in D dimensions fit as representations of the coset $G_{d'}/H_{d'}$.

Then, in quantum mechanics, the charge quantization restricts the duality group in a similar manner.

We consider a lattice of charges for the theory on T^d, made up of electric and magnetic charges and wrapping charges.

In particular, for the most relevant example of M theory on T^7, down to $D = 4$, where we can have only particle charges (electric and magnetic), but no extended objects (so no wrapping charges), since a particle is Poincaré dual to another particle, we have the usual Dirac–Schwinger–Zwanziger quantization condition,

$$m^i n'_i - m'^i n_i \in (2\pi\hbar)\mathbb{Z}, \tag{20.47}$$

where m^i are electric charges, and n_i are magnetic charges, for $i = 1, ..., 28$, relation that is invariant under an $Sp(56;\mathbb{Z})$ transformation, under which (m^i, n_i) is a vector.

Therefore, the duality group can be at most

$$E_{7(7)}(\mathbb{Z}) \in E_{7(7)}(\mathbb{R}) \cap Sp(56;\mathbb{Z}), \tag{20.48}$$

which gives the condition for $E_{d(d)}(\mathbb{Z})$ for $d \leq 7$.

Then, the U-duality group is

$$E_{d'(d')}(\mathbb{Z}) = SO(d'-1, d'-1;\mathbb{Z}) \rtimes Sl(d';\mathbb{Z}). \tag{20.49}$$

We calculate this and find the following table of U-duality groups:

D	d'	$E_{d'(d')}(\mathbb{R})$	$E_{d'(d')}(\mathbb{Z})$ U-duality group
10	1	1	1
9	2	$Sl(2;\mathbb{R})$	$Sl(2;\mathbb{Z}) \times \mathbb{Z}_2$
8	3	$Sl(3;\mathbb{R}) \times Sl(2;\mathbb{R})$	$Sl(3;\mathbb{Z}) \times Sl(2;\mathbb{Z})$
7	4	$Sl(5;\mathbb{R})$	$Sl(5;\mathbb{Z})$
6	5	$SO(5,5;\mathbb{R})$	$SO(5,5;\mathbb{Z})$
5	6	$E_{6(6)}(\mathbb{R})$	$E_{6(6)}(\mathbb{Z})$
4	7	$E_{7(7)}(\mathbb{R})$	$E_{7(7)}(\mathbb{Z})$
3	8	$E_{8(8)}(\mathbb{R})$	$E_{8(8)}(\mathbb{Z})$

The extremality relation "$Q = M$," after KK reduction on T^d depends on scalars = moduli, since for instance, in four dimensions the charges are dimensionless, whereas the masses have dimensions. More precisely, it should be $Z = M$, with Z a central charge. The general relation between the central charges, equal to the mass, and the quantized charges under the gauge fields, with charge vector m, is generically

$$Z = v \cdot m, \tag{20.50}$$

where $v \in H_{d'} \backslash G_{d'}$ is the matrix of the moduli (scalars), belonging to the (left) coset with isometry group $G_{d'}$ and local Lorentz $H_{d'}$.

For instance, for the particle multiplet obtained for M theory compactified on T^7, we obtain an $E_{7(7)}(\mathbb{Z})$ (U-duality group for $D = 4$ dimensions)-invariant mass formula.

The formula is written in terms of the following scalars: the compact metric components g_{IJ}, and the compact components of the 11-dimensional 3-form $A_{(3)}$, so A_{KLM}, and its Poincaré dual, $\tilde{A}_{(6)}$, so $\tilde{A}_{MNPQRS}$ (so we are in the "democratic formalism"), which gives am explicit parametrization of the upper triangular scalar matrix v in terms of physical compactification parameters.

Then the mass formula is written as (here $I, J = 0, 1, ..., 10$ in general, though we consider only the compact indices)

$$M^2 = (\tilde{m}_{(1)})^2 + \frac{1}{2!l_P^6}(\tilde{m}^{(2)})^2 + \frac{1}{5!l_P^{12}}(\tilde{m}^{(5)})^2 + \frac{1}{7!l_P^{18}}(\tilde{m}^{(1;7)})^2, \qquad (20.51)$$

where

$$\tilde{m}_I = m_I + \frac{1}{2}A_{JKI}m^{JK} + \left(\frac{1}{4!}A_{JKL}A_{MNI} + \frac{1}{5!}\tilde{A}_{JKLMNI}\right)m^{JKLMN}$$

$$+ \left(\frac{1}{3!4!}A_{JKL}A_{MNP}A_{QRI} + \frac{1}{2!5!}A_{JKL}A_{MNPQRI}\right)m^{J;KLMNPQR}$$

$$\tilde{m}^{IJ} = m^{IJ} + \frac{1}{3!}A_{KLM}m^{KLMIJ} + \left(\frac{1}{4!}A_{KLM}A_{NPQ} + \frac{1}{5!}\tilde{A}_{KLMNPQ}\right)m^{K;LMNPQIJ}$$

$$\tilde{m}^{IJKLM} = m^{IJKLM} + \frac{1}{2}A_{NPQ}m^{N;PQIJKLM}$$

$$\tilde{m}^{I;JKLMNPQ} = m^{I;JKLMNPQ}, \qquad (20.52)$$

and where, for instance,

$$(\tilde{m}_{(1)})^2 = \tilde{m}_I g^{IJ} \tilde{m}_J, \quad (\tilde{m}^{(2)})^2 = \tilde{m}^{IJ} g_{IK} g_{JL} \tilde{m}^{KL}, \text{ and so on} \qquad (20.53)$$

and $g_{IJ} = R^I R^J$, so $g^{IJ} = 1/(R^I R^J)$.

The above mass formula has *manifest* $Sl(d'; \mathbb{Z})$ symmetry, but it is also invariant under $SO(d'-1, d'-1; \mathbb{Z})$ T-duality.

Here $m_{(1)}, m^{(2)}, m^{(5)}$, and $m^{(1;7)}$ are for KK momentum, M2-brane, M5-brane, and KK6 monopole wrappings, respectively, with m_I the vector of momentum charge, m^{IJ} the 2-cycle charges, m^{IJKLM} the 5-cycle charges and $m^{I;JKLMNPQ}$ the (7;1)-cycle charges, so for instance, m^{IJ} is the state of the M2-brane wrapped on directions of radii R^I, R^J, with *undeformed* mass $R^I R^J / l_P^3$.

Finally, to complete the U-duality groups, we can write a similar classification for the heterotic string (again, when compactifying, it doesn't matter which heterotic string it is, since they are T-dual to each other) compactified to D dimensions. We obtain

D	Sugra duality group	String T-duality group	Full U-duality group
10	$O(16) \times SO(1,1)$	$O(16; \mathbb{Z})$	$O(16; \mathbb{Z}) \times \mathbb{Z}_2$
9	$O(1,17) \times SO(1,1)$	$O(1,17; \mathbb{Z})$	$O(1,17; \mathbb{Z}) \times \mathbb{Z}_2$
8	$O(2,18) \times SO(1,1)$	$O(2,18; \mathbb{Z})$	$O(2,18; \mathbb{Z}) \times \mathbb{Z}_2$
7	$O(3,19) \times SO(1,1)$	$O(3,19; \mathbb{Z})$	$O(3,19; \mathbb{Z}) \times \mathbb{Z}_2$
6	$O(4,20) \times SO(1,1)$	$O(4,20; \mathbb{Z})$	$O(4,20; \mathbb{Z}) \times \mathbb{Z}_2$
5	$O(5,21) \times SO(1,1)$	$O(5,21; \mathbb{Z})$	$O(5,21; \mathbb{Z}) \times \mathbb{Z}_2$
4	$O(6,22) \times SO(1,1)$	$O(6,22; \mathbb{Z})$	$O(6,22; \mathbb{Z}) \times Sl(2; \mathbb{Z})$
3	$O(8,24)$	$O(7,23; \mathbb{Z})$	$O(8,24; \mathbb{Z})$

Important concepts to remember

- A continuous symmetry group for supergravity compactification is reduced to a discrete one in the quantum theory (= string theory), by the need to keep the mass spectrum invariant as well.
- U-duality combines T-dualities and S-dualities into a larger group (containing both).
- M theory is string theory at large coupling, acting as an 11th dimension, $R = g_s l_s$.
- All five string theories are obtained from M theory, and their states are matched with M theory states.
- Type IIA and type IIB are T-dual, and IIA is M theory on a circle. The heterotic $E_8 \times E_8$ string theory is M theory on an interval ending in M9-branes, and is T-dual to the $SO(32)$ heterotic string, which is itself S-dual to type I $SO(32)$.
- The 11-dimensional supergravity compactified on $T^{d'}$ continuous symmetry groups are $E_{d'(d')}$, a noncompact form of $E_{d'}$, that includes $SO(d'-1, d'-1; \mathbb{R}) \times Sl(d'; \mathbb{R})$.
- In the quantum theory (string theory), the U-duality group is $E_{d'(d')}(\mathbb{Z}) = SO(d'-1, d'-1; \mathbb{Z}) \ltimes Sl(d'; \mathbb{Z})$, which leaves invariant the BPS mass formulas.

References and further reading

For U-duality, see the original paper [64], as well as the review [65]. M theory was defined by Witten in [59].

Exercises

(1) (Hull's duality) Show that *massive* type IIA supergravity, with string frame action

$$
\frac{1}{\kappa_{N,10}^2} \int d^{10}x \sqrt{-g} \left(e^{-2\phi} R + \tilde{M}^2 + ... \right) = \frac{1}{\kappa_N^2} \int d^{10}x \sqrt{-g} \left(e^{-2(\phi-\phi_0)} R + g_s^2 \tilde{M}^2 + ... \right),
$$
(20.54)

where $\tilde{M} = m/l_s$ is the supergravity mass parameter, can be obtained from M theory on a three-dimensional space $B(A, R_3)$, composed of a two-dimensional torus of modulus depending on the third coordinate, $\tau(x_3)$, with the area $A \to 0$ and $R_3 \to 0$, and metric

$$
ds_3^2 = R_3^2 dx_3^2 + \frac{A}{\text{Im}(\tau)} |dx_1 + \tau(x_3)dx_2|^2 = R_3^2 dx_3^2 + R_2^2 dx_2^2 + R_1^2 (dx_1 + mx_3 dx_2)^2, \quad (20.55)
$$

with all the radii going to 0, and periodicities $x_i \sim x_i + 1$, keeping fixed

$$
g_s^A = \frac{l_s}{\text{Im}\tau_0 R_3} = \frac{R_1 l_s}{R_2 R_3} = \text{fixed}, \quad l_s = \frac{l_P^{3/2}}{R_1^{1/2}} = \text{fixed}, \quad m = \text{fixed}. \quad (20.56)
$$

Hint: Consider the T-duality to type IIB theory on x_3, with a "Sherck-Schwarz compactification," namely fields defined modulo an element of $Sl(2;\mathbb{Z})$ (in the quantum theory), and so with $\tau(x, x_3) = \tau(x) + mx_3/R_3$.

(2) Use the ansatz at exercise 1 to find the M theory solution that corresponds to a D8-brane (massive type IIA solution charged under the mass parameter).

(3) Show that the states of the heterotic string are obtained from the states of M theory, just like the type IIA states are obtained in the text.

(4) Rewrite the mass formula for the BPS particle multiplet in $D = 4$ in terms of type IIA 10-dimensional quantities.

Gravity duals: Decoupling limit and Penrose limits on solutions and algebras

In this chapter, we will describe the supergravity limit of the AdS/CFT correspondence and, more generally of gauge/gravity duality, focusing on the gravity duals, decoupling limits for obtaining them, and the Penrose limits for the gravity duals (solutions) and their symmetry algebras.

21.1 The AdS/CFT correspondence

We will first describe the original heuristic derivation of the correspondence, for the standard case.

The derivation is based on having two points of view for the Hawking process, in a low-energy limit. The points of view are related to the equality of the string theory D-branes, as objects defined abstractly, with the extremal p-brane solutions of supergravity, which equality we have already described.

(1) In the first point of view, the string theory one, two open strings on a D-brane can collide, creating a closed string, that then peels off from the D-brane, understood as an abstract string theory object, the end of the open strings, and moves into the bulk.

(2) In the second point of view, the D-brane is replaced by the extremal p-brane solution of supergravity. Such a solution, when perturbed a bit away from extremality, to be a black p-brane of nonzero temperature and with a horizon, emits Hawking radiation, via an interaction between the gravitational bulk and its boundary, situated at the D-brane's position.

In the original D3-brane case, we have the following:

Point of view nr. 1: The theory contains the following:

– Open strings on the D3-branes. In the low energy $\alpha' \to 0$ limit (α' is the string length squared), the theory reduces to the theory of the massless open strings only, which is $\mathcal{N} = 4$ SYM.

– Closed strings in the bulk. In the low-energy limit, understood as $\delta r \to \infty$, we obtain just the theory of gravitons (massless closed strings).

– Interactions between the two, with $S_{\text{int}} \propto \kappa_N \sim g_s \alpha'^2$. Then for $\alpha' \to 0$, $S_{\text{int}} \to 0$, and also $\kappa_N \to 0$, so we obtain actually just free (super)gravity.

Point of view nr. 2: We have a supergravity solution, with $g_{00} = H_3^{-1/2}$. Consider the energy measured at point p in it, E_p, and consider a point situated at $r \to 0$. Then we have

$$E_p \to i\frac{d}{d\tau} = \frac{i}{\sqrt{-g_{00}}}\frac{d}{dt} \to \frac{1}{\sqrt{-g_{00}}}E, \tag{21.1}$$

where E is the energy measured at infinity, where $H_3 \to 1$. Since

$$H_3 = 1 + \frac{R^4}{r^4}, \quad R^4 = 4\pi g_s N\alpha'^2, \quad Q = g_s N, \tag{21.2}$$

we have

$$E = H_3^{-1/4}E_p \propto rE_p. \tag{21.3}$$

We must consider fixed E_p (energy in gravity), which means that as $r \to 0$, the energy at infinity $E \to 0$, so we have low energy.

To consider the process of Hawking radiation in the supergravity solution, we must move a bit away from the extremal limit $Q = M$, and add a small $\delta M \to 0$, giving a small Hawking temperature $T \to 0$. In this case, we obtain that the theory contains the following:

– At large distances in the gravitational background, $\delta r \to \infty$, or low energy, we obtain free gravity as before, since the effective (dimensionless, G_N has dimension) coupling of gravity is $G_N E^2 \to 0$.
– At $r \to 0$, we also obtain low-energy gravitational modes.
– Again, there is no interaction.

Identifying the two points of view, in the low-energy limit, we have two sets of modes that must be equal, but one of them is the same: free gravity at large distances. This means that the other must also be the same, and their identification leads to the definition of the AdS/CFT correspondence:

"$\mathcal{N} = 4\ SYM$" = gravity at $r \to 0$ in the D3-brane gravitational background, if we have $\alpha' \to 0$.

The $r \to 0$, $\alpha' \to 0$ limit is called a decoupling limit, as the interaction between the two modes of the theory vanishes: we decouple the *boundary* D-brane field theory from the *bulk* gravity theory, through $S_{\text{int}} \propto \alpha'^2 \to 0$.

Now we try to define better the decoupling limit. If $r \to 0$, then

$$H_3 \simeq \frac{R^4}{r^4} \Rightarrow ds^2 \simeq \frac{r^2}{R^2}(-dt^2 + d\vec{x}^2) + \frac{R^2}{r^2}dr^2 + R^2 d\Omega_3^2. \tag{21.4}$$

As $\alpha' \to 0$, we need that the string energy to be fixed in string units (energy has dimensions, so it is strictly speaking meaningless to say that an energy is fixed by itself), so $E_p\sqrt{\alpha'}$ fixed, while also we want the energy at infinity, E, to be fixed, but

$$E = E_p H_3^{-1/4} \propto E_p \frac{r}{\sqrt{\alpha'}}, \tag{21.5}$$

which implies that we need that

$$U = \frac{r}{\alpha'} = \text{fixed}. \tag{21.6}$$

This will be the energy scale in the field theory (indeed, it has dimensions of energy).

Finally then, taking into account that $R^4 = 4\pi g_s N \alpha'^2$, the full gravitational solution is

$$ds^2 = \alpha' \left[\frac{U^2}{\sqrt{4\pi g_s N}} (-dt^2 + d\vec{x}_3^2) + \sqrt{4\pi g_s N} \left(\frac{dU^2}{U^2} + d\Omega_5^2 \right) \right],$$

$$F_5 = (1 + *) dt \wedge dx_1 \wedge dx_2 \wedge dx_3 \wedge d(H_3^{-1})$$

$$\rightarrow (1 + *) \frac{4r^3}{R^4} dt \wedge dx_1 \wedge dx_2 \wedge dx_3 \wedge dr$$

$$= 4R^4 (1 + *) \epsilon_{(5)} = 16\pi g_s \alpha'^2 N (1 + *) \epsilon(5), \qquad (21.7)$$

where

$$\epsilon_{(5)} = \frac{\sqrt{-g}}{R^5} dx^0 \wedge ... \wedge dx^3 \wedge dr \qquad (21.8)$$

is the volume form on AdS_5, which integrated gives 1 (note that $\sqrt{-g} \propto R/R^4 = 1/R^3$, so the powers of R work out).

Although we will not be interested in this here, as we will only work with the gravity side of the AdS/CFT correspondence, we note here that the map to $\mathcal{N} = 4$ SYM is given by

$$4\pi g_s = g_{YM}^2 \qquad (21.9)$$

and N (the number of D3-branes, or charge) becomes the rank of the gauge group $SU(N)$.

The original statement of AdS/CFT, in the original paper of Maldacena in 1997, is that one considers the $\alpha' \rightarrow 0$ limit, but together with the condition for no string *worldsheet* corrections, meaning that the string length parameter must be much smaller than the curvature length of the space, $R \gg \sqrt{\alpha'}$, and no string *quantum* corrections, so $g_s \rightarrow 0$. Later, it was found that the AdS/CFT correspondence is actually valid at all g_s and N (so R), but as we are only interested in supergravity, we will not consider this here.

In this limiting case, we have only free classical supergravity, so we only need to consider the supergravity backgrounds. In general, these are called "gravity duals," and are obtained in a decoupling limit.

We also note that the case described in this section is the "top-down" construction, in which we start with a certain brane background in string theory and take a decoupling limit. But in principle, we can also use a "bottom-up," or phenomenological one, where we assume that the holographic map is valid, and invent a background that gives the correct properties (that we seek) for the nonperturbative, strongly coupled, dual field theory, and then use it to calculate more properties of the field theory. However, we will not be interested in that here. Here we will only deal with the supergravity backgrounds obtained in the decoupling limits of brane theories.

Besides the case described in this section, original AdS/CFT case, other cases with maximal supersymmetry are as follows.

21.2 M theory cases

Within M theory, we have three cases of AdS/CFT with maximal supersymmetry.

From N extremal M5-branes, we obtain $AdS_7 \times S^4$.

The N M5-branes solution is

$$ds^2 = H_5^{-1/3}(r)(-dt^2 + d\vec{x}_{(5)}^2) + H_5^{2/3}(r)(dr^2 + r^2 d\Omega_4^2),$$
$$F_{(4)} = * \left(dt \wedge dx^1 \wedge ... \wedge dx^5 \wedge d \left(H_5^{-1} \right) \right),$$
$$H_5 = 1 + \frac{\pi N l_P^3}{r^3}. \tag{21.10}$$

The gravity decoupling limit in M theory is with $l_P \to 0$ (the only dimensional parameter in M theory, replacing α'), and the same near-horizon limit $r \to 0$ (for the extremal brane, there is a horizon at $r = 0$), but now we must keep

$$U^2 \equiv \frac{r}{l_P^3} \tag{21.11}$$

fixed instead, as we now show. By the same logic from before, we must have

$$E = E_p \sqrt{-g_{00}} = E_p H_5^{-1/6} \sim E_p \sqrt{\frac{r}{l_P}}, \tag{21.12}$$

and keep this fixed, while also keeping the gravitational energy in gravitational units, $E_p l_p$ fixed, so we obtain that the above $U = \sqrt{r/l_P^3}$ is fixed.

Then, in the decoupling limit, the gravitational background becomes

$$ds^2 = l_P^2 \left[\frac{U^2}{(\pi N)^{1/3}} (-dt^2 + d\vec{x}_{(5)}^2) + 4(\pi N)^{2/3} \frac{dU^2}{U^2} + (\pi N)^{2/3} d\Omega_4^2 \right],$$
$$F_{(7)} = \frac{3r^2}{\pi N l_P^3} dx^0 \wedge dx^1 \wedge ... \wedge dx^5 \wedge dr \Rightarrow,$$
$$F_{S^4} = \frac{3}{\sqrt{4!7!}} l_P^3 \pi N \epsilon_{S^4}, \tag{21.13}$$

where $R_{S^4} = l_P (\pi N)^{1/3}$ and $R_{AdS_7} = 2 R_{S^4}$.

From N extremal M2-branes, we obtain $AdS_4 \times S^7$.
The N M2-branes solution is

$$ds^2 = H_2^{-2/3}(-dt^2 + dx^2 + dy^2) + H_2^{1/3}(dr^2 + r^2 d\Omega_7^2),$$
$$F_{(4)} = d(H_2^{-1}) \wedge dt \wedge dx \wedge dy,$$
$$H_2(r) = 1 + \frac{32\pi^2 N l_P^6}{r^6}. \tag{21.14}$$

The decoupling limit is again $r \to 0, l_P \to 0$, but now with

$$U = \frac{r^2}{l_P^3} \tag{21.15}$$

fixed, as we now show. As before, we have

$$E = E_p \sqrt{-g_{00}} = E_p H_2^{-1/3} \sim E_p \frac{r^2}{l_P^2} \tag{21.16}$$

fixed, but $E_p l_P$ also fixed, which means that, indeed, $U = r^2/l_P^3$ is fixed.

In the decoupling limit, the solution becomes

$$ds^2 = l_P^2 \left[\frac{U^2}{(32\pi N)^{2/3}} (-dt^2 + dx^2 + dy^2) + (32\pi^2 N)^{1/3} \left(\frac{1}{4} \frac{dU^2}{U^2} + d\Omega_7^2 \right) \right],$$

$$F_{(4)} = \frac{6r^5}{32\pi^2 N l_P^6} dr \wedge dt \wedge dx \wedge dy = \frac{\pi \sqrt{N}}{2} l_P^3 \epsilon_{(4)}. \tag{21.17}$$

The Aharony–Bergman–Jafferis–Maldacena (ABJM) model and its $AdS_4 \times \mathbb{CP}^3$ dual
The ABJM field theory is an $\mathcal{N} = 6$ supersymmetric, three-dimensional CS gauge field theory with gauge group $SU(N) \times SU(N)$.

It is obtained from N M2-branes, which were considered above in flat space, but now we consider them instead near a $\mathbb{C}^4/\mathbb{Z}_k$ singularity, and moreover we take the IR limit, which means we obtain a conformal field theory (at the fixed point).

For the N M2-branes near the $\mathbb{C}^4/\mathbb{Z}_k$ singularity, we obtain in the decoupling limit, similarly to what was done in flat space, an $AdS_4 \times S^7/\mathbb{Z}_k$ space. Indeed, the S^7 is described in $\mathbb{C}^4$ (the embedding eight-dimensional flat space) by

$$\sum_{i=1}^4 |Z_i|^2 = 1, \tag{21.18}$$

but the S^7 is a Hopf fibration over $\mathbb{CP}^3$ with fiber S^1, defined by

$$Z_i = e^{i\alpha} Z_i', \quad \forall i. \tag{21.19}$$

Then Z_i' defines the $\mathbb{CP}^3$, with coordinates

$$\zeta_l = \frac{Z_l}{Z_4}, \quad l = 1, 2, 3, \tag{21.20}$$

where the sphere constraint becomes just

$$\sum_{l=1}^3 |\zeta_l|^2 + 1 = \frac{1}{|Z_4|^2}, \tag{21.21}$$

which just defines Z_4, outside of $\mathbb{CP}^3$.

Finally, the action of the $\mathbb{Z}_k$ factor is just on the S^1 fiber, reducing it k times. Therefore, in the $k \to \infty$ limit, we are left with just the $AdS_4 \times \mathbb{CP}^3$ gravity dual, a solution of type IIA string theory (the circle reduction of 11-dimensional M theory, or supergravity).

21.3   Gravity dual of N extremal Dp-branes

For N Dp-branes in general, for $p \neq 3$, we have no AdS factor, so the dual field theory is not conformal. Indeed, the isometry of AdS_{d+1} is $SO(d, 2)$, which equals the conformal group in d dimensions, meaning that whenever we have an AdS factor in gravity, we have a dual conformal field theory, but if not, like here, the theory is not conformal, meaning it is energy-dependent.

The N extremal Dp-branes solution is

$$ds^2 = H_p^{-1/2}(-dt^2 + d\vec{x}_p^2) + H_p^{1/2}(dr^2 + r^2 d\Omega_{8-p}^2),$$

$$e^{(\phi-\phi_\infty)} = H_p^{\frac{3-p}{4}}, \quad g_s = e^{\phi_\infty},$$

$$A_{01...p} = -\frac{1}{2}(H_p^{-1} - 1),$$

$$H_p = 1 + \frac{d_p(2\pi)^{p-2}g_s N(\sqrt{\alpha'})^{7-p}}{r^{7-p}}. \tag{21.22}$$

But now, the field theory YM coupling has dimensions, so it's being fixed means that

$$g_{YM}^2 = (2\pi)^{p-2}g_s\alpha'^{\frac{p-3}{2}} \tag{21.23}$$

must be fixed. Then, defining $U = r/\alpha'$ as before, we find

$$H_p = 1 + \frac{d_p g_{YM}^2 N}{U^{7-p}\alpha'^2}, \tag{21.24}$$

so that we have, as in the previous cases,

$$E = E_p H_p^{-1/4} \sim E_p\sqrt{\alpha'}\frac{U^{\frac{7-p}{4}}}{(g_{YM}^2 N)^{1/4}}, \tag{21.25}$$

and if it is fixed, g_{YM} is fixed, and moreover $E_p\sqrt{\alpha'}$ fixed as before, then again U is fixed, as for $p = 3$.

But then the decoupling limit, $\kappa_N^2 \propto g_s\alpha'^2 \to 0$, is correct if $p < 7$ (since $g_{YM}^2 = (2\pi)^{p-2}[g_s\alpha'^2]\alpha'^{\frac{p-7}{2}}$ is fixed), but then $g_s \to 0$ as well only if $p \le 3$. If instead we have $p > 3$, then $g_s \to \infty$ so, if available, we must use an S-dual description, with $\tilde{g}_s \to 0$, and hope that we can still work with it.

Since the supergravity description is valid only when the field theory is nonperturbative, we must see what the latter means. In $p+1$ dimensions, in general (for $p \ne 3$), the coupling has dimensions, so the effective (dimensionless) gauge theory coupling is

$$g_{\text{eff}}^2 = g_{YM}^2 N U^{p-3}, \tag{21.26}$$

which means that the condition for nonperturbative SYM, $g_{\text{eff}}^2 \gg 1$, so we can trust the dual supergravity description, implies a range of field theory energies U,

$$U = \begin{cases} (g_{YM}^2 N)^{\frac{1}{3-p}}, & p < 3 \\ (g_{YM}^2 N)^{\frac{1}{p-3}}, & p > 3 \end{cases}. \tag{21.27}$$

Then, in the decoupling limit, the supergravity solution becomes

$$ds^2 = \alpha'\left[\frac{U^{\frac{7-p}{2}}}{g_{YM}\sqrt{d_p N}}(-dt^2 + d\vec{x}_p^2) + \frac{g_{YM}\sqrt{d_p N}}{U^{\frac{7-p}{2}}} + g_{YM}\sqrt{d_p N}U^{\frac{p-3}{2}}d\Omega_{8-p}^2\right],$$

$$e^\phi = (2\pi)^{2-p}g_{YM}^2\left(\frac{g_{YM}^2 d_p N}{U^{7-p}}\right)^{\frac{3-p}{4}} \sim \frac{g_{\text{eff}}^{\frac{7-p}{2}}}{N}. \tag{21.28}$$

This means that we can trust supergravity, for $g_s = e^\phi \ll 1$ and

$$\alpha' \mathcal{R} \sim \frac{1}{g_{\text{eff}}} = \sqrt{\frac{U^{3-p}}{g_{YM}^2 N}} \ll 1, \tag{21.29}$$

in a limited range of g_{eff}^2,

$$1 \ll g_{\text{eff}}^2 \ll N^{\frac{4}{7-p}}. \tag{21.30}$$

21.3.1 Observations on supersymmetry

Dp-branes and M-branes all preserve only half of supersymmetry, as we saw in Chapter 20, via the constraint $\Gamma^{01...p} \epsilon_0 = \pm \epsilon_0$. On the other hand, $AdS_5 \times S^5$, $AdS_4 \times S^7$, and $AdS_7 \times S^4$, obtained as limits of them, are maximally (fully) supersymmetric. This is part of a general story, that when we take a limit we cannot lose supersymmetry, since the supersymmetry generators may be at most rescaled and recombined. Supersymmetry can only increase, via the appearance of new Killing spinors.

In fact, there is a theorem by Kowalski–Glikman in 10-dimensional type IIB, and a similar one for 11-dimensional supergravity, that the only maximally supersymmetric backgrounds are Minkowski space (10-dimensional and 11-dimensional, respectively), $AdS_5 \times S^5$ in IIB and $AdS_4 \times S^7$ and $AdS_7 \times S^4$ in 11 dimensions, as well as the maximally supersymmetric pp waves that are obtained as their limits, as we will show next.

21.4 Penrose limits of AdS/CFT dualities

The Penrose limit is defined by the *Penrose theorem*, which states that in the neighborhood of a null geodesic space becomes a pp wave. This is seen as follows. Consider the null geodesic defined by $V = 0 = Y^i$, where V is a null coordinate, and Y^i are coordinates transverse to the motion of propagation (we can always choose coordinates such that the null geodesic is $V = 0 = Y^i$). Moreover, the motion is in $U = \tau$, also a null coordinate.

Then, part of the theorem is that we can always put the metric into the form

$$R^{-2} ds^2 = dV \left(2dU + \alpha dV + \sum_i \beta_i dY^i \right) + \sum_{i,j} C_{ij} dY^i dY^j. \tag{21.31}$$

Then, to go near the null geodesic, we consider the rescaling by R,

$$U = u, \quad V = \frac{v}{R^2}, \quad Y^i = \frac{y^i}{R}, \tag{21.32}$$

and then take $R \to \infty$, obtaining the pp wave in Rosen coordinates,

$$ds^2 = 2dudv + g_{ij}(u)dy^i dy^j, \tag{21.33}$$

where $g_{ij}(u) = C_{ij}(U, V = 0, Y^i = 0)$.

There is then a coordinate transformation that takes the pp wave into the standard form, in terms of Brinkmann coordinates,

$$ds^2 = 2dx^+dx^- + H(x^+, x^a)(dx^+)^2 + d\vec{x}^2, \qquad (21.34)$$

and where

$$H(x^+, x^a) = \sum_{a,b} A_{ab}(x^+)dx^a dx^b \qquad (21.35)$$

in this case (for a general pp wave, $H = H(x^+, x^a)$).

The interpretation of the previously discussed limiting procedure is as follows. We boost along x, while taking the scale of the metric, identified with the boost parameter, to ∞, so

$$x' - t' = e^{-\beta}(x-t), \quad x' + t' = e^{+\beta}(x+t), \quad e^\beta = R \to \infty. \qquad (21.36)$$

In the pp-wave limit of AdS/CFT, we can in fact obtain *strings* on the pp wave from the point of view of the field theory, not just supergravity modes, which we couldn't do otherwise. These strings on the pp wave are dual to a large charge sector of SYM, later called the "BMN sector," as we have shown in the "BMN" paper. However, here, as we are interested only in supergravity, we only focus on the pp-wave solutions obtained (supergravity solutions).

In the previously discussed analysis, we have shown that we can always obtain the pp wave first in Rosen coordinates. However, sometimes that is not necessary, if we do not start with the metric in the Penrose form (21.31), and we can obtain directly the Brinkmann pp wave.

Penrose limit of $AdS_5 \times S^5$

This is the case in the standard situation, of the Penrose limit of $AdS_5 \times S^5$, along a certain null geodesic. Indeed, by the Kowalski–Glikman theorem, and because the amount of susy cannot decrease, after the Penrose limit of the maximally supersymmetric $AdS_5 \times S^5$ near some null geodesic, we can only obtain either Minkowski space or the maximally supersymmetric pp wave. Here we show how to obtain the latter case, as it is more interesting.

Consider $AdS_5 \times S^5$ in global coordinates, with metric

$$ds^2 = R^2(-\cosh^2\rho\, d\tau^2 + d\rho^2 + \sinh^2\rho\, d\Omega_3^2) + R^2(\cos^2\theta\, d\psi^2 + d\theta^2 + \sin^2\theta\, d\Omega_3'^2), \qquad (21.37)$$

and the null geodesic sitting in the middle of AdS_5, at $\rho = 0$, and moving along an equator of S^5, at $\theta = 0$, with motion in τ, ψ. Then, near the geodesic we find

$$ds^2 \simeq R^2\left[-(1+\rho^2)d\tau^2 + d\rho^2 + \rho^2 d\Omega_3^2\right] + R^2\left[(1-\theta^2)d\psi^2 + d\theta^2 + \theta^2 d\Omega_3'^2\right]. \qquad (21.38)$$

Define the null coordinates $\tilde{x}^\pm = (\tau \pm \psi)/\sqrt{2}$ and then rescale as in the Penrose theorem,

$$\tilde{x}^+ = x^+, \quad \tilde{x}^- = \frac{x^-}{R^2}, \quad \rho = \frac{r}{R}, \quad \theta = \frac{y}{R}. \qquad (21.39)$$

With an additional rescaling $x^+ \to \sqrt{2}\mu x^+, x^- \to x^-/(\sqrt{2}\mu)$ needed to go to the usual form of the metric, we obtain

$$ds^2 = -2dx^+dx^- - \mu^2(\vec{r}^2 + \vec{y}^2)(dx^+)^2 + d\vec{y}^2 + d\vec{r}^2. \qquad (21.40)$$

We also obtain the self-dual 5-form

$$F_{+1\ldots4} = F_{+5\ldots8} = 4\mu. \tag{21.41}$$

Penrose limits of $AdS_4 \times S^7$ and $AdS_7 \times S^4$

The cases of $AdS_4 \times S^7$ and $AdS_7 \times S^4$ are similar, and give the same result for the pp wave (since actually, there is a unique maximally supersymmetric pp wave in 11 dimensions), and the analysis is similar to the $AdS_5 \times S^5$ case, as well as any other $AdS_p \times S^q$ cases.

For $AdS_4 \times S^7$, for the metric in global coordinates, for the same null geodesic at $\rho = 0$ and $\theta = 0$, we obtain near it

$$ds^2 \simeq R^2 \left[-(1 + \rho^2)d\tau^2 + d\rho^2 + \rho^2 d\Omega_2^2 \right] + 4R^2 \left[(1 - \theta^2)d\psi^2 + d\theta^2 + \theta^2 d\Omega_5^2 \right]. \tag{21.42}$$

Making the rescalings $2\psi = \tilde{\psi}, 2\theta = \tilde{\theta}$, defining the lightcone coordinates $\tilde{x}^\pm = (\tau \pm \tilde{\psi})/\sqrt{2}$ and then rescaling as in the Penrose theorem,

$$\tilde{x}^+ = x^+, \quad \tilde{x}^- = \frac{x^-}{R^2}, \quad \rho = \frac{r}{R}, \quad \tilde{\theta} = \frac{y}{R}, \tag{21.43}$$

we obtain

$$ds^2 = -2dx^+ dx^- - \mu^2 \left(\vec{r}^2 + \frac{1}{4}\vec{y}^2 \right)(dx^+)^2 + d\vec{y}^2 + d\vec{r}^2. \tag{21.44}$$

With the additional rescaling $\mu \to \mu/3$ we arrive at the standard form of the metric of this maximally supersymmetric pp wave,

$$ds^2 = -2dx^+ dx^- - \mu^2 \left(\sum_{i=1}^{3} \frac{x_i^2}{9} + \sum_{j=4}^{9} \frac{y_i^2}{36} \right)(dx^+)^2 + d\vec{y}^2 + d\vec{r}^2. \tag{21.45}$$

For $AdS_7 \times S^4$, again for the metric in global coordinates, and the same null geodesic at $\rho = 0$ and $\theta = 0$, we obtain near it

$$ds^2 \simeq 4R^2 \left[-(1 + \rho^2)d\tau^2 + d\rho^2 + \rho^2 d\Omega_5^2 \right] + R^2 \left[(1 - \theta^2)d\psi^2 + d\theta^2 + \theta^2 d\Omega_2^2 \right]. \tag{21.46}$$

Making the rescalings $2\theta = \tilde{\theta}, 2\rho = \tilde{\rho}$, then defining the lightcone coordinates $\tilde{x}^\pm = (\tilde{\tau} \pm \psi)/\sqrt{2}$ and then rescaling as in the Penrose theorem,

$$\tilde{x}^+ = x^+, \quad \tilde{x}^- = \frac{x^-}{R^2}, \quad \tilde{\rho} = \frac{r}{R}, \quad \theta = \frac{y}{R}, \tag{21.47}$$

we obtain the same pp wave,

$$ds^2 = -2dx^+ dx^- - \mu^2 \left(\vec{r}^2 + \frac{1}{4}\vec{y}^2 \right)(dx^+)^2 + d\vec{y}^2 + d\vec{r}^2, \tag{21.48}$$

which after the same $\mu \to \mu/3$ gives the same standard form of the metric of the maximally supersymmetric pp wave,

$$ds^2 = -2dx^+ dx^- - \mu^2 \left(\sum_{i=1}^{3} \frac{x_i^2}{9} + \sum_{j=4}^{9} \frac{y_i^2}{36} \right)(dx^+)^2 + d\vec{y}^2 + d\vec{r}^2. \tag{21.49}$$

The pp wave for the ABJM model in type IIA

The $\mathbb{CP}^3$ and AdS_4 space metrics are written as

$$R^{-2}ds_{\mathbb{CP}_3}^2 = d\xi^2 + \cos^2 \xi \sin^2 \xi \left(d\psi + \frac{\cos \theta_1}{2}d\phi_1 - \frac{\cos \theta_2}{2}d\phi_2 \right)^2$$

$$+ \frac{1}{4} \cos^2 \xi \left(d\theta_1^2 + \sin^2 \theta_1 d\phi_1 \right)^2 + \frac{1}{4} \sin^2 \xi \left(d\theta_2^2 + \sin^2 \theta_2 d\phi_2 \right)^2,$$

$$R^{-2}ds_{AdS_4}^2 = -\cosh^2 \rho d\tau^2 + d\rho^2 + \sinh^2 \rho d\Omega_2^2. \tag{21.50}$$

We consider the null geodesic at $\rho = 0, \theta_1 = \theta_2 = 0$, and $\xi = \pi/4$. We therefore define the coordinates

$$\tilde{\psi} = \psi + \frac{\phi_1 - \phi_2}{2}, \quad \tilde{x}^+ = \frac{\tau + \tilde{\psi}}{2}, \quad \tilde{x}^- = \frac{\tau - \tilde{\psi}}{2}, \tag{21.51}$$

considering motion in $\tilde{\psi}$, and then we rescale the coordinates like in the Penrose theorem,

$$\tilde{x}^+ = x^+, \quad \tilde{x}^- = \frac{x^-}{R^2}, \quad \rho = \frac{r}{R}, \quad \theta_i = \frac{\sqrt{2}y_i}{R}, \quad \xi = \frac{\pi}{4} + \frac{y_3}{2R}, \tag{21.52}$$

to obtain the metric

$$ds^2 = -4dx^+dx^- - (r^2 + y_3^2)(dx^+)^2 + dx^+(-y_1^2d\phi_1 + y_2^2d\phi_2)$$

$$+ dr^2 + r^2 d\Omega_3^2 + dy_1^2 + y_1^2 d\phi_1^2 + dy_2^2 + y_2^2 d\phi_2^2 + dy_3^2. \tag{21.53}$$

But, as we can see, the metric has unwanted off-diagonal components involving dx^+, so we must get rid of them. We do this by redefining

$$\tilde{\phi}_1 = \phi_1 - \frac{x^+}{2}, \quad \tilde{\phi}_2 = \phi_2 - \frac{x^+}{2}, \tag{21.54}$$

as well as going to Cartesian coordinates, and obtain

$$ds_{IIA}^2 = -4dx^+dx^- - \left(\sum_{i=1}^4 x_i^2 + \frac{1}{4}\sum_{i=5}^8 x_i^2 \right)(dx^+)^2 + \sum_{i=1}^8 dx_i^2. \tag{21.55}$$

Besides the metric, we also obtain a nontrivial 2-form and 4-form field strengths,

$$F_{+4} = \frac{k}{2\tilde{R}}, \quad F_{+123} = \frac{3}{2}\frac{k}{\tilde{R}}. \tag{21.56}$$

The previously discussed pp-wave solution is also unique in type IIA, due to the maximal supersymmetry.

21.4.1 General theory

Until now, we have used the metric of AdS in global coordinates, leading us to the pp wave in Brinkmann coordinates directly, so there was no need for a general theory. However, now we consider it.

First, one has to define the null geodesic, parametrized by λ. Then its equation of motion is

$$\frac{D^2 x^\mu}{d\lambda^2} = \frac{d^2 x^\mu}{d\lambda^2} + \Gamma^\mu_{\nu\rho}\frac{dx^\nu}{d\lambda}\frac{dx^\rho}{d\lambda} = 0, \quad \forall \mu. \tag{21.57}$$

If, moreover, we have motion in a direction x^k only, then we have the condition

$$\Gamma^\mu{}_{kk} = 0, \quad \forall \mu. \tag{21.58}$$

If further, x^k is an isometry, so $\partial_k g_{\mu\nu} = 0$, then the condition is simply

$$\partial^\mu g_{kk} = 0. \tag{21.59}$$

But generally, the motion is not in a single direction x^k. Then, we must construct the Lagrangian for the massless particle moving on the directions we want,

$$L = \frac{1}{2} g_{\mu\nu} \dot{x}^\mu \dot{x}^\nu, \quad \dot{x}^\mu \equiv \frac{dx^\mu}{d\lambda}. \tag{21.60}$$

Then, we also need the condition $L = 0$, which guarantees that we have a *null* geodesic.

Further, usually the metric has some isometries, $\partial g_{\mu\nu} / \partial x^a = 0$, which for the Lagrangian means that there are cyclic coordinates (L depends only on $\dot{x}^a$, not on x^a). The cyclic conditions are

$$\frac{\partial L}{\partial \dot{x}^a} = \text{const.} \equiv x_0^a, \tag{21.61}$$

which can be integrated to give $x^a(\lambda)$, $a = 0, 1, ..., n$.

Assume that the motion is in directions $x^m = \{t, \psi, x^q\}$, where (t and) ψ is an isometry direction. This covers the cases we are interested in. Then we make a change of variables to new coordinates, $\{\lambda, \beta, \phi^p\}$, with differential form

$$dt = \frac{dt}{d\lambda} d\lambda + \frac{dt}{d\beta} d\beta + \sum_{p=1}^{n-1} \frac{dt}{d\phi^p} d\phi^p,$$

$$d\psi = \frac{d\psi}{d\lambda} d\lambda + \left(\frac{d\psi}{d\beta} d\beta \right) + \sum_{p=1}^{n-1} \frac{d\psi}{d\phi^p} d\phi^p,$$

$$dx^q = \frac{x^q}{d\lambda} d\lambda + \left(\frac{dx^q}{d\beta} d\beta \right) + \sum_{p=1}^{n-1} \frac{dx^q}{d\phi^p} d\phi^p, \quad q = 1, ..., n-1. \tag{21.62}$$

We choose this transformation such that $g_{\lambda\beta} = 1$ and $g_{\lambda\phi^p} = 0$, which puts the metric in the Penrose theorem form (which we know that can always be done)

$$R^{-2} ds^2 = 2 d\lambda d\beta + \alpha d\beta^2 + \sum_{i=1}^{d-2} \gamma_i d\beta d\phi_i + \sum_{i,j=1}^{d-2} C_{ij} d\phi_i d\phi_j. \tag{21.63}$$

Then, the rescaling and limit lead to the pp wave in Rosen coordinates, which then can be changed into Brinkmann coordinates.

Example 1   *AdS$_5 \times S^5$ in Poincaré coordinates*.

If we consider the $AdS_5 \times S^5$ solution, but in Poincaré coordinates for AdS instead of global ones,

$$ds^2 = R^2 \left(\frac{-dt^2 + d\vec{x}_3^2 + dz^2}{z^2} \right) + R^2 (d\psi^2 + \sin^2 \psi \, d\Omega_4^2), \tag{21.64}$$

it is less clear what is the correct geodesic to give the pp wave. The geodesic still has to be in the ψ direction, giving an equator of S^5, and in time t, but has to move also in z, as it turns out.

So we consider the massless (null) geodesic Lagrangian

$$R^{-2}L = -z^{-2}\dot{t}^2 + z^{-2}\dot{z}^2 + \dot{\psi}^2. \tag{21.65}$$

The cyclic conditions for the coordinates t and ψ (the Lagrangian is independent on them) are

$$R^{-2}\frac{\partial L}{\partial \dot{t}} = -2E = \text{const.}, \quad R^{-2}\frac{\partial L}{\partial \dot{\psi}} = 2\mu = \text{const.}, \tag{21.66}$$

which give

$$\dot{t} = z^2 E, \quad \dot{\psi} = \mu, \tag{21.67}$$

together with the null condition $L = 0$, giving

$$\dot{z}^2 = \dot{t}^2 - z^2\dot{\psi}^2. \tag{21.68}$$

Then the transformation of coordinates from $\{t, \psi, z\}$ to $\{\lambda, \beta, \phi\}$ is, substituting Eq. 21.68, and including the condition to have $g_{\lambda\beta} = 1, g_{\lambda\phi} = 0$,

$$dt = \frac{dt}{d\lambda}d\lambda - \frac{d\beta}{E} + \mu\phi = z^2 E d\lambda - \frac{d\beta}{E} + \mu d\phi,$$

$$d\psi = \frac{d\psi}{d\lambda}d\lambda + E d\phi = \mu d\lambda + E d\phi,$$

$$dz = \frac{dz}{d\lambda}d\lambda = z\sqrt{z^2 E^2 - \mu^2}d\lambda. \tag{21.69}$$

Substituting this in the metric, we obtain the metric in the Penrose theorem form,

$$R^{-2}dz^2 = 2d\lambda d\beta - \frac{d\beta^2}{E^2 z^2} + \frac{2\mu}{E}\frac{d\beta d\phi}{z^2} + d\phi^2\left(E^2 - \frac{\mu^2}{z^2}\right) + \frac{d\vec{x}^2}{z^2} + \sin^2(\psi_\lambda + E\phi)d\Omega_4^2, \tag{21.70}$$

where $\psi_\lambda = \int \dot{\psi}d\lambda$, here $= \mu\lambda$.

Example 2   The Abelian T-dual to the $AdS_5 \times S^5$.

This is a case in which, although the null geodesic moves in several directions, nevertheless we obtain the Brinkmann form of the pp wave directly, through a clever choice of the transformation of coordinates.

The Abelian T-dual to $AdS_5 \times S^5$ has metric

$$ds^2 = 3\left(-\cosh^2\rho\, d\tau^2 + d\rho^2 + \sinh^2\rho\, d\Omega_3^2\right) + 4\left(d\alpha^2 + \sin^2\alpha\, d\gamma^2\right)$$

$$+ \frac{d\psi^2}{\cos^2\alpha} + \cos^2\alpha\left(d\chi^2 + \sin^2\chi\, d\xi^2\right). \tag{21.71}$$

We look for a null geodesic at $\alpha = 0, \chi = \pi/2, \rho = 0$, for motion in $\{\tau, \psi, \xi\}$ (so, unlike before the T-duality, motion in τ, ψ is not enough, it has to be supplanted by motion in ξ). The Lagrangian is then

$$L = \frac{R^2}{2}\left(-4\dot{\tau}^2 + \dot{\psi}^2 + \dot{\xi}^2\right), \tag{21.72}$$

and all three coordinates are cyclic, giving the conditions

$$\frac{\partial L}{\partial \dot{\tau}} = -4R^2 \dot{\tau} = -R^2, \quad \frac{\partial L}{\partial \dot{\xi}} = R^2 \dot{\xi} = -JR^2, \quad \frac{\partial L}{\dot{\psi}} = R^2 \dot{\psi} = \text{const.,} \qquad (21.73)$$

where we have put $E = 1$ by a rescaling, and where we don't need the last condition, since we already must supplant it with the null condition, that also derives $\dot{\psi}$, as

$$L = 0 \Rightarrow \dot{\psi} = \frac{1}{4}(1 - 4J^2). \qquad (21.74)$$

Expand around the null geodesic position as

$$\rho = \frac{\bar{r}}{2R}, \quad \alpha = \frac{x}{2R}, \quad \chi = \frac{\pi}{2} + \frac{z}{R}, \qquad (21.75)$$

and write the coordinate transformation to $\{\lambda, \beta, \phi\}$ as

$$d\tau = C_1 d\lambda,$$

$$d\xi = C_2 d\lambda + C_3 \frac{d\phi}{R},$$

$$d\psi = C_4 d\lambda + C_5 \frac{d\phi}{R} + C_6 \frac{d\beta}{R^2}, \qquad (21.76)$$

which amounts to including already the scaling $\tilde{\lambda} = \lambda, \tilde{\beta} = \beta/R^2$, and $\tilde{\phi} = \phi/R$ (so that λ is like u or x^+ and β is like v or x^-), and where

$$C_1 = \frac{1}{4} = \dot{\tau}, \quad C_2 = -J = \dot{\xi}, \quad C_4 = \frac{1}{2}\sqrt{1 - 4J^2} = \dot{\psi}. \qquad (21.77)$$

But now also impose that the $\mathcal{O}(R)$ terms in the metric vanish, which gives

$$C_2 C_3 + C_4 C_5 = 0, \qquad (21.78)$$

that $g_{\lambda\beta} = 1$, which gives

$$C_4 C_6 = 1, \qquad (21.79)$$

and that $g_{\phi\phi} = 1$, which gives

$$C_3^2 + C_5^2 = 1, \qquad (21.80)$$

to find the solution

$$C_3 = \sqrt{1 - 4J^2}, \quad C_5 = 2J, \quad C_6 = \frac{2}{\sqrt{1 - 4J^2}}. \qquad (21.81)$$

Note that the vanishing of the $\mathcal{O}(R)$ terms (after the rescaling) was the condition that $g_{\lambda\phi} = 0$, which previously we imposed in order to have the form of the metric in the Penrose theorem.

Then, after the transformation of coordinates, we find the pp wave in Brinkmann coordinates

$$ds^2 = 2d\lambda d\beta + d\bar{r}^2 + \bar{r}^2 d\Omega_3^2 + dz^2 + dx^2 + x^2 d\gamma^2 + d\phi^2$$

$$- \left[\frac{\bar{r}^2}{16} + \frac{8J^2 - 1}{16}x^2 + J^2 z^2 \right] d\lambda^2. \qquad (21.82)$$

21.4.2 Penrose limit on the isometry group and algebra

Finally, we note that the isometry group of the pp wave in the Penrose limit is obtained as a limit of the isometry group of the original background.

To understand this, we consider the case of $AdS_5 \times S^5$.

For $AdS_5 \times S^5$, the bosonic isometry group is $SO(4,2) \times SO(6)$, where $SO(4,2)$ is the isometry group of AdS_5 (equal to the conformal group in four dimensions) and $SO(6) = SU(4)$ is the isometry group of S^5. The bosonic isometry group is extended to $PSU(2,2|4)$, which has the maximal supersymmetry, for 32 supercharges.

Part of the Penrose limit is the limit of infinite scale of the metric, $R \to \infty$. In the Penrose limit, this is coupled to the infinite boost, giving a nontrivial result.

But let us understand the effect of the $R \to \infty$ limit on the symmetry algebra in the simpler case of limit that gives just flat space (with no boost). The idea of taking a limit on the background coupled to a limit on the isometry algebra is known as Wigner–Inönü contraction.

The bosonic isometry of $SO(6)$ is described by generators with indices in the Cartesian (and Euclidean) embedding of the sphere S^5, namely

$$J_{MN} = X_M \partial_N - X_N \partial_M. \tag{21.83}$$

Commuting these generators, we find the algebra

$$[J_{MN}, J_{PQ}] = \delta_{MP} J_{NQ} - \delta_{MQ} J_N - \delta_{NP} J_{MQ} + \delta_{NQ} J_{MP}. \tag{21.84}$$

The $SO(4,2)$ algebra has the same generators, but with coordinates on a space with (4,2) signature, leading to an algebra with δ_{MP}, and so on replaced with $\tilde{\eta}_{MP}$, and so on, the invariant tensor for (4,2) signature.

Then the Wigner–Inönü contraction of AdS_5 to $Minkowski_5$ symmetry algebras is obtained from $R_{AdS_5} \to \infty$, leading to $SO(4,2)$ going over into $ISO(4,1)$, the Poincaré group. From the generators J_{MN}, with $M, N = 0, 1, ..., 4, 5$, we select the Lorentz generators $J_{\mu\nu}$, for $\mu, \nu = 0, 1, ..., 4$, and define

$$J_{MN} = (J_{\mu\nu}, J_{\mu 0} = \tilde{P}_\mu), \tag{21.85}$$

and then rescale

$$\tilde{P}_\mu = R P_\mu, \tag{21.86}$$

where P_μ is the translation generator in five dimensions.

Then the Lorentz algebra for $J_{\mu\nu}$ is unchanged, and the commutator of $J_{\mu\nu}$ with $J_{\rho 0}$ gives

$$[J_{\mu\nu}, P_\rho] = \eta_{\mu\rho} P_\nu - \eta_{\nu\rho} P_\mu, \tag{21.87}$$

since the R cancels from the two sides, while the only commutator that changes is

$$[J_{\mu 0}, J_{\nu 0}] = \eta_{00} J_{\mu\nu}, \tag{21.88}$$

which becomes

$$[P_\mu, P_\nu] = \frac{1}{R^2} J_{\mu\nu} \to 0, \tag{21.89}$$

as it should be for the Poincaré group.

For the pp wave, we could do a similar Wigner–Inönü contraction. But rather, we can consider directly the symmetry algebra of the pp wave, which was found by Figueroa–O'Farrill and Papadopoulos. The symmetry generators for the 10-dimensional type IIB maximally supersymmetric pp wave are

$$h = e^+ = -\frac{1}{\mu}\partial_+, \quad e = e^- = -\mu\partial_-,$$

$$e_i = -\cos(\mu x^+)\partial_i - \mu\sin(\mu x^+)\tilde{y}^i\partial_-, \quad i = 1, ..., 8,$$

$$e_i^* = -\sin(\mu x^+)\partial_i + \mu\cos(\mu x^+)\tilde{y}^i\partial_-,$$

$$M_{ij} = x_i\partial_j - x_j\partial_i, \quad i,j = 1, ..., 4 \text{ OR } 5, ..., 8. \tag{21.90}$$

Then their algebra gives

$$[e_i, e_j^*] = e\delta_{ij},$$

$$[h, e_i] = e_i^*,$$

$$[h, e_i^*] = -e_i,$$

$$[M_{ij}, e_k] = -\delta_{ik}e_j + \delta_{jk}e_i,$$

$$[M_{ij}, e_k^*] = -\delta_{ik}e_j^* + \delta_{jk}e_i^* \tag{21.91}$$

and the Lorentz algebra for the two $SO(4)$ terms ($i,j = 1, ..., 4$ and $i,j = 5, ..., 8$). Note that e appears as a central charge (it commutes with everything and only appears on the right-hand side).

We note that we can define

$$a_i = \frac{e_i + ie_i^*}{\sqrt{2}}, \quad a_i^\dagger = \frac{e_i - ie_i^*}{\sqrt{2}}, \tag{21.92}$$

which means that

$$M_{ij} = i(a_i^\dagger a_j - a_j^\dagger a_i). \tag{21.93}$$

If, moreover, we fix $e = (e^- = e_+ = -\partial_+) = i$, then we also obtain the usual creation and annihilation algebra

$$[a_i, a_j^\dagger] = \delta_{ij}, \tag{21.94}$$

and the Hamiltonian takes the standard form,

$$H = i\sum_i a_i^\dagger a_i. \tag{21.95}$$

We have considered here only the bosonic algebra, but the fermionic one can also be represented in terms of fermionic creation and annihilation operators that can then be added to the previously discussed Hamiltonian.

Then, we can create representations of the pp-wave symmetry algebra in the usual way, in the Wigner representation (familiar for $SU(2)$ from quantum mechanics, but valid for all algebras), by acting with the creation operators on a vacuum (or lowest weight) state $|0\rangle$.

Finally, we also observe that the pp wave has 32 supercharges: as we said, the number of supersymmetries can only increase, not decrease during a limit like the Penrose limit. Since we already had the maximum number before the limit, we keep it after the limit.

Important concepts to remember

- The AdS/CFT correspondence relates the theory on a decoupled system of branes, with string theory in the background curved by the branes. In the original case, $\mathcal{N} = 4$ SYM in four dimensions vs. string theory in $AdS_5 \times S^5$.
- The relation is holographic, the field theory living on the boundary of the space, and the radial direction acts as energy, $U = r/\alpha'$ in the original case.
- Although originally derived for the case of $4\pi g_s = g_{YM}^2 \to 0$ and $N \to \infty$, with $g_{YM}^2 N = R^4/(\alpha')^2$ fixed and large, in which case there are no α' corrections and no g_s corrections on the string side, and we have just free classical gravity (and nonperturbative large N field theory on the other side), the duality is assumed to be valid at all g_s (all g_{YM}^2) and N, so at all α' and g_s.
- The construction in the previous three bullet points is a "top-down" construction, but there is also a "bottom-up" one, where one makes up a gravitational theory in AdS background that would, via the standard holographic map, have the correct field theory properties we want.
- Other susy cases of AdS/CFT are the following: $AdS_7 \times S^4$ vs. the six-dimensional (0,2) theory, $AdS_4 \times S^7$ vs. the three-dimensional theory on M2-branes, and its nice limit, $AdS_4 \times \mathbb{CP}^3$ vs. the ABJM model. The Dp-branes give also susy holographic backgrounds, but they are nonconformal.
- In the Penrose limit for AdS/CFT, on the gravity side, we obtain pp waves, and on the field theory side, we obtain large charge operators, forming a sector called the "BMN sector," on which we have spin chains in a "dilute gas" approximation.
- The Penrose theorem says that in the Penrose limit, going near a null geodesic in a spacetime, we find a pp (parallel plane) wave.
- The pp-wave symmetry algebra for the maximally supersymmetric case is a limit (Wigner–Inönü contraction) of the algebra of the spacetime. In general, we can never lose charges, only gain them, so in particular supersymmetry can only increase.
- For the maximally supersymmetric pp wave, the algebra can be written in terms of bosonic and fermionic creation and annihilation operators, so we can construct representations using $a^\dagger$'s, in the usual Wigner representation.

References and further reading

For more about the AdS/CFT correspondence, see my book about it [66]. Also Maldacena's original paper [67], and for general Dp-branes, [68]. For the Penrose limit of gravity duals giving pp waves, see the original BMN paper, [69]. For the pp wave for the ABJM case, see [70]. For the T-dual example for general Penrose limit, see [71]. For the symmetry algebras of pp waves, see [72, 73].

Exercises

(1) Consider the NS5-brane of type IIB theory. Write a decoupling limit and a resulting gravity dual coming from it.

(2) Is there a decoupling limit for D7-branes? What about for D7-branes intersecting with D3-branes in a supersymmetric configuration?

(3) Calculate the Penrose limit for a null geodesic moving solely in the AdS_5 part of $AdS_5 \times S^5$. Should it depend on the particular geodesic you took?

(4) Do the Wigner–Inönü contraction for $SO(6)$, the symmetry of S^5, to go into flat space.

22

Supersymmetric AdS/CFT gravity dual pairs and their deformations (susy, marginal, integrable)

In this chapter, we will consider the common supersymmetric AdS/CFT maps in three and four dimensions, which are found to be integrable, and then consider deformations of those gravity dual pairs that preserve some supersymmetry and are marginal (so, preserve conformal invariance) and/or intgrability.

22.1 Supersymmetric AdS/CFT maps

We are interested in the supergravity limit, which corresponds to the nonperturbative region of the field theory.

$\mathcal{N} = 4$ SYM vs. $AdS_5 \times S^5$

The original and the most important example of AdS/CFT map is for $\mathcal{N} = 4$ SYM vs. $AdS_5 \times S^5$.

The $SO(4, 2)$ conformal invariance of $\mathcal{N} = 4$ SYM (that includes Lorentz invariance) is mapped to the same isometry of AdS_5, and the $SO(6)$ R-symmetry of $\mathcal{N} = 4$ SYM is mapped to the same isometry of S^5. But these are just the bosonic symmetry groups; the full symmetry group of $\mathcal{N} = 4$ SYM (and so isometry of $AdS_5 \times S^5$ string theory) is the supergroup $PSU(2, 2|4)$.

$\mathcal{N} = 4$ SYM is the unique theory in four dimensions with maximal, namely $\mathcal{N} = 4$ supersymmetry for spins ≤ 1. As such, this is a standard four-dimensional example, or prototype, or toy model, for field theories that we might be interested in. We will always use it as a zeroth-order approximation.

The action of the model, using the trace normalization $\text{Tr}[T^a T^b] = -\frac{1}{2}\delta^{ab}$, is

$$
\begin{aligned}
S_{4d,\mathcal{N}=4\text{ SYM}} &= (-2) \int d^4x \ \text{Tr}\left[-\frac{1}{4}F_{\mu\nu}^2 - \frac{1}{2}\bar{\psi}_i \slashed{D}\, \psi^i - \frac{1}{2}D_\mu \phi_{ij}D^\mu \phi^{ij} \right.\\
&\qquad \left. -g\bar{\psi}^i[\phi_{ij}, \psi^j] - \frac{g^2}{4}[\phi_{ij}, \phi_{kl}][\phi^{ij}, \phi^{kl}] \right]\\
&= (-2) \int d^4x \ \text{Tr}\left[-\frac{1}{4}F_{\mu\nu}^2 - \frac{1}{2}\bar{\psi}_i \slashed{D}\, \psi^i - \frac{1}{2}D_\mu \phi_m D^\mu \phi^m \right.\\
&\qquad \left. -g\bar{\psi}^i[\phi_n, \psi^j]\tilde{\gamma}^n_{[ij]} - \frac{g^2}{4}[\phi_m, \phi_n][\phi^m, \phi^n] \right],
\end{aligned}
\tag{22.1}
$$

where $i, j = 1, ..., 4$ are fundamental indices for the $SU(4)$ R-symmetry (so in the spinor representation of $SO(6)$), and $m, n = 1, ..., 6$ are indices in the fundamental representation of $SO(6)$, equal to the antisymmetric of $SU(4)$, for which the $\tilde{\gamma}^m_{[ij]}$ are Clebsch–Gordan coefficients,

$$\phi_{[ij]} = \phi_m \tilde{\gamma}^m_{[ij]}. \tag{22.2}$$

The scalars ϕ_{ij} obey the reality condition

$$\phi^\dagger_{ij} = \phi^{ij} \equiv \frac{1}{2}\epsilon^{ijkl}\phi_{kl}, \tag{22.3}$$

so that there are only six real independent components in them, as there should be.

Since there is no covariant $\mathcal{N} = 4$ superspace formalism (only a lightcone one), it is worth describing it in $\mathcal{N} = 1$ language: we have the gauge multiplet plus three chiral multiplets Φ_l ($l = 1, 2, 3$), which appear symmetrically, with a superpotential

$$W = \text{Tr}(\Phi_1[\Phi_2, \Phi_3]) = \frac{1}{3}\epsilon_{lmn}\text{Tr}[\Phi^l\Phi^m\Phi^n]. \tag{22.4}$$

The AdS/CFT map relates gauge invariant operators $\mathcal{O}^{I_n}_n(\vec{x}_{(4)})$, having sources $\phi_{(n)}$, with fields in AdS_5 (after reducing on S^5, to obtain KK towers of fields in AdS_5), for which $\phi_{(n)}$ is a boundary source, or value. For example, the gauge invariant operator $\text{Tr}[F_{\mu\nu}F^{\mu\nu}]$ corresponds to the dilaton φ in AdS_5.

The ABJM model vs. $AdS_4 \times \mathbb{CP}^3$

The standard example in three dimensions, or prototype or toy model for any theory we might be interested in is the ABJM model. It has $\mathcal{N} = 6$ supersymmetry (note that this is not maximal in three dimensions, the maximal amount is $\mathcal{N} = 8$), with CS gauge fields for $SU(N)_k \times SU(N)_{-k}$, where k and $-k$ refer to the CS level or quantized coefficient of the CS term. The action is

$$\begin{aligned}
S_{\text{ABJM}} = \int d^3x \Bigg[&\frac{k}{4\pi}\epsilon^{\mu\nu\lambda}\text{Tr}\left(A_\mu\partial_\nu A_\lambda + \frac{2i}{3}A_\mu A_\nu A_\lambda - \hat{A}_\mu\partial_\nu\hat{A}_\lambda - \frac{2i}{3}\hat{A}_\mu\hat{A}_\nu\hat{A}_\lambda \right) \\
&- \text{Tr}\left(D_\mu C^\dagger_I D^\mu C^I \right) - i\text{Tr}\left(\psi^{I\dagger}\gamma^\mu D_\mu\psi_I \right) \\
&+ \frac{4\pi^2}{3k^2}\text{Tr}\left(C^I C^\dagger_I C^J C^\dagger_J C^K C^\dagger_K + C^\dagger_I C^I C^\dagger_J C^J C^\dagger_K C^K \right. \\
&\left. + 4C^I C^\dagger_J C^K C^\dagger_I C^J C^\dagger_K - 6C^I C^\dagger_J C^J C^\dagger_I C^K C^\dagger_K \right) \\
&+ \frac{2\pi i}{k}\text{Tr}\left(C^\dagger_I C^I \psi^{J\dagger}\psi_J - \psi^{\dagger J} C^I C^\dagger_I \psi_J - 2C^\dagger_I C^J \psi^{\dagger I}\psi_J + 2\psi^{\dagger J} C^I C^\dagger_J \psi_I \right. \\
&\left. + \epsilon^{IJKL}\psi_I C^\dagger_J \psi_K C^\dagger_L - \epsilon_{IJKL}\psi^{\dagger I} C^J \psi^{\dagger K} C^L \right) \Bigg]. \tag{22.5}
\end{aligned}$$

Here the covariant derivative, for instance on the scalars C^I (and the same on the fermions ψ_I) acts like

$$D_\mu C^I = \partial_\mu C^I + i\left(A_\mu C^I - C^I \hat{A}_\mu \right),$$

and so A_μ is in the first $SU(N)$ (with coefficient $k/(4\pi)$ for the CS term in the action), and $\hat{A}_\mu$ is in the second $SU(N)$ (with coefficient $-k/(4\pi)$ for the CS term in the action), while, as we can see, C^I and ψ_I, $I, J = 1, ..., 4$ are in the bifundamental representation of $SU(N) \times SU(N)$ (action from the left for one, and from the right for the other).

The model has $U(1) \times SU(4) = U(1) \times SO(6)$ R-symmetry, which is the correct one for six supersymmetry charges in three dimensions (the spinors can be thought to be real), which matches the isometry group of $\mathbb{CP}^3$. C^I and ψ_I are in the fundamental of $SU(4)$ and of charge $+1$ under $U(1)$.

Just as $g_{YM}^2 N$ is large for the supergravity limit, but g_{YM}^2 is small in the case of $\mathcal{N} = 4$ SYM, in the case of ABJM, k (playing the role of $1/g_{YM}^2$) is large, but N/k is also large. Note that maximal $\mathcal{N} = 8$ supersymmetry is recovered only for $N = 2$ or $k = 1$, but these are outside the supergravity limit, consistent with the fact that $AdS_4 \times \mathbb{CP}^3$ has only $\mathcal{N} = 6$ supersymmetry.

22.2 Supersymmetric and integrable deformations

Integrability, classically means that there are as many integrals of motion (that can be used to completely solve the theory) as degrees of freedom. In quantum mechanics, that becomes a bit more complicated, as there will be a notion of operators replacing the classical integrals of motion, but essentially the same picture is correct. In quantum field theory, the number of degrees of freedom becomes also infinite.

Both $\mathcal{N} = 4$ SYM and the ABJM model are integrable, though the proof is not completely mathematically rigorous. On the gravity side, the question is a bit more nuanced. String theory is defined only perturbatively, and only in some backgrounds, so the notion of nonperturbative string theory is not something well defined. But one can take the point of view that AdS/CFT is correct at all couplings, and then define nonperturbative string theory in $AdS_5 \times S^5$ as just $\mathcal{N} = 4$ SYM. From that point of view, the gravity side is also integrable.

But save that, we have only a genus by genus definition on the worldsheet of strings. In that, classical string picture, we do have also integrability on the gravity side: on the worldsheet of classical strings.

The question we can then ask is what are the possible deformations of $\mathcal{N} = 4$ SYM and ABJM, and of their corresponding gravity duals, that preserve: (at least $\mathcal{N} = 1$ supersymmetry and) conformal invariance, so are *marginal* deformations and/or integrability? These will be studied next.

22.2.1 β deformation of $\mathcal{N} = 4$ SYM

This deformation was found by Leigh and Strassler. It preserves $\mathcal{N} = 1$ supersymmetry and is marginal. Moreover, it has an $U(1) \times U(1)$, non-R symmetry.

It is defined by the star product (note that a particular type of star product, the Moyal product, is used for the most rigorous noncommutative deformation of field theory, equivalent to a field theory on a noncommutative space)

$$f * g = e^{i\pi\gamma(Q_f^1 Q_g^2 - Q_f^2 Q_g^1)} fg, \tag{22.6}$$

where f and g are fields, and (Q^1, Q^2) are charges under $U(1) \times U(1)$.

Then the superpotential of $\mathcal{N} = 4$ SYM in the $\mathcal{N} = 1$ formulation is changed by the deformation as

$$W = \text{Tr}(\Phi_1\Phi_2\Phi_3 - \Phi_1\Phi_3\Phi_2) \rightarrow \text{Tr}\left(e^{i\pi\gamma}\Phi_1\Phi_2\Phi_3 - e^{-i\pi\gamma}\Phi_1\Phi_3\Phi_2\right). \tag{22.7}$$

The gravity dual is in type IIB supergravity that, as we said, is obtained from M theory on a torus of modulus τ, that becomes $\tau = B_{12} + i\sqrt{g}$ in type IIB (note that $\sqrt{g}$ is the volume). τ is acted upon by the $Sl(2;\mathbb{R})$ symmetry of the supergravity equations of motion.

Then, the beta deformation of $\mathcal{N} = 4$ SYM, found by Lunin and Maldacena, with a parameter γ as above, corresponds to the transformation

$$\tau \to \tau_\gamma = \frac{\tau}{1 + \tau\gamma}, \tag{22.8}$$

which acts as a solution-generating transformation.

The action above is obtained from a T-duality on one of the torus circles, then a change of coordinates (shift), and another T-duality.

If we have an $\mathbb{R}^4$ in two polar coordinates,

$$ds^2 = d\rho_1^2 + d\rho_2^2 + \rho_1^2 d\varphi_1^2 + \rho_2^2 d\varphi_2^2, \tag{22.9}$$

we can formally consider the torus made by the two angles φ_1, φ_2, with $\sqrt{g} = \rho_1\rho_2$, so $\tau = i\rho_1\rho_2$, thus obtaining the beta transformed solution

$$ds^2 = d\rho_1^2 + d\rho_2^2 + \frac{1}{1 + \gamma^2\rho_1^2\rho_2^2}(\rho_1^2 d\varphi_1^2 + \rho_2^2 d\varphi_2^2)$$
$$B_{12} = \frac{\gamma\rho_1^2\rho_2^2}{1 + \gamma^2\rho_1^2\rho_2^2}$$
$$e^{2\Phi} = e^{2\Phi_0}\frac{1}{1 + \gamma^2\rho_1^2\rho_2^2}. \tag{22.10}$$

Then similarly, on $AdS_5 \times S^5$ (with action only on S^5, since the beta deformation is marginal, so preserves conformal invariance, and conformal invariance is mapped to an AdS space in gravity), the beta deformed metric and dilaton are

$$ds^2_{\text{string}} = R^2\left[ds^2_{AdS_5} + \sum_i(d\mu_i^2 + G\mu_i^2 d\phi_i^2) + \hat{\gamma}G\mu_1^2\mu_2^2\mu_3^2\left(\sum_i d\phi\right)^2\right]$$
$$G^{-1} = 1 + \hat{\gamma}^2(\mu_1^2\mu_2^2 + \mu_2^2\mu_3^2 + \mu_3^2\mu_1^2)$$
$$\hat{\gamma} = R^2\gamma, \quad R^4 = 4\pi e^{\Phi_0}N$$
$$e^{2\Phi} = e^{2\Phi_0}G, \tag{22.11}$$

as well as B^{NS}, C_2, C_4, and F_5 forms.

22.2.2 γ deformation (generalization with three parameters)

This generalization, found by Frolov, has three parameters, but corresponds to a *nonsupersymmetric*, yet still marginal, deformation of $\mathcal{N} = 4$ SYM.

Before describing it, we note that the Lunin–Maldacena gravity dual to the beta deformation is obtained as a TsT transformation, acting only on the S^5.

One describes the S^5 in the six-dimensional Euclidean space with three polar coordinate pairs,

$$ds^2 = \sum_{i=1}^{3}(dr_i^2 + r_i^2 d\phi_i^2), \tag{22.12}$$

with the constraint $\sum_i r_i^2 = 1$.

We then redefine

$$\phi_1 = \varphi_3 - \varphi_2$$
$$\phi_3 = \varphi_3 - \varphi_1$$
$$\phi_2 = \varphi_3 + \varphi_1 + \varphi_2. \tag{22.13}$$

(a) Doing a T-duality on φ_1, we obtain

$$\hat{g}_{11} = \frac{1}{r_2^2 + r_3^2}, \quad \hat{g}_{12} = \hat{g}_{13} = 0$$

$$\hat{g}_{22} = \frac{r_1^2 r_2^2 + r_1^2 r_3^2 + r_2^2 r_3^2}{r_2^2 + r_3^2}$$

$$\hat{g}_{33} = 1 - \frac{(r_2^2 - r_3^2)^2}{r_2^2 + r_3^2}$$

$$\hat{b}_{12} = \frac{r_2^2}{r_2^2 + r_3^2}, \quad \hat{b}_{13} = \frac{r_2^2 - r_3^2}{r_2^2 + r_3^2}, \quad \hat{b}_{23} = 0. \tag{22.14}$$

(b) Doing the shift $\varphi_2 \to \varphi_2 + \gamma \varphi_1$, we obtain

$$\hat{G}_{11} = \hat{g}_{11} + \gamma^2 \hat{g}_{22} = \frac{G^{-1}}{r_2^2 + r_3^2}$$

$$\hat{G}_{12} = \hat{\gamma}\hat{g}_{22}, \quad \hat{G}_{13} = \hat{\gamma}\hat{g}_{23}, \quad \hat{G}_{23} = \hat{g}_{23}, \quad \hat{g}_{22} = \hat{g}_{22}, \quad \hat{G}_{33} = \hat{g}_{33}. \tag{22.15}$$

(c) After the final T-duality back on φ_1, we obtain the beta deformed solution (22.11).

To generalize the construction we consider three TsT transformations:

(a) T duality on ϕ_1, shift by $\hat{\gamma}_3$ on ϕ_2, T duality back on ϕ_1.
(b) T duality on ϕ_2, shift by $\hat{\gamma}_1$ on ϕ_3, T duality back on ϕ_2.
(c) T duality on ϕ_3, shift by $\hat{\gamma}_2$ on ϕ_1, T duality back on ϕ_3.

Then, one obtains the solution

$$ds^2_{\text{string}} = R^2 \left[ds^2_{AdS_5} + \sum_i (dr_i^2 + Gr_i^2 d\phi_i^2) + Gr_1^2 r_2^2 r_3^2 \left(\sum_i \hat{\gamma}_i d\phi_i \right)^2 \right]$$

$$G^{-1} = 1 + \hat{\gamma}_3^2 r_1^2 r_2^2 + \hat{\gamma}_1^2 r_2^2 r_3^2 + \hat{\gamma}_2^2 r_3^2 r_1^2$$

$$\hat{\gamma}_i = \gamma_i \sqrt{\lambda} = \gamma_i \sqrt{4\pi e^{\Phi_0} N}$$

$$e^{2\Phi} = e^{2\Phi_0} G, \tag{22.16}$$

as well as B^{NS}, C_2, C_4, and F_5 forms. The proof is left as an exercise.

The deformation of $\mathcal{N} = 4$ SYM is marginal but not supersymmetric, so can be expressed in terms of the potential only (not in terms of a superpotential),

$$V = \text{Tr}\left[|\Phi_1\Phi_2 - e^{-2\pi i\gamma_3}\Phi_2\Phi_1|^2 + |\Phi_2\Phi_3 - e^{-2\pi i\gamma_1}\Phi_3\Phi_2|^2 + |\Phi_3\Phi_1 - e^{-2\pi i\gamma_2}\Phi_1\Phi_3|^2\right]$$
$$+ \text{Tr}\left[\left([\Phi_1, \bar{\Phi}_1] + [\Phi_2, \bar{\Phi}_2] + [\Phi_3, \bar{\Phi}_3]\right)^2\right]. \tag{22.17}$$

Note that when we put all deformation parameters to be equal, $\gamma_i = \gamma$, we get the beta deformation.

22.3 η and λ deformations

While in the previous cases, we had a well-defined field theory deformation, and we found a gravity dual corresponding to it, we now move on to a case where the only thing we know is the integrable deformation for the string worldsheet moving in the gravity dual background. These are the η and λ deformations.

The deformations are for the string worldsheet action written in WZW form and are classically integrable.

We have not yet described the string worldsheet actions, we will do so in two chapters, so it seems strange to already write their deformations. However, here we will write the actions completely algebraically, without bothering about spacetime interpretations, so the goal is just to show that such integrable deformations exist.

The cases of interest for the deformation are (1) $AdS_5 \times S^5$, with symmetry supergroup $PSU(2,2|4)$; (2) $AdS_3 \times S^3 \times T^4$, with symmetry supergroup $[PSU(1,1|2)]^2$, and (3) $AdS_2 \times S^2 \times T^6$, with symmetry supergroup $PSU(1,1|2)$.

In these cases, there is a $\mathbb{Z}_4$ grading of the algebra of the supergroup, extending the simple $\mathbb{Z}_2$ one that divides generators into bosons and fermions. We need to have the Poincaré group $ISO(1, d-1)$, with generators P_a and J_{ab}, and two types of supercharges, Q^1 and Q^2, such that the $\mathbb{Z}_4$ action on the generators can be described in terms of roots of unity as (note that $i = e^{i\pi/2}, -1 = i^2 = e^{i\pi}, -i = i^3 = e^{3i\pi/2}$)

$$J_{ab} \rightarrow J_{ab}, \quad P_a \rightarrow -P_a, \quad Q^1 \rightarrow iQ^1, \quad Q^2 \rightarrow -iQ^2. \tag{22.18}$$

Another way of saying this is that the generators $T_A = \{P_a, J_{ab}, Q_\alpha^I\}$, $I = 1, 2$ and $\alpha = 1, ..., N$ split according to $\mathbb{Z}_4$ projectors $P^{(m)}$, $m = 0, 1, 2, 3$ as

$$P^{(0)}(T_A) = J_{ab}, \quad P^{(1)}(T_A) = Q_\alpha^1, \quad P^{(2)}(T_A) = P_a, \quad P^{(3)}(T_A) = Q_\alpha^2. \tag{22.19}$$

We also define the supertrace

$$\text{Str}(T_A T_B) \equiv K_{AB}. \tag{22.20}$$

Next, define the operators

$$\hat{d} = P^{(1)} + 2\hat{\eta}^{-2}P^{(2)} - P^{(3)}$$
$$\hat{d}^T = -P^{(1)} + 2\hat{\eta}^{-2}P^{(2)} + P^{(3)}, \tag{22.21}$$

so that

$$\hat{d} + \hat{d}^T = 4\hat{\eta}^{-2}P^{(2)}, \tag{22.22}$$

in terms of a deformation parameter η, related to $\hat{\eta}$ by

$$\hat{\eta} = \sqrt{1 - c\eta^2}. \tag{22.23}$$

We next define the further operators

$$\mathcal{O}_+ = 1 + \eta R_g \hat{d}^T$$
$$\mathcal{O}_- = 1 - \eta R_g \hat{d}, \tag{22.24}$$

where

$$R_g = (\text{Adj}_g)^{-1} R(\text{Adj}_g), \tag{22.25}$$

and where R, with $R^T = -R$, called the R-matrix, acts on the algebra $\mathcal{G}$. Moreover, we define

$$R(X) = r^{ij} T_i \text{Str}(T_j X), \quad T_i, X \in \mathcal{G} \tag{22.26}$$

as the action of the R-matrix on X. Then, the R-matrix satisfies the (modified) classical Yang–Baxter equations, or (m)CYBE, related to integrability,

$$[R(X), R(Y)] - R([R(X), Y] + [X, R(Y)]) = c[X, Y], \quad \forall X, Y \in \mathcal{G}, \tag{22.27}$$

sometimes also rewritten as (to emphasize the matricial nature)

$$[RM, RN] - R([RM, N] + [M, RN]) = c[M, N]. \tag{22.28}$$

The case $c = 0$ is the classical Yang–Baxter equation (CYBE), whereas $c = \pm 1$ is the mCYBE.

Note that in integrability in quantum theory, defined by scattering, the usual *Yang–Baxter equation* (*YBE*) relates the 3-point scattering to 2-point scatterings, happening in two possible ways (so that the 3-point S-matrix is not independent, but is completely defined by the 2-point S-matrix),

$$S_{123}^{(3)} = S_{12}(\lambda_1 - \lambda_2) S_{13}(\lambda_1 - \lambda_3) S_{23}(\lambda_2 - \lambda_3)$$
$$= S_{23}(\lambda_2 - \lambda_3) S_{13}(\lambda_1 - \lambda_3) S_{12}(\lambda_1 - \lambda_2), \tag{22.29}$$

where the λ's are complex versions of the rapidities for the scatterings. More generally, in the algebraic approach, one writes a relation in $\mathcal{A} \otimes \mathcal{A} \otimes \mathcal{A}$,

$$\mathcal{R}_{12} \mathcal{R}_{13} \mathcal{R}_{23} = \mathcal{R}_{23} \mathcal{R}_{13} \mathcal{R}_{23}, \tag{22.30}$$

where $\mathcal{R}_{12} = \mathcal{R} \otimes \mathbb{1} \in \mathcal{A} \otimes \mathcal{A} \otimes \mathcal{A}$, $\mathcal{R}_{23} = \mathbb{1} \otimes \mathcal{R} \in \mathcal{A} \otimes \mathcal{A} \otimes \mathcal{A}$, and $\mathcal{R}_{13}$ has the identity stuck in the middle. Then, for some representation of the YBE, one finds the *Yang–Baxter algebra*

$$R_{12}(u - v)(L_j(u) \otimes \mathbb{1})(\mathbb{1} \otimes L_j(v)) = (\mathbb{1} \otimes L_j(v))(L_j(v) \otimes L_j(u)) R_{12}(u - v). \tag{22.31}$$

The relation to the CYBE is not easy to describe, however.

Finally, the η *deformed string worldsheet model* is, in this algebraic WZW formulation,

$$\mathcal{L}_{\text{wsh}} = -\frac{(1 + c\eta)^2}{4(1 - c\eta^2)} (\gamma^{ij} - \epsilon^{ij}) \text{Str} \left[(g^{-1} \partial_i g) \hat{d} \mathcal{O}_-^{-1} (g^{-1} \partial_j g) \right], \tag{22.32}$$

where note that $A_i = g^{-1}\partial_i g \in \mathcal{G}$ is a gauge field (is in the algebra of the group).

Similarly, the λ *deformed string worldsheet model* is

$$\mathcal{L}_{\text{wsh}} = -\frac{k}{2\pi}(\gamma^{ij} - \epsilon^{ij})\text{Str}\left[(g^{-1}\partial_i g)\left(1 + \hat{B}_0 - 2\tilde{\mathcal{O}}_-^{-1}\right)(g^{-1}\partial_j g)\right], \qquad (22.33)$$

where

$$\begin{aligned}
\tilde{\mathcal{O}}_+ &= (\text{Adj}_g)^{-1} - \Omega^T \\
\tilde{\mathcal{O}}_- &= 1 - (\text{Adj}_g)^{-1}\Omega \\
\Omega &= P^{(0)} + \lambda^{-1}P^{(1)} + \lambda^{-2}P^{(2)} + \lambda P^{(3)} \\
\Omega^T &= P^{(0)} + \lambda P^{(1)} + \lambda^{-2}P^{(2)} + \lambda^{-1}P^{(3)} \\
1 - \Omega\Omega^T &= 1 - \Omega^T\Omega = (1 - \lambda^{-4})P^{(2)} \\
d\hat{B}_0 &= \frac{1}{3}\text{Str}\left(g^{-1}dg \wedge g^{-1}dg \wedge g^{-1}dg\right),
\end{aligned} \qquad (22.34)$$

and λ is the deformation parameter.

The λ *deformation corresponds in spacetime to a solution of the supergravity constraints.*

On the other hand, the η deformation in general doesn't, but satisfies some Generalized Supergravity Equations (GSE), to be described a bit later.

The case when the η deformation satisfies the usual supergravity equations in spacetime is when

$$r^{ij}[T_i, T_j] = 0, \qquad (22.35)$$

where the R-matrix is

$$R = \frac{1}{2}r^{ij}T_i \wedge T_j, \qquad (22.36)$$

and the abelian case corresponds to $[T_i, T_j] = 0$, $T_i \in \mathcal{G}$, while for the nonabelian case $[T_i, T_j] \neq 0$.

Note then that some cases of deformations are obtained from $AdS_5 \times S^5$ via TsT transformations, but there is (as of yet) no general recipe when this happens.

22.4 Yang–Baxter deformations

A more general theory has been developed, which includes η and λ deformations, known as Yang–Baxter deformations.

It is based on the Metsaev–Tseytlin supercoset action for $AdS_5 \times S^5$, where the supercoset description of the supersymmetric $AdS_5 \times S^5$ is as the supercoset

$$\frac{PSU(2,2|4)}{SO(1,4) \times SO(5)}. \qquad (22.37)$$

Indeed, $PSU(2,2|4)$ is the isometry group of the supersymmetric $AdS_5 \times S^5$, and $SO(1,4) \times SO(5)$ the local Lorentz group.

One writes then the undeformed action as

$$S = -\frac{T}{2} \int d\tau d\sigma P_-^{\alpha\beta} \mathrm{Str}\left[A_\alpha d_- A_\beta\right], \tag{22.38}$$

where $A = g^{-1} dg$ is in the algebra, $g \in SU(2,2|4)$, $P_\pm^{\alpha\beta}$ are projectors on the worldsheet, and $d_\pm$ are written in terms of the $\mathbb{Z}_4$ grading projectors,

$$P^{\alpha\beta} = \frac{\gamma^{\alpha\beta} - \epsilon^{\alpha\beta}}{2}$$
$$d_\pm = \mp P^{(1)} + 2P^{(2)} \pm P^{(3)}, \tag{22.39}$$

and the gauge field A is decomposed according to the $\mathbb{Z}_4$ grading as

$$A = A^{(0)} + A^{(1)} + A^{(2)} + A^{(3)}, \quad A^{(i)} = P^{(i)} A. \tag{22.40}$$

Then the undeformed action is rewritten as

$$S = \frac{T}{2} \int d\tau d\sigma \, \mathrm{Str}(A^{(2)} \wedge *_\gamma A^{(2)} - A^{(1)} \wedge A^{(3)}). \tag{22.41}$$

Moreover, define

$$gT_i g^{-1} \equiv \left[\mathrm{Adj}_g\right]_i^{\ j} T_j. \tag{22.42}$$

Then, the Killing vectors of AdS_5 are

$$\hat{T}_i = \hat{T}_i^m \partial_m = \left(\mathrm{Adj}_{g_{\mathrm{bos}}^{-1}}\right)_i^{\ a} e_a^m \partial_m, \tag{22.43}$$

which leads to the representation of the isometry group of AdS_5 (equal to the conformal group in four dimensions)

$$\hat{P}_\mu = \partial_\mu$$
$$\hat{M}_{\mu\nu} = x_\mu \partial_\nu - x_\nu \partial_\mu$$
$$\hat{P} = x^\mu \partial_\mu + z \partial_z$$
$$\hat{K}_\mu = (x^\nu \partial_\nu + z^2) \partial_\mu - 2x_\mu (x^\nu \partial_\nu + z \partial_z). \tag{22.44}$$

Finally, the Yang–Baxter deformation of the worlsheet string in $AdS_5 \times S^5$ is

$$S_{\mathrm{YB}} = -\frac{T(1 - \tilde{c}^2 \eta)}{2} \int d\sigma d\tau P_-^{\alpha\beta} \mathrm{Str}\left[A_\alpha \hat{d}_i \circ \mathcal{O}_-^{-1} A_\beta\right], \tag{22.45}$$

where $-\tilde{c}^2 = +c$ from before, and

$$\hat{d}_\pm = \mp P^{(1)} + 2\hat{\eta}^{-2} P^{(2)} \pm P^{(3)}$$
$$\mathcal{O}_\pm = 1 \pm \eta R_g \circ \hat{d}_\pm$$
$$\hat{\eta} = \sqrt{1 + \tilde{\chi}^2} \eta$$
$$r = \frac{1}{2} r^{ij} T_i \wedge T_j, \quad T_i \in \mathcal{G}. \tag{22.46}$$

22.5 Generalized supergravity equations

The GSE obtained for Yang–Baxter deformations (and η deformations) are

$$R_{mn} - \frac{1}{4} H_{mpq} H_n{}^{pq} + 2D_m \partial_n \phi + D_m U_n + D_n U_m = T_{mn}$$

$$- \frac{1}{2} D^k H_{kmn} + \partial_k \phi H^k{}_{mn} + U^k H_{kmn} + D_m I_n - D_n I_m = K_{mn}$$

$$R - \frac{1}{2} |H_3|^2 + 4D^m \partial_m \phi - 4|\partial \phi|^2 - 4\left(I_m I^m + U_m U^m + 2U^m \partial_\mu \phi - D_m U^m\right) = 0$$

$$d * \hat{F}_n - H_3 \wedge *\hat{F}_{n+2} - i_I B_2 \wedge *\hat{F}_n - i_I * \hat{F}_{n-2} = 0, \tag{22.47}$$

where i_I stands for the projection onto the Killing vector I, we have the usual notation

$$|\alpha_p|^2 \equiv \frac{1}{p!} \alpha_{m_1 \ldots m_p} \alpha^{m_1 \ldots m_p}, \tag{22.48}$$

and the definitions

$$T_{mn} = \frac{e^{2\phi}}{4} \sum_p \left[\frac{1}{(p-1)!} \hat{F}_{(m}^{k_1 \ldots k_{p-1}} \hat{F}_{n)k_1 \ldots k_{p-1}} - \frac{1}{2} g_{mn} |\hat{F}_p|^2 \right]$$

$$K_{mn} = \frac{e^{2\phi}}{4} \sum_p \frac{1}{(p-2)!} \hat{F}_{k_1 \ldots k_{p-2}} \hat{F}_{mn}^{k_1 \ldots k_{p-2}}$$

$$\hat{F}_p = d\hat{A}_{p-1} + H_3 \wedge \hat{A}_{p-3} - i_I B_2 \wedge \hat{A}_{p-1} - i_I \hat{A}_{p+1}. \tag{22.49}$$

As we see, we have that the GSEs are defined by the presence of a Killing vector I,

$$I = I^m \partial_m, \quad I_m U^m = 0, \tag{22.50}$$

with

$$\mathcal{L}_I g_{mn} = 0, \quad \mathcal{L}_I \phi = 0, \quad \mathcal{L}_I B_2 + d(U - i_I B_2) = 0. \tag{22.51}$$

Then when $I = 0$ we obtain the usual supergravity equations, and in particular this happens when $r^{ij}[T_i, T_j] = 0$, with $T_i \in SU(2, 2|4)$.

Note that in the case of $c^2 = 0$ in the CYBE and the Yang–Baxter deformations, these reduce to the β transformations, and the $O(d, d)$ transformations studied in Chapter 17. In this case,

$$g'_{mn} + B'_{mn} = \left[(G^{-1} + \beta)\right]^{-1}_{mn}, \quad d' = d, \quad e^{-2d} = \sqrt{|g|} e^{-2\phi}, \tag{22.52}$$

and then the β matrix, corresponding to the β transformation, is

$$\beta^{mn}(x) = -r^{mn}(x) = 2\eta r^{ij} \hat{T}_i^m(x) \hat{T}_j^n(x), \tag{22.53}$$

where $\hat{T}_i^m(x)$ is the Killing vector associated with the generator T_i.

For instance, in the case of the r-matrix r^{ij} of the type

$$r = \frac{1}{2}(\mu_3 h_1 \wedge h_2 + \mu_1 h_2 \wedge h_3 + \mu_2 h_3 \wedge h_1), \tag{22.54}$$

we obtain the gamma deformations of Frolov, with the β^{mn} matrix

$$\beta = 2\eta(\mu_3 \partial_{\phi_1} \wedge \partial_{\phi_2} + \mu_1 \partial_{\phi_2} \wedge \partial_{\phi_3} + \mu_2 \partial_{\phi_3} \wedge \partial_{\phi_1}). \tag{22.55}$$

We leave the details as an exercise.

Important concepts to remember

- The standard supersymmetric AdS/CFT maps are $AdS_5 \times S^5$ vs. $\mathcal{N} = 4$ SYM in four dimensions, $AdS_4 \times \mathbb{CP}^3$ vs. the ABJM model in three dimensions ($\mathcal{N} = 6$ susy, CS gauge theory with $SU(N) \times SU(N)$ gauge group).
- The beta deformation of $\mathcal{N} = 4$ SYM, preserving $\mathcal{N} = 1$ susy and being marginal, corresponds to the star product $f * g = e^{i\pi\gamma(Q_f^1 Q_g^2 - Q_f^2 Q_g^1)} fg$, and deforms the superpotential as $W = \text{Tr}(\Phi_1\Phi_2\Phi_3 - \Phi_1\Phi_3\Phi_2) \rightarrow \text{Tr}\left(e^{i\pi\gamma}\Phi_1\Phi_2\Phi_3 - e^{-i\pi\gamma}\Phi_1\Phi_3\Phi_2\right)$, and in the gravity dual is a TsT deformation acting on the S^5.
- The gamma deformation deforms with three γ's, in three different superpotential directions, and corresponds to three TsT transformations.
- The classically integrable string worldsheet in $AdS_5 \times S^5$ can be deformed, preserving integrability, into the η and λ deformations.
- The λ deformation is defined in terms of a solution to the (m)CYBE, and corresponds in spacetime to a solution of the supergravity equations of motion.
- The η deformation is defined in terms of a solution to the (m)CYBE, and it corresponds to a supergravity equation in spacetime when the corresponding r-matrix satisfies $r^{ij}[T_i, T_j] = 0$.
- The general Yang–Baxter deformations generalize the η and λ deformations, and satisfy GSE, that generalize the supergravity equations of motion through the introduction of a Killing vector I. The YB deformations reduce to β deformations, $O(d, d)$ transformations and η and λ deformations in particular cases.

References and further reading

The gravity dual of the β deformation was found in [74] and of the γ deformations in [75]. A good review for the η and λ deformations is in [76], while the original η deformation was obtained in [77], and the original λ deformation in [78] and [79]. A review of the Yang–Baxter deformations is in [80].

Exercises

(1) Calculate the forms B^{NS}, C_2, C_4, and F_5 for the gravity dual to the beta deformation.
(2) Show that for the three TsT transformations corresponding to the gamma deformation, one obtains the gravity dual with

$$ds^2_{\text{string}} = R^2\left[ds^2_{AdS_5} + \sum_i(dr_i^2 + Gr_i^2 d\phi_i^2) + Gr_1^2 r_2^2 r_3^2\left(\sum_i \hat{\gamma}_i d\phi_i\right)^2\right]$$

$$G^{-1} = 1 + \hat{\gamma}_3^2 r_1^2 r_2^2 + \hat{\gamma}_1^2 r_2^2 r_3^2 + \hat{\gamma}_2^2 r_3^2 r_1^2$$

$$\hat{\gamma}_i = \gamma_i\sqrt{\lambda} = \gamma_i\sqrt{4\pi e^{\Phi_0}N}$$

$$e^{2\Phi} = e^{2\Phi_0}G. \tag{22.56}$$

(3) Show that the vanishing of the Killing vector I in the GSEs implies that $r^{ij}[T_i, T_j] = 0$.

(4) Show that indeed, for the r-matrix

$$r = \frac{1}{2}(\mu_3 h_1 \wedge h_2 + \mu_1 h_2 \wedge h_3 + \mu_2 h_3 \wedge h_1),\tag{22.57}$$

we obtain the gamma deformations of Frolov.

Extremal black holes, the attractor mechanism, and holography

In this chapter, we describe the attractor mechanism, which refers to extremal black holes. Moreover, we will consider it in AdS background, making a connection with holography in this case.

23.1 The attractor mechanism

The attractor mechanism was first observed in four-dimensional $\mathcal{N} = 2$ supersymmetric theories, but is more general than that, applying to $\mathcal{N} = 4$ and $\mathcal{N} = 8$ supersymmetric theories, as well as theories in higher dimensions ($d \geq 5$, with $\mathcal{N} \geq 1$) and gauged supergravities, with AdS backgrounds.

The idea is the following. In theories with scalar fields, that require no energy to change, so "moduli," there are black hole solutions depending on them. Then these moduli change, or "flow" between their values at infinity, and their values at the horizon, which horizon values are fixed by the charges, and are found by minimizing some function.

The relevant black hole solutions are *extremal*, meaning they obey

$$M = |Z|, \tag{23.1}$$

where Z is a central charge, depending on the moduli. The fact that the horizon values of the moduli are found by minimizing some function tells us that the horizon values are fixed points of an attractor (in dynamical systems, due to the nonlinearity of the equations, it happens that for a large number of initial conditions, called the basin of attraction, the system evolves to the same final values, called the attractor), so there is a basin of attraction for the scalar field values at infinity.

Note that, even though until now we acted as if these two things are the same (and called them both BPS saturated), extremality does not necessarily imply supersymmetry, there are relevant counterexamples, and we will encounter one of them later on in the chapter.

Thus the mass M and entropy S of these extremal black holes are found from the values of Z at the horizon, called Z_{fix}, which will depend solely on the charges of the black hole.

23.1.1 Four-dimensional $\mathcal{N} = 2$ case (the original one)

In four-dimensional $\mathcal{N} = 2$ theories, we have seen that there is so-called special geometry. We can choose "symplectic sections" (L^Λ, M_Λ), with $\Lambda = 0, 1, ..., n$ on the geometry, obeying the constraint

$$i\left(\bar{L}^\Lambda M_L - L^\Lambda \bar{M}_\Lambda\right) = 1, \tag{23.2}$$

and these symplectic sections are functions of the coordinates on the moduli space (special geometry), denoted either by $(\phi_i, \bar{\phi}_i)$ or, as before, by $(z_i, \bar{z}_i)$, so $L^\Lambda(\phi_i, \bar{\phi}_i)$, $M_\Lambda(\phi_i, \bar{\phi}_i)$. The symplectic sections are related by the "period matrix" $\mathcal{N}_{\Lambda\Sigma}$,

$$M_\Lambda = \mathcal{N}_{\Lambda\Sigma} L^\Sigma \Rightarrow D_{\bar{\imath}} \bar{M}_\Lambda = \mathcal{N}_{\Lambda\Sigma} D_{\bar{\imath}} \bar{L}^\Sigma. \tag{23.3}$$

We can express these symplectic sections in terms of holomorphic ones (X^Λ, F_Λ), using the Kähler potential K, via

$$L^\Lambda = e^{K/2} X^\Lambda, \quad M_\Lambda = e^{K/2} F_\Lambda, \quad \partial_{\bar{\imath}} X^\Lambda = \partial_{\bar{\imath}} F_\Lambda = 0. \tag{23.4}$$

The Kähler potential is then obtained as (using the constraint on the symplectic sections)

$$K = -\ln i\left(\bar{X}^\Lambda F_\Lambda - X^\Lambda \bar{F}_\Lambda\right). \tag{23.5}$$

We can define inhomogeneous coordinates, or "special coordinates,"

$$Z^\Lambda = \frac{X^\Lambda(\phi_i)}{X^0(\phi_i)}, \quad Z^0 = 1, \tag{23.6}$$

which are therefore also holomorphic,

$$\partial_{\bar{\imath}} Z^\Lambda(\phi_i, \bar{\phi}_i) = 0 \Rightarrow Z^\Lambda = Z^\Lambda(\phi_i). \tag{23.7}$$

Moreover, it follows that we can then identify $Z^i = \phi_i = z_i$, with $Z^0 = 1$. Here

$$F_\Lambda = \partial_\Lambda F, \tag{23.8}$$

where F is called the prepotential.

Consider now the $\mathcal{N} = 2$ supersymmetric black holes that have an ADM mass M depending only on the (q, p) charges, associated with X^Λ, F_Λ, and the moduli ϕ_i, through $X^\Lambda(\phi), F_\Lambda(\phi)$, as

$$M^2 = |Z|^2,$$
$$Z(z_i, \bar{z}_i, q, p) = e^{\frac{K(z, \bar{z})}{2}} \left(X^\Lambda(z_i) q_\Lambda - F_\Lambda(z_i) p^\Lambda\right)$$
$$= L^\Lambda q_\Lambda - M_\Lambda p^\Lambda. \tag{23.9}$$

Here z_i are the moduli at infinity: this is in flat space, where we measure the ADM mass. Then

$$M_{\text{ADM}}^2 = |Z|^2 = M_{\text{ADM}}^2(z_i, \bar{z}^i, p, q). \tag{23.10}$$

The area of the horizon of these extremal black hole solutions depends only on p, q and has no moduli dependence, so $A_H = A_H(p, q)$.

On the other hand, extremality, which in this case means unbroken supersymmetry, so minimum of the potential, means that the Kähler covariant derivative of the scalars vanishes (as we saw when discussing the general supersymmetric theories with chiral multiplets),

$$D_i Z \equiv \left(\partial_i + \frac{1}{2}\partial_i K\right) Z(z_j, \bar{z}_j, p, q) = 0. \tag{23.11}$$

But since in general, from the form (23.9) of the central charge, we find that it is also Kähler covariantly holomorphic, so $\bar{Z}$ is covariantly anti-holomorphic,

$$D_i \bar{Z} \equiv \left(\partial_i - \frac{1}{2} \partial_i K \right) \bar{Z} = 0, \tag{23.12}$$

we obtain

$$D_i(Z\bar{Z}) = 0 = (D_i Z)\bar{Z} + Z D_i(\bar{Z}) = (\partial_i Z)\bar{Z} + Z \partial_i \bar{Z} = \partial_i(Z\bar{Z}) \Rightarrow \partial_i |Z| = 0. \tag{23.13}$$

This means that we obtain

$$\frac{\partial}{\partial z_i} |Z| = 0 \quad \text{at} \quad Z = Z_{\text{fix}} = Z(L^\Lambda(p,q), M_\Lambda(p,q)), \tag{23.14}$$

and then the Hawking entropy of the black hole is written in terms of this fixed point central charge,

$$S = \frac{A}{4} = \pi |Z_{\text{fix}}|^2. \tag{23.15}$$

But then, at the horizon, we can express the horizon area in terms of the Bertotti–Robinson mass,

$$\frac{A}{4\pi} = M_{\text{BR}}^2, \tag{23.16}$$

so that the latter doesn't have a dependence on the moduli,

$$M_{\text{BR}}^2 = |Z_{\text{fix}}|^2 = M_{\text{BR}}^2(p,q). \tag{23.17}$$

Only in the absence of scalars, the ADM and Bertotti–Robinson mass are equal (the first is for an asymptotically flat space, and the second is defined directly for the curved space at the horizon of the Bertotti–Robinson form).

The solution for these extremal black holes in the near horizon, $r \to 0$, is of the general type

$$ds^2 = -e^{2U} dt^2 + e^{-2U} d\vec{x}^2, \tag{23.18}$$

where $r^2 = \vec{x}^2$, and e^{-U} is harmonic,

$$\Delta e^{-U} = 0, \tag{23.19}$$

meaning that we can choose (since we drop the 1 in the near-horizon limit, and the three-dimensional harmonic function is $\propto 1/r$, $e^{-U} \propto 1/r$, $e^{-2U} \propto 1/r^2$)

$$e^{-2U} = \frac{A_H}{4\pi r^2}. \tag{23.20}$$

(note that we can put $e^{-2U} = C/r^2$, but then the horizon transverse metric is $C d\Omega_2^2$, so the horizon area is $4\pi C$).

Finally then, the metric takes the Bertotti–Robinson form,

$$ds^2 = ds_{\text{BR}}^2 = -\frac{r^2}{A_H/(4\pi)} dt^2 + \frac{A_H/(4\pi)}{r^2} d\vec{x}^2, \tag{23.21}$$

if we identify $M_{\text{BR}}^2 = A_H/(4\pi)$. With the further coordinate change $\rho = M_{\text{BR}}^2/r$, and $\rho^2 = \vec{y}^2$, we get

$$ds_{\text{BR}}^2 = \frac{M_{\text{BR}}^2}{\rho^2}(-dt^2 + d\vec{y}^2). \tag{23.22}$$

Because of the extremum of the moduli of the central charge, $\partial_i Z = 0$, giving Z_{fix}, we obtain an extremum of the Bertotti–Robinson mass and entropy over the same,

$$\frac{\partial}{\partial z_i} M_{\text{BR}} = 0, \quad M^2 = |Z_{\text{fix}}|^2, \quad S = \frac{A_H}{4} = \pi |Z_{\text{fix}}|. \tag{23.23}$$

This is the essence of the attractor mechanism in four dimensions.

Similar relations hold in $d = 5$ and higher, just that then we have different power laws, for instance, in $d = 5$, we have

$$S = \frac{A}{4} \sim |Z_{\text{fix}}|^{3/2}. \tag{23.24}$$

23.2 Interpretation and Sen's entropy function formalism

Sen's entropy function formalism provides a physical interpretation for the attractor mechanism previously described. Moreover, it is valid even in the case of higher derivative gravity, as was already noted in Sen's original paper.

Also, the formalism only depends on the $AdS_2 \times S^{D-2}$ metric at the horizon, which comes in fact from extremality, not necessarily from supersymmetry, as in the previously discussed case. As we said, there are in fact non-supersymmetric but extremal examples of relevance.

We therefore consider a theory with Abelian gauge fields $A_\mu^{(i)}$, neutral scalars $\{\phi_s\}$, and with $SO(2,1) \times SO(3)$ invariance at the horizon in four dimensions, easily generalized (we will do this next) to $SO(2,1) \times SO(d-1)$ in d dimensions.

Under these conditions, the ansatz (most general one) for the solution at the horizon is

$$ds_{\text{near-horizon}}^2 = v_1\left(-r^2 dt^2 + \frac{dr^2}{r^2}\right) + v_2(d\theta^2 + \sin^2\theta d\phi^2),$$

$$\phi_s = u_s,$$

$$F_{rt}^{(i)} = e_i,$$

$$F_{\theta\phi}^{(i)} = \frac{p_i}{4\pi}\sin\theta, \tag{23.25}$$

where u_s are the moduli at the horizon, e_i are the electric fields, and p_i are the magnetic charges. These form our unknowns. The metric takes the form of $AdS_2 \times S^2$, with radii squared v_1 and v_2.

We can calculate the Riemann tensor components for AdS_2 and S^2 on the ansatz as

$$R_{\alpha\beta\gamma\delta} = -\frac{1}{v_1}(g_{\alpha\gamma}g_{\beta\delta} - g_{\alpha\delta}g_{\beta\gamma}), \quad AdS_2,$$

$$R_{mnpq} = \frac{1}{v_2}(g_{mp}g_{nq} - g_{mq}g_{np}), \quad S^2. \tag{23.26}$$

Define the following function as the integral over the sphere of the Lagrangian (g is the metric on the sphere),

$$f(\vec{u}, \vec{v}, \vec{e}, \vec{p}) \equiv \int d\theta d\phi \sqrt{-\det g}\mathcal{L}. \tag{23.27}$$

Then we can check that the scalar and the metric (Einstein) equations of motion reduce to just the extremization with respect to u_s (the scalars at the horizon) and v_i (the radii, the only parameters in the near horizon metric),

$$\frac{\partial f}{\partial u_s} = 0, \quad \frac{\partial f}{\partial v_i} = 0. \tag{23.28}$$

The gauge field equations of motion and Bianchi identities become

$$\partial_r \left(\frac{\partial \sqrt{-\det g}\mathcal{L}}{\partial F_{rt}^{(i)}} \right) = 0, \quad \partial_r F_{\theta\phi}^{(i)} = 0, \tag{23.29}$$

which are satisfied by the ansatz (23.25), in which e_i are the electric fields and p_i are the magnetic charges. Since the electric charges are conjugate to the electric fields, we have

$$\frac{\partial f}{\partial e_i} = q_i. \tag{23.30}$$

Then, given the charges $\vec{q}, \vec{p}$, we have to solve equations (23.28) and (23.30) for the unknowns u_s, v_i, e_i, in the same number as the equations, so we can solve them completely in terms of $\vec{q}, \vec{p}$, as we did in the normal attractor mechanism before.

Next, we consider the more general *Wald entropy* formula that equals with the Hawking entropy in the standard cases in general relativity, but is also valid for higher derivative gravity theories (in fact, one of the reasons it was invented was to deal with these cases), which in this background becomes

$$S_{\text{BH}} = 8\pi \frac{\partial \mathcal{L}}{\partial R_{rtrt}} g_{rr} g_{tt} A_H, \tag{23.31}$$

where in the derivative with respect to the Riemann tensor component R_{rtrt}, we are instructed to ignore covariant derivatives D's and treat $R_{\mu\nu\rho\sigma}$ components as independent variables. One can then further define the quantity

$$\delta\mathcal{L} = \frac{\partial \mathcal{L}}{\partial R_{\mu\nu\rho\sigma}} R_{\mu\nu\rho\sigma}, \tag{23.32}$$

which helps us define the λ derivative of the function $f_\lambda(\vec{u}, \vec{v}, \vec{e}, \vec{p})$, which is obtained by multiplying only the component R_{rtrt} by λ in the Lagrangian.

Indeed, then we can find that

$$\frac{\partial f_\lambda(\vec{u}, \vec{v}, \vec{e}, \vec{p})}{\partial \lambda} = \int d\theta d\phi \sqrt{-\det g} R_{\alpha\beta\gamma\delta} \frac{\partial \mathcal{L}}{\partial R_{\alpha\beta\gamma\delta}}, \tag{23.33}$$

where $\alpha, \beta, \gamma, \delta$ are AdS_2 indices, so take only the values t, r, and by symmetries we can reduce the Riemann tensor to R_{rtrt} in this sector.

Moreover, due to the form of the Riemann tensor in AdS_2, we find

$$\frac{\partial \mathcal{L}}{\partial R_{\alpha\beta\gamma\delta}} = -v_1^2 (g^{\alpha\gamma} g^{\beta\delta} - g^{\alpha\delta} g^{\beta\gamma}) \frac{\partial \mathcal{L}}{\partial R_{rtrt}}, \tag{23.34}$$

which we can reverse, and use the object appearing in $\partial f_\lambda / \partial \lambda$ to relate to it, so

$$\frac{\partial \mathcal{L}}{\partial R_{rtrt}} A_H = \frac{1}{4} v_1^{-2} \left. \frac{\partial f_\lambda(\vec{u}, \vec{v}, \vec{e}, \vec{p})}{\partial \lambda} \right|_{\lambda=1}. \tag{23.35}$$

Also, we can calculate

$$\lambda g^{rr} g^{tt} R_{rtrt} = -\frac{\lambda}{v_1},$$

$$\sqrt{-g^{rr} g^{tt}} F^{(i)}_{rt} = \frac{e_i}{v_i},$$

$$\sqrt{-\det g} = v_1, \tag{23.36}$$

which means that we can write a scaling relation for f_λ in terms of a function g with v_1 appearing only in scaling relations as in Eq. (23.36),

$$f_\lambda(\vec{u}, \vec{v}, \vec{e}, \vec{p}) = v_1 g\left(\vec{u}, v_2, \vec{p}, \frac{\lambda}{v_1}, \frac{\vec{e}}{v_1}\right), \tag{23.37}$$

which in turn means we find that f_λ satisfies the differential relation

$$\left(\lambda \frac{\partial}{\partial \lambda} + v_1 \frac{\partial}{\partial v_1} + e_i \frac{\partial}{\partial e_i} - 1\right) f_\lambda(\vec{u}, \vec{v}, \vec{e}, \vec{p}). \tag{23.38}$$

Putting $\lambda = 1$ and substituting the derivative with respect to λ obtained from the differential relation in the Wald entropy formula, we obtain the black hole (Wald) entropy

$$S_{\text{BH}} = 2\pi \left(e_i \frac{\partial f}{\partial e_i} - f\right). \tag{23.39}$$

But, since $\partial f / \partial e_i = q_i$, we can define the function

$$F(\vec{u}, \vec{v}, \vec{q}, \vec{p}) = 2\pi \left[e_i \frac{\partial f(\vec{u}, \vec{v}, \vec{e}, \vec{p})}{\partial e_i} - f(\vec{u}, \vec{v}, \vec{e}, \vec{p})\right] = 2\pi [e_i q_i - f(\vec{u}, \vec{v}, \vec{e}, \vec{p})], \tag{23.40}$$

so the Legendre transform of f with respect to e_i.

The relation $\partial f / \partial e_i = q_i$ becomes now one more extremization,

$$\frac{\partial F}{\partial e_i} = 0, \tag{23.41}$$

besides the ones with respect to the scalars,

$$\frac{\partial F}{\partial u_s} = 0, \quad \frac{\partial F}{\partial v_i} = 0, \tag{23.42}$$

where now F is called *Sen's entropy function*, and it equals the entropy at the extremum,

$$S_{\text{BH}}(\vec{q}, \vec{p}) = F(\vec{u}, \vec{v}, \vec{q}, \vec{p})_{\text{extr}}. \tag{23.43}$$

We can easily generalize the previously discussed analysis to higher dimensions, where the horizon geometry of the extremal black holes is $AdS_2 \times S^{D-2}$, so

$$ds^2_{\text{near-horizon}} = v_1 \left(-r^2 dt^2 + \frac{dr^2}{r^2}\right) + v_2 d\Omega^2_{D-2},$$

$$\phi_s = u_s,$$

$$F^{(i)}_{rt} = e_i,$$

$$H^{(a)}_{l_1 \ldots l_{D-2}} = p_a \epsilon_{l_1 \ldots l_{D-2}} \frac{\sqrt{\det h^{(D-2)}}}{\Omega_{D-2}}. \tag{23.44}$$

Note that now we have defined the magnetic field through the Poincaré dual $D - 2$-form H.

Then we define as before

$$f(\vec{u}, \vec{v}, \vec{e}, \vec{p}) = \int d^{D-2}\vec{x}\sqrt{-\det g}\mathcal{L}, \tag{23.45}$$

which implies the extremization equations

$$\frac{\partial f}{\partial u_s} = 0, \quad \frac{\partial f}{\partial v_i} = 0, \quad \frac{\partial f}{\partial e_i} = q_i, \tag{23.46}$$

and the entropy of the extremal black hole equals the extremum value of the Legendre transform of the above function,

$$S_{\text{BH}} = F_{\text{extremum}} = 2\pi(e_i q_i - f)_{\text{extremum}}, \tag{23.47}$$

or, in terms of $F = e_i q_i - f$,

$$\frac{\partial F}{\partial u_s} = 0, \quad \frac{\partial F}{\partial v_i} = 0, \quad \frac{\partial F}{\partial e_i} = 0. \tag{23.48}$$

23.3 Attractors in five-dimensional gauged supergravity and holography

The five-dimensional gauged supergravity has an AdS background, which is holographic, so this generalization of the attractor mechanism is worth exploring.

23.3.1 Five-dimensional gauged supergravity attractors

Since the theory is gauged, we consider the five-dimensional action for gravity coupled to the scalars ϕ^i, with a nonlinear sigma model kinetic term with moduli space metric g_{ij}, and a potential V, as well as the nonabelian gauge fields with field strengths $F^A_{\mu\nu}$, with a coupling function $f_{AB}(\phi)$,

$$S[G_{\mu\nu}, \phi^i, A^I_\mu] = \frac{1}{2\kappa_N^2} \int_M d^5x\sqrt{-G}[R - g_{ij}(\phi)\partial_\mu\phi^i\partial^\mu\phi^j$$
$$- f_{AB}(\phi)F^A_{\mu\nu}F^{B\,\mu\nu} + V(\phi)], \tag{23.49}$$

where we have $\kappa_N^2 = 8\pi G_N$.

The ansatz for the solutions we want is

$$ds^2 = -a(r)^2 dt^2 + a(r)^{-2}dr^2 + b(r)^2 d\Omega_3^2, \tag{23.50}$$

where

$$d\Omega_3^2 = d\theta^2 + \sin^2\theta\, d\phi^2 + \cos^2\theta\, d\psi^2. \tag{23.51}$$

The Bianchi identities and equations of motion for the gauge fields are solved by the ansatz

$$F^A = \frac{1}{b^3} f^{AB} Q_B \, dt \wedge dr, \tag{23.52}$$

where Q_A are constants that determine the electric charges carried by the gauge fields F^A and f^{AB} is the matrix inverse of f_{AB}.

We observe that on the ansatz, an important quantity that appears is the "effective potential"

$$V_{\text{eff}}(\phi) = f^{AB}(\phi) Q_A Q_B, \tag{23.53}$$

and the action reduces to the one-dimensional action

$$S = \frac{1}{2\kappa_N^2} \int dr \left(6b + 6ab^2 a'b' + 6a^2 bb'^2 - b^3 V(\phi) - a^2 b^3 (\phi_i')^2 - \frac{2}{b^3} V_{\text{eff}}(\phi_i) \right). \tag{23.54}$$

Again we have an extremal Reissner–Nordstrom (charged) AdS black hole, with *constant scalar fields* and

$$b(r) = r,$$
$$a^2(r) = 1 + \frac{r^2}{l^2} - \frac{m}{r^2} + \frac{q^2}{r^4} = \frac{1}{l^2 r^4}(r - r_H)^2 (r + r_H)^2 (r^2 + 2r_H^2 + l^2), \tag{23.55}$$

where $r = r_H$ is a degenerate horizon (the two horizons of the Reissner–Nordstrom black hole degenerate into a single one for the case of an extremal solution), and the parameters are given in terms of it by

$$m = 2r_H^2 \left(1 + \frac{3}{2} \frac{r_H^2}{l^2} \right),$$
$$q^2 = r_H^4 \left(1 + 2 \frac{r_H^2}{l^2} \right). \tag{23.56}$$

These mass and charge parameters are related to the (asymptotic) ADM mass and charge M and Q by

$$M = \frac{3\pi}{8G_N} m, \qquad Q = \sqrt{3} q, \tag{23.57}$$

and then the electric field becomes

$$F = \frac{1}{2} F_{\mu\nu} dx^\mu \wedge dx^\nu = \frac{Q}{r^3} dr \wedge dt. \tag{23.58}$$

In the near-horizon limit, $\rho = r - r_H \to 0$, we obtain

$$a(\rho) = \frac{4}{l^2 r_H^2}(3r_H^2 + l^2)\rho^2 = \frac{1}{v_1}\rho^2, \tag{23.59}$$

where as before v_1 is the radius squared of AdS_2. With the further rescaling $t = v_1 \tau$, we obtain the $AdS_2 \times S^3$ geometry in the usual form in the near horizon, as we expected.

Note that this solution is *extremal*, since it has a degenerate horizon and an $AdS_2 \times S^3$ near-horizon geometry and no Hawking temperature, $T = 0$, but is *not supersymmetric*.

The susy bound is saturated for $M = 2Q$, but in that case, we obtain a naked singularity. Observe that this happens only because the space is asymptotically AdS. In the asymptotically flat space, this could not happen.

Near the horizon, the metric and electric field become

$$ds^2 = v_1 \left(-\rho^2 d\tau^2 + \frac{1}{\rho^2} d\rho^2 \right) + v_2 d\Omega_3^2,$$

$$F^A = e^A d\tau \wedge d\rho. \tag{23.60}$$

Then a calculation finds for the entropy function (in the convention that $16\pi G_N = 1$)

$$F(u^i, v_1, v_2, e^A, Q_A) = 2\pi [Q_A e^A - f(u^i, v_1, v_2, e^A)], \tag{23.61}$$

$$f(u^i, v_1, v_2, e^A) = 2\pi^2 \left[-2v_2^{3/2} + 6v_1\sqrt{v_2} + 2\frac{v_2^{3/2}}{v_1} f_{AB} e^A e^B - v_1 v_2^{3/2} V(\phi) \right].$$

The attractor equations mean extremizing the above function over all its variables, but keeping fixed the charges Q_A, plus extremization over e^A (which gives the Legendre transform equation).

By combining the two variations (equations of motion) with respect to v_i, we get

$$\frac{4}{v_2} - \frac{1}{v_1} - V(\phi) = 0, \tag{23.62}$$

while from $\partial F / \partial e^A = 0$, we obtain the relation between charge and electric field,

$$Q_A = 8\pi^2 \frac{v_2^{3/2}}{v_1} f_{AB} e^B. \tag{23.63}$$

Replacing the extremum values inside the entropy function, we find its value at the extremum,

$$F_{\text{extremum}} = 8\pi^3 v_2^{3/2}, \tag{23.64}$$

which must therefore equal the entropy of the extremal black hole in the convention $16\pi G_N = 1$.

The variations of the entropy function with respect to u^i, $\partial F / \partial u^i = 0$, give

$$2\frac{\partial f_{AB}}{\partial u^i} e^A e^B = -v_1^2 \frac{\partial V}{\partial u^i}. \tag{23.65}$$

In the case that the potential is constant, $V(\phi) = $ constant, as it happens in AdS space, the right-hand side of Eq. (23.65) is zero, so the equations reduce to finding the critical points of the "effective potential" V_{eff} in (23.53) at the horizon,

$$\partial_i V_{\text{eff}} = 0. \tag{23.66}$$

This is one of the conditions needed to have an attractor mechanism, the second being the condition that there should be no unstable directions around this minimum, so the matrix of second derivatives of the effective potential at the critical point,

$$M_{ij} = \frac{1}{2} \partial_i \partial_j V_{\text{eff}}(\phi_0^k), \tag{23.67}$$

should have no negative eigenvalues.

23.3.2 Embedding in string theory and AdS/CFT

The extremal black holes above described previously were considered just in a generic theory in five dimensions, so even though they have AdS asymptotics, we are not guaranteed to have the full AdS/CFT technology available, and we would only know that the theory is holographic.

Because of that, we try to see whether we can embed them in string theory, and moreover obtain them from a certain decoupling limit. That will in fact be the case, and this guarantees that we can use standard AdS/CFT techniques.

We will find that the extremal Reissner–Nordstrom AdS black hole is a consistent truncation in the near-horizon limit of a system of rotating D3-branes.

Thus extremal RNAdS in five dimensions is a special case of a three-charge black hole solution, for the case $H_1 = H_2 = H_3 = H$. The general (and nonextremal) solution is

$$
\begin{aligned}
ds_5^2 &= -(H_1 H_2 H_3)f \, dt^2 + (H_1 H_2 H_3)^{1/3}(f^{-1}dr^2 + r^2 d\Omega_{3,k}^2), \\
X_i &= H_i^{-1}(H_1 H_2 H_3)^{1/3}, \\
A^i &= \sqrt{k}(1 - H_i^{-1}) \coth \beta_i dt,
\end{aligned}
\tag{23.68}
$$

where f is a "blackening factor" (non-extremality), and

$$
\begin{aligned}
f &= k - \frac{\mu}{r^2} + g^2 r^2 (H_1 H_2 H_3), \\
H_i &= 1 + \frac{\mu \sinh^2 \beta_i}{kr^2},
\end{aligned}
\tag{23.69}
$$

and $k = 1, 0, -1$ corresponds to having S^3, T^3, or H^3 foliations for the "sphere" (although we have put a sphere there, it is worth noticing that the solution exists for all three cases).

The extremal RNAdS solution is obtained for $\beta_i = \beta$, so $H_i = H$, and X_i=constant=X. In order to obtain the precise form of the metric, we consider the change of coordinates

$$
\begin{aligned}
\tilde{r}^2 &= Hr^2 = r^2 + \mu \sinh^2 \beta, \\
H(r)^{-2}f(r) &= a(r),
\end{aligned}
\tag{23.70}
$$

obtained for

$$
g^2 = \frac{1}{l^2};
$$

$$
\sinh^2 \beta = -\frac{1}{2} + \frac{1}{2}\sqrt{1 + \frac{1 + 2r_H^2/l^2}{r_H^2/l^2(1 + 9r_H^2/4l^2)}}; \quad \mu = \frac{r_H^2(2 + 3r_H^2/l^2)}{\sqrt{1 + \frac{1 + 2r_H^2/l^2}{r_H^2/l^2(1+9r_H^2/4l^2)}}}.
\tag{23.71}
$$

The embedding in 10 dimensions is done using the (consistent, as it can be proven) KK reduction ansatz

$$
ds_{10}^2 = \sqrt{\tilde{\Delta}} ds_5^2 + \frac{1}{g^2\sqrt{\tilde{\Delta}}} \sum_{i=1}^{3} X_i^{-1}(d\mu_i^2 + \mu_i^2(d\phi_i + gA_i)^2),
\tag{23.72}
$$

where

$$\tilde{\Delta} = \sum_{i=1}^{3} X_i \mu_i^2,$$

$$d\Omega_5^2 = \sum_{i=1}^{3} d\mu_i^2 + \mu_i^2 d\phi_i^2;$$

$$\mu_1 = \sin\theta; \quad \mu_2 = \cos\theta \sin\psi; \quad \mu_3 = \cos\theta \cos\psi. \tag{23.73}$$

For our solution, $X_i = X = \text{constant}$, $\tilde{\Delta} = X$, and $1/g^2 = l^2$, and putting $X = 1$, we get

$$ds_{10}^2 = ds_5^2 + l^2 \sum_{i=1}^{3} [d\mu_i^2 + \mu_i^2 (d\phi_i + gA_i)^2]. \tag{23.74}$$

So the first step toward using AdS/CFT, of embedding of the solution in string theory, is there, but moreover we want to show that this is obtained as a *brane system in a decoupling limit*, as is the case for the $AdS_5 \times S^5$ background in which the oxidized RN AdS black hole lives.

We find that the solution is a rotating D3-brane solution in a decoupling limit.

The solution for D3-branes rotating with three angular momenta is

$$ds^2 = H^{-1/2} \left[-\left(1 - \frac{2m}{r^4 \Delta}\right) dt^2 + dx_1^2 + dx_2^2 + dx_3^2 \right] + H^{1/2} \left[\frac{\Delta dr^2}{H_1 H_2 H_3 - 2m/r^4} \right.$$

$$\left. +r^2 \sum_{i=1}^{3} H_i (d\mu_i^2 + \mu_i^2 d\phi_i^2) - \frac{4m \cosh\alpha}{r^4 H \Delta} dt \sum_{i=1}^{3} l_i \mu_i^2 d\phi_i + \frac{2m}{r^4 H \Delta} \left(\sum_{i=1}^{3} l_i \mu_i^2 d\phi_i \right)^2 \right], \tag{23.75}$$

where

$$\Delta = H_1 H_2 H_3 \sum_{i=1}^{3} \frac{\mu_i^2}{H_i};$$

$$H = 1 + \frac{2m \sinh^2\alpha}{r^4 \Delta};$$

$$H_i = 1 + \frac{l_i^2}{r^2}. \tag{23.76}$$

To find our case, we take $l_1 = l_2 = l_3 \equiv l_0$, which gives $H_1 = H_2 = H_3 \equiv h$, and $\Delta = h^2 \sum_i \mu_i^2$. Making again the change of variables $\tilde{r}^2 = r^2 + l_0^2$, we get

$$ds^2 = H^{-1/2} \left[-\left(1 - \frac{2m}{\tilde{r}^4}\right) dt^2 + dx_1^2 + dx_2^2 + dx_3^2 \right] + H^{1/2} \left[\frac{d\tilde{r}^2}{1 - \frac{2m}{\tilde{r}^4} + \frac{2ml_0^2}{\tilde{r}^6}} \right.$$

$$\left. +\tilde{r}^2 d\Omega_5^2 - \frac{4ml_0 \cosh\alpha}{\tilde{r}^4 H} dt \sum_{i=1}^{3} \mu_i^2 d\phi_i + \frac{2ml_0^2}{\tilde{r}^4 H} \left(\sum_{i=1}^{3} \mu_i^2 d\phi_i \right)^2 \right], \tag{23.77}$$

where $H = 1 + 2m \sinh^2\alpha \, r^{-4}$.

We define the decoupling limit via a parameter $\epsilon \to 0$, which is introduced by rescalings in

$$m = \epsilon^4 m'; \quad \sinh \alpha = \epsilon^{-2} \sinh \alpha'; \quad r = \epsilon r'; \quad x^\mu = \epsilon^{-1} x'^\mu; \quad l_i = \epsilon l'_i, \qquad (23.78)$$

after which we drop the primes. One then indeed gets our extremal RN AdS black hole solution oxidized to 10 dimensions, with the identifications

$$d\Omega_{3,k}^2 = d\vec{y} \cdot d\vec{y}; \quad \vec{y} = g\vec{x};$$
$$\frac{1}{g^2} = \sqrt{2m} \sinh \alpha; \quad \mu = 2mg^2; \quad l_i^2 = \mu \sinh^2 \beta_i. \qquad (23.79)$$

We see that we are interested in the case where the "sphere" becomes flat.

Then our solution interpolates between AdS_5 in the UV, with metric

$$ds_5^2 \simeq -\frac{r^2}{l^2} dt^2 + l^2 \frac{dr^2}{r^2} + r^2 d\Omega_{3,k=1}^2, \qquad (23.80)$$

and the space $AdS_2 \times S^3$ in the IR (the horizon, at $r \to r_H$), with

$$ds^2 = ds_{AdS_2}^2 (\rho = r - r_H) + r_H^2 d\Omega_3^2. \qquad (23.81)$$

The interpolation between r^2 and r_H^2 for the coefficient of $d\Omega_3^2$ is done via the function $b(r)^2$, and it is dual to a holographic flow between two field theory RG flow fixed points.

The flow between fixed points is known to be characterized by a "c-function," due to a "c-theorem," introduced and proven in the two-dimensional case by Zamolodchikov (the four-dimensional case is considerably more complicated and was only proven in 2011 by Zohar Komargodski and Adam Schwimmer), that is a function that is non-increasing (decreasing or constant) along the RG flow from UV to IR, and in the UV and IR takes the values of the corresponding conformal field theory central charges c, that measure the (effective) number of degrees of freedom, always higher in the UV than in the IR.

In fact, for the slightly more general metric

$$ds^2 = -a(r)^2 dt^2 + \frac{dr^2}{c(r)^2} + b(r) d\Omega_3^2, \qquad (23.82)$$

the holographic c-function is

$$C(r) = C_0 \frac{a^3}{b'^3 c^2} \to C_0 \frac{1}{b'^3}, \qquad (23.83)$$

so it is only a function of $b'(r)$ in the case that $a(r) = c(r)$, as we had until now, which explains our statement that $b(r)$ is dual to the holographic RG flow between two fixed points.

We should also note that the UV of the theory is dual to the field theory, while the IR is described via the attractor mechanism, that gives the entropy (so the number of degrees of freedom) in the IR. So at the horizon we have at least some partial information about the theory (in some cases, for a "membrane paradigm," we can describe more general properties of the theory at the horizon only, such as transport)

Important concepts to remember

- The attractor mechanism says that for *extremal* black holes with scalar fields, these change between (arbitrary) values at infinity and given (attractor) values at the horizon of the black hole.
- The horizon values are fixed by the charges of the black holes and are found by extremizing a certain function. The extremal black holes, usually correspond to BPS ones, with $M = |Z|$, but are more generally defined by the AdS horizons.
- In the original case for the attractor mechanism, for $\mathcal{N} = 2$ susy black holes in four dimensions, the mass and entropy are found in terms of an extremal value for Z, Z_{fix}.
- Sen's entropy function formalism states that for extremal black holes, defined by an $AdS_2 \times S^{d-2}$ horizon, defining the function $f(\vec{u}, \vec{v}, \vec{e}, \vec{p}) \equiv \int d\theta d\phi \sqrt{-\det g}\mathcal{L}$, with u scalar moduli at the horizon, v radii at the horizon, e electric fields, and p magnetic charges, Sen's entropy function $F(\vec{u}, \vec{v}, \vec{q}, \vec{p}) = 2\pi \left[e_i \frac{\partial f(\vec{u}, \vec{v}, \vec{e}, \vec{p})}{\partial e_i} - f(\vec{u}, \vec{v}, \vec{e}, \vec{p}) \right] = 2\pi[e_i q_i - f(\vec{u}, \vec{v}, \vec{e}, \vec{p})]$ has a minimum equal to the black hole entropy, and its extremization fixes u, v, e in terms of the charges q, p.
- In five-dimensional gauged supergravity, the attractor mechanism reduces to finding extrema of the "effective potential," $F^A = \frac{1}{b^3}f^{AB}Q_B \; dt \wedge dr$, whose matrix of second derivatives should have no negative eigenvalues.
- One can embed the extremal Reissner–Nordstrom–AdS five-dimensional black holes into string theory, as a system of rotating D3-branes in a decoupling limit. That allows us to use AdS/CFT and also define a "c-function," that gives the central charge at fixed points, which are the UV and IR limits of the (RG) flow corresponding to the gravity solution.

References and further reading

For the attractor mechanism, see the papers in which it was originally developed, [81] and [82, 83] (though there are many other relevant papers around that time). For Sen's entropy function formalism, see Sen's original paper [84]. For the attractor mechanisms in five-dimensional gauged supergravity with AdS background, and the relation to holography, see [85].

Exercises

(1) Given the argument about extremal $\mathcal{N} = 2$ supersymmetric black holes in four dimensions, what can you say about $\mathcal{N} = 4$ ones and $\mathcal{N} = 8$ ones?

(2) Given the ansatz (23.25), check explicitly that the equations of motion of the Lagrangian reduce to

$$\frac{\partial f}{\partial u_s} = 0, \quad \frac{\partial f}{\partial v_i} = 0, \quad \frac{\partial f}{\partial e^i} = q_i, \tag{23.84}$$

and that the gauge field equations of motion and Bianchi reduce to

$$\partial_r \left(\frac{\partial \sqrt{-\det g} \mathcal{L}}{\partial F_{rt}^{(i)}} \right) = 0, \quad \partial_r F_{\theta\phi}^{(i)} = 0, \tag{23.85}$$

which are satisfied by the ansatz.

(3) Check that

$$b(r) = r,$$

$$a^2(r) = 1 + \frac{r^2}{l^2} - \frac{m}{r^2} + \frac{q^2}{r^4} = \frac{1}{l^2 r^4}(r - r_H)^2(r + r_H)^2(r^2 + 2r_H^2 + l^2) \tag{23.86}$$

is a solution of the one-dimensional action

$$S = \frac{1}{2\kappa_N^2} \int dr \left(6b + 6ab^2 a'b' + 6a^2 bb'^2 - b^3 V(\phi) - a^2 b^3 (\phi_i')^2 - \frac{2}{b^3} V_{eff}(\phi_i) \right). \tag{23.87}$$

(4) Check that the entropy function

$$F(u^i, v_1, v_2, e^A, Q_A) = 2\pi [Q_A e^A - f(u^i, v_1, v_2, e^A)], \tag{23.88}$$

$$f(u^i, v_1, v_2, e^A) = 2\pi^2 \left[-2v_2^{3/2} + 6v_1 \sqrt{v_2} + 2\frac{v_2^{3/2}}{v_1} f_{AB} e^A e^B - v_1 v_2^{3/2} V(\phi) \right]$$

has the extremum

$$F_{extremum} = 8\pi^3 v_2^{3/2}. \tag{23.89}$$

24

Supersymmetric string (NS-R, GS, Berkovits) and supergravity on the worldsheet vs. spacetime supergravity

In this chapter, I will finally describe a few things about (super)string theory, in its three formulations, Neveu–Schwarz–Ramond (NSR), Green–Schwarz (GS), and Berkovits (pure spinor), and make the connection with supergravity in spacetime, as well as supergravity on the worldsheet of the string.

String theory is a theory of strings, generalizing particle theory, but in its less familiar, worldline formulation. Note that quantum field theory can be completely described as a theory on the worldline of the particle, though the formulation is a bit unyielding, and is less used. That is why we will start with an analysis of the relativistic particle action on the worldline.

24.1 Particle actions

When considering a particle, we begin with the nonrelativistic case. For a free particle, the action is

$$S_{\text{NR}} = \int dt\, L = \int dt \frac{m\dot{\vec{x}}^2}{2}. \tag{24.1}$$

This generalizes easily to the relativistic particle, whose action is m times the length of the worldline, that is, the integral of the proper time,

$$S_{\text{Rel}} = S_1 = -m(c^2) \int d\tau = -mc^2 \int d\tau \sqrt{-\dot{X}^\mu \dot{X}^\nu \eta_{\mu\nu}}$$

$$= -mc^2 \int dt \sqrt{1 - \frac{v^2}{c^2}}, \tag{24.2}$$

where we have reexpressed, *in a form that is only equivalent on-shell for* $X^\mu(\tau)$, $d\tau$ as $d\tau \sqrt{-\dot{X}^\mu \dot{X}^\nu \eta_{\mu\nu}}$, since on-shell we have $-\dot{X}^\mu \dot{X}^\nu \eta_{\mu\nu} = -ds^2/d\tau^2 = 1$. We see that by expanding the square root in the final form of the action in v/c, we get the nonrelativistic result minus the constant potential = rest energy mc^2.

The equation of motion of the action, when varying with respect to $X^\mu(\tau)$, is the free motion of the particle,

$$\frac{d}{d\tau}p^\mu = \frac{d}{d\tau}\left(m\frac{dX^\mu}{d\tau}\right) = 0, \tag{24.3}$$

as expected.

This seems trivial, but we can generalize this action to the case when we have interactions with external fields:

– The simplest generalization is to consider the interaction with a gravitational field, replacing $\eta_{\mu\nu}$ with $g_{\mu\nu}$. This leads to the generalization of the free particle motion to the geodesic motion in a gravitational field,

$$\frac{D}{d\tau}\left(m\frac{dX^\mu}{d\tau}\right) = 0. \tag{24.4}$$

– We can also couple the particle to an external electromagnetic field, via the worldline action term

$$\int d\tau A_\mu(X^\rho(\tau))q\frac{dX^\mu}{d\tau} = \int d^4x A_\mu(X^\rho)j^\mu(X^\rho), \tag{24.5}$$

where we have rewritten the term on the worldline as the usual spacetime electromagnetic coupling, via the current

$$j^\mu(X^\rho) = q\delta^3(X^\rho - X^\rho(\tau))\frac{dX^\mu}{d\tau}. \tag{24.6}$$

But the above action is highly nonlinear, due to the square root, so it is not very nice to quantize. However, we know how to deal with this situation: For the above "second-order action," we write a "first-order form," by introducing an auxiliary field. In particular, it is clear what is the auxiliary field here: We must couple the worldline action to independent gravity on the worldline, described via an einbein ("viel" bein, but there is only one component now, so "ein" bein) $e(\tau) = \sqrt{-\det\gamma(\tau)}$ (where, of course, the determinant is moot in one dimension, but we put it there so that the origin of the term is clear: It is the integration measure in curved space). Then the action in first-order formulation becomes

$$S_p = \frac{1}{2}\int d\tau\left[e^{-1}(\tau)\frac{dX^\mu}{d\tau}\frac{dX^\nu}{d\tau}\eta_{\mu\nu} - em^2\right]. \tag{24.7}$$

It is also clear how we wrote the two terms: The first term is just the massless scalar action in one dimension, since $e^{-1}(\tau) = \sqrt{-\det\gamma}\,\gamma^{ab}$, while the second term is just a constant, since $e = \sqrt{-\det\gamma}$. The coefficient of $-m^2$ of the second term is chosen such that the action becomes equivalent with the above second-order action:

Considering the equation of motion for $e(\tau)$,

$$\frac{\delta S}{\delta e(\tau)} = 0 \Rightarrow e^2(\tau) = -\frac{\dot{X}^\mu\dot{X}_\mu}{m^2}, \tag{24.8}$$

and by substituting that back into S_p, we obtain back S_1.

24.2 Bosonic string

We can now easily generalize that to the case of the (bosonic) string. According to the general idea, the worldline is generalized to a *worldsheet*, for the spatial direction along the string σ, plus the proper time direction along the motion of the string τ, and the (mass times) length of the worldline becomes (tension times) area of the worldsheet. In general, for a brane, the action would be the volume of the worldvolume. So

$$S_{\text{string}} = -T \int dA = -\frac{1}{2\pi\alpha'} \int d\sigma \int d\tau \sqrt{-\det\gamma_{ab}}, \qquad (24.9)$$

where γ_{ab} is the metric on the worldsheet. But what kind of metric?

24.2.1 Actions and equations of motion

–Second-order formalism

In a second-order formalism, γ^{ab} stands for the *induced* metric from spacetime to the worldsheet,

$$h_{ab}(\xi^a) = \partial_a X^\mu \partial_b X^\nu g_{\mu\nu}(X^\rho(\xi^a)). \qquad (24.10)$$

Then, we obtain the Nambu–Goto action,

$$S_{\text{NG}} = -\frac{1}{2\pi\alpha'} \int d\sigma \int d\tau \sqrt{-\det h_{ab}(X^\rho(\xi^a))}. \qquad (24.11)$$

–First-order formalism

The first-order formulation, paralleling what we did for the particle, is the *Polyakov action* (even though it was found by Brink, DiVecchia, Howe, Deser, and Zumino). In flat spacetime (for $g_{\mu\nu} = \eta_{\mu\nu}$), it is as follows: couple the action to gravity on the worldsheet, via the independent metric γ_{ab} on the worldsheet, acting as an auxiliary field, so

$$S_P[X^\rho, \gamma_{ab}] = -\frac{1}{4\pi\alpha'} \int d\sigma\, d\tau \sqrt{-\gamma}\, \gamma^{ab} \partial_a X^\mu \partial_b X^\nu \eta_{\mu\nu}. \qquad (24.12)$$

We want to notice that, besides the usual diffeomorphism invariance on the worldsheet and Poincaré invariance in spacetime, the action is also Weyl invariant on the worldsheet, namely invariant under a transformation for which X^μ are invariant, but the metric changes by a conformal factor (a common factor depending on the worldsheet coordinates)

$$\gamma'_{ab} = e^{2\omega(\sigma,\tau)}\gamma_{ab}. \qquad (24.13)$$

The variation of the action with respect to the independent metric γ^{ab} gives

$$\delta S_P = -\frac{1}{4\pi\alpha'} \int d\sigma\, d\tau \sqrt{-\gamma}\, \delta\gamma^{ab} \left[\partial_a X^\mu \partial_b X^\nu \eta_{\mu\nu} - \frac{1}{2}\gamma_{ab} \left(\gamma^{cd} \partial_c X^\mu \partial_d X^\nu \eta_{\mu\nu} \right) \right] = 0, \qquad (24.14)$$

which gives

$$\gamma_{ab} = h_{ab} \equiv \partial_a X^\mu \partial_b X^\nu \eta_{\mu\nu}, \qquad (24.15)$$

and so again, when substituting in the Polyakov action, we get back the Nambu–Goto action.

The energy–momentum tensor on the worldsheet of the string is usually multiplied by a conventional 2π, to make formulas nicer (the overall constant doesn't matter, since in two dimensions, gravity is trivial, so its coupling to T_{ab} is irrelevant), giving

$$T^{ab} = -4\pi \frac{1}{\sqrt{-\gamma}} \frac{\delta S_P}{\delta\gamma^{ab}} = \frac{1}{\alpha'} \left(\partial^a X^\mu \partial^b X^\nu - \frac{1}{2}\gamma^{ab} \partial_c X^\mu \partial^c X^\nu \right) \eta_{\mu\nu}. \qquad (24.16)$$

Note that, since there is no kinetic term for γ^{ab} (or rather, it gives no contribution to the equations of motion, being topological), the equation of motion for it is just

$$T_{ab} = 0. \tag{24.17}$$

As usual, the energy–momentum tensor is conserved,

$$\nabla_a T^{ab} = 0, \tag{24.18}$$

which comes from the Noether theorem.

But, now we have also a Weyl invariance of the action, which can be expressed as (since it multiplies all components of the metric by the same factor)

$$\gamma^{ab} \frac{\delta S}{\delta \gamma^{ab}} = 0 \Rightarrow T^a{}_a = 0, \tag{24.19}$$

just that this tracelessness is *off-shell*.

Finally then, the equation of motion for X^μ is just the free wave equation,

$$\frac{\delta S}{\delta X^\mu} = 0 \Rightarrow \nabla^2 X^\mu = 0. \tag{24.20}$$

The equation of motion must be supplemented with boundary conditions, which should vanish the boundary terms in the general variation of the action. The possibilities are:

– *closed strings*, which means periodic boundary conditions on the worldsheet,

$$X^\mu(\tau, \sigma + l) = X^\mu(\tau, \sigma), \quad \gamma^{ab}(\tau, \sigma + l) = X^\mu(\tau, \sigma). \tag{24.21}$$

– *open strings*, which can have either:
– Neumann boundary conditions,

$$\partial^\sigma X^\mu(\tau, 0) = \partial^\sigma X^\mu(\tau, l), \tag{24.22}$$

– or Dirichlet boundary conditions,

$$\delta X^\mu(\tau, 0) = \delta X^\mu(\tau, l) = 0. \tag{24.23}$$

We should fix a gauge for the general coordinate invariance, with two local parameters $\xi^a(\sigma, \tau)$, and for the Weyl invariance, with one local parameter $\omega(\sigma, \tau)$, which means that we can completely fix the form of the metric, obtaining the *unit gauge* (sometimes also called conformal gauge, though that is not quite correct),

$$h_{ab} = \eta_{ab}. \tag{24.24}$$

As usual when we fix a gauge, the equation of motion of the object that was fixed becomes a constraint (in electromagnetism, for instance, in Coulomb gauge $A_0 = 0$, we obtain the Gauss constraint $\vec{\nabla} \cdot \vec{E} = 0$). Here then, having fixed completely γ_{ab}, the constraint becomes its equation of motion, so

$$T_{ab} = 0. \tag{24.25}$$

In this case, we finally obtain the free action in conformal (unit) gauge,

$$S = \frac{T}{2} \int d^2\sigma \, \eta^{ab} \partial_a X^\mu \partial_b X^\nu \eta_{\mu\nu}. \tag{24.26}$$

We next solve the free wave equation in this gauge, with each of the particular boundary conditions. Since the free wave equation can be written as

$$(\partial_\sigma + \partial_\tau)(\partial_\sigma - \partial_\tau)X^\mu = 0, \tag{24.27}$$

we can have a constant and a linear term in τ, describing the center of mass motion, plus oscillator terms that depend on $\sigma + \tau$ (left movers) and ones that depend on $\sigma - \tau$ (right movers). Then, we obtain:

– for closed strings,

$$X^\mu(\sigma,\tau) = x^\mu + \alpha' p^\mu \tau + \frac{i\sqrt{2\alpha'}}{2} \sum_{n\neq 0} \frac{1}{n} \left[\alpha_n^\mu e^{-in(\tau-\sigma)} + \tilde{\alpha}_n^\mu e^{-in(\tau+\sigma)} \right], \tag{24.28}$$

– for Neumann open strings, where we basically identify α_n^μ with $\tilde{\alpha}_n^\mu$,

$$X^\mu(\sigma,\tau) = x^\mu + 2\alpha' p^\mu \tau + i\sqrt{2\alpha'} \sum_{n\neq 0} \frac{1}{n} \alpha_n^\mu e^{-in\tau} \cos n\sigma. \tag{24.29}$$

We will not describe the Dirichlet open strings here.

24.2.2 Constraints, oscillators, and quantization

We still have to solve the constraints, which will give conditions on α_n^μ and $\tilde{\alpha}_n^\mu$, and among them is the condition that the worldsheet Hamiltonian vanishes, $H = 0$. Here we obtain

$$H_{\text{open}} = \frac{1}{\alpha'} \sum_{n=-\infty}^{+\infty} \alpha_{-n}^\mu \alpha_n^\mu$$

$$H_{\text{closed}} = \frac{2}{\alpha'} \sum_{n=-\infty}^{+\infty} \left(\alpha_{-n}^\mu \alpha_n^\mu + \tilde{\alpha}_{-n}^\mu \tilde{\alpha}_n^\mu \right). \tag{24.30}$$

From the Lagrangian, we can calculate the canonically conjugate momentum to $X^\mu(\sigma,\tau)$, $P^\mu(\sigma,\tau)$, and define Poisson brackets, such that

$$\{X^\mu(\sigma,\tau), P^\nu(\sigma',\tau)\}_{P.B.} = \delta(\sigma - \sigma')\eta^{\mu\nu},$$

$$\{X^\mu(\sigma,\tau), X^\nu(\sigma',\tau)\}_{P.B.} = 0 = \{P^\mu(\sigma,\tau), P^\nu(\sigma',\tau)\}_{P.B.} = 0. \tag{24.31}$$

Substituting the expansion in terms of α_n^μ and $\tilde{\alpha}^\mu$, we obtain their commutators,

$$\{\alpha_m^\mu, \alpha_n^\nu\}_{P.B.} = \{\tilde{\alpha}_m^\mu, \tilde{\alpha}_n^\nu\}_{P.B.} = -im\delta_{m+n}\eta^{\mu\nu}. \tag{24.32}$$

The next step is then to quantize these Poisson brackets, with the usual replacement $\{,\}_{P.B.} \to \frac{1}{i\hbar}[,]$. However, we have an extra symmetry to fix, a residual gauge symmetry from the unit gauge, conformal invariance. This is fixed, for instance, via the lightcone gauge

$$X^+(\sigma,\tau) = x^+ + p^+\tau, \tag{24.33}$$

which says that X^+ is not independent and has no nontrivial oscillators. We must also consider the constraints, which fix also the X^- oscillators, as

$$\alpha_n^- = \frac{\sqrt{2\alpha'}}{2p^+} \sum_{m\in\mathbb{Z}} \alpha_{n-m}^i \alpha_m^i, \tag{24.34}$$

so they are dependent functions of α_n^i. This means that only α_n^i are independent variables and are quantized in lightcone gauge.

We only showed explicitly the Hamiltonian $H = 0$ constraint, since this is the one that gives the mass spectrum. Indeed, from it, we find

$$M^2 \equiv -p^\mu p_\mu = 2p^+ p^- - p^i p^i = \frac{1}{\alpha'} \left(\sum_{n\geq 1} \alpha_{-n}^i \alpha_n^i - a \right) = \frac{1}{\alpha'} \left(\sum_{n\geq 1} n a_n^{\dagger i} a_n^i - a \right), \quad (24.35)$$

where we have rescaled the oscillators,

$$\alpha_m^\mu = \sqrt{m} a_m^\mu, \quad \alpha_{-m}^\mu = \sqrt{m} a_m^{\dagger\mu}, \quad (24.36)$$

in order to obtain creation and annihilation operators,

$$[a_n^i, a_m^{\dagger j}] = \delta_{ij} \delta_{mn}. \quad (24.37)$$

Here a is a quantum ordering constant: When going to the quantum theory, a_n^i and $a_n^{\dagger i}$ are noncommuting operators, so the order in the formula for M^2 matters: When we write it as $a^\dagger a$, we obtain (in zeta function regularization, where $\sum_n n = \zeta(-1) = \frac{1}{12}$)

$$a = \frac{D-2}{24}. \quad (24.38)$$

But we will shortly see that we must have $a = 1$, which singles out $D = 26$: Bosonic strings are only quantum mechanically consistent in $D = 26$ dimensions.

The spectrum is obtained as usual in the case we have quanta for some field (photons, phonons, etc.): We act with the creation operator on some vacuum. The only difference now is that the modes of the string create different kinds of particles, and so the vacuum, corresponding to a tachyonic particle, also has a momentum, thus $|0;\vec{k}\rangle$. In particular, for open strings, we obtain (for $n = 1$)

$$a_1^{\dagger i} |0;\vec{k}\rangle, \quad (24.39)$$

which looks like a gauge field, A_i, hence must be massless. But the mass is $M^2 = (1-a)/\alpha'$, so that fixes $a = 1$, as we said.

Note that we have shown here how to quantize in lightcone gauge, but there are other possibilities: There is a (modern) covariant quantization, Gupta–Bleuler, or BRST quantization.

24.2.3 Background fields

Considering next nontrivial background (spacetime) fields, the string action in NS-NS background is (as seen in Chapter 18)

$$S = -\frac{1}{4\pi\alpha'} \int d^2\sigma \left[\sqrt{-\gamma} \gamma^{ab} \partial_a X^\mu \partial_b X^\nu g_{\mu\nu}(X^\rho(\sigma)) \right.$$
$$\left. + \alpha' \epsilon^{ab} \partial_a X^\mu \partial_b X^\nu B_{\mu\nu}(X^\rho(\sigma)) - \alpha' \sqrt{-\gamma} \mathcal{R}^{(2)} \Phi(X^\rho(\sigma)) \right], \quad (24.40)$$

in terms of the metric $g_{\mu\nu}$, the (NS-NS) Kalb–Ramond field $B_{\mu\nu}$ and the dilaton Φ, where $\mathcal{R}^{(2)}$ is the two-dimensional Ricci tensor. However, in two dimensions,

$$\frac{1}{4\pi} \int d^2\sigma \sqrt{-\gamma} \mathcal{R}^{(2)} = \chi, \quad (24.41)$$

which is a topological invariant, known as the Euler character, related to the genus g (the unique topological invariant of bosonic two-dimensional surfaces) by $\chi = 2(1 - g)$.

But then, if $\Phi(X^\rho(\sigma))$ is constant, it goes outside the integral, and we form simply the constant term $\Phi\chi = 2(1 - g)\Phi$ in the action.

For quantum consistency of string theory, like for any gauge theory, the local symmetries present at the classical level should be maintained at the quantum level. In the case of the string, we have conformal invariance, which should be maintained at the quantum level. But the nontrivial NS-NS (and supergravity, more generally) backgrounds can be thought of as quantum corrections to flat spacetime: There are "vertex operators" which create particles = quanta of the corresponding fields $(g_{\mu\nu}, B_{\mu\nu}, \Phi)$, which can be shown to simply add a contribution to the action such that $g_{\mu\nu}, B_{\mu\nu}, \Phi$ are changed (note that, since these fields are bosons, we can have Bose–Einstein condensation that creates a nontrivial background from many particles).

Then the quantum consistency dictates that the backgrounds are such that the conformal invariance is preserved. This means that all the beta functions must be zero (since conformal invariance includes scale invariance, and the beta functions correspond to quantum changes with the scale). But the couplings of the two-dimensional action are exactly the spacetime fields $g_{\mu\nu}, B_{\mu\nu}, \Phi$, so we have the equations

$$\beta_\Phi = 0, \quad \beta_{g_{\mu\nu}} = 0, \quad \beta_{B_{\mu\nu}} = 0, \tag{24.42}$$

which give just the supergravity equations of motion in zeroth order, plus α' corrections in nontrivial orders.

24.3 Supersymmetric strings

We now want to supersymmetrize the bosonic string action. But that could mean:

- to supersymmetrize on the worldsheet, which is useful for worldsheet calculations. This is the Neveu–Schwarz–Ramond (NSR) superstring or "spinning string" formalism.
- to supersymmetrize in spacetime, which is what we really want, certainly in the low-energy supergravity. This is the Green–Schwarz (GS) superstring formalism.

But, while the NSR string is easily quantized, the GS string can be easily quantized only in a gauge, which breaks spacetime Lorentz invariance, and where it is shown to be equivalent to the NSR formalism.

So we would like to have a *quantum* covariant superstring formulation. That is the Berkovits formalism, using pure spinors. However, that is considerably harder to use, despite having the best of both worlds.

24.3.1 The superparticle

As in the bosonic case, before we embark on the superstring analysis, we start with the superparticle, the supersymmetric generalization of the particle action, more precisely of the first-order S_p action.

We introduce N spacetime spinors and worldsheet scalars $\theta^{A\alpha}$, with $A = 1, ..., N$ and α the spacetime spinor index in 10 dimensions.

We generalize $dX^\mu/d\tau$ to a manifestly supersymmetric quantity,

$$\Pi^\mu_\tau = \frac{dX^\mu}{d\tau} + \bar{\theta}^A \Gamma^\mu \frac{d\theta^A}{d\tau}, \tag{24.43}$$

which is invariant under N supersymmetries,

$$\delta\theta^A = \epsilon^A, \quad \delta\bar{\theta}^A = \bar{\epsilon}^A, \quad \delta e = 0$$
$$\delta X^\mu = -\bar{\epsilon}^A \Gamma^\mu \theta^A. \tag{24.44}$$

Here Γ^μ are $SO(9,1)$ gamma matrices, so $\{\Gamma^\mu, \Gamma^\nu\} = 2\eta^{\mu\nu}$, decomposed in terms of $SO(8)$ gamma matrices γ^i, so $\{\gamma^i, \gamma^j\} = 2\delta^{ij}$, as (an example of decomposition, or of the representation)

$$\Gamma^0 = -i\sigma_2 \otimes \mathbb{1}_{16}, \quad \Gamma^i = \sigma_1 \otimes \gamma^i, \quad i = 1, ..., 8, \quad \Gamma^9 = \sigma_3 \otimes \mathbb{1}_{16}. \tag{24.45}$$

The superparticle action, with spacetime supersymmetry, is

$$S = \frac{1}{2} \int d\tau e^{-1}(\tau) \Pi^\mu_\tau \Pi^\mu_\tau. \tag{24.46}$$

However, now that we wrote the action, we note that it has an additional symmetry, known as kappa symmetry, with parameters $\kappa^{A\alpha}$, which are N spacetime spinors, and transformation rules

$$\delta\theta^A = -\Gamma_\mu \Pi^\mu_\tau \kappa^{A\alpha},$$
$$\delta X^\mu = -2\bar{\theta}^A \Gamma^\mu \delta\theta^A,$$
$$\delta e = 4e\dot{\bar{\theta}}^A \kappa^A. \tag{24.47}$$

The invariance of the action under kappa symmetry is obtained from

$$\delta\Pi^\mu_\tau = -2\dot{\bar{\theta}}^A \Gamma^\mu \delta\theta^A \Rightarrow$$
$$\delta\Pi^2_\tau = -4\dot{\bar{\theta}}^A \Gamma^\mu \Pi^\mu_\tau \delta\theta^A = 4\Pi^2_\tau \dot{\bar{\theta}}^A \kappa_A,$$
$$\delta e^{-1} = -4e^{-1}\dot{\bar{\theta}}^A \kappa_A. \tag{24.48}$$

24.3.2 The GS superstring

Moving on to the string, to obtain the GS superstring, with spacetime supersymmetry, we supersymmetrize the Polyakov action via introducing manifestly supersymmetric objects Π^μ_a like Π^μ_τ, but now in σ, τ. Thus, we define

$$\Pi^\mu_a = \partial_a X^\mu + \bar{\theta}^A \Gamma^\mu \partial_a \theta^A, a = \sigma, \tau. \tag{24.49}$$

These are manifestly invariant under N (global) spacetime supersymmetries,

$$\delta\theta^A = \epsilon^A, \quad \delta\bar{\theta}^A = \bar{\epsilon}^A, \quad \delta X^\mu = -\bar{\epsilon}^A \Gamma^\mu \theta^A. \tag{24.50}$$

Thus the naive guess for the GS superstring action gives a kinetic term that we call

$$S_1 = S_{\text{kin}} = -\frac{1}{4\pi\alpha'} \int d^2\sigma \sqrt{-\gamma} \gamma^{ab} \Pi^\mu_a \Pi^\nu_b \eta_{\mu\nu}, \tag{24.51}$$

with the obvious generalization to a supergravity background of replacing $\eta_{\mu\nu}$ by $g_{\mu\nu}$.

But this action has no kappa symmetry, and kappa symmetry will be gauge fixed to give worldsheet susy, and thus a supersymmetric spectrum, which is needed!

This means that we must include a second term, which will absorb the variation under a certain kappa symmetry of the first term.

This second term is, written for $\mathcal{N} = 2$ susies,

$$S_2 = \frac{1}{2\pi\alpha'} \int d^2\sigma \left[\epsilon^{ab} \partial_a X^\mu (\bar\theta^1 \Gamma_\mu \partial_b \theta^1 - \bar\theta^2 \Gamma^\mu \partial_b \theta^2) \right.$$
$$\left. - \epsilon^{ab} (\bar\theta^1 \Gamma^\mu \partial_a \theta^1)(\bar\theta^2 \Gamma_\mu \partial_b \theta^2) \right], \tag{24.52}$$

and it can be rewritten as a super-Wess–Zumino (WZ) term in superspace, in the super-geometric approach, as

$$S_2 = S_{\rm WZ} = -\frac{1}{4\pi} \int d^2\sigma \epsilon^{ab} \Pi_a^\Lambda \Pi_b^\Sigma B_{\Lambda\Sigma}, \tag{24.53}$$

where $\Lambda = (\mu, \alpha)$ is a superspace index, with α a spacetime spinor index, and

$$\Pi^\Lambda = (\Pi^\mu, d\theta^\alpha) \tag{24.54}$$

are basis one-forms in superspace for the coset $\frac{N=2 \text{ susies}}{SO(9,1)}$.

Of course, $B_{\mu\nu}$ is a background, which is $= 0$ in flat spacetime, but now, even in flat spacetime, there are fermionic components that are nonzero.

As we discussed earlier, this WZ term is only for $\mathcal{N} = 2$ susies, but one can also write one with $\mathcal{N} = 1$ for an $\mathcal{N} = 1$ invariant action.

Proof

The proof that S_2 can be rewritten as a WZ term is as follows (following Henneaux and Mezincescu). A WZ term is an integrated total derivative, so we look for 3-forms in the $\mathcal{N} = 2$ superspace $= \frac{N=2 \text{ susies}}{SO(9,1)}$, meaning the translation generators and the susy generators, with basis Π^Λ. Consider then a generic 3-form

$$\Omega_3 = f_{\Lambda\Pi\Sigma} \Pi^\Lambda \wedge \Pi^\Pi \wedge \Pi^\Sigma. \tag{24.55}$$

For correct (Lorentz) transformation properties, the $f_{\Lambda\Pi\Sigma}$ must be the structure constants of the susy algebra (which is the only three-index antisymmetric quantity with good properties),

$$[T_\Lambda, T_\Pi\} = f_{\Lambda\Pi\Sigma} T_\Sigma. \tag{24.56}$$

But the relevant $\mathcal{N} = 2$ algebra on superspace, so for P_μ (translation generator) and Q_α (susy generator) is

$$\{Q_\alpha, \bar{Q}_\beta\} = (C\gamma^\mu)_{\alpha\beta} P_\mu, \quad [Q_\alpha, P_\mu] = [\bar{Q}_\alpha, P_\mu] = [P_\mu, P_\nu] = 0, \tag{24.57}$$

which means that the most general 3-form must be

$$\Omega_3 = \Pi^\mu \wedge d\bar\theta^{\alpha A} (C\gamma^\mu)_{\alpha\beta} d\theta^{\beta B} a_{AB}, \tag{24.58}$$

where a_{AB} is a symmetric matrix, which can be diagonalized. Since, moreover, we want $d\Omega_3 = 0$ (as we want $\Omega_3 = d\Omega_2$ to write our WZ term), it follows that we want a_{AB} to be traceless, so finally a_{AB} is diagonalized to

$$a_{AB} = a \begin{pmatrix} 1 & 0 \\ 0 & -1 \end{pmatrix}. \tag{24.59}$$

Then

$$\begin{aligned}
\Omega_3 &= a \left(\Pi^\mu \wedge d\bar{\theta}^1 \gamma_\mu \wedge d\theta^1 - \Pi^\mu \wedge d\bar{\theta}^2 \gamma_\mu \wedge d\theta^2 \right) \\
&= a \left(dX^\mu - \bar{\theta}^c (C\gamma^\mu) d\theta^c \right) \wedge \left(d\bar{\theta}^1 (C\gamma_\mu) \wedge d\theta^1 - d\bar{\theta}^2 (C\gamma_\mu) \wedge d\theta^2 \right),
\end{aligned} \tag{24.60}$$

so

$$\begin{aligned}
a^{-1}\Omega_3 &= dX^\mu \wedge \left(d\bar{\theta}^1 (C\gamma_\mu) \wedge d\theta^1 - d\bar{\theta}^2 (C\gamma_\mu) \wedge d\theta^2 \right) \\
&\quad - \bar{\theta}^c (C\gamma^\mu) d\theta^c \left(d\bar{\theta}^1 (C\gamma_\mu) \wedge d\theta^1 - d\bar{\theta}^2 (C\gamma_\mu) \wedge d\theta^2 \right) \\
&= d \left[dX^\mu \wedge (\bar{\theta}^1 \gamma_\mu d\theta^1 - \bar{\theta}^2 \gamma_\mu d\theta^2) \right. \\
&\quad \left. - (\bar{\theta}^1 \gamma^\mu d\theta^1 + \bar{\theta}^2 \gamma^\mu d\theta^2) \wedge (\bar{\theta}^1 \gamma_\mu \wedge d\theta^1 - \bar{\theta}^2 \gamma_\mu \wedge d\theta^2) \right].
\end{aligned} \tag{24.61}$$

But

$$\gamma_\mu d\theta^1 \wedge d\bar{\theta}^1 \wedge \gamma^\mu d\theta^1 \equiv 0, \tag{24.62}$$

and similarly for $1 \leftrightarrow 2$, so we obtain

$$\begin{aligned}
a^{-1}\Omega_3 &= d \left[dX^\mu \wedge (\bar{\theta}^1 \gamma_\mu d\theta^1 - \bar{\theta}^2 \gamma_\mu d\theta^2) \right] \\
&\quad + \bar{\theta}^1 \gamma^\mu d\theta^1 \wedge d\bar{\theta}^2 \gamma_\mu \wedge d\theta^2 + d\bar{\theta}^1 \gamma_\mu \wedge d\theta^1 \wedge \bar{\theta}^2 \gamma^\mu \wedge d\theta^2 \\
&= d \left[dX^\mu \wedge (\bar{\theta}^1 \gamma_\mu d\theta^2) + (\bar{\theta}^1 \gamma^\mu d\theta^1) \wedge (\bar{\theta}^2 \wedge \gamma_\mu d\theta^2) \right].
\end{aligned} \tag{24.63}$$

So, as we wanted, we have that $\Omega_3 = d\Omega_2$, where

$$a^{-1}\Omega_2 = dX^\mu \wedge (\bar{\theta}^1 \gamma_\mu d\theta^2) + (\bar{\theta}^1 \gamma^\mu d\theta^1) \wedge (\bar{\theta}^2 \wedge \gamma_\mu d\theta^2). \tag{24.64}$$

But then, using the general Stokes theorem, we have

$$\int_{\mathcal{M}} \Omega_3 = \int_{\mathcal{M}} d\Omega_2 = \int_{\partial\mathcal{M}} \Omega_2, \tag{24.65}$$

and we choose $\mathcal{M}$ as the three-dimensional "disk" or, more precisely, cylinder, embedded in superspace (rather, $\mathcal{M}$ is a submanifold such that there is a map between the disk and $\mathcal{M}$) and such that $\partial\mathcal{M}$ is a two-dimensional (closed) surface for the worldsheet parametrized by (σ, τ).

Then, written in (σ, τ) space, $\int_{\partial\mathcal{M}} \Omega_2$ is just S_2. q.e.d.

Denote then Ω_3 by H and write it as

$$H = H_{\Lambda\Pi\Sigma} \Pi^\Lambda \wedge \Pi^\Pi \wedge \Pi^\Sigma; \tag{24.66}$$

it follows that

$$S_2 = \int_{\mathcal{M}} H = \int_{\mathcal{M}} H_{\Lambda\Pi} \Pi^\Lambda \wedge \Pi^\Pi \wedge \Pi^\Sigma = \int_{\partial\mathcal{M}} B_{\Lambda\Sigma} \Pi^\Lambda \wedge \Pi^\Sigma. \tag{24.67}$$

Moreover, then

$$H = dB = (\gamma_\mu)_{\alpha\beta} \Pi^\alpha \wedge \Pi^\beta \wedge \Pi^\mu. \tag{24.68}$$

Finally then, the total GS action is

$$S_{\text{GS}} = S_{\text{kin}} + S_{\text{WZ}} = S_1 + S_2 \tag{24.69}$$

and it also has, besides the spacetime susy invariance, an invariance under a kappa symmetry with parameters $\kappa^{Aa\alpha}$, $A = 1, 2$, which are worldsheet vectors with index a and spacetime spinors with index α, given by

$$\delta_\kappa \theta^A = -2\Gamma_\mu \Pi_a^\mu \kappa^{Aa}$$
$$\delta_\kappa X^\mu = -\bar{\theta}^A \Gamma^\mu \delta\theta^A$$
$$\delta_\kappa (\sqrt{-\gamma}\gamma^{ab}) = -16\sqrt{-\gamma} \left(P_-^{ac} \bar{\kappa}^{1b} \partial_c \theta^1 + P_+^{ac} \bar{\kappa}^{2b} \partial_c \theta^2 \right), \tag{24.70}$$

where we have defined the (anti)self-dual projectors

$$P_\pm^{ab} = \frac{1}{2} \left(\gamma^{ab} \pm \frac{\epsilon^{ab}}{\sqrt{-\gamma}} \right). \tag{24.71}$$

In a general supersymmetric background, the GS action is written as (in the supergeometric approach)

$$S_{\text{GS}} = -\frac{1}{4\pi\alpha'} \int d^2\sigma \sqrt{-\gamma}\gamma^{ab} \Pi_a^\Lambda \Pi_b^\Pi G_{\Lambda\Pi} + \int_{\partial\mathcal{M}} B_{\Lambda\Pi} \Pi^\Lambda \wedge \Pi^\Pi. \tag{24.72}$$

One observation is that, if the θ^A have the same chirality, we obtain the type IIB superstring action, while if they have opposite chiralities, we obtain the type IIA superstring action.

We can fix the kappa symmetry, by introducing an extra lightcone gauge condition for the fermions (on top of the usual lightcone gauge condition for the bosons),

$$\Gamma^+ \theta^1 = \Gamma^+ \theta^2, \quad \Gamma^\pm = \frac{\Gamma^0 \pm \Gamma^9}{\sqrt{2}}. \tag{24.73}$$

Imposing these two lightcone gauge conditions (for the bosons and the fermions) in flat spacetime turns the GS action in the lightcone action

$$S_{\text{lc}} = -\frac{1}{4\pi\alpha'} \int d^2\sigma \left[\partial_a X^i \partial^a X^i + 2\alpha' \bar{S}^m \gamma^a \partial_a S^m \right], \tag{24.74}$$

where the gauge-fixed worldsheet scalars $\theta^{A\alpha}$ have been regrouped into two-component Majorana (real) worldsheet spinors S^m, with m a spinor index of $SO(8)$, so that S^m now transform under the two-dimensional Lorentz transformations.

The above lightcone action has now both spacetime and worldsheet supersymmetry.

The lightcone action can be quantized, but we cannot find a covariant quantization formulation. For that, we must consider the Berkovits formalism.

24.3.3 NSR (Spinning) string

We next move on to the case with manifest worldsheet supersymmetry, the NSR string, or spinning string.

It is written in terms of fermionic variables $\psi^\mu(\sigma, \tau)$ that are worldsheet spinors and spacetime vectors. The action is

$$S = -\frac{1}{4\pi\alpha'} \int d^2\sigma \left[\partial_a X^\mu \partial^a X_\mu - \bar{\psi}^\mu \gamma^a \partial_a \psi_\mu \right], \tag{24.75}$$

where the two-dimensional gamma matrices are

$$\gamma^0 - \begin{pmatrix} 0 & -1 \\ 1 & 0 \end{pmatrix} = -i\sigma_2, \quad \gamma^1 = \begin{pmatrix} 0 & 1 \\ 1 & 0 \end{pmatrix} = \sigma_1. \tag{24.76}$$

The action has worldsheet supersymmetry

$$\delta X^\mu = \bar{\epsilon}\psi^\mu, \quad \delta\psi^\mu = \gamma^a \partial_a X^\mu \epsilon. \tag{24.77}$$

This is nothing but the susy transformation of the two-dimensional WZ model, for each value of μ.

When varying the action, besides the usual bosonic boundary term (whose vanishing gives the possible boundary conditions), we also get now a fermionic boundary term,

$$\psi_+ \delta\psi_+ - \psi_- \delta\psi_- |_0^\pi . \tag{24.78}$$

Its vanishing imposes the fermionic boundary conditions

$$\psi_+ = \pm\psi_- \tag{24.79}$$

at each endpoint. However, we can put $\psi_+^\mu(0, \tau) = \psi_-^\mu(0, \tau)$ (at one end) by redefining the fermions; however, then at the other endpoint, there are two independent possibilities,

$$\psi_+(\pi, \tau) = \pm\psi_-(\pi, \tau). \tag{24.80}$$

The case with $+$ is the Ramond (R) sector and gives rise to spacetime fermions, and the case with $-$ is the Neveu–Schwarz (NS) case and gives rise to spacetime bosons.

For the closed string, we have independent left and right movers, so we must consider the boundary conditions and corresponding sectors for each, obtaining NS-NS (Bose–Bose) and R-R (Fermi–Fermi) fields that are bosonic, and NS-R (Bose–Fermi) and R-NS (Fermi–Bose) fields that are fermionic.

In lightcone gauge, the NSR string is equivalent to the GS superstring, showing that there is in fact a single superstring theory (of course, each is of type IIA, IIB, I, or heterotic, and these are then unified in M theory).

If we put the NSR string in worldsheet curved space, the supersymmetry on the worldsheet means that in fact we must (for consistency) couple to supergravity, not just to gravity.

Introduce then the vielbein $e_a^m(\sigma, \tau)$, where a is a curved index and m a flat index, with the worldsheet metric

$$\gamma_{ab} = e_a^m e_b^n \eta_{mn}. \tag{24.81}$$

The worldsheet supergravity multiplet is (γ^{ab}, χ_a^A), where χ_a^A is a gravitino, that is, a vector-spinor, with A a spinor index, $A = 1, 2$ and a the vector index, $a = 1, 2$. We need to couple this multiplet to the matter one, (X^μ, ψ^μ).

But in two dimensions the Einstein–Hilbert action is topological, as we saw, so non-dynamical, while the gravitino action *vanishes*. This means that we can consider the coupling to external (or non-dynamical) worldsheet supergravity fields.

Considering also off-shell supersymmetry, so besides the on-shell matter (WZ) multiplet (X^μ, ψ^μ) we consider also an auxiliary field F^μ, we write the full action with off-shell worldsheet susy as

$$\mathcal{L} = \frac{1}{2\pi\alpha'} \left[-\frac{e}{2}\gamma^{ab}\partial_a X^\mu \partial_b X^\nu - \frac{e}{2}\bar{\psi}^\mu \gamma^a \partial_a \psi^\nu + \frac{e}{2}F^\mu F^\nu \right.$$
$$\left. + e(\bar{\chi}_a \gamma^b \gamma^a \psi^\mu)(\partial_b X^\nu) + \frac{e}{4}(\bar{\psi}^\mu \psi^\nu)(\bar{\chi}_a \gamma^b \gamma^a \chi_b) \right] \eta_{\mu\nu}. \qquad (24.82)$$

Here $\bar{\chi}^a = \chi_a^T C = \chi_a^T i\gamma^0$, with χ_a real (two-dimensional Majorana).

Besides the usual invariances (Einstein, local Lorentz since we are using the vielbein-spin connection formulation, and Weyl invariance), there is a conformal supersymmetry (partner to the Weyl invariance), with

$$\delta_{cs} X^\mu = 0, \quad \delta_{cs} e_a^m = 0, \quad \delta_{cs} \psi^\mu = 0, \quad \delta_{cs} F^\mu = 0,$$
$$\delta_{cs} \chi_a = \gamma_a \epsilon_{\text{conf}}. \qquad (24.83)$$

The local susy, or supergravity, transformations are

$$\delta_s e_a^m = 2\bar{\epsilon}_s \gamma^m \chi_a$$
$$\delta_s \chi_a = D_a(\hat{\omega})\epsilon_s$$
$$\delta_s X^\mu = \bar{\epsilon}_s \psi^\mu$$
$$\delta_s \psi^\mu = \gamma^a \hat{D}_a \epsilon_s + F^\mu \epsilon_s$$
$$\delta F^\mu = \bar{\epsilon}_s \gamma^a \hat{D}_a \psi^\mu, \qquad (24.84)$$

where the first two lines are the standard transformations of the supergravity multiplet and the next three the standard transformations of the WZ multiplet, and

$$D_a(\hat{\omega})\psi = \partial_a \psi + \frac{1}{4}\hat{\omega}_a^{mn} \gamma_{mn} \psi \qquad (24.85)$$

is the supercovariant derivative with $\hat{\omega}_a^{mn}$ the supercovariant spin connection, such that $\delta\hat{D}_a X$ contains no $\partial_a \epsilon_s$ terms.

24.3.4 Berkovits formalism

The Berkovits formalism is, as we said, a way to covariantly quantize the superstring, so it is based on a reformulation of the GS string.

The left mover action in the GS formalism (for the heterotic case, but also in type II) is, in complex coordinates,

$$S_L = \int d^2z \left[\frac{1}{2}\Pi^\mu \Pi_\mu + \frac{1}{4}\Pi_\mu \theta^\alpha (\gamma^\mu)_{\alpha\beta} \bar{\partial}\theta^\beta - \frac{1}{4}\bar{\Pi}_\mu \theta^\alpha (\gamma^\mu)_{\alpha\beta} \partial\theta^\beta \right], \qquad (24.86)$$

where, as before,

$$\Pi^\mu = \partial X^\mu + \frac{1}{2}\theta^\alpha(\gamma^\mu)_{\alpha\beta}\partial\theta^\beta$$
$$\bar\Pi^\mu = \bar\partial X^\mu + \frac{1}{2}\theta^\alpha(\gamma^\mu)_{\alpha\beta}\bar\partial\theta^\beta. \tag{24.87}$$

Since the canonically conjugate momentum to θ^α doesn't appear in the action, we have the Dirac constraint (i.e., calculating p_α from the action)

$$p_\alpha = \frac{1}{2}\left(\Pi_\mu - \frac{1}{4}\theta\gamma_\mu\partial_1\theta\right)(\gamma^\mu\theta)_\alpha, \tag{24.88}$$

meaning that the constraint is

$$d_\alpha = p_\alpha - \frac{1}{2}\left(\Pi_\mu - \frac{1}{4}\theta\gamma_\mu\partial_1\theta\right)(\gamma^\mu\theta)_\alpha = 0. \tag{24.89}$$

We can calculate the algebra of constraints using the Poisson brackets, and find

$$\{d_\alpha, d_\beta\} = i(\gamma^\mu)_{\alpha\beta}\Pi^\mu, \tag{24.90}$$

where $\Pi^\mu\Pi_\mu = 0$. This means that there are eight (half) first-class constraints (for which on the right-hand side of the anticommutator, we have zero or another constraint) and eight second-class constraints (for which we don't).

Siegel then introduced an action with an independent canonically conjugate momentum to θ^α,

$$S = \int d^2z\left[\frac{1}{2}\partial X^\mu\bar\partial X_\mu + p_\alpha\bar\partial\theta^\alpha\right]. \tag{24.91}$$

But now

$$d_\alpha = p_\alpha - \frac{1}{2}\left(\partial X^\mu + \frac{1}{4}\theta\gamma^\mu\partial\theta\right)(\gamma_\mu\theta)_\alpha \tag{24.92}$$

is not zero anymore.

Next, we introduce (bosonic) *pure spinors* as objects satisfying the constraint

$$\lambda^\alpha(\gamma^\mu)_{\alpha\beta}\lambda^\beta = 0, \quad \mu = 0, 1, ..., 9. \tag{24.93}$$

This is a Lorentz invariant constraint (like the Lorenz gauge $\partial^\mu A_\mu = 0$ of electromagnetism), so we are covariantly quantizing the superstring using pure spinors.

As in general *BRST quantization*, we define a BRST charge Q_{BRST}, and physical states are obtained as the objects in the cohomology of Q_{BRST}, that is, Q-closed modulo Q-exact states ($Q_{\text{BRST}}|\psi\rangle = 0$, where $|\psi\rangle \sim |\psi\rangle + Q_{\text{BRST}}|\chi\rangle$).

The BRST charge is defined as

$$Q_{\text{BRST}} = \int \lambda^\alpha d_\alpha, \tag{24.94}$$

and physical states as ghost number 1 states in the cohomology of Q_{BRST}.

Introducing also a canonically conjugate momentum to λ called w_α, so a term $w_\alpha\bar\partial\lambda^\alpha$ in the action, and considering both left movers and right movers, the open superstring action in flat spacetime is then

$$S_0 = -\frac{1}{\alpha'}\int d^2\sigma\left[\frac{1}{2}\partial X^\mu\bar\partial X_\mu + p_\alpha\bar\partial\hat\theta^\alpha + \hat p_\alpha\partial\hat\theta^\alpha + w_\alpha\bar\partial\lambda^\alpha + \hat w_\alpha\partial\hat\lambda^\alpha\right]. \tag{24.95}$$

The BRST transformations of the fields are found by applying the Poisson brackets with them, obtaining

$$\delta_Q X^\mu = \lambda \gamma^\mu \theta$$
$$\delta_Q \theta^\alpha = \lambda^\alpha$$
$$\delta_Q d_\alpha = -\Pi^\mu (\gamma_\mu \lambda)_\alpha$$
$$\delta_Q w_\alpha = d_\alpha. \tag{24.96}$$

This is similar to the kappa symmetry, which had $\delta X^\mu = \kappa \gamma^\mu \theta$ and $\delta \theta^\alpha = \kappa^\alpha$.

Important concepts to remember

- The worldline particle action is $S_1 = -mc^2 \int d\tau \sqrt{-\dot{X}^\mu \dot{X}^\nu \eta_{\mu\nu}}$, with generalization to $\eta_{\mu\nu} \to g_{\mu\nu}$, to which we can add couplings to other fields, and the first-order form is $S_p = \frac{1}{2} \int d\tau \left[e^{-1}(\tau) \frac{dX^\mu}{d\tau} \frac{dX^\nu}{d\tau} \eta_{\mu\nu} - em^2 \right]$, in terms of an independent worldline einbein $e(\tau)$.

- For a string, we write an action on the worldsheet that minimizes the area spanned in spacetime, giving the second-order Nambu–Goto action $S_{\text{NG}} = -\frac{1}{2\pi\alpha'} \int d\sigma \int d\tau \sqrt{-\det h_{ab}(X^\rho(\xi^a))}$, and the first-order Polyakov action, $S_P[X^\rho, \gamma_{ab}] = -\frac{1}{4\pi\alpha'} \int d\sigma d\tau \sqrt{-\gamma} \gamma^{ab} \partial_a X^\mu \partial_b X^\nu \eta_{\mu\nu}$, in terms of an independent worldsheet metric γ_{ab}.

- Fixing the gauge $\gamma_{ab} = \eta_{ab}$, the equation of motion for γ_{ab} becomes the constraint $T_{ab} = 0$. $T^a{}_a$ is zero off-shell due to Weyl invariance.

- The equation of motion for X^μ is the wave equation, solved with boundary conditions: closed string, open Neumann or open Dirichlet string, giving a sum of oscillators (left and right).

- Quantization of the string modes, turning Poisson brackets into commutators, can be done in lightcone gauge, where the gauge fixes $X^+ = x^+ + p^+ \tau$ and the constraints fix the oscillators in X^-, leaving only X^i as independent. Consistency of the quantization requires $D = 26$ dimensions for the bosonic case, and $D = 10$ for the supersymmetric case.

- The mass spectrum is obtained by acting with the oscillators = creating operators on a vacuum state, of given momentum.

- For quantum consistency, the quantum corrected worldsheet action must have the same symmetries as in the classical case, which restricts to backgrounds satisfying one-loop worldsheet beta functions equal to zero, giving corrected supergravity equations of motion.

- The Green–Schwarz superstring (with spacetime supersymmetry) has kinetic term $S_1 = S_{\text{kin}} = -\frac{1}{4\pi\alpha'} \int d^2\sigma \sqrt{-\gamma} \gamma^{ab} \Pi_a^\mu \Pi_b^\nu \eta_{\mu\nu}$, but also, for the existence of a fermionic kappa symmetry, a WZ term, $S_2 = S_{\text{WZ}} = -\frac{1}{4\pi} \int d^2\sigma \epsilon^{ab} \Pi_a^\Lambda \Pi_b^\Sigma B_{\Lambda\Sigma}$, with $\Pi_a^\mu = \partial_a X^\mu + \bar{\theta} \Gamma^\mu \partial_a \theta^A$ and $\Pi^\Lambda = (\Pi^\mu, d\theta^\alpha)$.

- When fixing a gauge for kappa symmetry, a fermionic lightcone gauge $\Gamma^+ \theta^1 = \Gamma^+ \theta^2$, the spacetime scalars and worldsheet fermions $\theta^{A\alpha}$ are regrouped into worldsheet spinors S^m, giving the same action as the NSR action for the spinning string, with worldsheet supersymmetry, $S = -\frac{1}{4\pi\alpha'} \int d^2\sigma \left[\partial_a X^\mu \partial^a X_\mu - \bar{\psi}^\mu \gamma^a \partial_a \psi_\mu \right]$.

- For the NSR string, imposing $\psi_+^\mu(0,\tau) = \psi_-^\mu(0,\tau)$, we are left with two choices at the other end, $\psi_+(\pi,\tau) = \pm\psi_-(\pi,\tau)$, with $+$ being the spacetime fermionic Ramond sector and $-$ the spacetime bosonic Neveu–Schwarz sector.
- The NSR spinning string can be coupled off-shell to supergravity (local supersymmetry), which extends the formulation with an independent worldsheet metric γ_{ab} to supersymmetry.
- The Berkovits formalism introduces independent momenta p_α conjugate to θ^α, and similarly for right movers, and pure spinors, defined by the covariant constraint $\lambda^\alpha (\gamma^\mu)_{\alpha\beta} \lambda^\beta = 0$, $\mu = 0, 1, ..., 9$. The flat spacetime action is $S_0 = -\frac{1}{\alpha'} \int d^2\sigma \left[\frac{1}{2}\partial X^\mu \bar{\partial} X_\mu + p_\alpha \bar{\partial}\theta^\alpha + \hat{p}_\alpha \partial \hat{\theta}^\alpha + w_\alpha \bar{\partial}\lambda^\alpha + \hat{w}_\alpha \partial \hat{\lambda}^\alpha \right]$, the BRST charge is $Q_{\text{BRST}} = \int \lambda^\alpha d_\alpha$, and physical states are ghost number one states in its cohomology.

References and further reading

For an introduction to string theory, see the classic books by Green, Schwarz, and Witten [86] and Polchinski [87]. For a short version, see my AdS/CFT book [66]. The proof of the WZ term being a superspace WZ term is by Henneaux and Mezincescu [88]. The Berkovits formalism was introduced in [89] and is reviewed in [90].

Exercises

(1) Prove that you get the geodesic equation (24.4) from the relativistic particle action (24.2) with $\eta_{\mu\nu} \rightarrow g_{\mu\nu}$, and specialize to the Newton limit to obtain motion in Newtonian gravity.

(2) Defining $L_0 = \int_0^{2\pi} d\sigma \, T_{--}$, $\tilde{L}_0 = \int_0^{2\pi} d\sigma \, T_{++}$ for the closed string, and $L_0 = \int_0^\pi d\sigma \, (T_{++} + T_{--})$, show that $H_{\text{open}} = L_0$ and $H_{\text{closed}} = L_0 + \tilde{L}_0$ leads to the Hamiltonians in (24.30).

(3) Prove explicitly (without making use of the superspace description) the supersymmetry of the WZ term of the superstring.

(4) Prove the kappa symmetry formulas (24.70) and therefore the kappa symmetry of the GS action for the superstring $S_{\text{kin}} + S_{\text{WZ}}$.

25 Kappa symmetry and spacetime supergravity equations of motion; superembedding formalism

In this chapter, we extend what we learned in Chapter 24 about the superstring and construct kappa symmetric brane actions. We will find that their consistency conditions are related to the supergravity equations of motion. Finally, we will consider the superembedding formalism, for embedding the worldvolume supermanifold into the spacetime supermanifold, and find that the consistency conditions for that, in many cases, lead to equations of motion for backgrounds.

25.1 WZ terms and superstrings

Before we get into it, let us review a bit of information about WZ terms. Consider the Lie algebra-valued object

$$g^{-1}dg = T^a e^\alpha_a \partial_\alpha \phi^i dx_i. \tag{25.1}$$

Then the WZ term for the string is written in the three-dimensional form (with the string worldsheet at its boundary) as

$$\int d^3x \, \mathrm{Tr}\left[(g^{-1}dg)^3\right], \tag{25.2}$$

where

$$f_{abc} = \mathrm{Tr}(T_a T_b T_c), \tag{25.3}$$

so the WZ term is, explicitly,

$$f^{abc} \int d^3x \, e^\alpha_a e^\beta_b e^\gamma_c \partial_\alpha \phi^i \partial_\beta \phi^j \partial_\gamma \phi^j \epsilon_{ijk}. \tag{25.4}$$

In general (for a general dimension, here $d = 2n$, with the term written in $d = 2n + 1$, where the $2n$-dimensional space is at its boundary), the WZ term is

$$S_{\mathrm{WZ}} = \int d^{2n+1}x \, \mathrm{Tr}[(g^{-1}dg)...(g^{-1}dg)] = \int \Omega_{2n+1}, \tag{25.5}$$

and we consider it to be valid for both bosonic and fermionic generators.

In particular, for the super-Poincaré algebra, superspace is $\frac{\mathcal{N}=2 \, SUSY}{SO(1,9)}$, with generators $G_A = \{P_\mu, Q_\alpha\}$. For the superstring case, we obtain

$$S_{\mathrm{WZ,string}} = \int (g^{-1}dg)^A \wedge (g^{-1}dg)^B \wedge (g^{-1}dg)^C f_{ABC}, \tag{25.6}$$

where

$$g^{-1}dg^A = e^A_M dz^M,$$
$$e^m = dx^m - \theta^A \gamma^m d\theta,$$
$$e^\alpha = d\theta^\alpha, \tag{25.7}$$

so

$$(g^{-1}dg)^m = dx^m - \bar{\theta}\gamma^m d\theta,$$
$$(g^{-1}dg)^\alpha = d\theta^\alpha. \tag{25.8}$$

Since the superalgebra is $\{Q^A_\alpha, Q^B_\beta\} = (\gamma^m)_{\alpha\beta} P_m \delta^{AB}$, and the rest are trivial, we have

$$f^m{}_{\alpha\beta} = (\gamma^m)_{\alpha\beta}, \quad \text{rest } 0. \tag{25.9}$$

However, technically Eq. (25.9) would need to have lowered the index m with the Killing metric on the superalgebra,

$$\gamma_{\alpha\beta} = f_{\alpha p}{}^q f_{\beta q}{}^p, \tag{25.10}$$

but

$$\gamma_{mm'} = f_{mA}{}^B f_{nB}{}^A = 0 \Rightarrow f_{m\alpha\beta} = 0. \tag{25.11}$$

But we can guess that the coefficient f'_{ABC} must be $(\gamma_m)_{\alpha\beta} a_{AB}$, where a_{AB} can be diagonalized, so that finally

$$S_{\text{WZ,string}} = \int_{\mathcal{M}} (\gamma_m)_{\alpha\beta} (dx^m - \bar{\theta}\gamma^m d\theta) \wedge d\theta^\alpha \wedge d\bar{\theta}^\beta$$
$$= \int d^3 x \partial_\alpha \bar{\theta}^A \gamma_m \partial_\beta \theta^B \left[\partial_\gamma X^m - \sum_{c=1,2} \bar{\theta}^c \gamma^m \partial_\gamma \theta^c \right] \epsilon^{\alpha\beta} a_{AB}. \tag{25.12}$$

In general then, we would get that the derivative of the WZ term (so dH) is given by (since $dg^{-1} = -g^{-1}dgg^{-1}$, from $dg^{-1}g + g^{-1}dg = 0$)

$$d \operatorname{Tr}[(g^{-1}dg)....(g^{-1}dg)] = - \operatorname{Tr}[(g^{-1}dg)...(g^{-1}dg)(g^{-1}dg)], \tag{25.13}$$

so with a minus and an extra $g^{-1}dg$.

In Chapter 24, we have learned that the kinetic action for the (GS) superstring was written as

$$S_1 = S_{\text{kin}} = -\frac{1}{4\pi\alpha'} \int d^2\sigma \sqrt{-\gamma} \gamma^{ab} \Pi^\mu_a \Pi^\nu_b \eta_{\mu\nu}, \tag{25.14}$$

but in order to have a kappa symmetric action, we needed to add a second term, a Wess–Zumino (WZ) one,

$$S_2 = S_{\text{WZ}} = \int_{\partial\mathcal{M}} B = \int_{\mathcal{M}} H = \int H_{\text{ABC}} \Pi^A \wedge \Pi^B \wedge \Pi^C, \tag{25.15}$$

where A, B, C were curved superspace indices. Moreover, in flat spacetime background, the 3-form field strength was fixed in terms of Π^A (the supervielbeins),

$$H = dB = (\gamma_\mu)_{\alpha\beta} \Pi^\alpha \Pi^\beta \Pi^\mu. \tag{25.16}$$

25.2 Super-*p*-branes

25.2.1 Actions

We now want to generalize that to a (super)*p*-brane (supersymmetric "walls" extended in *p* space dimensions and time) and find the action for motion in flat spacetime background. We start with the kinetic term, which is generalized as

$$S_1 = S_{\text{kin}} = \int d^{p+1}\xi \frac{1}{2}\sqrt{-\gamma}\left[\gamma^{ij}\Pi_i^\mu\Pi_j^\nu\eta_{\mu\nu} - (p-1)\right]. \tag{25.17}$$

Note that the constant term $(p-1)$ is needed in order that the equation of motion for the independent metric γ^{ij} gives the embedding condition, in terms of the (super)vielbein Π_i^μ, which will be renoted (because we will need a lot of indices and vielbeins later on) as E_i^a, such that the condition is

$$\gamma_{ij} = E_i^a E_j^b \eta_{ab}. \tag{25.18}$$

Indeed then, varying S_1 with respect to γ_{kl}, we obtain

$$\frac{\delta S_1}{\delta \gamma_{kl}} = -\frac{1}{2}\frac{-\gamma}{\sqrt{-\gamma}}\gamma^{kl}\left[\gamma^{ij}E_i^a E_j^b \eta_{ab} - (p-1)\right] + \sqrt{-\gamma}\gamma^{ik}\gamma^{jl}E_i^a E_j^b \eta_{ab} = 0 \Rightarrow$$

$$\frac{(p-1)}{2}\gamma_{kl} = E_i^a E_j^b \eta_{ab}\left[\frac{1}{2}\gamma_{kl}\gamma^{ij} - \delta_k^i\delta_l^j\right]. \tag{25.19}$$

By multiplying with γ^{kl} and then substituting back the result in the equation, we obtain (since $i, j = 0, 1, ..., p$)

$$\frac{(p-1)(p+1)}{2} = \frac{p-1}{2}\delta_k^k = E_i^a E_j^b \eta_{ab}\left[\frac{1}{2}(p+1)\gamma^{ij} - \gamma^{ij}\right] \Rightarrow$$

$$\gamma_{kl} = E_k^a E_l^b \eta_{ab}, \tag{25.20}$$

as advertised.

On the other hand, $S_2 = S_{\text{WZ}}$ easily generalizes as a WZ term on the worldvolume,

$$S_2 = S_{\text{WZ}} = \int_{\partial\mathcal{M}=\text{worldvolume}} B = \int_{\mathcal{M}} H, \tag{25.21}$$

where the $p+2$ form is (using again the notation with supervielbeins Π^A, in order to see more clearly the analogy with the superstring case)

$$H = dB = (\gamma_{\mu_1...\mu_p})_{\alpha\beta}\Pi^\alpha \wedge \Pi^\beta \wedge \Pi^{\mu_1} \wedge ... \wedge \Pi^{\mu_p}. \tag{25.22}$$

Since we want that $H = dB$ (exact form), it should also be closed, $dH = 0$, which in general is possible only if we have a consistency condition. Indeed, since $d\Pi^\alpha = 0$ ($\Pi^\alpha = d\theta^\alpha$) and

$$d\Pi^\mu = d[dX^\mu - \bar\theta\gamma^\mu d\theta] = -d\bar\theta\gamma^\mu \wedge d\theta = -\Pi^\alpha(\gamma^\mu)_{\alpha\beta}\Pi^\beta, \tag{25.23}$$

we obtain that

$$dH = -p(\gamma_{\mu_1...\mu_p})_{\alpha\beta}\Pi^\alpha \wedge \Pi^\beta \wedge \Pi^\gamma(\gamma^{\mu_1})_{\gamma\delta} \wedge \Pi^\delta \wedge \Pi^{\mu_2} \wedge ... \wedge \Pi^{\mu_p}$$

$$= -p(\gamma_{\mu_1\mu_2...\mu_p})_{(\alpha\beta}\gamma_{\gamma\delta)}^{\mu_1}\Pi^\alpha \wedge \Pi^\beta \wedge \Pi^\gamma \wedge \Pi^\delta \wedge \Pi^{\mu_2} \wedge ... \wedge \Pi^{\mu_p}, \tag{25.24}$$

where we have put explicitly the antisymmetrization sign in $(\alpha\beta\gamma\delta)$ because it multiplies the antisymmetric object $\Pi^\alpha \wedge \Pi^\beta \wedge \Pi^\gamma \wedge \Pi^\delta$. Now we consider the fact that the Π^α's and Π^μ's are arbitrary so can be peeled off, and from $dH = 0$ we obtain the consistency condition

$$(\gamma_{\mu_1 \ldots \mu_p})(\alpha\beta\gamma^{\mu_1}{}_{\gamma\delta}) = 0. \tag{25.25}$$

25.2.2 Brane scan and kappa symmetry

This is a gamma matrix identity, depending on the worldvolume dimension $d = p + 1$ and spacetime dimension D that needs to hold. It gives a *brane-scan* of possibilities in the D vs. d plot.

By "endless" Fierzing we can compute for each possibility that we have Bose–Fermi matching, with $D - d$ scalars on the worldvolume being the p-brane's coordinates.

Consider that the dimension of a spacetime spinor is M and that we have N supersymmetries.

Now comes the important point: we require kappa symmetry, by analogy with the case of the string, where it was needed for worldsheet susy in a kappa symmetry gauge, which gave the supersymmetric spectrum equal to the NSR string one. Here kappa symmetry is also required in order to have worldsheet susy, but the requirement for the latter is less motivated, unless in the special cases of M2-branes and M5-branes, as we will see.

The kappa symmetry generalizes the one of the superstring,

$$\delta\Pi^\mu = 0,$$
$$\delta\Pi^\alpha = (\mathbb{1} + \Gamma)^\alpha{}_\beta \kappa^\beta,$$
$$\delta\gamma_{ij} = 2\left[X_{ij} - \frac{\gamma_{ij}X^k{}_k}{p-1}\right], \tag{25.26}$$

where

$$\Gamma^\alpha{}_\beta = \frac{\eta}{(p+1)!\sqrt{-\gamma}}\epsilon^{i_1 \ldots i_{p+1}}\Pi^{\mu_1}_{i_1}\ldots\Pi^{\mu_{p+1}}_{i_{p+1}}(\gamma_{\mu_1 \ldots \mu_{p+1}})^\alpha{}_\beta,$$
$$\eta = (-1)^{\frac{(p+1)(p-2)}{4}}, \tag{25.27}$$

and X_{ij} is a very long expression, which we will not need.

Then the kappa symmetry gives the "brane scan." It halves the number of fermionic degrees of freedom, obtaining $MN/2$. If we go on-shell, we lose another half, for $MN/4$.

But, as for the string, when we fix kappa symmetry, we obtain worldvolume supersymmetries. Denoting by m the dimension of the minimal spinor on the worldvolume and by n the number of worldvolume supersymmetries after fixing the kappa symmetry, we have on-shell (which halves the number of fermionic components) $mn/2$.

Thus we have the condition for supersymmetry (both spacetime and worldsheet) for $D - 2$ transverse scalars (ones transverse to the direction of motion of the center of mass) on-shell

$$D - 2 = \frac{1}{2}mn = \frac{1}{4}MN. \tag{25.28}$$

The one exception is actually the $\mathcal{N} = 1$ or heterotic string, with $d = p + 1 = 2$, in which case θ_L and θ_R are independent, and we require matching independently for the left and right moving fermions. Then we have

$$D - 2 = \frac{1}{2}mn = \frac{1}{2}MN. \tag{25.29}$$

But for type II strings we have again $MN/4$, which refers now to $\theta_L + \theta_R$ fermions.

The resulting brane scan in the D vs d space has four *diagonal* lines that start at $d = 2$ and go up in dimension: the "real" sequence from $d = 2, D = 3$ to $d = 3, D = 4$, the "complex" sequence from $d = 2, D = 4$ to $d = 4, D = 6$, the "quaternionic" sequence from $d = 2, D = 6$ to $d = 6, D = 10$, and the "octonionic" sequence, from $d = 2, D = 10$ to $d = 3, D = 11$.

The brane scan obtained from (25.28) is consistent with the gamma–gamma condition (25.25). This is partly because some of the conditions for kappa symmetry are just that $H = dB$ in the previous WZ term (remember that the gamma–gamma condition (25.25) was obtained from the $dH = 0$ condition).

So we have that **kappa symmetry $\Rightarrow$ worldsheet susy $\leftrightarrow$ spacetime susy**.

And we saw that the **gamma matrix condition $\leftrightarrow$ kappa symmetry condition**, which will be soon shown to be also $\leftrightarrow$ **spacetime susy condition**.

25.2.3 Curved superspace and sugra equations from kappa symmetry

We now go back to the kappa symmetry condition and the (super)p-brane action. While previously we worked in flat superspace, we now generalize the action to *curved* superspace. The kinetic term is trivial to generalize, by just replacing $\eta_{\mu\nu}$ with $g_{\mu\nu}$, so

$$S_1 = S_{\text{kin}} = \int d^{p+1}\xi \frac{1}{2}\sqrt{-\gamma} \left[\gamma^{ij}E_i^\mu E_j^\nu g_{\mu\nu} - (p - 1)\right]. \tag{25.30}$$

The WZ term is written in terms of general supervielbeins as

$$S_2 = S_{\text{WZ}} = \int_{\mathcal{M}} H = \int_{\partial\mathcal{M}} B, \tag{25.31}$$

with

$$H = dB = \left[\frac{\eta(-1)^p}{(p+1)!}\right](\gamma_{a_1\ldots a_p})_{\alpha\beta}E^{a_1}\ldots E^{a_p}E^\alpha E^\beta + \frac{\eta}{p!}\Lambda_\beta(\gamma_{a_1\ldots a_{p+1}})^\beta{}_\alpha E^{a_1}\ldots E^{a_{p+1}}E^\alpha. \tag{25.32}$$

In the particular (and very relevant) case of the *11-dimensional on-shell supergravity* in superspace (in the super-geometric approach), we have, as we have already mentioned,

– the usual supervielbein and super-spin connection variables E_M^A and $\Omega_{M\,A}{}^B$, with

$$E^A = E_M^A dz^M, \quad \Omega^A{}_B = \Omega_{M\,A}{}^B dz^M, \tag{25.33}$$

which define the torsion

$$T^A = \frac{1}{2}E^C E^B T_{BC}{}^A = DE^E = dE^A + E^B\Omega_B{}^A, \tag{25.34}$$

and curvature

$$R_A{}^B = d\Omega_A{}^B + \Omega_A{}^C \Omega_C{}^B, \tag{25.35}$$

satisfying the Bianchi identities

$$DT^A = E^B R_B{}^A, \quad DR_A{}^B = 0, \tag{25.36}$$

but now we also have
— the superspace gauge 3-form

$$X = \frac{1}{3} E^C E^B E^A X_{ABC}, \tag{25.37}$$

with gauge transformation

$$\delta X = dY, \quad Y = \frac{1}{2} E^B E^A Y_{AB}, \tag{25.38}$$

and field strength

$$H = dX = \frac{1}{4!} E^D E^C E^B E^A H_{ABCD}, \quad dH = 0. \tag{25.39}$$

Then there are nontrivial constraints for $T_{AB}{}^C$ and H_{ABCD} that constrain them to only constants:

$$\begin{aligned}
T_{\alpha\beta}{}^c &= -i(\Gamma^c)_{\alpha\beta}, \\
T_{\alpha\beta}{}^\gamma &= T_{ab}{}^c = 0, \\
H_{\alpha\beta\gamma\delta} &= H_{\alpha\beta\gamma c} = H_{\alpha bcd} = 0, \\
H_{\alpha\beta cd} &= i(\Gamma_{cd})_{\alpha\beta}.
\end{aligned} \tag{25.40}$$

But after Fierzing, one finds that the kappa symmetry condition for a p-brane in a general background *is only satisfied if we satisfy the constraints*

$$\begin{aligned}
(T^a)_{\alpha\beta} &= (\gamma^a)_{\alpha\beta}, \\
T_{(ca} T^c{}_{b)\alpha} &= \eta_{ab} \Lambda_\alpha, \\
H_{\alpha a_{p+1} \ldots a_1} &= \frac{\eta}{\psi!} \Lambda_\beta (\gamma_{a_1 \ldots a_{p+1}})^\beta{}_\alpha, \\
H_{\alpha\beta a_p \ldots a_2} &= \frac{\eta(-1)^p}{(p+1)!} (\gamma_{a_1 \ldots a_p})_{\alpha\beta}, \\
H_{\alpha\beta\gamma A_1 \ldots A_{p-1}} &= 0,
\end{aligned} \tag{25.41}$$

for all Bose or Fermi $A_1, .., A_{p-1}$.

These constraints on the torsion T and the H field, found from kappa symmetry, are satisfied, *at least for $D = 11$ and $d = 3$ (for the 11-dimensional supermembrane)* by the on-shell constraints of 11-dimensional supergravity!!

This means that the *supermembrane only propagates consistently in the supergravity background*, just like the superstring only propagates consistently (for different reasons, though!!) in the background of supergravity plus α' corrections.

But this kappa symmetry (so worldsheet supersymmetry, in a gauge) requirement for $H = dB$ fixes the form of B in $S_{WZ} = S_2$, in terms of the superspace map from the worldvolume coordinate ξ^i to the superspacetime one (x^μ, θ^α), so it is not an independent variable anymore.

Another observation is that the brane scan we have derived is a *fundamental* brane scan only: Dp-branes don't enter in it, neither do KK monopoles and M5-branes. The relevant objects that we obtain are the following: in $D = 11$, only the fundamental M2-brane, and in 10 dimensions, the F1 fundamental string and the NS5-brane.

25.3 Superembedding formalism

The superembedding formalism refers to embedding the superspace on the worldvolume into the superspace in spacetime, as well as the similar case, for branes ending on branes, where we embed the boundary superspace of the ending brane into the superspace of the brane it ends on.

We will see that, from the consistency of this superembedding, in various cases, we obtain the equations of motion for various backgrounds.

Denoting spacetime with an underline, and worldvolume without it, we have

$$\mathcal{M}_{p+1} \subset \underline{\mathcal{M}_D}. \tag{25.42}$$

25.3.1 Superembedding conditions

Bosonic case

To better understand the issues, we start with the simpler bosonic case. Denoting by a flat worldvolume indices and $\underline{a}$ spacetime flat indices, and with e worldvolume vielbeins and with E spacetime vielbeins, we have

$$e^a(\xi^m) = d\xi^m e_m^a(\xi), \quad E^{\underline{a}} = dx^{\underline{m}} E_{\underline{m}}^{\underline{a}}(x). \tag{25.43}$$

Pulling back the spacetime vielbein onto the worldvolume, we obtain

$$E^{\underline{a}}(x(\xi)) = d\xi^m \partial_m x^{\underline{m}} E_{\underline{m}}^{\underline{a}} = e^a E_a^{\underline{a}}. \tag{25.44}$$

But we want to break the local Lorentz invariance under $SO(1, D-1)$ to $SO(1, p) \times SO(D - p - 1)$ as usual (in, say, KK compactification). We do this by the gauge choice

$$E^a = e^a, \quad E^i = 0. \tag{25.45}$$

Then the pullback in terms of the metric is

$$g_{mn}(\xi) = \partial_m x^{\underline{m}} \partial_n x^{\underline{n}} g_{\underline{mn}}(x(\xi)). \tag{25.46}$$

The minimal volume embedding (like, for instance, minimizing the area of the string) gives the brane equations of motion: the equations of motion of the fields on the worldvolume (in the case of the string, the coordinates X^μ and the metric γ^{ab}).

Supersymmetric case

Going to the supersymmetric case, with spacetime supermanifold $\mathcal{M}_{D,2n}$ described by coordinates

$$Z^{\underline{M}} = (X^{\underline{m}}, \Theta^{\underline{\mu}}), \quad \underline{\mu} = 1, ..., 2n, \tag{25.47}$$

and worldvolume supermanifold (with half as many spinors: the brane breaks half supersymmetries) $\mathcal{M}_{p+1,n}$, described by coordinates

$$z^M = (\xi^m, \eta^\mu), \quad \mu = 1, ..., n. \tag{25.48}$$

Writing the flat indices with $\underline{A}$ and A, respectively, we have the supervielbeins

$$E^{\underline{A}}(Z) = dZ^{\underline{M}} E^A_{\underline{M}} = (E^a, E^\alpha),$$
$$e^a(z) = dz^M e^A_M = (e^a, e^{\alpha q}), \tag{25.49}$$

where $\underline{\mu}, \underline{\alpha}$ are spinor indices of $SO(1, D-1)$ (curved and flat, respectively), μ, α are spinor indices of $SO(1, p)$, and q is a spinor index of $SO(D - p - 1)$.

Then the *superembedding conditions* (actually, pullback conditions) are

$$E^a = e^a(z), \quad E^i = 0, \tag{25.50}$$

with the first one being the basic superembedding condition, rewritten explicitly as

$$E^a_{\alpha q} = e^M_{\alpha q} \partial_M Z^{\underline{M}} E^a_{\underline{M}} = 0. \tag{25.51}$$

In general, we would write

$$E^{\underline{a}} = e^A e^M_A \partial_M Z^{\underline{M}} E^{\underline{a}}_{\underline{M}} = e^a E^{\underline{a}}_a + e^{\alpha q} E^{\underline{a}}_{\alpha q},$$
$$E^{\underline{\alpha}} = e^A e^M_A \partial_M Z^{\underline{M}} E^{\underline{\alpha}}_{\underline{M}} = e^a E^{\underline{\alpha}}_a + e^{\alpha q} E^{\underline{\alpha}}_{\alpha q}. \tag{25.52}$$

By a rotation, we would then choose

$$E^a = E^{\underline{b}} u^a_{\underline{b}}(z) = e^a(z). \tag{25.53}$$

25.3.2 Particle embedding

As an example, which we will follow until almost the end, we will consider the particle. Strings and branes are considerably more complicated, but the logic is the same.

Bosonic case

The worldline coordinates are $x^{\underline{m}}(\tau)$, and the action (in what before we called first-order form) with also a worldline einbein $e(\tau)$ is (as we have already seen)

$$S = \int d\tau \frac{1}{2e(\tau)} \dot{x}^{\underline{m}} \dot{x}_{\underline{m}}. \tag{25.54}$$

Reparametrization invariance on the worldline of the above action, or equivalently the equation of motion of $e(\tau)$, gives

$$\frac{1}{2} \dot{x}^{\underline{m}} \dot{x}_{\underline{m}} = 0 \Leftrightarrow p^{\underline{m}} p_{\underline{m}} = 0, \tag{25.55}$$

so the momentum of the particle is null: the massless on-shell condition for the particle.

We can write a true first-order action (first order in derivatives), by introducing the momentum as a further auxiliary field, obtaining

$$S = \int d\tau \left(p_{\underline{m}} \dot{x}^{\underline{m}} - \frac{e}{2} p_{\underline{m}} p^{\underline{m}} \right).$$ (25.56)

Spinning particle

Next, we consider the "spinning particle," with worldline supersymmetry. Introducing the worldline spinor parameter η, the coordinate superfield is

$$X^{\underline{m}}(\tau, \eta) = x^{\underline{m}}(\tau) + i\eta \chi^{\underline{m}}(\tau).$$ (25.57)

The supereinbein and momentum superfields are expanded in η as

$$E^{\underline{m}}(\tau, \eta) = e(\tau) + i\eta \psi(\tau),$$
$$p_{\underline{m}}(\tau, \eta) = p_{\underline{m}}(\tau) + i\eta \rho_{\underline{m}}(\tau).$$ (25.58)

Then the bosonic momentum is found to be

$$p_{\underline{m}} = \frac{1}{e} \left(\dot{x}_{\underline{m}} - \frac{i}{e} \psi \chi_{\underline{m}} \right).$$ (25.59)

The action of the spinning particle in superspace is

$$S = -\int d\tau d\eta \frac{i}{2E} DX^{\underline{m}} \partial_\tau X_{\underline{m}}.$$ (25.60)

In components, it becomes

$$S = \int d\tau \left[\frac{1}{2e} \left(\dot{x}^{\underline{m}} \dot{x}_{\underline{m}} + i\dot{\chi}^{\underline{m}} \chi_{\underline{m}} \right) - \frac{i}{e^2} \psi \chi_{\underline{m}} \dot{x}^{\underline{m}} \right].$$ (25.61)

The constraints, from (super)reparametrization invariance, or equations of motion for $e(\tau)$ and its superpartner ψ, give

$$p^{\underline{m}} p_{\underline{m}} = 0, \quad \chi^{\underline{m}} p_{\underline{m}} = 0.$$ (25.62)

Superparticle

Next, consider the superparticle, with target space (spacetime) supersymmetry, under

$$\delta\theta^{\underline{\mu}} = \epsilon^{\underline{\mu}}, \quad \delta x^{\underline{m}} = i\bar{\theta} \Gamma^{\underline{m}} \delta\theta.$$ (25.63)

The fermion is a minimal spinor in D dimensions, either Majorana, Weyl, or Majorana–Weyl.

The supervielbein is written as

$$\mathcal{E}^{\underline{A}} = (\mathcal{E}^{\underline{a}}, \mathcal{E}^{\underline{\alpha}}),$$
$$\mathcal{E}^{\underline{a}} = \delta^{\underline{a}}_{\underline{m}} (dx^{\underline{m}} - i d\bar{\theta} \Gamma^{\underline{m}} \theta),$$
$$\mathcal{E}^{\underline{\alpha}} = d\theta^{\underline{\mu}} \delta^{\underline{\alpha}}_{\underline{\mu}}.$$ (25.64)

The pullback (embedding) is

$$\mathcal{E}^{\underline{a}}_{\underline{\alpha}}(Z(\tau)) \equiv d\tau \mathcal{E}^{\underline{a}}_{\underline{\alpha}\tau}(Z(\tau)).$$ (25.65)

The action is

$$S = \int d\tau \frac{1}{2e(\tau)} \mathcal{E}^{\underline{a}}_\tau \mathcal{E}^{\underline{b}}_\tau \eta_{\underline{ab}},$$ (25.66)

and in a first-order form it is

$$S = \int d\tau \left(p_{\underline{a}} \mathcal{E}_\tau^a - \frac{e}{2} p_{\underline{a}} p^{\underline{a}} \right),$$ (25.67)

where

$$\mathcal{E}_\tau^a = \partial_\tau x^{\underline{a}} - i \partial_\tau \bar{\theta} \Gamma^a \theta.$$ (25.68)

It has kappa symmetry under

$$\delta\theta^{\underline{\mu}} = i\not{p}\kappa = ip_{\underline{m}}(\Gamma^{\underline{m}})^{\underline{\mu}}_{\ \underline{\nu}}\kappa^{\underline{\nu}}(\tau),$$
$$\delta x^{\underline{m}} = i\bar{\theta}\Gamma^{\underline{m}}\delta\theta,$$
$$\delta e = 4\dot{\theta}\kappa,$$
$$\delta p_{\underline{m}} = 0.$$ (25.69)

The fermionic constraint is

$$\Pi_{\underline{\mu}}(\tau)\left(p_{\underline{m}}\Gamma^{\underline{m}}\right)^{\underline{\nu}}_{\ \underline{\nu}} = 0,$$ (25.70)

where $\Pi_{\underline{\mu}}$ is the momentum canonically conjugate to $\theta^{\underline{\mu}}$.

Then we have

$$D_{\underline{\mu}} = \Pi_{\underline{\mu}} - i\bar{\theta}_{\underline{\nu}}\left(p_{\underline{m}}\Gamma^{\underline{m}}\right)^{\underline{\nu}}_{\ \underline{\mu}} = 0.$$ (25.71)

Spinning superparticle and the doubly supersymmetric formalism

Next, we consider the "spinning superparticle," that is, the particle with both worldline and spacetime supersymmetry, or the "doubly supersymmetric formalism."

In this case, the embedding is through

$$Z^{\underline{M}}(z^M), \quad Z^{\underline{M}} = (X^{\underline{m}}, \Theta^{\underline{\mu}}), \quad z^M = (\tau, \eta),$$ (25.72)

and the spacetime spinor superfield is

$$\Theta^{\underline{\mu}}(\tau, \eta) = \theta^{\underline{\mu}}(\tau) + \eta\lambda^{\underline{\mu}}(\tau).$$ (25.73)

The wordline superfield symmetries are

$$\delta\theta^{\underline{\mu}} = -a(\tau)\dot{\theta}^{\underline{\mu}} - \frac{1}{2}\alpha(\tau)\lambda^{\underline{\mu}},$$
$$\delta\lambda^{\underline{\mu}} = -a(\tau)\dot{\lambda}^{\underline{\mu}} - \frac{1}{2}\dot{a}(\tau)\lambda^{\underline{\mu}} - \frac{i}{2}\alpha(\tau)\dot{\theta}^{\underline{\mu}},$$ (25.74)

where $a(\tau)$ is a bosonic reparametrization and $\alpha(\tau)$ is a local supersymmetry, or supergravity (fermionic reparametrization), and so

$$\delta\tau = \Lambda(\tau, \eta) - \frac{1}{2}\eta D\Lambda,$$
$$\delta\eta = -\frac{i}{2}D\Lambda,$$
$$\Lambda(\tau, \eta) = a(\tau) + i\eta\alpha(\tau),$$ (25.75)

with Λ being the superreparametrization parameter.

The worldline supervielbein is

$$e^A(\tau, \eta) = (e^\tau, e^\eta),$$ (25.76)

with its flat spacetime version

$$e_0^A(\tau, \eta) = (e_0^\tau, e_0^\eta) = (d\tau + i\eta d\eta, d\eta).$$ (25.77)

The spacetime supervielbeins are

$$\mathcal{E}^{\underline{\alpha}}(Z(z^M)) = dz^M \mathcal{E}_M^{\underline{\alpha}} = e_0^\tau \partial_\tau \Theta^{\underline{\alpha}} + e_0^\eta D\Theta^{\underline{\alpha}},$$
$$\mathcal{E}^{\underline{a}}(Z(z^M)) = dz^M \mathcal{E}_M^{\underline{a}} = e_0^\tau (\partial_\tau X^{\underline{a}} - i\partial_\tau \bar{\Theta} \Gamma^{\underline{a}} \Theta) + e_0^\eta (DX^{\underline{a}} - iD\bar{\Theta}\Gamma^{\underline{a}}\Theta),$$ (25.78)

where $DX^{\underline{a}} - iD\bar{\Theta}\Gamma^{\underline{a}}\Theta = \mathcal{E}_\eta^{\underline{a}}$.

Then the spinning superparticle action is

$$S = -\int d\tau d\eta \frac{1}{2E} \mathcal{E}_\eta^{\underline{a}} D\mathcal{E}_\eta^{\underline{a}}$$ (25.79)

and has the spectrum equal to the tensor product of the spectra of the spinning particle and of the superparticle!

25.3.3 Examples

Example 1 $\mathcal{N} = 1, D = 3$ doubly supersymmetric (spinning superparticle)
As before, writing a truly first-order action in terms of the momentum (now a superfield),

$$S = \int d\tau d\eta \left[P_{\underline{a}} \mathcal{E}_\eta^{\underline{a}} - \frac{E}{2} P_{\underline{a}} P^{\underline{a}} \right],$$ (25.80)

we can actually drop the second term, since it vanishes on the constraint, $P_{\underline{a}} P^{\underline{a}} = 0$. The action then has the following:

- Wordline supersymmetry
- Target space (spacetime) supersymmetry
- Super-kappa symmetry, which however is redundant, as we will see

Imposing the matching of the Bose–Fermi degrees of freedom, having D bosons, and $2^{[D/2]}$ fermions and $n = 2^{[D/2]-1}$ worldline fermions, and n kappa symmetries imposes the constraint that

$$\mathcal{N} = 1, \quad D = 3, 4, 6, 10.$$ (25.81)

So the doubly supersymmetric formalism only occurs, for the particle, in $D = 3, 4, 6$ or 10 dimensions.

But, as we see, for $D = 3$, we have $n = 2^{[D/2]-1} = 1$, so this case is special (there are no spinor indices for the fermions in the transverse – non-worldline – dimensions) and will be treated separately.

The action is then

$$S = \int d\tau d\eta P_{\underline{a}} \mathcal{E}_\eta^{\underline{a}} = \int d\tau d\eta P_{\underline{a}} (DX^{\underline{a}} - iD\bar{\theta}\Gamma^{\underline{a}}\theta),$$ (25.82)

and in components on the worldline, it becomes

$$S = \int d\tau p_{\underline{a}}(\dot{x}^{\underline{a}} - i\dot{\bar{\theta}}\Gamma^{\underline{a}}\theta - \bar{\lambda}\Gamma^{\underline{a}}\lambda) + i\int d\tau \rho_{\underline{a}}(\chi^{\underline{a}} - \bar{\lambda}\Gamma^{\underline{a}}\lambda). \tag{25.83}$$

The equations of motion for χ, ρ, and $p_{\underline{a}}$ are, respectively,

$$\rho_{\underline{a}} = 0,$$
$$\chi^{\underline{a}} = \bar{\lambda}\Gamma^{\underline{a}}\lambda,$$
$$\mathcal{E}^{\underline{a}}_{\tau}\big|_{\eta=0} = \dot{x}^{\underline{a}} - i\bar{\theta}\Gamma^{\underline{a}}\theta = \bar{\lambda}\Gamma^{\underline{a}}\lambda, \tag{25.84}$$

stating that ρ and χ are auxiliary, while the third defines the Cartan–Penrose, or twistor, representation of the lightlike vector $p_{\underline{a}}$.

Further, the λ, x, and θ equations of motion are, respectively (using the equations in Eq. (25.84) for λ),

$$p_{\underline{a}}(C\Gamma^{\underline{a}})_{\underline{\alpha\beta}}\lambda^{\underline{\beta}} = 0,$$
$$\partial_{\tau}p_{\underline{a}} = 0,$$
$$(p_{\underline{a}}\Gamma^{\underline{a}})^{\underline{\alpha}}{}_{\underline{\beta}}\partial_{\tau}\theta^{\underline{\beta}} = 0, \tag{25.85}$$

where the last two equations are the *superparticle equations of motion*!

Using a *local supersymmetry (supergravity)* with

$$\alpha(\tau) = -\frac{4i}{e}\bar{\lambda}_{\underline{\alpha}}\kappa^{\underline{\alpha}}, \tag{25.86}$$

we have a kappa symmetry obtained from the local supersymmetry (so, as we said, the kappa symmetry is redundant, as it can be obtained from the local supersymmetry).

The spinning particle equations of motion (see the previous spinning particle action),

$$\chi^{\underline{m}}p_{\underline{m}} \equiv \chi^{\underline{m}}\left(\dot{x}_{\underline{m}} - \frac{i}{e}\psi\chi_{\underline{m}}\right) = 0,$$
$$\dot{\chi}^{\underline{m}} = \frac{1}{e}\psi\dot{x}^{\underline{m}} + \frac{\dot{e}}{2e}\chi^{\underline{m}}, \tag{25.87}$$

where we have defined a new $p_{\underline{m}}$ as previously (consistent with the spinning particle action) are related to our equations for χ and $p_{\underline{a}}$,

$$p^{\underline{m}} = \frac{1}{e}\bar{\lambda}\Gamma^{\underline{m}}\lambda,$$
$$\chi^{\underline{m}} = \bar{\theta}\Gamma^{\underline{m}}\lambda, \tag{25.88}$$

(where the second equation is the superpartner of the first) via the Cartan–Penrose (twistor) representation of the lightlike vector $p_{\underline{a}}$ defined previously.

The superembedding condition comes from varying the action

$$S = \int d\tau d\eta P_{\underline{a}}\mathcal{E}^{\underline{a}}_{\eta} \tag{25.89}$$

with respect to $P_{\underline{a}}$,

$$\mathcal{E}^{\underline{a}}_{\eta} = DX^{\underline{a}} - iD\bar{\theta}\Gamma^{\underline{a}}\theta = 0. \tag{25.90}$$

Example 2 $\mathcal{N} = 1, D = 4, 6, 10$ doubly supersymmetric (spinning superparticle)
Next, we consider the other dimensions for the propagation of the spinning superparticle, which means that now there are indices for the transverse (non-worldline) spinor, which will be denoted, as in the general theory, by q (such that the full D-dimensional spinor index is $\alpha q = \eta q$).

Because of supersymmetry, there are now (as much as the on-shell bosons) $n = D - 2 > 1$ worldline fermions, with the index q.

The superembedding condition is now

$$\mathcal{E}_q^{\underline{a}}(Z(z)) = D_q X^{\underline{a}} - iD_q \bar{\Theta} \Gamma^{\underline{a}} \Theta = 0. \tag{25.91}$$

Note that, what was before simply D is now D_q (for the spinor of $SO(D - p - 1)$), and correspondingly the superfield superreparametrization parameter Λ has now three such indices, Λ^{qrs}.

The action is, generalizing the $D = 3$ dimensional action,

$$S = i \int d\tau d^{D-2}\eta P_{\underline{a}}^q \mathcal{E}_{\underline{a}}^q$$
$$= -i \int d\tau d^{D-2}\eta P_{\underline{a}}^q \left[D_q X^{\underline{a}} - iD_q \bar{\Theta} \Gamma^{\underline{a}} \Theta \right]. \tag{25.92}$$

From the superembedding condition in its most general form, we have *in curved superspace*

$$E^{\underline{A}}(Z(z)) = e^\tau(z)\partial_\tau Z^M E_{\underline{M}}^{\underline{A}} + e^a(z)D_q Z^M E_{\underline{M}}^{\underline{A}} = e^B E_B^{\underline{A}} \Rightarrow,$$
$$E_q^{\underline{a}}(Z(z)) \equiv D_q Z^M E_{\underline{M}}^{\underline{a}} = 0, \tag{25.93}$$

and the action

$$S = i \int d\tau d^{D-2}\eta P_{\underline{a}}^q E_{\underline{a}}^q, \tag{25.94}$$

where we used the notation with E of the general supersymmetric case in curved superspace to distinguish from the doubly supersymmetric vielbein $\mathcal{E}$ in flat superspace.

Then in *curved superspace*, the target space (super)torsion constraints are obtained (in a gauge) from the invariance under

$$\delta P_{\underline{b}}^q = \left(\delta_{\underline{b}}^{\underline{a}} D_r + \Omega_{r\underline{b}}^{\phantom{r\underline{b}}\underline{a}} \right) \left(\Lambda^{qrs} \Gamma_{\underline{a}} E_s \right). \tag{25.95}$$

Note that in flat superspace, the above invariance is written as

$$\delta P_{\underline{a}}^q = D_r \left(\Lambda^{qrs} \Gamma_{\underline{a}} D_s \Theta \right). \tag{25.96}$$

25.3.4 Superstrings

Having mostly talked about (super)particles, we now just write a few equations for the (super)string case, to see that we can indeed generalize.

Besides the term generalizing the particle one, we have an extra term, generalizing the superstring WZ term in the GS formulation, so

$$S = -i \int d^2\xi \, d^n\eta \, P_{\underline{a}}^{-q} \mathcal{E}_{-q}^{\underline{a}} + \int d^2\xi \, d^n\eta \, P^{MN} \mathcal{F}_{MN}^{(2)}, \tag{25.97}$$

where

$$\mathcal{F}_{MN}^{(2)} = \mathcal{E}_{[M}^q e_{N]}^{++} \mathcal{E}_{++q} + B_{MN}^{(2)} + \partial_{[M} A_{N]}$$
$$B^{(2)} = i \, dx^{\underline{a}} d\bar{\theta} \, \Gamma_{\underline{a}} \theta, \tag{25.98}$$

and $+$ and $-$ refers to lightcone gauge indices.

In a nontrivial (curved) spacetime supergravity background, we exchange $\mathcal{E}$ with the general E, so we write the action

$$S = -i \int d^2\xi \, d^n\eta \, P_{\underline{a}}^{-q} E_{-q}^{\underline{a}} + \int d^2\xi \, d^n\eta \, P^{MN} \mathcal{F}_{MN}^{(2)}. \tag{25.99}$$

From invariance of δP^{MN} (in a gauge) we obtain the torsion constraints, as well as the constraints

$$H_{\underline{\alpha}\underline{\beta}\underline{a}} = 2ie^{\Phi(z)} (C\Gamma_{\underline{a}})_{\alpha\beta}, \quad H_{\underline{\alpha}\underline{\beta}\underline{\gamma}} = 0. \tag{25.100}$$

p-branes

In the case of *p*-branes:

- The superembedding integrability condition gives the torsion constraints *sometimes*, and also sometimes the equations of motion.
- In the case of the bosonic embedding, the minimal embedding condition (embedding with minimal volume) gives the (Nambu–Goto-like or Polyakov-like) brane equations of motion (for X's and γ's).
- For the superparticle and superstrings, $E_{\alpha q}^{\underline{a}} = 0$ is dynamical, so we get no equations of motion.
- In the case of the M2-branes and M5-branes, we obtain the supergravity equations of motion for the M2-brane case, and for M5-branes, we get the M5-brane equations of motion. For D*p*-branes, we obtain the D*p*-brane equations of motion.
- For open branes, specifically the most relevant cases, M2-branes ending on M5-branes, where the boundary of M2 is an M1 manifold inside M5, we obtain the M5-brane equations of motion. For F1 strings ending on D*p*-branes, again from the F0 manifold at the boundary inside D*p*, we obtain the D*p* equations of motion. Also from D*p* ending on D($p+2$), from the boundary manifold inside D($p+2$), we obtain the D($p+2$) equations of motion.

Important concepts to remember

- The WZ term is generically of the type $\int d^{2n+1}x \, \text{Tr}[(g^{-1}dg)...(g^{-1}dg)]$, and in the case of the superstring it is written as $S_2 = S_{\text{WZ}} = \int_{\partial\mathcal{M}} B = \int_{\mathcal{M}} H = \int H_{ABC} \Pi^A \wedge \Pi^B \wedge \Pi^C$, where in flat background $H = dB = (\gamma_\mu)_{\alpha\beta} \Pi^\alpha \Pi^\beta \Pi^\mu$.
- For the super-*p*-brane, $S_1 = S_{\text{kin}} = \int d^{p+1}\xi \frac{1}{2}\sqrt{-\gamma} \left[\gamma^{ij} \Pi_i^\mu \Pi_j^\nu \eta_{\mu\nu} - (p-1) \right]$ and $S_2 = S_{\text{WZ}} = \int_{\partial\mathcal{M}=\text{worldvolume}} B = \int_{\mathcal{M}} H$, with $H = dB = (\gamma_{\mu_1...\mu_p})_{\alpha\beta} \Pi^\alpha \wedge \Pi^\beta \wedge \Pi^{\mu_1} \wedge ... \wedge \Pi^{\mu_p}$.
- The consistency condition for the super-*p*-brane, from the closure of the above H, $dH = 0$, is the gamma matrix identity $(\gamma_{\mu_1...\mu_p})_{(\alpha\beta} \gamma^{\mu_1})_{\gamma\delta)} = 0$. It gives a brane scan of possibilities for D vs. d.

- The condition for existence of kappa symmetry gives $D - 2 = \frac{1}{2}mn = \frac{1}{4}MN$, where M, N are in spcetime and m, n on the worldvolume, with $MN/2$ instead for $\mathcal{N} = 1$ or heterotic strings, which also gives a brane scan, consistent with the previous one.
- Thus the gamma matrix condition is the condition of worldsheet susy, related to the condition of kappa symmetry, which is needed in order to obtain worldsheet susy in addition to spacetime susy, and thus obtain the correct spectrum.
- For motion in curved superspace, in the super-geometric approach, we need a certain number of constraints on the background superspace to be satisfied (such that we have consistent propagation). In the case of the supermembrane, the constraints are the on-shell constraints of 11-dimensional supergravity.
- The superembedding formalism is about embedding worldvolume superspace into target (spacetime) superspace consistently. The superembedding conditions (or pullback conditions in superspace) are $E^a = e^a(z), E^i = 0$.
- For a particle, we can have a bosonic particle, a spinning particle, a superparticle, and a spinning superparticle, or doubly supersymmetric one, which has as spectrum the tensor product of the spectra of the spinning particle and the superparticle.
- For consistent propagation in the doubly supersymmetric formalism, we find we need to satisfy the spacetime (super)torsion constraints.
- So from the integrability condition of superembedding, we sometimes find the torsion constraints and/or equations of motion. Also for branes ending on branes, we find the equations of motion of the brane one ends on.

References and further reading

For the superembedding formalism, see the reviews [91] and [92].

Exercises

(1) Check that the gamma–gamma condition,

$$(\gamma_{\mu_1...\mu_p})_{(\alpha\beta}\gamma^{\mu_1})_{\gamma\delta)} = 0, \tag{25.101}$$

and the kappa symmetry condition

$$D - 2 = \frac{1}{2}mn = \frac{1}{2}MN \tag{25.102}$$

are satisfied for the 11-dimensional supermembrane and 10-dimensional type II superstring.

(2) Check the kappa symmetry,

$$\delta\theta^{\underline{\mu}} = i\not{p}\kappa = ip_{\underline{m}}(\Gamma^m)^{\underline{\mu}}_{\ \underline{\nu}}\kappa^{\underline{\nu}}(\tau),$$

$$\delta x^{\underline{m}} = i\bar{\theta}\Gamma^m\delta\theta,$$

$$\delta e = 4\dot{\bar{\theta}}\kappa,$$

$$\delta p_{\underline{m}} = 0, \tag{25.103}$$

of the superparticle action

$$S = \int d\tau \left(p_{\underline{a}} \mathcal{E}_\tau^{\underline{a}} - \frac{e}{2} p_{\underline{a}} p^{\underline{a}} \right).$$ (25.104)

(3) Check the superreparametrization invariance,

$$\delta\tau = \Lambda(\tau, \eta) - \frac{1}{2} \eta D\Lambda,$$

$$\delta\eta = -\frac{i}{2} D\Lambda,$$

$$\Lambda(\tau, \eta) = a(\tau) + i\eta\alpha(\tau),$$ (25.105)

of the spinning superparticle action

$$S = -\int d\tau d\eta \frac{1}{2E} \mathcal{E}_\eta^{\underline{a}} D\mathcal{E}_\eta^{\underline{a}}.$$ (25.106)

(4) Check that the $D = 3$ spinning superparticle action,

$$S = \int d\tau d\eta P_{\underline{a}} \mathcal{E}_\eta^{\underline{a}} = \int d\tau d\eta P_{\underline{a}} (DX^{\underline{a}} - iD\bar{\theta}\Gamma^{\underline{a}}\theta),$$ (25.107)

becomes in components

$$S = \int d\tau p_{\underline{a}} (\dot{x}^{\underline{a}} - i\dot{\bar{\theta}}\Gamma^{\underline{a}}\theta - \bar{\lambda}\Gamma^{\underline{a}}\lambda) + i \int d\tau \rho_{\underline{a}} (\chi^{\underline{a}} - \bar{\lambda}\Gamma^{\underline{a}}\lambda).$$ (25.108)

26 Supergravity and cosmological inflation models

In this chapter we will study cosmological inflation from the point of view of supergravity.

Inflation is the most popular cosmological inflation model for the beginning of the universe. It is important to emphasize that it is the most popular, in the sense that it is the most successful at solving the problems of Big Bang cosmology in a generic way, but by no means the only model. There are also other models, such as the ekpyrotic and cyclic models, the string gas cosmology, and holographic cosmology.

Inflation is usually described by scalar fields. Again, this is not the only choice, we can have other fields, such as vector fields, just that those are more complicated, because usually they break Lorentz invariance, unless one chooses a scalar-like combination, which amounts also to an effective scalar. The simplest such model is then with a single scalar field.

Moreover, in the simplest incarnation of inflation with a single scalar field, the field "slowly rolls" down a potential for ϕ, $V(\phi)$. For this "slow-roll" inflation, the most popular and simplest models are as follows.

- the power–law potential $V(\phi) = \lambda\phi^p/p!$, however, this is disfavored by recent data on the cosmological perturbation parameters ϵ, η, and r (to be defined later).
- the "new inflation" model, which is favored by the data: there is a plateau in the potential, followed by a drop, and a minimum, where "reheating" of the universe happens: the energy of the potential is transferred into kinetic (thermal) energy of standard ("low energy") particles. After reheating, the usual radiation-dominated (RD) cosmology begins.

As an observation, among the new inflation models, the first inflationary model to be defined, of Starobinsky, is the one that sits in the center of the error bars. This is described as a model with gravity action $R + aR^2$, but can be rewritten as a model of usual gravity, R, plus a scalar field with an exponential potential, as we will describe at the end of the chapter.

It turns out, for reasons we will try to explain shortly, to be very hard to embed inflation in string theory. In fact, there is no good, completely rigorous, string theory model for inflation yet. But here we try to embed it in supergravity, yet working with string theory in mind. So we don't need to worry about string theory embedding, which comes with its complicated constraints, but only consider supergravity models.

Moreover, for phenomenological reasons, we will consider only $\mathcal{N} = 1$ models: already $\mathcal{N} = 2$ has no complex representations (the chiral and antichiral superfields come in pairs), while we know that the Standard Model has chiral fields. This means that, in order to see something useful at accelerators, the $\mathcal{N} = 2$ and higher supersymmetry must be broken at a much higher scale, perhaps close to the Planck scale, while the $\mathcal{N} = 1$ can be broken at reachable (or rather, testable) scales.

We will then consider in this chapter $\mathcal{N} = 1$ supergravity coupled to chiral multiplets and/or vector multiplets.

26.1 $\mathcal{N} = 1$ supergravity coupled with a single chiral superfield Φ

The potential for $\mathcal{N} = 1$ supergravity coupled to chiral superfields is

$$V = e^{k/M_{\text{Pl}}^2}\left[g^{i\bar{j}} D_i W \overline{D_j W} - \frac{3}{M_{\text{Pl}}^2} |W|^2 \right], \tag{26.1}$$

where $k(\phi^i, \bar{\phi}^{\bar{j}})$ is the Kähler potential, giving the kinetic term, and $W(\phi^i)$ is the superpotential, which is a holomorphic function. Moreover, the Kähler-covariant derivative is

$$D_i W = \partial_i W + \frac{\partial_i k}{M_{\text{Pl}}^2} W. \tag{26.2}$$

As we said, we will consider only one Φ.

The *simplest* assumption (not necessarily true! see later) is for K to be Taylor-expandable at zero,

$$K(\phi, \bar{\phi}) = K_0 + \partial_\phi \partial_{\bar{\phi}} K\big|_0 \, \bar{\phi}\phi + ... \tag{26.3}$$

Note that this is not the most general expansion: we could have linear terms, ϕ and $\bar{\phi}$, but these give zero contribution to the metric on the moduli space, since $\partial_i \partial_{\bar{j}}(\alpha\phi + \beta\bar{\phi}) = 0$. However, we will see that by ignoring these, we will miss an important case.

Then the kinetic term is

$$- (\partial_\phi \partial_{\bar{\phi}} K)\big|_0 \, \partial_\mu \phi \partial^\mu \bar{\phi}, \tag{26.4}$$

and the scalar potential is

$$V = V_0 \left(1 + g_{\phi\bar{\phi}} \frac{\bar{\phi}\phi}{M_{\text{Pl}}^2} + ... \right) + ..., \tag{26.5}$$

independently on the form of the superpotential W, only due to the term e^{k/M_{Pl}^2}.

On the other hand, the slow-roll condition (for slow-roll inflation) is $\epsilon, \eta \ll 1$, where if the field is slowly rolling, ϵ and η depend only on the potential, namely one has

$$\epsilon = \frac{M_{\text{Pl}}^2}{2}\left(\frac{V'}{V} \right)^2, \quad \eta = M_{\text{Pl}}^2 \frac{V''}{V}. \tag{26.6}$$

This condition is satisfied at least for (slow-roll) new inflation.

But in string theory we have the so-called "eta problem," which is: after stabilizing the moduli (in string theory there are many moduli, or scalars with no potential, at least perturbatively, but since we see no scalars other than the Higgs in the real world, these scalar moduli need to be "stabilized," that is, fixed at the minimum of a nonperturbative potential, with a large enough mass so that we don't see them; examples of such moduli are shape and volume moduli of the compact space, moduli associated with the Kalb–Ramond B field, etc.), we create (quantum) contributions to the potential for the inflation that spoil $\eta \ll 1$.

Quantum corrections also give corrections to the mass of the dilaton, which are of the order of the Hubble scale,

$$\Delta m_\phi \sim H = \frac{\dot{a}}{a}, \tag{26.7}$$

where $a(t)$ is the common scale factor of Friedmann–Lemaître–Robertson–Walker (FLRW) cosmology. But for FLRW cosmology, from the Einstein equations, one can prove the Friedmann equation,

$$V_0 = 3M_{\mathrm{Pl}}^2 H^2. \tag{26.8}$$

Together, the scalar potential is roughly

$$V \sim 3M_{\mathrm{Pl}}^2 H^2 + \Delta m_\phi^2 \frac{\phi^2}{2} = V_0 + \Delta m_\phi \frac{\phi^2}{2}, \tag{26.9}$$

giving a contribution to the η of

$$\Delta \eta = \frac{\Delta m_\phi^2}{3H^2} \sim 1. \tag{26.10}$$

Note that the same thing is obtained here, from (26.5), though here we are at the classical level only:

$$\Delta \eta = g_{\phi\bar{\phi}} \sim \mathcal{O}(1). \tag{26.11}$$

Note that

$$V_0 = e^{k_0/\mathcal{M}_{\mathrm{Pl}}^2} \left[(\partial_\phi \partial_{\bar{\phi}} K)^{-1} D_\phi W \overline{D_\phi W} - \frac{3}{M_{\mathrm{Pl}}^2} |W|^2 \right]. \tag{26.12}$$

Having in mind string theory, which has quantum corrections at the Planck scale M_{Pl}, we Taylor expand the superpotential as

$$W = M_{\mathrm{Pl}}^3 \sum_p c_p \left(\frac{\phi}{M_{\mathrm{Pl}}} \right)^p, \tag{26.13}$$

where $c_p \sim \mathcal{O}(1)$. That gives also a quantum string contribution coming from W of

$$\Delta \eta \sim \mathcal{O}(1). \tag{26.14}$$

We note that we have shown three types of $\mathcal{O}(1)$ contributions to η, so we could of course have *cancellations* between them, but that would be *fine-tuning*, so not a true solution.

We see that perhaps if we didn't have a Taylor expansion for K at zero, we could solve the eta problem. For instance, consider the volume modulus φ of a CY_3 complex space for the six compact dimensions, which (we will describe it better in a later chapter) has a complex Kähler modulus combining φ with another scalar into a complex scalar ρ, with Kähler potential

$$K = -3 \ln [\rho + \bar{\rho}]. \tag{26.15}$$

Indeed, this K has no Taylor expansion around $\rho = 0$, so it would be fine. But, not surprisingly, such K would be generically spoiled in string theory by the moduli stabilization mechanism.

Another possibility for the η problem resolution is usually defined as follows: if we have a *softly broken* symmetry, such that the resulting potential is nearly flat. The symmetry is obvious: shift symmetry, under $\phi \to \phi$+constant. If the symmetry is softly broken by power laws, then

$$\mathcal{L} = -\frac{1}{2}(\partial_\mu \phi)^2 - \lambda_p \frac{\phi^p}{M_{\text{Pl}}^p}, \tag{26.16}$$

but with $\lambda_p \ll 1$, which indeed results in $\eta \ll 1$. Note that $\lambda_p = 0, \forall p$ has the exact shift symmetry.

However, it is worth noting that, except if we give a *reason* why $\lambda_p \ll 1$ in the fundamental theory, this is not a *solution* of the eta problem, but rather a *parametrization* of the same.

26.2 D-term inflation

Another interesting case is when we add vector superfields V^a, where a is in the adjoint of some gauge group G, and the auxiliary field is called D^a. Then, replacing the solution for the auxiliary field D^a in the potential for this "D-term,"

– assuming a canonical kinetic term for the gauge fields, $-\frac{1}{4g^2}F^a_{\mu\nu}F^{a\mu\nu}$, the D-term potential is

$$V_D = \frac{g^2 D^a D^a}{2}, \tag{26.17}$$

which is supposed to be added to the F-term (which in *rigid* susy is $F^i F^{\dagger i}/2$). Solving for D^a, we find

$$D^a = \phi^\dagger T^a \phi = \phi^{\dagger i}(T^a)_i^{\ j}\phi_j. \tag{26.18}$$

– if the kinetic term for the gauge fields has a function depending on the chiral fields, $-\frac{1}{4g_a^2}f_a(\phi)F^a_{\mu\nu}F^{a\mu\nu}$, then

$$D^a = \phi_i(T^a)^i_{\ j}\frac{\partial K}{\partial \phi_j} \Rightarrow V_d = \sum_a \left[\text{Re }f_a(\phi)\right]^{-1}\frac{g_a^2 D_a^2}{2}. \tag{26.19}$$

We can add a Fayet–Iliopoulos term (FI) with FI parameter ξ,

$$\int d^2\theta d^2\bar{\theta}\xi^a V^a, \tag{26.20}$$

which effectively replaces $D^a \to D^a + \xi^a$ in the D-term. Then, since there is no more minimum at zero energy, $E_0 \neq 0$, we have spontaneous supersymmetry breaking (as we remember, this was because the susy algebra starts with $\{Q, Q\} = H + ...$, so for a supersymmetric vacuum, $Q|\psi\rangle = 0$, it follows that $H|\psi\rangle = 0$, so $E_0 = 0$). Conversely, $E_0 \neq 0$ implies spontaneous (in the vacuum) supersymmetry breaking.

26.2.1 Example: the original FI model

1. The original FI model has one abelian ($U(1)$ gauge group) vector multiplet, and 2 chiral multiplets of $U(1)$ charges ± 1. Then the (unique renormalizable and $U(1)$ invariant) superpotential is

$$W = m\Phi_+\Phi_-. \qquad (26.21)$$

With the canonical kinetic terms, so with

$$K = \bar{\Phi}_+\Phi_+ + \bar{\Phi}_-\Phi_-, \qquad (26.22)$$

the potential becomes

$$V = m^2|\phi_+|^2 + m^2|\phi_-|^2 + \left(\xi + e^2|\phi_+|^2 - e^2|\phi_-|^2\right)^2. \qquad (26.23)$$

We see that its only extremum is at $\phi_+ = \phi_- = 0$, where $V_0 = \xi^2 \neq 0$, so we have spontaneous supersymmetry breaking. On the other hand, $\phi_+ = \phi_- = 0$ at this minimum, so there is no Higgs mechanism (remember that the complex fields are charged under the $U(1)$, so if the VEVs would be nonzero, we would have a Higgs mechanisms). This means that the $U(1)$ symmetry is unbroken.

2. The above model was for rigid susy, but of course we want models coupled to supergravity.

Generalizing the above to inflation in $\mathcal{N} = 1$ supergravity, we add the abelian vector multiplet as above, also the two charged chiral multiplets of charges ± 1, $\Phi_\pm$, but now also a neutral scalar S, which takes the place of the mass m via its VEV, so the superpotential is

$$W = S\Phi_+\Phi_-, \qquad (26.24)$$

while the Kähler potential is canonical,

$$K = \bar{S}S + \bar{\Phi}_+\Phi_+ + \bar{\Phi}_-\Phi_-. \qquad (26.25)$$

Then the same kind of analysis follows.

26.3 Field redefinitions

When we consider potential models for inflation, in particular supergravity models, we have to be aware that field redefinitions don't change the theory, but rather give an equivalent theory.

In the case of rigid susy, we saw that there was a "Kähler transformation," that was a transformation on the Kähler potential, of the type

$$K \to K + f_1(\phi) + f_2(\bar{\phi}). \qquad (26.26)$$

Under it, the metric on the scalar field space, $g_{i\bar{j}} = \partial_i\partial_{\bar{j}}K$, in our case $g_{\phi\bar{\phi}} = \partial_\phi\partial_{\bar{\phi}}K$, is preserved. If we want to maintain K real (though that is not needed, only the observables need to be real, and K is not an observable), we could impose that $f_1 = \bar{f}_1 = f$.

In rigid susy, this transformation of K is also a symmetry of the potential. But in the supergravity case, because $V = e^{k/M_{\text{Pl}}^2}[|DW|^2 - 3|W|^2/M_{\text{Pl}}^2]$, it is not. To obtain a symmetry of the potential, in the case $f_1 = f_2 = f$, we can also transform the superpotential as

$$W \to e^{-f/(2M_{\text{Pl}}^2)}W, \tag{26.27}$$

This transformation then also preserves supersymmetry, since we have transformed the $\mathcal{N} = 1$ quantities K and W, but not the component fields themselves.

On the other hand, we can also consider field redefinitions that do not preserve supersymmetry, for instance just redefining the scalars ϕ^i, which would break susy. One such example is when we go to the canonical kinetic term,

$$- g_{i\bar{j}}\partial_\mu \phi^i \partial^\mu \bar{\phi}^{\bar{j}} \to -\partial_\mu \tilde{\phi}^i \partial^\mu \bar{\tilde{\phi}}^{\bar{j}}. \tag{26.28}$$

26.4   Example of supergravity model with ϵ, $\eta \ll 1$

26.4.1   First try

To find our first example of slow-roll inflation in an $\mathcal{N} = 1$ supergravity model coupled to a single scalar superfield, consider first the canonical Kähler potential (now without the constant term,

$$K = \bar{\Phi}\Phi, \tag{26.29}$$

so considered from the beginning, not as a Taylor expansion), and consider also the *simplest* superpotential, which is actually the linear one,

$$W = A\Phi. \tag{26.30}$$

Then we obtain

$$g_{\phi\bar{\phi}} = 1, \quad DW = A\left(1 + \frac{\bar{\phi}\phi}{M_{\text{Pl}}^2}\right), \tag{26.31}$$

so the scalar potential becomes

$$\begin{aligned}
V &= e^{\bar{\phi}\phi/M_{\text{Pl}}^2}\left[|A|^2\left|1 + \frac{\bar{\phi}\phi}{M_{\text{Pl}}^2}\right|^2 - 3\frac{|A|^2}{M_{\text{Pl}}^2}|\phi|^2\right] \\
&= |A|^2 e^{\bar{\phi}\phi/M_{\text{Pl}}^2}\left[-1\frac{|\phi|^2}{M_{\text{Pl}}^2} + \frac{|\phi|^4}{M_{\text{Pl}}^4}\right].
\end{aligned} \tag{26.32}$$

For small fields, if $|\phi|/M_{\text{Pl}} \ll 1$, the mass terms for ϕ cancel between the bracket and the one coming from the exponential, and we find

$$V \simeq |A|^2\left[1 + \frac{1}{2}\frac{|\phi|^4}{M_{\text{Pl}}^4} + \mathcal{O}\left(\frac{|\phi|^6}{M_{\text{Pl}}^6}\right)\right]. \tag{26.33}$$

If moreover, we consider that the inflaton is the real part of ϕ, $\sigma = \text{Re}\,\phi$, then we obtain

$$\epsilon \simeq 2\left(\frac{\sigma}{M_{\text{Pl}}}\right)^6, \quad \eta \simeq \frac{6\sigma^2}{M_{\text{Pl}}^2}. \tag{26.34}$$

This means that we can put $\epsilon \ll 1, \eta \ll 1$ just by having $\sigma/M_{Pl} \ll 1$.

However, even if we have $\epsilon, \eta \ll 1$, we still don't have new inflation, since *there is no end to inflation*. Indeed, note that there is a plateau for $V(\sigma)$, but after it the field *increases* instead of decreasing.

26.4.2 Modified example

We can modify the above example such that the potential actually decreases after inflation.

We modify the superpotential by including a subleading term in ϕ,

$$W = a\Phi \left(1 - b\Phi^n\right), \quad n \geq 3, \tag{26.35}$$

while keeping the canonical Kähler potential, $K = \bar{\Phi}\Phi$. Then the Kähler-covariant derivative is

$$DW = A\left[1 - (n+1)b\phi^n + \frac{\bar{\phi}\phi}{M_{Pl}^2}(1 - b\phi^n)\right], \tag{26.36}$$

and so the scalar potential becomes

$$V = e^{\bar{\phi}\phi/M_{Pl}^2}|A|^2 \left[\left|1 - (n+1)b\phi^n + \frac{\bar{\phi}\phi}{M_{Pl}^2}(1 - b\phi^n)\right|^2 - \frac{3|\phi|^2}{M_{Pl}^2}|1 - b\phi^n|^2\right]. \tag{26.37}$$

At small fields, $|\phi|/M_{Pl} \ll 1$, we obtain

$$V \simeq |A|^2 e^{|\phi|^2/M_{Pl}^2}\left[1 - \frac{|\phi|^2}{M_{Pl}^2} - (n+1)b(\phi^n + \bar{\phi}^n) + \frac{|\phi|^4}{M_{Pl}^4} + \mathcal{O}\left(\frac{|\phi|^5}{M_{Pl}^5}\right)\right]. \tag{26.38}$$

Then, in the particular cases of:

- For $n = 3$ and the inflation is $\sigma = \text{Re } \phi$, we obtain ($b' = b/M_{Pl}^3$ is adimensional)

$$V \simeq |A|^2 \left[1 - \frac{8b'\sigma^3}{M^3} + \mathcal{O}\left(\frac{\sigma^4}{M_{Pl}^4}\right)\right]. \tag{26.39}$$

- For $n = 4$ and the same inflaton $\sigma = \text{Re } \phi$, we obtain (now $b' = b/M_{Pl}^4$ is adimensional)

$$V \simeq |A|^2 \left[1 = (10b' - 1/2)\frac{\sigma^4}{M_{Pl}^4} + \mathcal{O}\left(\frac{\sigma^5}{M_{Pl}^5}\right)\right]. \tag{26.40}$$

We see that in both cases, there are no mass terms (they have been canceled), so as before, we obtain $\epsilon, \eta \ll 1$ just by choosing $\sigma/M_{Pl} \ll 1$.

However, as we see from the form of the approximate potential, now the leading term goes *down*, not up, so we do obtain good new inflation. This model was found by Yzawa and Yanagida.

In fact, the two models above are the only possibilities for inflation with the standard $K = \bar{\Phi}\Phi$.

Possible modification of K

We have motivated (though we have not quite done it this way) the Kähler potential from the Taylor expansion around zero. If the above models are all there in this case, it stands to reason that we can modify this assumption.

Specifically, consider the case when there is no Taylor or Laurent expansion for K and W, which means that there is a singularity at $\phi = 0$ that is not of finite-order pole form, that is, it is an essential singularity.

In particular, in string theory, one finds the Kähler potential

$$K = -3\log(\Phi + \bar{\Phi}), \tag{26.41}$$

and consider the inflaton as $\sigma = \mathrm{Re}\,\phi$. This case can arise in a KK compactification, if σ is a volume modulus for the compact space. Specifically, the metric on the compact space $g_{mn}(x, y)$ is written in terms of the tensor spherical harmonic $g_{mn}(y)$ as

$$\bar{g}_{mn}(x, y) = e^{2u(x)} g_{mn}(y). \tag{26.42}$$

In that case, the actual volume modulus is

$$\mathrm{Re}\,\phi = e^{cu(x)}, \tag{26.43}$$

where c is a number. One then finds for it the kinetic term

$$\frac{\alpha}{(\phi + \bar{\phi})^2} \partial_\mu \phi \partial^\mu \bar{\phi}, \tag{26.44}$$

which is easily found to come from the Kähler potential

$$K = -\alpha \ln(\Phi + \bar{\Phi}). \tag{26.45}$$

In the case of the most standard compact six-dimensional space in string theory, CY_3 (which we will describe better in a later chapter), we find $\alpha = 3$, as written above.

However, for this strange-looking K, inflation is only achieved if we add terms inside the log, depending on D3-brane moduli = positions z_α, $\alpha = 1, 2, 3$ in the CY_3, so if we have

$$K = -3 \ln \left[\rho + \bar{\rho} - \Delta k(z_\alpha, \bar{z}_\alpha) \right]. \tag{26.46}$$

As explained before, while this looks fine as it is, nonperturbative contributions that fix, or "stabilize," the moduli (leaving only the inflaton), spoil this form.

26.5   Special embedding of inflationary potentials in supergravity

But there is a way to embed (almost) any potential inside $\mathcal{N} = 1$ supergravity plus one chiral superfield, albeit with an unusual Kähler potential,

$$K = -3M_{\mathrm{Pl}}^2 \ln \left(1 + \frac{\bar{\Phi} + \Phi}{\sqrt{3}M_{\mathrm{Pl}}} \right), \tag{26.47}$$

which, despite its similarity with the above string theory case, it is very hard (no examples so far) to embed in string theory. But, since we are interested only in supergravity, so we don't need to consider all the constraints coming from string theory, this is acceptable.

Taylor expanding the above Kähler potential around 0 (which is now possible!), we find

$$
K \simeq -\sqrt{3}M_{\text{Pl}}(\Phi + \bar{\Phi}) + \frac{1}{2}(\Phi + \bar{\Phi})^2 + \ldots
$$

$$
= \bar{\Phi}\Phi + \left[\frac{1}{2}\Phi^2 - \sqrt{3}M_{\text{Pl}}\Phi\right] + \ldots
$$

$$
+ \left[\frac{1}{2}\bar{\Phi}^2 - \sqrt{3}M_{\text{Pl}}\bar{\Phi}\right] + \ldots \tag{26.48}
$$

We see that *to first order* the above K is Kähler-equivalent to $\bar{\Phi}\Phi$, since the terms in brackets are, respectively, an $f(\Phi)$ and an $f(\bar{\Phi})$.

Observe that we have a linear term, $\propto \Phi + \bar{\Phi}$, which we neglected before (see now, that it was not quite right to do so) and, more importantly, the equivalence was only in first order. In higher orders, we have mixings of Φ and $\bar{\Phi}$ that are asymmetric, so actually, we cannot use the Kähler transformation to get rid of $\Phi + \bar{\Phi}$ terms (rather, it serves no purpose). So we see that, while we thought we wrote the more general Taylor expansion model, we didn't, and this model slipped through the cracks.

Also, another important observation is that now, unlike the previous cases, we want the inflaton to be the *imaginary part of ϕ, not the real one*, with canonical field (the square root of 2 is because $-\partial_\mu \phi \partial^\mu \phi$ lacks the 1/2 in front for the canonical form)

$$
\phi_{\text{can}} = \sqrt{2}\text{Im }\phi. \tag{26.49}
$$

But this means that, if we would just substitute Re $\phi = 0$ from the beginning, we would get $K \equiv 0$. Instead, we must first take the derivatives and then put the real part of ϕ to 0. Then the kinetic term becomes

$$
-\partial_\phi \partial_{\bar{\phi}} K \partial_\mu \phi \partial^\mu \bar{\phi} = -\left.\frac{\partial_\mu \phi \partial^\mu \bar{\phi}}{\left(1 + \frac{\phi + \bar{\phi}}{\sqrt{3}M_{\text{Pl}}}\right)^2}\right|_{\text{Re}\phi = 0} = -\frac{1}{2}\partial_\mu \phi_{\text{can}} \partial^\mu \phi_{\text{can}}. \tag{26.50}
$$

The Kähler-covariant derivative, on the constraint Re $\phi = 0$, becomes

$$
DW = \partial_\phi W + \frac{\sqrt{3}}{1 + \frac{\phi + \bar{\phi}}{\sqrt{3}M_{\text{Pl}}}} \frac{W}{M_{\text{Pl}}} \rightarrow DW|_{\phi_{\text{can}}} = W'\left(\sqrt{2}i\text{Im }\phi\right) + \frac{\sqrt{3}}{M_{\text{Pl}}}W\left(\sqrt{2}i\text{Im }\phi\right).
$$

$$
\tag{26.51}
$$

Then, *IF* W is a real function (meaning, with real coefficients), the first term in the covariant derivative above is imaginary, and the second is real, so in $|DW|^2$ they don't interfere, and the second term cancels the $-3|W|^2/M_{\text{Pl}}^2$ term. Note that $g_{\phi\bar{\phi}} = 1$ and $K = 0$ for Re $\phi = 0$. Thus,

$$
V = e^{K/M_{\text{Pl}}^2}\left[|DW|^2 - \frac{3}{M_{\text{Pl}}^2}|W|^2\right]
$$

$$
= \left|W'\left(\sqrt{2}i\text{Im }\phi\right)\right|^2 \equiv \hat{W}'(\phi_{\text{can}})^2, \tag{26.52}
$$

where we have defined the *real function* $\hat{W}$ by

$$
W(\phi) = \frac{1}{\sqrt{2}}\hat{W}(-\sqrt{2}i\phi). \tag{26.53}
$$

Since we obtain a $V \geq 0$, written as a square of $\hat{W}'$, for *any positive single scalar inflationary potential* V, we can embed it into $\mathcal{N} = 1$ supergravity (take the square root, find $\hat{W}'$, then $\hat{W}$, then W).

The only problem of this construction is that it is hard to see where it could come from in string theory, but in supergravity it is fine.

26.6   Kallosh and Linde's "α attractors" in $\mathcal{N} = 1$ supergravity

Finally, consider the models known as "α attractors" defined by Renata Kallosh, Andrei Linde and others.

Consider a simple generalization of the Kähler potential from string theory KK compactification, with an additional α in front, so

$$K = -3\alpha \log(T + \bar{T}), \tag{26.54}$$

which leads to a canonical field (the real part only, the inflaton, neglect the imaginary part, which is an axion χ) φ defined by

$$T = \exp\left(\sqrt{\frac{2}{3\alpha}}\varphi\right) + i\chi. \tag{26.55}$$

This gives the kinetic term for T

$$\frac{3\alpha}{(T + \bar{T})^2} \partial_\mu T \partial^\mu \bar{T}. \tag{26.56}$$

Another way to parametrize this variable T is in terms of the "disk variable" Z, defined by

$$Z = \frac{T - 1}{T + 1} \Rightarrow T = \frac{1 + Z}{1 - Z}, \tag{26.57}$$

in terms of which the Kähler potential is

$$K = -3\alpha \log(1 - \bar{Z}Z). \tag{26.58}$$

In terms of the canonical scalar φ, at $\chi = 0$, Z becomes

$$Z|_{\chi=0} = \tanh\left(\frac{\varphi}{\sqrt{6\alpha}}\right). \tag{26.59}$$

Then, we must give a superpotential W, which can be described as a function of $Z = \tanh\left(\varphi/\sqrt{6\alpha}\right)$, or as a function of $T = \exp\left(\varphi\sqrt{2/(3\alpha)}\right)$.

The advantage of this is that the object $T = \exp\left(\varphi\sqrt{2/(3\alpha)}\right)$, for a particular α, is something that appears in the Starobinsky model of inflation, which is known to be preferred by data.

One can then write (among other possibilities) the natural potentials:

– "T-models," with

$$V = Z^{2n} = \tanh^{2n}\left(\frac{\varphi}{\sqrt{6\alpha}}\right). \tag{26.60}$$

– "E-models," with

$$V = V_0 \left(1 - e^{-\sqrt{2}3\alpha\varphi} \right)^{2n}.$$ (26.61)

Then the Starobinsky model corresponds to the E-model with $\alpha = 1, n = 1$.

One can also embed various models of inflation in $\mathcal{N} = 1$ supergravity via a formalism of a "Kähler function" (that we defined before), that puts together K and W into a single function

$$\mathcal{G} \equiv K + \log W + \log \bar{W},$$ (26.62)

which is real if ϕ is real. The advantage of this is that now the scalar potential depends solely on $\mathcal{G}$,

$$V = e^{\mathcal{G}} \left[\mathcal{G}^{\alpha\bar{\beta}} \mathcal{G}_\alpha \mathcal{G}_\beta - 3 \right],$$ (26.63)

where, as is natural, $\mathcal{G}_\alpha \equiv \partial\mathcal{G}/\partial\Phi^\alpha$ and $\mathcal{G}_{\alpha\bar{\beta}} = \partial^2\mathcal{G}/\partial\Phi^\alpha\partial\bar{\Phi}^{\bar{\beta}}$.

Then we can find $\mathcal{G}$ from V in an easier manner, and then from it deduce K and W. This works in most useful cases (beyond even a single scalar superfield).

However, note that choosing to write things in terms of Z, or T and then finding simple forms for W is not too powerful, it simply parametrizes the result we want to obtain, *unless we explain why such a coordinate is natural in a more fundamental theory*. Otherwise, we might as well work with the canonical field φ and invent (super)potentials for it.

Important concepts to remember

- Inflation is the most popular model of cosmology, and single scalar slow-roll inflation the most common for it. Power–law inflation is disfavored by data, and new inflation (motion on a plateau, followed by a drop and reheating) is favored.

- It is very difficult to embed inflation in string theory *rigorously*: currently there is no such model.

- In inflation, we define the slow-roll parameters $\epsilon = M_{\mathrm{Pl}}^2/2(V'/V)^2$ and $\eta = M_{\mathrm{Pl}}^2 V''/V$, which should be $\ll 1$.

- In string theory, we have the "eta problem," that we get generically $\eta \sim 1$ either classically, or quantum mechanically, when we stabilize the moduli. To avoid it would require fine-tuning, for cancellations between effects.

- In supergravity, in particular, $\mathcal{N} = 1$ supergravity with one chiral superfield, it is easier to embed inflation. One simple model, $K = \bar{\Phi}\Phi$ and $W = A\Phi$, with inflaton $\sigma = \mathrm{Re}\,\phi$, seems to work, but gives a *rising* potential after the plateau, so no end of inflation.

- A modified example introduces a subleading term to the superpotential, $W = a\Phi(1 - b\Phi^n)$, $n \geq 3$, and gives a correct plateau followed by a drop, and is the only good example with canonical Kähler potential.

- One can achieve inflation with the Kähler potential modified by D3-brane moduli, $K = -3\ln[\rho + \bar{\rho} - \Delta k(z_\alpha, \bar{z}_\alpha)]$, but moduli stabilization spoils inflation.

- There is a special embedding of inflationary models into $\mathcal{N} = 1$ supergravity with one chiral superfield, for the special Kähler potential $K = -3M_{\mathrm{Pl}}^2 \ln\left(1 + \frac{\bar{\phi}+\phi}{\sqrt{3}M_{\mathrm{Pl}}}\right)$, with inflation the *imaginary part* of ϕ, and superpotential defined by $V = \hat{W}'(\phi_{\text{can}})^2$, $W(\phi) = \hat{W}(-\sqrt{2}i\phi)/\sqrt{2}$.

- Alpha attractors are defined by a Kähler potential $K = -3\alpha \log(T + \bar{T})$, for a T related to the canonical field φ by $T = \exp\left(\sqrt{\frac{2}{3\alpha}}\varphi\right) + i\chi$, and superpotentials written in terms of T or $Z = (T-1)/(T+1)$. The Starobinsky model, understood also as gravity with an R^2 correction, is obtained as a particular case.

References and further reading

For a more in detail discussion of supergravity inflation and the problems of embedding it in string theory, see [93] and the paper [94], and the reviews [95, 96]. For alpha attractors, see for instance [97].

Exercises

(1) Calculate the slow-roll parameters ϵ and η for $\mathcal{N} = 1$ supergravity plus one chiral superfield, with

$$K = -3\ln(\Phi + \bar{\Phi}), \quad W = A\rho. \tag{26.64}$$

(2) Calculate the potential and its equations of motion for the supergravity modification of the original FI model, with V^a coupled to $\Phi_\pm$ (with charge ± 1 under $U(1)$) and neutral S, with FI term $\xi^a V^a$ and

$$W = S\Phi_+\Phi_-, \quad K = |S|^2 + |\Phi_+|^2 + |\Phi_-|^2. \tag{26.65}$$

(3) Embed "chaotic inflation," with $V = \lambda_p \phi^p / p!$, into $\mathcal{N} = 1$ supergravity coupled to one chiral superfield, via the special embedding.

(4) For the Yzawa–Yanagida model, with

$$K = \bar{\Phi}\Phi, \quad W = \alpha\Phi(1 - b\Phi^n), \tag{26.66}$$

calculate the potential exactly, and then expand it in $\sigma = \text{Re }\phi$ to find, for $n = 3$ and $n = 4$, (26.39) and (26.40), as well as for the $n = 1$ and $n = 2$ cases. Repeat the exercise for $K = -3\ln[\Phi + \bar{\Phi}]$.

Maldacena–Núñez and supergravity no-go theorems; loopholes

In this chapter, we will describe no-go theorems in supergravity, starting with the standard one of Maldacena and Núñez, and then considering more recent generalizations leading to a conjecture, the "swampland conjecture."

The Maldacena–Núñez no-go theorem is for (warped) KK compactification of a D-dimensional supergravity theory to $\mathbb{R}^d$ (Minkowski) or dS^d (de Sitter), for $d \geq 2$, and with a finite Newton's constant in the lower dimensions, $\kappa_{N,d}$. Of course, the case of most relevance is $d = 4$, but the theorem is more general.

The argument will be very general and will only be based on the equation of motion for the warp factor.

As usual, we will use $M, N, \dots$ for D-dimensional indices, $\mu, \nu, \dots$ for d-dimensional ones, and $m, n, \dots$ for compact ones.

27.1 No-go theorem

D-dimensional gravity ($D > 2$), compactified to d dimensions has no $\mathbb{R}^d$ or dS^d solutions IF:

(1) the gravity theory has no higher curvature corrections.
(2) the potential is nonpositive, $V \leq 0$. This is not true for the case of massive type IIA supergravity in 10 dimensions, which has a cosmological constant, but this case will be treated separately, and the same result will be obtained. The potential can thus be $V = -V_0$ or $V(\phi)$, but in the range of values explored for the solution considered, it must be nonpositive.
(3) The theory has massless fields with positive kinetic terms, with n-form field strengths $F_{M_1 \dots M_n} \times n = 1$ corresponds to scalars, $n = 2$ to Maxwell fields, or nonabelian ones, if the metric on the group space is positive definite, so the kinetic terms are still positive. Then $n < D$, since $n = D$ is just a contribution to the potential, considered at (2).
(4) $\kappa_{N,d}$ is finite.

Proof. Consider the Einstein equations (note that these are with no higher derivative corrections to the gravity part!), putting $\kappa_{N,D} = 1$,

$$R_{MN} = T_{MN} - \frac{1}{D-2} g_{MN} T^P{}_P. \tag{27.1}$$

Consider the (warped! there is a warp factor Ω depending on the compact coordinates only) ansatz

$$ds_D^2 = \Omega^2(y)(d\vec{x}_d^2 + \hat{g}_{mn}(y)dy^m dy^n), \tag{27.2}$$

where $d\vec{x}^2 = \eta_{\mu\nu}dx^\mu dx^\nu$, but where "$\eta_{\mu\nu}$" stands either for Minkowski or for the de Sitter (also maximally symmetric space) metric. Note that we have put the warp factor common also to the compact space metric, though we could have absorbed it in a redefinition of the compact metric and/or coordinates.

Then, for $(MN) = (\mu\nu)$, we obtain

$$R_{\mu\nu} = R_{\mu\nu}(\eta) - \eta_{\mu\nu}\left[\hat{\nabla}^2 \log \Omega + (D-2)(\hat{\nabla} \log \Omega)^2\right]$$
$$= T_{\mu\nu} - \frac{1}{D-2}\Omega^2 \eta_{\mu\nu}T^L{}_L, \tag{27.3}$$

where $\hat{\nabla}$ means with the metric $\hat{g}_{mn}$, and $R_{\mu\nu}(\eta)$ is the Ricci tensor for the $\eta_{\mu\nu}$ metric (so $=0$ for the Minkowski case, and $\neq 0$ for de Sitter). In the second line we have used the Einstein equations.

Take the trace of Eq. (27.3) with the $\eta^{\mu\nu}$ metric and obtain

$$\hat{\nabla}^2 \log \Omega + (D-2)\left(\hat{\nabla} \log \Omega\right)^2 = \frac{1}{(D-2)\Omega^{D-2}}\hat{\nabla}^2\Omega^{D-2}$$
$$= R(\eta) + \Omega^2\left(-T^\mu{}_\mu + \frac{d}{D-2}T^P{}_P\right). \tag{27.4}$$

Define

$$\tilde{T} \equiv -T^\mu{}_\mu + \frac{d}{D-2}T^P{}_P. \tag{27.5}$$

We will first prove that $\tilde{T} \geq 0$, considering separately the potential and n-form cases:

- potential, $T_{MN} \propto -Vg_{MN}$, in which case $T^M{}_M \propto -DV$ and $T^\mu{}_\mu \propto -dV$, so

$$\tilde{T} \propto -Vd\left(-1 + \frac{D}{D-2}\right) = -\frac{2d}{D-2}V \geq 0, \tag{27.6}$$

since we have assumed that $-V \geq 0$.
- n-form. Now

$$T_{MN} \propto F_{MP_1...P_{n-1}}F_N{}^{P_1...P_{n-1}} - \frac{1}{2n}g_{MN}F^2, \tag{27.7}$$

so

$$\tilde{T} \propto -F^{\mu P_1...P_{n-1}}_{\mu P_1...P_{n-1}} + \frac{d}{D-2}\left(1 - \frac{1}{n}\right)F^2. \tag{27.8}$$

But, in order to preserve the $\mathbb{R}^d$ (Minkowski) or dS^d (de Sitter) isometries, the indices could be completely along the internal dimensions, $F_{m_1...m_n}$, or (if $n \geq d$) d indices along $\mathbb{R}^d$ and the rest internal, $F_{\mu_1...\mu_d m_{d+1}...m_n}$. So if we have a μ index, we must be in the second case, which means that in the contraction with one μ index we actually have

$$F_{\mu P_1...P_{n-1}}F^{\mu P_1...P_{n-1}} = \frac{d}{n}\tilde{F}^2, \tag{27.9}$$

where $\tilde{F}^2$ means just the case with d indices in $\mathbb{R}^d$. Then, since we have one 0 component among the $\mu_1...\mu_d$, and the contraction is with $g^{00} < 0$, it means that $\tilde{F}^2 < 0$.

The contribution $\hat{F}^2$, for the contraction of the terms with only internal indices, is positive, since it has no g^{00} contraction, and it contributes to $\tilde{T}$ with a positive coefficient, so in the end

$$\tilde{T} = -\tilde{F}^2 \frac{d(D-1-n)}{n(D-2)} + \frac{d(n-1)}{(D-2)n}\hat{F}^2 \geq 0. \tag{27.10}$$

Then, multiplying (27.4) with $\Omega^{2(D-2)}(D-2)$, we obtain

$$\Omega^{D-2}\hat{\nabla}^2\Omega^{D-2} = \Omega^{D-2}(D-2)R(\eta) + \Omega^{2D-2}\tilde{T} \geq 0, \tag{27.11}$$

since $R(\eta) \geq 0$ and $\tilde{T} \geq 0$, with equality only for the Minkowski case.

But since

$$\frac{1}{2\kappa_{N,d}^2} = \frac{2}{2\kappa_{N,D}^2}\int d^{D-d}y\sqrt{\hat{g}}\Omega^{D-2}, \tag{27.12}$$

where we have rewritten the D-dimensional factor $\sqrt{-G}G^{MN} = \sqrt{\hat{g}}\Omega^D\sqrt{-\eta}\frac{1}{\Omega^2}\eta^{\mu\nu}$ in terms of the measure of integration over the internal manifold, $\sqrt{\eta}\eta^{\mu\nu}$ (to be used in d dimensions) and Ω^{D-2}, it means that *if Ω is bounded above and below in the internal manifold*, the (assumed at point 4) finiteness of $\kappa_{N,d}$ implies that the internal manifold should actually be *compact*.

Then integrating by parts the inequality (27.11), we don't obtain boundary terms, so we get

$$-\int d^{D-d}y\sqrt{-\hat{g}}\left(\hat{\nabla}\Omega^{D-2}\right)^2 \geq 0, \tag{27.13}$$

which is only possible if Ω is constant. But then from (27.4) we have $R(\eta) + \Omega^2\tilde{T} = 0$ and, since $\tilde{T} \geq 0$, if follows that $R(\eta) = 0$, so *de Sitter space is excluded*. Moreover, we actually need $\tilde{T} = 0$, which is only possible if we only have scalars ($n = 1$-forms) turned on.

We can relax the condition that Ω is bounded both above and below, and consider only the case where Ω can vanish (so Ω is still bounded above), and regularize the internal manifold ($\mathcal{R} \to \mathcal{R}_\epsilon$) by staying at some ϵ away from them ($\Omega > \epsilon$), which means that $\vec{\nabla}\Omega$ is either 0 (constant Ω) or points inward at the boundary, so $\vec{n} \cdot \vec{\nabla}\Omega < 0$. Then by considering the boundary term when we partially integrate (27.11), we get

$$\int_{\mathcal{R}_\epsilon}\left(\hat{\nabla}\Omega^{D-2}\right)^2 \leq \int_{\partial\mathcal{R}_\epsilon}\left(\vec{n} \cdot \vec{\nabla}\Omega^{D-2}\right)\Omega^{D-2} \leq 0, \tag{27.14}$$

so again we obtain that Ω must be constant. That then also *excludes the Randall–Sundrum type compactifications* where the warp factor Ω goes to zero at a singularity = boundary of the space.

Note that the Randall–Sundrum background is given by a slice of AdS metric, with warp factor decreasing from a "Planck brane" at y_{max} to an "IR brane" at y_{min}, where particle physics (our world) is located,

$$ds^2 = e^{ky}d\tilde{x}_4^2 + dy^2. \tag{27.15}$$

q.e.d.

What are then possible loopholes to the previously discussed theorem?

We can introduce higher derivative corrections to the Einstein equations, as it happens, for instance, in string theory, for example, by introducing "orientifolds" (which effectively are non-dynamical, negative tension branes).

We can also introduce singularities such that $\kappa_{N,d} \to \infty$, which can also be achieved via orientifolds.

27.2   No Minkowski or de Sitter compactifications of massive type IIA supergravity

As promised, we now consider, separately, the case of massive type IIA supergravity. We consider compactification on smooth manifolds without boundaries down to $\mathbb{R}^d$ or dS^d, as before.

The Einstein equations and dilaton equation are given by

$$
R_{MN} = \frac{m^2}{16} e^{-5\phi} g_{MN} + 2\partial_M \partial_N \phi + 2^{2\phi} \left(H_M{}^{PQ} H_{NPQ} - \frac{1}{12} g_{MN} H_{(3)}^2 \right)
$$
$$
+ 2m^2 e^{-3\phi} \left(F_M{}^P F_{NP} - \frac{1}{16} g_{MN} F_{(2)}^2 \right) + \frac{e^{-\phi}}{3} \left(F_M{}^{PQR} F_{NPQR} - \frac{3}{32} g_{MN} F_{(4)}^2 \right).
$$
$$
\Box \phi + \frac{5m^2}{8} e^{-5\phi} + \frac{1}{48} e^{-\phi} F_{(4)}^2 - \frac{1}{6} e^{2\phi} H_{(3)}^2 + \frac{3m^2}{4} e^{-3\phi} F_{(2)}^2 = 0, \tag{27.16}
$$

where

$$
F_{(n)}^2 \equiv F_{M_1 \dots M_n} F^{M_1 \dots M_n}. \tag{27.17}
$$

As before, on the same compactification ansatz (but considering the particular case of the dilaton, $n = 1$ form field strength from the previous discussion, and then, as we said, we can only have $F_m = \partial_m \phi$, so $\phi = \phi(y)$ only) we find from the Einstein equation that

$$
\frac{1}{(D-2)\Omega^D} \hat{\nabla}^2 \Omega^{D-2} = \Omega^{-2} R(\eta) + d \left[-\frac{m^2}{16} e^{-5\phi} + \frac{1}{12} e^{2\phi} H_{(3)e}^2 + \frac{m^2}{2} e^{-3\phi} F_{(2)e}^2 + \frac{1}{32} e^{-\phi} F_{(4)e}^2 \right]
$$
$$
- d \left[\theta(3-d) \frac{e^{2\phi}}{4} H_{(3)l}^2 + \theta(2-d) \frac{7m^2}{8} F_{(2)l}^2 + \theta(4-d) \frac{5}{96} e^{-\phi} F_{(4)l}^2 \right], \tag{27.18}
$$

where $F_{(n)e}^2$ is the contraction with only indices in the internal dimension $(m, n, \dots)$ and $F_{(n)l}^2$ is the contraction with only indices in the d dimensions.

On the other hand, on the ansatz, we have

$$
\Box \phi = \frac{1}{\Omega^D} \hat{\nabla}_m \left(\Omega^{D-2} \hat{g}^{mn} \partial_n \phi \right). \tag{27.19}
$$

From (27.18) multiplied by $10\Omega^D$, plus the dilaton equation, with the above $\Box \phi$, multiplied by $d\Omega^d$, we find

$$\frac{10}{D-2}\hat{\nabla}^2\Omega^{D-2} + d\hat{\nabla}_m(\Omega^{D-2}\hat{g}^{mn}\partial_n\phi) = \Omega^{D-2}R(\eta) + d\Omega^D\left[\frac{2}{3}e^{2\phi}H^2_{(3)e} + 2m^2e^{-3\phi}F^2_{(2)e}\right.$$

$$\left. + \frac{1}{3}e^{-\phi}F^2_{(4)e} - \theta(3-d)\frac{8}{3}e^{2\phi}H^2_{(3)l} - \theta(2-d)8m^2F^2_{(2)l} - \theta(4-d)\frac{e^{-\phi}}{2}F^2_{(4)l}\right] \geq 0,$$

$$(27.20)$$

since $F^2_{(n)e} > 0$ by the same argument as in the previous theorem (this was called $\hat{F}^2$ before) and $F^2_{(n)l} < 0$ as well (this was called $\tilde{F}^2$ before).

But the left-hand side is now a total derivative (we have taken care to not multiply with extra Ω's like we did in the previous theorem, such that now we have a total derivative instead), hence by integrating over the internal manifold, we get zero (assuming that Ω is bounded above and below, so the internal space is compact and singularity free).

But then the right-hand side must be zero as well, which is only possible if both $R(\eta) = 0$ and $F_{(2)} = F_{(4)} = H_{(3)} = 0$. But in that case, we can integrate the dilaton equation, now reducing to $\Box\phi + \frac{5m^2}{8}e^{-\phi} = 0$, after multiplying with Ω^D, over the internal manifold, and we get a contradiction, since the $\Box\phi$ term integrates to a zero, as there are no boundaries.

This means that there are no non-singular Minkowski or de Sitter compactifications. Again, we can also assume that Ω is only bounded from above, so it can go to zero, and in that case the dilaton stays constant, we can include the same boundary term, and conclude that again we also *exclude the Randall–Sundrum type compactifications.*

27.3 No RS solution in $d = 5$ gauged supergravity

Having reached a conclusion on Randall–Sundrum-type solutions in general dimension, but in a precise sense for the construction, we can go back to a previous no-go theorem by Kallosh and Linde, about such solutions in the natural context for them, in $\mathcal{N} = 2$ $d = 5$ gauged supergravity with vector multiplets. Such theories would arise from a KK compactification of 10-dimensional supergravity on a nontrivial space and have AdS solutions in five-dimensions, like in the five-dimensional RS construction.

The RS solution would be one with ansatz

$$ds^2 = a^2(r)dx^\mu dx^\nu \eta_{\mu\nu} + dr^2, \qquad (27.21)$$

so a static solution with four-dimensional Poincaré invariance and a warp factor interpolating between two fixed points.

Substituting the ansatz in the Lagrangian, and because of the time independence the energy is minus the spatial integral of the Lagrangian density, $E = -\int d^3x\mathcal{L}$, so the energy functional on the ansatz becomes

$$E = \frac{1}{2}\int_{-\infty}^{+\infty} dr\, a^4 \left\{\left|e^a_i(\phi)' \mp 3e^{ai}(\phi)\partial_i W\right|^2 - 12\left[\frac{a'}{a} \pm W\right]^2\right\} \pm 3\int_{-\infty}^{+\infty} dr\frac{\partial}{\partial r}[a^4 W],$$

$$(27.22)$$

where prime refers to d/dr, W is the superpotential, and $\partial_i W = \partial W/\partial\phi^i$, $e^a_i(\phi)$ is the vielbein in the ϕ^i field space, with metric $g_{ij}(\phi) = e^a_i e^b_j \eta_{ab}$ corresponding to very special geometry.

One could consider the equations of motion coming from them (and we will, later), but these are second order in derivatives. For the first argument (no-go theorem), we will consider $\mathcal{N} = 1$ supersymmetric solutions (susy valid along the flow in $a(r)$), and for them the susy flow equations, which are only first order in derivatives (because $\{Q, Q\} \sim H$, the $Q|\psi\rangle = 0$ equations, or susy flow equations, are first order, while the $H|\psi\rangle = 0$ equations, or equations of motion for E, are second order),

$$(\phi^i)' = \pm 3 g^{ij} \partial_j W,$$

$$\frac{a'}{a} = \mp W. \tag{27.23}$$

Of course, they are also the equations that minimize the energy functional, which has been already put into a form that gives a BPS condition, with a sum of squares, plus a topological (boundary) term.

Defining the analog of the Hubble constant for motion in r instead of time t, $H_r \equiv a'/a$, we get that $H_r(\phi) = \mp W(\phi)$.

At the *critical points* (remember that we assume that the endpoints of the RS solution flow equations are fixed, or critical points), by definition

$$(\phi^i)_{\text{cr}} = 0 \Rightarrow (\partial_i W)_{\text{cr}} = 0, \tag{27.24}$$

and there calling $\phi = \phi_*$, $H_r(\phi_*) = \mp W(\phi_*)$ is constant.

Since

$$a \frac{\partial}{\partial a} \phi^i = \frac{a}{a'} \frac{\partial}{\partial r} \phi^i = \frac{1}{H_r} \phi^{i'}, \tag{27.25}$$

we obtain, using the susy flow equations and independently on the $\pm$ sign (that cancels),

$$a \frac{\partial}{\partial a} \phi^i = -\frac{3 g^{ij} \partial_j W}{W}. \tag{27.26}$$

On the other hand, for very special geometry near the critical point ϕ_*, we have the relation

$$(\partial_i \partial_j W)_{\text{cr}} = \frac{2}{3} g_{ij} W_{\text{cr}}. \tag{27.27}$$

Then, near the critical point, expanding the numerator of the right-hand side in (27.26), we get

$$a \frac{\partial}{\partial a} \phi^i \simeq -3 \frac{\partial_k (g^{ij} \partial_j W)(\phi^k - \phi_*^k)}{W}$$

$$= -2(\phi^i - \phi_*^i), \tag{27.28}$$

where in the second equality we used the relation (27.27) at the critical point.

Then the solution of the resulting equation is

$$\phi^i(a) \simeq \phi_*^i + \frac{c^i}{a^2}, \tag{27.29}$$

as we can easily check, which means that the *scale factor grows toward the fixed point*.

This means that there is no *decreasing* warp factor toward a fixed point, as needed to give an RS solution, where we need to have a minimum of the warp factor as the "physical brane" or "IR brane," where we live.

The previously discussed argument used the susy flow equations, so it only prohibits *supersymmetric* RS solutions, so we need a non-supersymmetric argument as well to eliminate also non-supersymmetric RS solutions.

From the minimum of the energy functional (27.22), we find the equations of motion

$$(\phi^i)'' + 4H_r(\phi^i)' + g^{ij}g_{jl,k}(\phi^k)'(\phi^l)' - 6g^{ij}\partial_j V = 0, \qquad (27.30)$$

where the potential is

$$V = -6\left[W^2 - \frac{3}{4}g^{ij}\partial_i W \partial_j W\right]. \qquad (27.31)$$

At the critical points, we have $V < 0$, and as in the susy case we have

$$(\partial_i \partial_j V)_{\text{cr}} = \frac{2}{3}g_{ij}V_{\text{cr}} < 0. \qquad (27.32)$$

If $H_r < 0$ at the critical point, as we want to happen for an RS-type solution (the warp factor to decrease toward it), and since by assumption $\phi \to \phi_*$ for large r (so that at large r we go to this critical point), meaning therefore that $g_{ij} \to$ constant, $g_{jl,k} \to 0$, $(\phi^k)' \to 0$, then in the equation of motion (27.30) the third term can be neglected with respect to the second, and therefore be ignored, the equation of motion (27.30) near the critical point becomes

$$(\delta\phi^i)'' - 4|H_r|(\delta\phi^i)' = -4|V|\delta\phi^i. \qquad (27.33)$$

This is clearly the equation for a harmonic oscillator with *negative* friction, so the solution blows up, contradicting our assumption of going to zero near the critical point. Thus this non-supersymmetric RS-type solution is also impossible.

27.4 The "swampland conjecture"

In 2018, Obied, Ooguri, Spodyneiko, and Vafa proposed a so-called swampland conjecture, which has been the subject of much debate and dispute, claiming that in string theory we have

$$|\nabla V| \geq cV, \qquad (27.34)$$

where V is the scalar potential and c is some constant. There are examples of subsets of string theory constructions that obey this but, if true, this would exclude the possibility of slow-roll inflation in string theory, including via the previously defined KKLT and KKLMMT constructions.

The part relevant to our discussion refers to certain "no-go theorems," generalizing the above by inclusion of some string theory effects.

27.4.1 First: scattered (sporadic) results

In one application, one considers an effective potential for the string moduli ρ that includes the overall volume (of the KK space) modulus and τ that includes the string dilaton (giving the coupling of string theory through its VEV) of the type

$$V_{\text{eff}} = \frac{A_{\mathcal{R}}}{\tau^2 \rho} + \frac{A_{\mathcal{O}}}{\tau^3 \rho^{(6-q)/2}} + \sum_i \frac{A}{\tau^{\alpha_i} \rho^{\beta_i}} \equiv V_{\mathcal{R}} + V_{\mathcal{O}} + \sum_i V_i. \qquad (27.35)$$

Here the first term is a curvature term, the second comes from orientifold O_q-planes, and the third from NS–NS fluxes, R–R fluxes, and D-branes.

For an arbitrary parameter (coefficient) a, and defining each term in V_{eff} as V_i as discussed previously, we define the derivative operators $\mathcal{D}(a)$ and their eigenvalues $C_i(a)$ on the terms in V_{eff} as

$$\mathcal{D}(a) = -a\tau \partial_\tau - \rho \partial_\rho,$$
$$C_i(a) = \frac{\mathcal{D}(a)V_i}{V_i}. \qquad (27.36)$$

Whenever

$$C_i^{(-)} \leq \min_j C_j^{(+)}, \ \forall i, \qquad (27.37)$$

where $C_i^{(-)}$ is the constant $C_i(a)$ (eigenvalue of the derivative operator $\mathcal{D}(a)$) corresponding to a negative contribution to the potential, and $C_i^{(+)}$ the same for a positive contribution to the potential, we have a no-go theorem.

The previously discussed condition leads to a set of inequalities for a, and then for a *given* a, we find that there can be no de Sitter minimum of the potential, and moreover there is a bound on $|\nabla V|/V$, with a certain constant c corresponding to it.

Acting with $\mathcal{D}(a)$ on the terms in the effective potential in (27.35), we find

$$\mathcal{D}(a)V_i = (a\alpha_i + \beta_i)V_i \equiv C_i V_i,$$
$$\mathcal{D}(a)V_{\mathcal{R}} = (2a + 1)V_{\mathcal{R}} \equiv C_{\mathcal{R}} V_{\mathcal{R}},$$
$$\mathcal{D}(a)V_{O_q} = \left(3a + \frac{6-q}{2}\right) V_{O_q} \equiv C_{O_q} V_{O_q}. \qquad (27.38)$$

We then want to choose an a such that, on the total V_{eff},

$$\mathcal{D}V_{\text{eff}} = kV_{\text{eff}} + \text{positive}, \qquad (27.39)$$

where $k > 0$ is a constant. Splitting as suggested before all the C_i's (including $C_{\mathcal{R}}$ and C_{O_q}) into positive and negative contributions to the potential, in order to satisfy (27.39), we need the condition (27.37) that, for a system of F_r RR fluxes, O_q orientifold planes, and Dp-branes, leads to the conditions

$$2a + 3 > 0, \ \ 4a + (r - 3) > 0, \ \ 3a - \frac{6-p}{2} > 0,$$
$$2a - q \leq 0, \ \ a + (r - 3) - \frac{6-q}{2} \geq 0, \ \ p - q \leq 0. \qquad (27.40)$$

Moreover, if the internal manifold is

- positively curved, we also need $2a + 1 > 0$, $a + \frac{4-q}{2} \leq 0$.
- negatively curved, we also need $2a + (r - 4) \geq 0$, $a + \frac{4-p}{2} \geq 0$.

Since for each a there is a no-go theorem, we find all possible (p, q, r), with $p = q$ (since orientifold planes must be compensated by D-branes of the same dimensionality) for which one can find an a that satisfies all the inequalities.

In each such case, there is a specific constant c, called c_*, such that

$$\frac{|\nabla V|^2}{V^2} \geq c_*^2. \tag{27.41}$$

So, as we see, these "no-go theorems" are actually only for sporadic string theory cases (corresponding to various systems, contributing to V_{eff}).

Among possible loopholes to even those discussed earlier, are the compactification on 10-dimensional supergravity on a *negatively curved* six-dimensional manifold. But also, of course, the examples of contributions to V_{eff} we took into account are just known ones, there are no guarantees that there are no others for which something like this is not possible.

27.4.2 Second: more general no-go theorem

This is a true generalization of the Maldacena–Núñez no-go theorem. But this *also implies no string (higher derivative to gravity) corrections, no singularities, and also assumes the strong energy condition (SEC),*

$$T_{00} + \frac{1}{D-2} T^M{}_M \geq 0. \tag{27.42}$$

Take the ansatz (note that, corresponding to the MN no-go theorem, here the Ω multiplies only the d-dimensional metric, and also that it, and g_{mn}, can also depend on time)

$$ds^2 = \Omega(t, y)^2 \left(-dt^2 + a(t)^2 d\vec{x}_{d-1}^2\right) + g_{mn}(y, t) dy^m dy^n. \tag{27.43}$$

Using the Einstein equation for the 00 component, and the SEC condition, we have

$$0 \leq T_{00} + \frac{1}{D-2} T^M{}_M = R_{00}, \tag{27.44}$$

and we calculate R_{00} on the ansatz.

Then, multiplying the result with Ω^{d-2} (note that the power is $d - 2$ and not $D - 2$, since the ansatz now has no Ω multiplying the compact space metric) and integrating over the compact space, and then using the d-dimensional FLRW equation

$$\frac{\ddot{a}}{a} = \left(\frac{\dot{a}}{a}\right)^2 \equiv H^2, \tag{27.45}$$

after some algebra, one finds the bound

$$(d-1)H^2 \leq -\int d^{D-d}y \Omega^{d-2} \sqrt{\det g} \frac{\ddot{\Omega}}{\Omega}. \tag{27.46}$$

But now we must relate this bound to the potential V. Consider the overall volume modulus τ; then, $\ddot{\tau}$ is the only contribution to the right-hand side of Eq. (27.46).

Separating the overall volume modulus τ in the metric ansatz, we write

$$ds^2 = e^{-2\tau\sqrt{\frac{D-d}{(D-2)(d-2)}}}\tilde{\Omega}(y,t)\left(-dt^2 + a(t)^2 d\vec{x}_{d-1}^2\right) + e^{2\tau\sqrt{\frac{d-2}{(D-d)(d-2)}}}\tilde{g}_{mn}(y,\Phi)dy^m dy^n,$$
(27.47)

where tilde indicates that we took out τ and Φ are all other fields.

The effective action from this ansatz becomes

$$S_{\text{eff}} = \int d^d x \sqrt{\det g_d}\left(\frac{R_d}{2\kappa_{N,d}^2} + \frac{\dot{\tau}^2}{2\kappa_{N,d}^2} - V(\tau,\Phi) + \text{other fields}\right).$$
(27.48)

At the point where $\dot{\tau} = 0, \dot{\Phi} = 0$, the d-dimensional Einstein equation (Friedmann equation, on the FLRW ansatz) and equation of motion for τ become

$$H^2 = \frac{2\kappa_{N,d}^2}{(d-1)(d-2)}V(\tau,\Phi),$$
$$\ddot{\tau} + \kappa_{N,d}^2 \partial_\tau V(\tau,\Phi) = 0,$$
(27.49)

and combining these with the bound (27.46), one finds the bound

$$-\frac{\partial_\tau V(\tau,\Phi)}{V(\tau,\Phi)} \geq 2\sqrt{\frac{D-2}{(D-d)(d-2)}},$$
(27.50)

so of the type of the conjecture, but with a very specific c_*.

But there are many loopholes to the previously discussed no-go theorem: first, *the SEC is actually violated in string theory*. Second, we already said that the usual assumptions of no higher derivative corrections to gravity (since we used the Einstein equation) and no singularities are actually broken in string theory.

For the SEC, we can replace it with the null energy condition (NEC), which is not violated in string theory, but only at the expense of some restriction on the geometry of the internal manifold.

Finally, we should say that, in the end, a counterexample was actually found to the original swampland conjecture of $|\nabla V|/V \geq c$.

That led Ooguri, Pati, Shiu, and Vafa to claim a modified conjecture that *either*

$$|\nabla V| \geq \frac{c}{M_{\text{Pl}}}V$$
(27.51)

OR

$$\min(\nabla_i \nabla_j V) \leq -\frac{c'}{M_{\text{Pl}}^2}V,$$
(27.52)

where both c and c' are $\sim \mathcal{O}(1)$, and the minimum of the Hessian matrix is an orthonormal frame.

In any case, we showed here the original swampland conjecture, since it was based on some concrete no-go theorems following the logic of Maldacena and Núñez.

Important concepts to remember

- The Maldacena–Núñez supergravity no-go theorem says that D-dimensional supergravity has no d-dimensional de Sitter (or Minkowski) solutions, if gravity has no higher-derivative (higher curvature) corrections, the potential is nonpositive (but massive IIA supergravity is included as a special case, treated separately), the massless fields have positive kinetic terms, and $\kappa_{N,d}$ is finite.
- The ansatz is warped, with warp factor $\Omega(y)$, and the theorem is based on an equation, leading to an integral inequality, on $\Omega(y)$.
- If we assume that Ω is only bounded from above, but $\Omega = 0$ singularities are allowed, then also RS-type compactifications, where the warp factor goes to zero at a singularity = boundary of the internal space, are not allowed. RS models have a warp factor in five dimensions, decreasing toward the physical brane.
- In massive type IIA supergravity, there are also no de Sitter or Minkowski solutions, nor RS compactifications.
- Loopholes include higher-order (higher-derivative) corrections and singularities (both realized in string theory, for instance, by orientifold planes).
- In $\mathcal{N} = 2$, five-dimensional gauged supergravity, with very special geometry for the scalars, there is no RS-type solution, with a warp factor interpolating between two fixed points, and decreasing toward the physical one. This is true for both supersymmetric and non-supersymmetric solutions (flows).
- The (original) swampland conjecture is that $|\nabla V|/V \geq c$, with c some constant, so in particular no de Sitter solutions, in string theory.
- We have some sporadic results, based on the form of the effective potential in string theory for the scalars ρ (including the overall volume) and τ (including the dilaton).
- We also have a more general no-go theorem, (which needs SEC which can be violated in string theory), again based on the now *time* evolution of the warp factor $\Omega(y, t)$, or NEC (satisfied in string theory), but with some restrictions on the geometry of the internal manifold.
- There are counterexamples to the original swampland conjecture; it was later modified.

References and further reading

The original Maldacena–Núñez no-go theorem was found in [98]. The Kallosh–Linde no-go for Randall–Sundrum was found in [99]. The swampland conjecture was proposed in [100], and the modified one in [101].

Exercises

(1) Show that on the warped ansatz for the MN no-go theorem we get

$$R_{\mu\nu} = R_{\mu\nu}(\eta) - \eta_{\mu\nu}\left[\hat{\nabla}^2 \log \Omega + (D-2)(\hat{\nabla} \log \Omega)^2\right]$$

$$= T_{\mu\nu} - \frac{1}{D-2}\Omega^2 \eta_{\mu\nu} T^L{}_L. \tag{27.53}$$

(2) Derive the equations of motion (27.30) from the energy functional (27.22).

(3) Show that the SEC bound (27.46), together with the equations of motion (27.49), give the bound

$$-\frac{\partial_\tau V(\tau, \Phi)}{V(\tau, \Phi)} \geq 2\sqrt{\frac{D-2}{(D-d)(d-2)}}. \tag{27.54}$$

(4) What happens if we allow singular Ω, going to zero, for the SEC bound (27.46)? Can it be salvaged by the introduction of the boundary term, as in the MN case?

28 Witten's positive energy theorem in general relativity and connection with supergravity

In this chapter, we will consider Witten's proof of the positive energy theorem in general relativity, namely the fact that the energy of an asymptotically flat spacetime is positive or zero. We will then make a connection with supergravity, following the work of Chris Hull.

28.1 Setup

Witten's proof is based on a simple idea, using spinors in the curved spacetime, and as such it is related to supergravity.

A simple, yet naive, argument for the connection with supergravity goes like the usual positive energy for a supersymmetric vacuum argument: consider the susy algebra, which generically is

$$\{Q_\alpha, Q_\beta\} = \hbar(C\gamma^\mu)_{\alpha\beta}P_\mu + ..., \tag{28.1}$$

where this is all for $\mathcal{N} = 1$ supergravity, and possible extra terms appear for $\mathcal{N} > 1$. Note that $\hbar$ appears due to dimensional reasons.

Then we can roughly say that

$$H = \frac{1}{\hbar}Q_\alpha^2, \tag{28.2}$$

so since $Q_\alpha^2 \geq 0$, it follows that $H \geq 0$.

Moreover, in the classical limit, $\hbar \to 0$, the spinors will vanish (there is no spinor background, for classical field theory), yet we still obtain $H \geq 0$, which means that we can use the supergravity argument, even if we are interested in a theory without supersymmetry.

Nevertheless, there are subtleties, so we need something other than this simple argument. But we will still use spinors, *in the classical limit*, to prove it.

One important assumption for the positive energy theorem is the *Dominant Energy Condition* (*DEC*), which is written as

$$T_{\mu\nu}U^\mu V^\nu \geq 0, \tag{28.3}$$

where U^μ, V^ν are non-spacelike (so null or timelike) vectors. This means that $T_{00} \geq 0$ in all Lorentz frames, or that

$$T_{00} \geq ||T_{0i}||. \tag{28.4}$$

Boundary conditions:

We still need to consider boundary conditions. We consider a spacelike foliation of space-time, with $g_{00} = -1$ and $g_{0i} = 0$, so with the time direction defined to be orthonormal to the spatial hypersurface foliation, on which coordinates are x^i, out of the total of x^μ.

Moreover, we need the spacetime to be asymptotically flat (otherwise there is no globally defined notion of energy), and moreover that at spacelike infinity on the hypersurface, $r = \sqrt{x^i x^i} \to \infty$, we have

$$g_{\mu\nu} = \eta_{\mu\nu} + \mathcal{O}\left(\frac{1}{r}\right)$$

$$\frac{\partial}{\partial x^k} g_{\mu\nu} = \mathcal{O}\left(\frac{1}{r^2}\right), \tag{28.5}$$

so the deviation from the flat metric are of order $1/r$, and in spatial derivatives of order $1/r^2$.

With these boundary conditions, we can define the ADM energy

$$E = \frac{1}{16\pi G_N} \int d^2 S^j \left(\partial_k g_{jk} - \partial_j g_{kk}\right). \tag{28.6}$$

28.2 Sketch of the proof

Before we give the proof, it is worth understanding the broad strokes, so that we don't lose sight of the big picture while we are preoccupied with details.

We will consider the "Witten condition," which is a particular type of Dirac equation,

$$\slashed{D}\epsilon = 0: \quad \gamma^i D_i \epsilon = 0, \tag{28.7}$$

where $i = 1, 2, 3$ (spatial indices), but D_i are four-dimensional covariant derivatives, constructed with x^μ.

We will find that the object S, defined as

$$S = \int d^2 S^k \epsilon^* D_k \epsilon = \int d^3 x \partial^k (\epsilon^* D_k \epsilon)$$

$$= \int d^3 x (D_i \epsilon^*) D_i \epsilon + \int d^3 x \left(T_{00} + \sum_i T_{0i} \gamma^0 \gamma^i\right) \epsilon \geq 0 \tag{28.8}$$

is positive due to being the sum of a square, and a second term positive by the DEC, $S \geq 0$.

Then, we will find that $\epsilon \to \epsilon_0$ at $r \to \infty$. Finally, we will express S in terms of this constant (and arbitrary!) ϵ_0 as

$$S = 4\pi G_N \left(\epsilon_0^* \epsilon_0 E + \sum_k \epsilon_0^* \gamma^0 \gamma^k \epsilon_0 P_k\right), \tag{28.9}$$

from which we find that $E \geq |P|$, which is the positive energy condition we are looking for.

Moreover, we will find that the equality only holds for $R_{\alpha\beta\gamma\delta} = 0$, which is flat space.

Finally, we will go back to the expression of S and find a better connection for it to supergravity, following Hull.

Example 1   To understand the meaning of the positive energy theorem, consider the case of a Schwarzschild metric,

$$ds^2 = -dt^2 \left(1 - \frac{2MG_N}{r} \right) + \frac{dr^2}{1 - \frac{2MG_N}{r}} + r^2 d\Omega_2^2. \tag{28.10}$$

The positive energy theorem says that $M \geq 0$, which means that the solution with $M \geq 0$ is fine.

On the other hand, for $M < 0$, there is an asymptotically flat spacelike surface that passes through the singular point $r = 0$, so that is not good.

28.3 Proof

We will divide the proof in four steps.

28.3.1 Step 1

We first show that there is no ϵ that is a solution to the Witten condition, which also goes to 0 at infinity, $\epsilon \to 0$ for $r \to \infty$.

We assume that such an ϵ exists, and we will find a contradiction. If we consider the Witten condition, and act again with $D\!\!\!/$, we obtain its integrability condition,

$$D\!\!\!/\epsilon = 0 \Rightarrow 0 = D\!\!\!/^2\epsilon = D_i D_i \epsilon + \frac{1}{4}[\gamma_i, \gamma_j][D_i, D_j]\epsilon. \tag{28.11}$$

But

$$[D_i, D_j]\epsilon = \frac{1}{8} R_{ij\alpha\beta}[\gamma^\alpha, \gamma^\beta]\epsilon, \tag{28.12}$$

which is found from specializing to ij, the general formula for anticommuting covariant derivatives, and

$$[\gamma^\mu, \gamma^\nu][\gamma^\alpha, \gamma^\beta] = 4\epsilon^{\mu\nu\alpha\beta}\gamma_5 - 4(g^{\mu\alpha}g^{\nu\beta} - g^{\mu\beta}g^{\nu\alpha})$$
$$+ 2\left(-g^{\mu\alpha}[\gamma^\nu, \gamma^\beta] + g^{\mu\beta}[\gamma^\nu, \gamma^\alpha] + g^{\nu\alpha}[\gamma^\mu, \gamma^\beta] - g^{\nu\beta}[\gamma^\mu, \gamma^\alpha]\right), \tag{28.13}$$

which is found by considering the three possible Lorentz structures consistent with the symmetries of the left-hand side and fixing the coefficients by various traces (multiplied with various structures).

Moreover, we have $\epsilon^{\mu\nu\alpha\beta}R_{\mu\nu\alpha\beta} = 0$, and, after using some of the symmetries of the Riemann tensor, the last term above gives $-\frac{1}{2}R^k{}_{0kj}\gamma^0\gamma^j\epsilon$, so we get the equation

$$0 = D^i D_i \epsilon - \frac{1}{4}R^{ij}{}_{ij}\epsilon - \frac{1}{2}R^k{}_{0kj}\gamma^0\gamma^k\epsilon. \tag{28.14}$$

But, using $g_{00} = -1$ and $g_{0i} = 0$, we find

$$R_{00} = R^i{}_{0i0}, \quad R = R^{ij}{}_{ij} + 2R^{0i}{}_{0i}. \tag{28.15}$$

Moreover, using also the Einstein equations (and the fact that the $R^{0i}{}_{0i}$ terms cancel between the two terms in the Einstein tensor), we find

$$R_{00} - \frac{1}{2} g_{00} R = \frac{1}{2} R^{ij}{}_{ij}$$
$$= 8\pi G_N T_{00}. \tag{28.16}$$

Further, also using $g_{00} = -1$ and $g_{0i} = 0$ and the Einstein equations, we find

$$R_{0j} = R^k{}_{0kj}$$
$$= 8\pi G_N T_{0j}, \tag{28.17}$$

Substituting this in the integrability condition, in the form (28.14), we find

$$D^i D_i \epsilon - 4\pi G_N \left(T_{00} + T_{0j} \gamma^0 \gamma^j \right) \epsilon = 0. \tag{28.18}$$

But, multiplying by ϵ^*, integrating with $\int d^3 x \sqrt{g}$ and integrating by parts, and noting that there is no boundary term since the solution of the Dirac equation on an asymptotically flat surface that vanishes at infinity does so at least as $1/r^2$ (claim by Witten, which is not very obvious), so $\int_{\Sigma_\infty} d^2 \Sigma^k \epsilon^* D_k \epsilon \to 0$, we obtain

$$\int d^3 x \sqrt{g} (D_i \epsilon^*)(D_i \epsilon) + 4\pi G_N \int d^3 x \sqrt{g} \epsilon^* \left[T_{00} + T_{0j} \gamma^0 \gamma^j \right] \epsilon = 0. \tag{28.19}$$

But $T_{00} \geq |T_{0j}|$ by the DEC, so the eigenvalues of the object in the square brackets are positive, hence so is the second term, while the first is a perfect square. This means that the only possibility for equality is if

$$D_i \epsilon = 0. \tag{28.20}$$

But then, if $\epsilon \neq 0$, it does not vanish at infinity, contrary to what we assumed. So there are no solutions to the Witten condition $\not{D}\epsilon = 0$ with $\epsilon \to 0$ at $r \to \infty$, contrary to the initial assumption. *q.e.d Step 1.*

28.3.2 Step 2

We now prove that there is an ϵ, solution to the Witten condition $\not{D}\epsilon = 0$, that goes to an (arbitrary) constant, $\epsilon \to \epsilon_0 + \mathcal{O}(1/r)$, at $r \to \infty$.

We write ϵ as a sum of a trial function ϵ_1 that satisfies the condition, $\epsilon_1 = \epsilon_0 + \mathcal{O}(1/r)$, and the rest is called ϵ_2,

$$\epsilon = \epsilon_1 + \epsilon_2. \tag{28.21}$$

We want to show that $\epsilon_2 \to \mathcal{O}(1/r)$, by writing ϵ_2 in terms of ϵ_1 via a Green's function. Moreover, we, in fact, can choose an ϵ_1 such that

$$\not{D}\epsilon_1 = \mathcal{O}\left(\frac{1}{r^3}\right). \tag{28.22}$$

This part of the proof will be skipped, so we just assume that it can be done. Then

$$\not{D}\epsilon_2 = -\not{D}\epsilon_1, \tag{28.23}$$

where $\epsilon_2 \to 0$.

We view the above equation as a Dirac equation with a given source (given by the trial function ϵ_1) $\not{D}\epsilon_2 = f$, whose solution is given in terms of the Green's function $S(x,y)$ for $\not{D}$ with boundary condition that $S \to 0$ as $x, y \to \infty$. The solution is thus

$$\epsilon_2(x) = -\int d^3y S(x,y)\not{D}\epsilon_1(y). \tag{28.24}$$

But then, we find that $S(x,y) \to 1/y^2$ as $y \to \infty$ (for the Dirac Green's function), while we said that we could choose $\not{D}\epsilon_1(y) \sim 1/y^3$, which means that the integral converges, so indeed $\epsilon_2(x)$ is finite.

Moreover, as $x \to \infty$, one has

$$S(x,y) \to \frac{1}{4\pi r^2}\gamma \cdot \hat{x} + \mathcal{O}\left(\frac{1}{r^3}\right), \tag{28.25}$$

so one finds

$$\epsilon_2(x) = -\frac{1}{4\pi r^2}\int dy\gamma \cdot \hat{x}\not{D}\epsilon_1(y) + \mathcal{O}\left(\frac{1}{r^3}\right). \tag{28.26}$$

q.e.d. Step 2.

28.3.3 Step 3

Now, consider this $\epsilon \to \epsilon_0 + \mathcal{O}(1/r)$, we repeat the calculation at step 1. There, we wanted to prove that there was no $\epsilon \to 0$, but here we just want to calculate what happens with the correct $\epsilon \to \epsilon_0$.

We obtain, as before,

$$D_iD_i\epsilon - 4\pi G_N\left(T_{00} + T_{0j}\gamma^0\gamma^j\right)\epsilon = 0. \tag{28.27}$$

Multiplying with ϵ^* and integrating over $\int d^3x\sqrt{g}$, and then integrating by parts as before, now we have a surface term, so we obtain

$$S \equiv \int_{\Sigma=\partial V} d^2 S^k\epsilon^* D_k\epsilon = \int_V d^3x\partial_k(\sqrt{g}\epsilon^* D_k\epsilon)$$
$$= \int_V d^3x\sqrt{g}(D_i\epsilon^*)(D_i\epsilon) + \int_V d^3x\sqrt{g}\epsilon^*\left(T_{00} + T_{0j}\gamma^0\gamma^j\right)\epsilon \geq 0, \tag{28.28}$$

where again the right-hand side is the sum of a perfect square term and a term that is positive by the DEC.

Now, however, the left-hand side is S, a quantity that we need to calculate. Since it is an invariant of the three-surface, and by the ADM theorem, the only possible invariants are E and P^k, it follows that S must be a linear combination of them. But, in order to prove the positive energy theorem, we need to calculate *which* combination.

Considering the asymptotically flat expansion of the metric and the vielbein,

$$g_{\mu\nu} = \eta_{\mu\nu} + h_{\mu\nu}, \quad e^i_\mu = \delta^i_\mu + \frac{1}{2}h^i_\mu, \tag{28.29}$$

we find for the corresponding spin connection matrix ($\Gamma_\mu \equiv \omega_\mu^{ab} \frac{1}{4} \gamma_{ab}$)

$$\Gamma_\mu = \frac{1}{16} (\partial_\beta h_{\alpha\mu} - \partial_\alpha h_{\beta\mu})[\gamma^\alpha, \gamma^\beta]. \tag{28.30}$$

Step 3a

First, consider the simple case of a single object at rest, which means the metric must be asymptotically Schwarzschild, with

$$h_{ij} = \frac{2G_N M}{r} \delta_{ij}, \quad h_{0i} = 0, \quad h_{00} = \frac{2G_N M}{r}. \tag{28.31}$$

Then we find

$$\Gamma_i = -\frac{1}{4} \frac{G_N M}{r^3} [\gamma_i, \vec{\gamma} \cdot \vec{x}] + \mathcal{O}\left(\frac{1}{r^3}\right) \tag{28.32}$$

and

$$\slashed{D} = \slashed{\partial} + \gamma^i \Gamma_i = \slashed{\partial} - \frac{G_N M}{r^3} \vec{\gamma} \cdot \vec{x}. \tag{28.33}$$

Moreover, we find that the condition $\slashed{D}\epsilon = 0$ with the boundary condition $\epsilon \to \epsilon_0 + \mathcal{O}(1/r)$ is solved by

$$\epsilon = \left(1 - \frac{G_N M}{r}\right) \epsilon_0 + \mathcal{O}\left(\frac{1}{r^2}\right), \tag{28.34}$$

in terms of the arbitrary ϵ_0 (it can also be derived from a conformal transformation of flat space, as Schwarzschild is conformally flat), which means that we obtain

$$S = \int_{\Sigma^\infty} dS^k \epsilon^* D_k \epsilon = \int_{\Sigma^\infty} d\Omega r^2 \epsilon^* D_r \epsilon$$
$$= \int_{\Sigma^\infty} d\Omega r^2 \epsilon_0^* \partial_r \left(1 - \frac{2G_N M}{r}\right) \epsilon_0 = 4\pi G_N M \epsilon_0^* \epsilon_0 \geq 0, \tag{28.35}$$

so from $S \geq 0$ and $\epsilon_0^* \epsilon_0 \geq 0$, we find $M \geq 0$.

Step 3b (General case)

Now, for the general case, with

$$g_{\mu\nu} = \eta_{\mu\nu} + h_{\mu\nu} = \eta_{\mu\nu} + \mathcal{O}\left(\frac{1}{r}\right), \quad \partial_\alpha h_{\mu\nu} \sim \mathcal{O}\left(\frac{1}{r^2}\right) \tag{28.36}$$

for $r \to \infty$, we calculate

$$S = \int_{\Sigma^\infty} d\Omega r^2 \epsilon^* D_r \epsilon. \tag{28.37}$$

But, as we see, more precisely we only need to calculate the $\mathcal{O}(1/r^2)$ term in $D_r \epsilon$, as it is the only one that contributes. In that case, we find

$$D_r \epsilon = \Gamma_r \epsilon_0 - \frac{1}{r^2} \tilde{\epsilon}(\theta, \phi), \tag{28.38}$$

where $\Gamma_r \sim \mathcal{O}(1/r^2)$, so

$$S = \int_{\Sigma^\infty} d\Omega r^2 \epsilon_0^* \Gamma_r \epsilon_0 - \int_{\Sigma^\infty} d\Omega \epsilon_0^* \tilde{\epsilon}(\theta, \phi), \tag{28.39}$$

but, after some algebra,

$$\int_{\Sigma^\infty} d\Omega \epsilon_0^* \tilde{\epsilon}(\theta, \phi) = \int_{\Sigma^\infty} d\Omega r^2 \epsilon_0^* \gamma^r \gamma^i \Gamma_i \epsilon_0, \qquad (28.40)$$

and then, using that $d\Omega r^2 \Gamma_r \to dS^k \Gamma_k$ to go back to this form in the first term of S, we obtain

$$S = \int_{\Sigma^\infty} dS^k \epsilon_0^* \left(\Gamma_k - \gamma_k \gamma^i \Gamma_k \right) \epsilon_0. \qquad (28.41)$$

Then we calculate that

$$\Gamma_k - \gamma_k \gamma^i \Gamma_i = \frac{1}{4} \left(\partial_i h_{ki} - \partial_k h_{ii} \right) + \frac{1}{8} (\partial_\beta h_{ki})[\gamma_i, \gamma_\beta]$$
$$- \frac{1}{8} \partial_k h_{\beta k}[\gamma^i, \gamma^\beta] - \frac{1}{8} \partial_\beta h_{ii}[\gamma_k, \gamma_\beta] + \frac{1}{8} \partial_i h_{\beta i}[\gamma_k, \gamma_\beta], \qquad (28.42)$$

and moreover find that only $\beta = 0$ contributes to the above. Then finally, we write S as

$$S = \frac{1}{4} \epsilon_0^* \epsilon_0 \int_{\Sigma^\infty} dS^j \left(\partial_i h_{ji} - \partial_j h_{ii} \right)$$
$$+ \frac{1}{4} \epsilon_0^* \gamma^0 \gamma^k \epsilon_0 \int_{\Sigma^\infty} dS^j \left(\partial_j h_{0k} - \partial_0 h_{jk} + \delta_{jk} \partial_0 h_{ii} - \delta_{jk} \partial_i h_{0i} \right). \qquad (28.43)$$

But we have the ADM formulae for the energy and momentum,

$$E = \frac{1}{16\pi G_N} \int_{\Sigma^\infty} dS^j \left(\partial_i h_{ji} - \partial_j h_{ii} \right)$$
$$P_k = \frac{1}{16\pi G_N} \int_{\Sigma^\infty} dS^j \left(\partial_j h_{0k} - \partial_0 h_{jk} + \delta_{jk} \partial_0 h_{ii} - \delta_{jk} \partial_i h_{0i} \right), \qquad (28.44)$$

so that we can write S as

$$S = 4\pi G_N \left(\epsilon_0^* \epsilon_0 E + \sum_k \epsilon_0^* \gamma^0 \gamma^k \epsilon_0 P_k \right). \qquad (28.45)$$

We can then *choose* (the otherwise arbitrary constant spinor) ϵ_0 to be the eigenvector of $\gamma^0 \vec{\gamma} \cdot \vec{P}$, with eigenvalue $-|\vec{P}|$, which means that

$$S \geq 0 \Rightarrow E \geq |\vec{P}|, \qquad (28.46)$$

which is what we wanted to prove.

q.e.d. Step 3 and q.e.d. theorem, part I.

Before we proceed to Step 4 and part II of the theorem, we note that the above formulae can be written in a covariant form. We have

$$S = \bar{\epsilon}_0 \gamma^\mu \epsilon_0 P_\mu \geq 0. \qquad (28.47)$$

Moreover, *on the Witten condition*, we have a covariant formulation of the definition of S, as

$$S = \int_{\Sigma_\infty^2 = \partial \Sigma^3} \frac{1}{2} d\Sigma_{\mu\nu} \left(\epsilon^{\mu\nu\rho\sigma} \bar{\epsilon} \gamma^5 \gamma_\rho D_\sigma \epsilon - (D_\sigma \bar{\epsilon}) \gamma^5 \gamma_\rho \epsilon \right), \qquad (28.48)$$

where the quantity in brackets is called $E^{\mu\nu}$. Note the difference from the previous (non-covariant) definition of $\int_{\Sigma_\infty^2} dS^k \epsilon^* D_k \epsilon$.

Using the Stokes theorem, we can write

$$S = \int_{\Sigma^3} D_\nu E^{\mu\nu} d\Sigma_\mu = \ldots =$$

$$= \int_{\Sigma^3} \left(R_{\mu\nu} - \frac{1}{2} g_{\mu\nu} R \right) \bar{\epsilon} \gamma^\mu \epsilon d\Sigma^\nu + \int_{\Sigma^3} (D_\mu \epsilon) \, (\gamma^\nu \sigma^{\mu\rho} + \sigma^{\mu\rho} \gamma^\nu) \, D_\rho \epsilon d\Sigma_\nu, \tag{28.49}$$

and in the first term, the Einstein tensor in the bracket equals $8\pi G_N T_{\mu\nu}$, so the terms are ≥ 0 by the DEC $T_{\mu\nu} U^\mu U^\nu \geq 0$, while the second term gives, in a spacelike foliation with the Σ^3 transverse (orthornormal to) the time direction,

$$2|D_j \epsilon|^2 - 2 \left| \sum_{j=1}^3 \gamma^j D_j \epsilon \right|^2 . \tag{28.50}$$

28.3.4 Step 4

Finally, the last step and part II of the theorem is to show that the equality is only achieved for Minkowski space.

If $E = 0$, then $|\vec{P}| = 0$, and in turn this means that $S = 0$, $\forall \epsilon_0$. But from (28.28), we see that this is possible only if $D_k \epsilon = 0$ *on the whole (initial in time) hypersurface* $V = \Sigma^3$. In turn, this means that (integrability condition)

$$[D_i, D_j]\epsilon = 0 \Rightarrow R_{ij\rho\sigma}[\gamma^\rho, \gamma^\sigma]\epsilon = 0, \quad \forall \epsilon_0. \tag{28.51}$$

But considering a few ϵ_0's, forming a basis, we find that

$$R_{ij\rho\sigma} = 0 \tag{28.52}$$

on the hypersurface Σ^3. But deforming this arbitrary hypersurface Σ^3 locally, we find that in fact

$$R_{\mu\nu\rho\sigma} = 0 \tag{28.53}$$

everywhere, which means that we are in Minkowski space, with $T_{\mu\nu} = 0$.

q.e.d. Step 4 and theorem.

28.4 Connection to supergravity

Finally, consider the connection to supergravity, in the formulation due to Hull, which was the reason to describe this theorem in a supergravity book.

As we said, this idea is to embed the theory into supergravity, and then obtain back the general relativity case by taking the classical limit $\hbar \to 0$. We will obtain the positive energy theorem from the variation of the supercharges under global supersymmetry. We will first describe the formalism.

Consider a fluctuation around a background $g_{\mu\nu}^{(b)}$ (flat or anti-de Sitter),

$$g_{\mu\nu} = g_{\mu\nu}^{(b)} + h_{\mu\nu}. \tag{28.54}$$

Moreover, consider the Killing spinors η^I, as generators of *local* supersymmetries, through the variation on the gravitino,

$$\delta_Q(\eta)\psi_\mu = (\hat{D}_\mu)\eta, \qquad (28.55)$$

where $\hat{D}_\mu$ is some differential operator: in the case of $\mathcal{N} = 1$ supergravity, this is just D_μ, but otherwise it can contain extra terms, usually of the type $F_{(n)}\Gamma^{(n)}$ (n-form field strengths, times some gamma functions).

Then, if the bosonic background has η as a Killing spinor, we have

$$\hat{D}_\mu^{(b)}\eta = 0, \qquad (28.56)$$

where $\hat{D}_\mu^{(b)}$ refers to the operator on the background solution.

The integrability condition for the Killing spinor is

$$\hat{D}_{[\mu}^{(b)}\hat{D}_{\nu]}^{(b)}\eta = 0. \qquad (28.57)$$

The gravitino field equation is

$$R^\mu = \epsilon^{\mu\nu\rho\sigma}\gamma^5\gamma_\nu\hat{D}_\rho^{(b)}\psi_\sigma, \qquad (28.58)$$

where $\gamma_\nu = e_\nu^{(b)a}\gamma_a$, contracted with the background vielbein and therefore we find

$$\hat{D}_\mu^{(b)}R^\mu = 0. \qquad (28.59)$$

Multiplying R^μ with the (commuting) Killing spinors η^I and the determinant of the background vielbein $e^{(b)}$, we find

$$\begin{aligned}
e^{(b)}\bar{\eta}^I R^\mu &= e^{(b)}\bar{\eta}^I \epsilon^{\mu\nu\rho\sigma}\gamma^5\gamma_\nu\hat{D}_\rho^{(b)}\psi_\sigma \\
&= e^{(b)}D_\rho^{(b)}(\bar{\eta}^I \epsilon^{\mu\nu\rho\sigma}\gamma_5\gamma_\nu\psi_\sigma) \\
&= \partial_\rho\left(e^{(b)}\bar{\eta}^I \epsilon^{\mu\nu\rho\sigma}\gamma_5\gamma_\nu\psi_\sigma\right),
\end{aligned} \qquad (28.60)$$

which implies (because of the symmetry of the partial derivatives) that

$$\partial_\mu(e^{(b)}\bar{\eta}^I R^\mu) = 0, \qquad (28.61)$$

which is a conservation equation for a *local* supersymmetry current (one can also find it via a generalized Noether procedure, where $(T_a)^i{}_j\psi^j$ is now replaced by $\hat{D}_\mu^{(b)}\eta$, as for a *local* YM current, with $\delta A_\mu = D_\mu\epsilon$, but note that it is a *fermionic* current, as needed), hence we find the *local supercharges* associated with the Killing spinors η^I as

$$Q^I = \int_{\Sigma^3} d^3 x e^{(b)}\bar{\eta}^I R^0 = \int_{\Sigma^3} d\Sigma_\mu\bar{\eta}^I \epsilon^{\mu\nu\rho\sigma}\gamma_\nu^{(b)}D_\rho^{(b)}\psi_\sigma, \qquad (28.62)$$

where the second form is a covariant form, and the first is the form in terms of a time direction (the 0 component of the current integrated over space) orthonormal to Σ^3.

Symmetry algebra: global case

Consider metrics that are either asymptotically flat (AF) or asymptotically anti-de Sitter (AAdS). The asymptotic symmetry generators are P^M, $M, N = 0, 1, 2, 3$, and J^{MN} in both cases.

Thus the $\mathcal{N} = 1$ global symmetry algebra in the AF case is (the Poincaré algebra and)

$$[P^M, Q^I] = 0$$
$$[J^{MN}, Q^I] = \hbar(\sigma^{MN})^I{}_J Q^J$$
$$\{Q^I, \bar{Q}^J\} = \hbar \Gamma^{IJ}_M P^M (+...), \tag{28.63}$$

where the dots are in case of $\mathcal{N} > 1$. In the AAdS case, besides the $[P^M, P^N] = J^{MN}$ turning the Poincaré into the AdS algebra, we also have an extra term in the susy algebra,

$$\{Q^I, \bar{Q}^J\} = \hbar \Gamma^{IJ}_M P^M + i\hbar(\sigma_{MN})^{IJ} J^{MN} (+...). \tag{28.64}$$

Corresponding to each of the Killing vectors k^A of the metric, there is a *global* symmetry generator $K^{(g)}_A$, standing for the generators above, so

$$K^{(g)}(k^M) = P^M, \quad K^{(g)}(k^{MN}) = J^{MN}, \quad K^{(g)}(\eta^I) = Q^{(g)}_I, \quad K^{(g)}(\bar{\eta}^I) = \bar{Q}^{(g)}_I. \tag{28.65}$$

They satisfy the (global symmetry =super)algebra, written together as

$$[K^{(g)}_A, K^{(g)}_B] = f_{AB}{}^C K^{(g)}_C. \tag{28.66}$$

Symmetry algebra: local case (supergravity)
In the local case, as we have described before, the algebra is an algebra of transformations, and it depends on the model. For the $\mathcal{N} = 1$ case, it is

$$[\delta_Q(\epsilon_1), \delta_Q(\epsilon_2)] = \delta_{\text{g.c.}}(\xi^\mu) + \delta_Q(-\xi^\mu \psi_\mu) + \delta_{1.L.}(\xi^\mu \omega^{ab}_\mu), \tag{28.67}$$

where

$$\xi^\mu = \frac{1}{2}\bar{\epsilon}_2 \gamma^\mu \epsilon_1. \tag{28.68}$$

Further, the Killing vectors are written as quadratic forms of the Killing spinors. In the AF case, we have

$$\bar{\eta}_I \gamma^\mu \eta_J = f_{IJ}{}^A k^\mu_A = \gamma^M_{IJ} k^\mu_M, \tag{28.69}$$

while in the AAdS case, we have

$$\bar{\eta}_I \gamma^\mu \eta_J = f_{IJ}{}^A k^\mu_A = \gamma^M_{IJ} k^\mu_M + i(\sigma^{MN})_{IJ} k^\mu_{MN}. \tag{28.70}$$

The (local) supercharges are the Q^I previously calculated,

$$Q^I = \int_{\Sigma^3} \bar{\eta}^I R^\mu d\Sigma_\mu. \tag{28.71}$$

Now the *global* supersymmetries are defined by the variation with respect to the Killing spinors (*constant* susy parameters, $\epsilon^I(x) \to \eta^I(x)$),

$$Q^{(g)}_I = \delta_{\bar{Q}}(\bar{\eta}_I), \quad \bar{Q}^{(g)}_I = \delta_Q(\eta_I), \tag{28.72}$$

and, as before, they satisfy the global susy algebra,

$$\{Q^{(q)}_I, \bar{Q}^{(g)}_J\} = f^M{}_{IJ} K^{(g)}_M. \tag{28.73}$$

We can define the transformation of the *local* charges K_B under *global* symmetries defined by $K_B^{(g)}$ and, not surprisingly, the charges are just "rotated" into each other by the global symmetry,

$$[K_A^{(g)}, K_B] = f_{AB}{}^C K_C. \tag{28.74}$$

In the particular case of supersymmetries, we obtain

$$\{Q_I^{(g)}, Q_J\} = \delta_Q(\eta^I) Q_J \equiv f_{IJ}{}^A K_A. \tag{28.75}$$

The logic of the positive energy theorem then is to express the mass (which is like K_A, specifically the zero component of P_M) as a function of the *change in the local supercharges under global supersymmetry*.

Linearized level

At the linearized level, for the AF case, we have

$$\delta_Q(\eta^I)\psi_\mu = D_\mu \eta^I + \mathcal{O}(\psi^2), \tag{28.76}$$

so, after some calculations, we find for the change of R_μ

$$\delta_Q(\eta^I) R^\mu = \dots = (G^\mu{}_a)_{\text{lin.}} \gamma^a \eta_I + \mathcal{O}(\psi^2), \tag{28.77}$$

where $(G^\mu{}_a)_{\text{lin.}}$ is the linearized Einstein tensor with one curved and one flat index, equal (by the Einstein equations) to $8\pi G_N \theta^\mu{}_a$, with $\theta^\mu{}_a$ being the linearized energy–momentum tensor.

Then, finally, the change under the global supersymmetry of the local supercharges is

$$\delta_Q(\eta^I) Q^J = \int_{\Sigma^3} \bar{\eta}^I \delta_Q(\eta^I) R^\mu d\Sigma_\nu + \mathcal{O}(\psi^2)$$

$$= \int_{\Sigma^3} \theta^{\mu\nu} \bar{\eta}^I \gamma_\mu \eta^J d\Sigma_\nu + \mathcal{O}(\psi^2)$$

$$= \gamma_M^{IJ} k_{IJ}^\mu P_\mu \equiv \gamma_M^{IJ} P_M, \tag{28.78}$$

where we have used (28.69) and where $P^\mu = \int_{\Sigma^3} \theta^{\mu\nu} d\Sigma_\nu$.

Full level

Consider the asymptotical Killing spinor (at $r \to \infty$) χ_0^I. We want to construct the change under $\delta_Q(\chi^I) = \bar{Q}_I^{(g)}$ (global susy) of $\chi_0^I Q_I$ (local supercharge).

Then, from

$$\{Q_I^{(g)}, Q_J\} = f_{IJ}{}^A K_A, \tag{28.79}$$

we obtain (after some algebra)

$$\{\delta_{\bar{Q}}(\bar{\eta}_I), \chi_0^J Q_J\} = (\bar{\chi}_0^J f_{IJ}{}^A \chi_0^J) K_A$$

$$= \int_{\Sigma^3} \epsilon^{\mu\nu\rho\sigma} \left[\bar{\chi} \gamma^5 \gamma_\rho \hat{D}_\nu \hat{D}_\sigma \chi + \overline{(\hat{D}_\nu \chi)} \gamma^5 \gamma_\rho \hat{D}_\sigma \chi \right] d\Sigma_\mu, \tag{28.80}$$

where the first term gives

$$\int_{\Sigma^3} T^{\mu\nu} (\bar{\chi} \gamma_\mu \chi) d\Sigma_\nu \tag{28.81}$$

and is ≥ 0 by the DEC, while the second is positive or zero by the Witten condition, as we have shown in (28.49):

Finally then, since K_A includes P_M, which includes $P_0 = M$, we find the bound on the mass,

$$(...)M = (\bar{\chi}_0^I f_{IJ}{}^A \chi_0^J)K_A = \int_{\Sigma^3} d^3x\, e(\hat{D}^j\chi)^\dagger (D_j\chi) - \int_{\Sigma^3} d^3x\, e(\gamma^i\hat{D}_i\chi)^\dagger(\gamma^j\hat{D}_j\chi), \quad (28.82)$$

and the first term is ≥ 0 as a sum of squares, while the second vanishes by the Witten condition.

Note that the physical interpretation of the Witten condition is that, in order to avoid the propagation of unphysical degrees of freedom (with negative energy), we need to impose the *spatial only* gamma-trace gauge condition

$$\sum_{j=1}^{3} \gamma^j \psi_j = 0, \qquad\qquad (28.83)$$

which is preserved by local supersymmetry (supergravity) if

$$\sum_{j=1}^{3} \gamma^j \hat{D}_j \epsilon = 0, \qquad\qquad (28.84)$$

that is, the Witten condition.

Finally, note that this proof of positivity of the energy was found in supergravity to follow Witten's but, as we said, and as it was explicitly shown by Hull, we can get back to classical gravity by just taking $\hbar \to 0$, and leaving only the bosonic background.

Important concepts to remember

- The positive energy theorem states that the energy of an asymptotically flat spacetime with only $1/r$ corrections to the metric and $1/r^2$ to the spatial derivatives of the metric, with matter obeying the DEC, $T_{\mu\nu}U^\mu V^\nu \geq 0$, with U^μ, V^ν non-spacelike, is positive or zero, with zero just for flat space.
- The theorem can be understood roughly as the $\hbar \to 0$ limit of the supersymmetric theorem arising from $Q_\alpha^2 = \hbar H$.
- Witten's proof is based on the existence of a spinor ϵ, solution to the "Witten condition" $\sum_{i=1}^{3} \gamma^i D_i\epsilon = 0$, with D_i four-dimensional covariant derivative with only spatial indices, such that $\epsilon \to \epsilon_0 + \mathcal{O}(1/r)$, as $r \to \infty$.
- From ϵ, we can form an object $S = \int d^2S^k \epsilon^* D_k\epsilon$, that is found to be a particular linear combination of E and P_k.
- For the connection to supergravity, in a supergravity theory, one defines the local supercharges $Q^I = \int_{\sigma^3} \bar{\eta}^I R^\mu d\Sigma_\mu$, and then their variation under *global* supersymmetry, by $\delta_{\bar{Q}}(\bar{\eta}^I)$, a rotation of the supercharge, is proportional to the mass M.
- Under the DEC, the above variation is a sum of a perfect square and a term vanishing by the Witten condition. The Witten condition preserves the gauge condition for the gravitino $\sum_{j=1}^{3} \gamma^j \psi_j = 0$, so spatial gamma-trace vanishes.
- In the $\hbar \to 0$ limit, the supergravity proof goes over to the proof for general relativity.

References and further reading

The positive energy theorem was proven in [102]. The connection with supergravity was made precise in [103].

Exercises

(1) Do the missing steps between (28.11) and (28.14).

(2) Show that

$$
\Gamma_k - \gamma_k \gamma^i \Gamma_i = \frac{1}{4} \left(\partial_i h_{ki} - \partial_k h_{ii} \right) + \frac{1}{8} (\partial_\beta h_{ki})[\gamma_i, \gamma_\beta]
$$
$$
- \frac{1}{8} \partial_k h_{\beta k}[\gamma^i, \gamma^\beta] - \frac{1}{8} \partial_\beta h_{ii}[\gamma_k, \gamma_\beta] + \frac{1}{8} \partial_i h_{\beta i}[\gamma_k, \gamma_\beta], \qquad (28.85)
$$

and moreover that only $\beta = 0$ contributes to the above.

(3) Show that the second term in (28.49) becomes (28.50).

(4) Do the missing steps to show that the two terms in (28.80) are positive.

Compactification of low-energy string theory

In this chapter, we will describe the compactification of string theory, mostly using its low-energy version, supergravity, the subject of this book, down to four dimensions. We will leave specific issues about the MSSM, and attempt to obtain it, for Chapter 30, here focusing on the mechanisms available. The few points about string theory will be stated without proof, and for the rest we will use supergravity.

29.1 Setup and conditions for compactification

As we have already described, string theory is a fundamental theory that hopefully will describe in a unified way the Standard Model together with gravity. At low energies, which means in particular energies much lower than the characteristic scale of string theory, $E \ll 1/\sqrt{\alpha'}$, string theory becomes supergravity. Here of course low energies is a relative term, that still means energies much larger than the compactification scale, which itself is usually larger than (or comparable to) the Grand Unified Theory (GUT) scale. String theory is quantum mechanically consistent only in 10 dimensions (where there are no quantum anomalies in several invariances – worldsheet conformal, Lorentz and BRST).

Also as we already mentioned, in 10 dimensions, there are 5 consistent perturbative string theories, called types IIA, IIB, I, and heterotic, the last of which itself comes in two variants. At the low energies they become, respectively, type IIA supergravity, IIB supergravity, and type I and heterotic become type I supergravity coupled to SYM. Here types II and I refer to the number of supersymmetries in 10 dimensions ($\mathcal{N} = 2$ is maximal in 10 dimensions), IIA has two supersymmetries of opposite chiralities, and IIB has two supersymmetries of the same chirality.

The great result of Michael Green and John Schwarz from 1985 that started the first superstring revolution was the complicated calculation showing that type IIA and IIB supergravities have no quantum gauge (local) anomalies (global anomalies are fine, and have physical implications, but local ones are not, they mean the theory is inconsistent), and type I coupled to SYM has no anomalies provided that the gauge group is one of three possible choices, $SO(32), SO(16), \times SO(16)$ or $E_8 \times E_8$ (technically, it is also possible to have $U(1)^{496}$ or $E_8 \times U(1)^{248}$, but these possibilities are trivial, and there are no known string theories for them). At the level of supergravity, $SO(16) \times SO(16)$ is bosonic (not supersymmetric); it actually can appear as some limit of string theories in some nonperturbative regime, but we will not discuss it here. Correspondingly, there are the type I string theory with gauge group $SO(32)$ and the heterotic string theory with gauge group $SO(32)$ or $E_8 \times E_8$.

One of the big results of the second superstring revolution from ~1995 was that this seemingly different superstring theories are in fact nonperturbatively related by superstring dualities and there is a single unifying picture, called M-theory, the -generically non-perturbative- superstring theory, for which various corners of parameter space appear as the five different string theories. But at the level of the perturbative theories, and in particular for their low-energy supergravity version that we study here, they look different.

In this chapter, we will focus on the type IIB and the heterotic $E_8 \times E_8$ supergravity theories, as they are the most appealing for phenomenology. The perturbative heterotic $E_8 \times E_8$ string was the first to be extensively studied, during the first superstring revolution. It has several appealing features: it has a gauge group already in 10 dimensions and is large enough so that it can accommodate not only the $SU(3) \times SU(2) \times U(1)$ of the Standard Model but also the common GUT groups $SU(5)$ and $SO(10)$; it is also perturbative. This means that we can use the low-energy supergravity to describe its compactification, without needing much information about string theory itself. Type IIB has become relevant in more recent years, due to the possibility of adding fluxes, which gives features useful for phenomenology.

Thus generically, we will compactify a 10-dimensional superstring theory on a compact space K_6, or $M_{10} = M_4 \times K_6$. For the resulting theory, we want to obtain $\mathcal{N} = 1$ supersymmetry in four dimensions, for phenomenological reasons that we already explained, but review here. The point is that in four-dimensional field theory it is very hard to break supersymmetry down to $\mathcal{N} = 1$, and we know that if we have supersymmetry at energies testable at accelerators (like the LHC), so as to solve the usual problems in the Standard Model (mass of the Higgs low, hierarchy of scales, gauge coupling unification), it can be at most $\mathcal{N} = 1$. The reason is that for $\mathcal{N} = 2$ and higher, there are no complex representations, in particular, fermion fields come in chiral pairs (a field of one chirality comes together with another field of opposite chirality, with the same properties as for Q and $\tilde{Q}$ in a hypermultiplet), and that contradicts experiments: in the Standard Model there are chiral fermions.

The condition of $\mathcal{N} = 1$ supersymmetry of the "vacuum" state (one with only low-energy fields) in four dimensions is $Q|\psi\rangle = 0$, where $|\psi\rangle$ is a vacuum state, which means only with low-energy fields. In terms of fields, when $Q \to \delta_Q$, we write the condition as $\delta_Q fields = 0$, but since $\delta_Q bosons \sim fermions$, and fermions have VEV = 0 (nonzero fermionic VEVs would spontaneously break local Lorentz invariance), the variation of the bosons is automatically satisfied, so we only need to satisfy $\delta_Q fermions = 0$.

In the $\mathcal{N} = 1$ heterotic supergravity, the fermionic variations are

$$\delta\psi_M = \frac{D_M\eta}{k_N} + \frac{k_N}{32g^2\phi}(\Gamma_M{}^{NPQ} - 8\delta_M^N\Gamma^{PQ})\eta H_{NPQ} + (\text{Fermi})^2,$$

$$\delta\chi^a = -\frac{1}{4g\sqrt{\phi}}\Gamma^{MN}F_{MN}^a\eta + (\text{Fermi})^2,$$

$$\delta\lambda = \frac{1}{\sqrt{2}\phi}(\Gamma \cdot \partial\phi)\eta + \frac{k_N}{8\sqrt{2}g^2\phi}\Gamma^{NPQ}\eta H_{NPQ} + (\text{Fermi})^2, \tag{29.1}$$

where ψ_M is the gravitino, χ^a is the gluino, the superpartner of the YM field (gluon) in the $\mathcal{N} = 1$ SYM multiplet, and λ is the dilatino (superpartner of the dilaton), $H = dB$ is the field strength of the antisymmetric tensor 2-form B_{MN} of string theory, the NS–NS B-field or Kalb–Ramond field (which couples to the string itself), ϕ is the dilaton, a scalar field whose VEV is related to the string coupling by $g_s = e^{\langle\phi\rangle}$, and F_{MN}^a are the YM fields. Here we didn't write the explicit form of the $(Fermi)^2$ terms since they vanish on the VEV.

We see that if we choose $H = d\phi = 0$ (the B-field is pure gauge and the dilaton is constant), the supersymmetry conditions reduce to

$$D_i\eta = 0,$$
$$\Gamma^{ij}F_{ij}\eta = 0, \tag{29.2}$$

where $i, j \in K_6$. Since we also want a *single supersymmetry in four dimensions*, it follows that we should have a single solution for η of Eq. (29.2). Note that the first equation is a Killing spinor equation. In the case of the sphere compactification, the Killing spinor equation had also a constant term (γ_μ), which came from a constant flux for the antisymmetric tensor on the sphere, but in this case there are no fluxes.

Before we can study these equations in detail, we need to learn a bit about topology.

29.2 Topology

29.2.1 Spinors and holonomies

If we have spinors in a curved space, we know that we can define parallel transport of the spinors along the curve, using $D\eta = 0$ (with the spin connection) to replace $d\eta = 0$ from flat space as the defining equation for the parallel transport. Along a closed path γ, the spinor will generically come back to a rotated version of itself,

$$\eta^\alpha \to U^\alpha{}_\beta\eta^\beta. \tag{29.3}$$

Here U is given by

$$U = P\exp\int_\gamma \omega \cdot dx, \tag{29.4}$$

where ω is the spin connection, belonging to the Lie algebra of $SO(n)$, and $P\exp$ stands for path-ordered exponential: since ω is in a nonabelian group, when we write the exponential, we have products of noncommuting objects, the ω's at various points on the curve, so their order matters. If we discretize the path $\gamma : x^\mu(t)$ to be x_i^μ, $i = 1, ..., M$, then path ordering means to order the objects in the product for increasing i, for instance,

$$P(\omega(x_2)\omega(x_5)\omega(x_3)) = \omega(x_2)\omega(x_3)\omega(x_5). \tag{29.5}$$

Then, since ω belongs to the Lie algebra of $SO(n)$, U will generically be in $SO(n)$, and is called a *holonomy*.

The group formed by all possible U's (holonomies) is called *holonomy group*, and is then a subgroup of $SO(n)$.

We saw that we want to have a single η such that $D_i\eta = 0$, that is, a covariantly constant spinor. There is a theorem that says that if there is a unique covariantly constant spinor η, then the holonomy group is $SU(n/2)$.

29.2.2 Kähler and Calabi–Yau manifolds

We can define a *complex manifold* in the following way. We say we have an *almost complex structure* if there is a matrix $J^i{}_j$ such that $J^2 = -1$ (a generalization of i for dimensions larger than two), and can be diagonalized at any point over $\mathbf{C}$. If an object called the Nijenhuis tensor (that will not be defined here) is zero when diagonalized (a statement analogous to having a zero Riemann tensor $R_{\mu\nu\rho\sigma}$ for metric $g_{\mu\nu} = \eta_{\mu\nu}$), then we say we have a *complex manifold*, and we can use coordinates z^i and $\bar{z}^{\bar{j}}$.

A complex manifold with holonomy $\subseteq U(N)$, where $N = n/2$, is called a *Kähler manifold*, in which case $J^i{}_j$ is covariantly constant. We had already seen an equivalent definition of a Kähler manifold, namely a complex manifold for which *locally* there is a function K, called the Kähler potential, such that

$$g_{i\bar{j}} = \partial_i\partial_{\bar{j}}K. \tag{29.6}$$

The spin connection ω on a Kähler manifold is a $U(N) \sim SU(N) \times U(1)$ gauge field. If the holonomy is actually $SU(N)$, it means that the $U(1)$ part is topologically trivial (it is pure gauge, with zero YM curvature), or equivalently, we say that the *first Chern class of K is zero*, $c_1(K) = 0$.

Let us explain a bit better this definition. If $dF = 0$, we can define the *cohomology* of F as the closed forms, that is, forms F satisfying $dF = 0$, modulo exact forms, that is, $F = dA$. Now if $dF = 0$, we can always *locally* write $F = dA$, that is, for A's defined on patches. The patches generically intersect, so that there are two different A's on the intersection of two patches, which obviously have to be related by a gauge transformation, known as a transition function. Therefore by defining objects satisfying $dF = 0$ modulo objects satisfying $F = dA$ (with such an equivalence class), we define a topological property. The cohomology class of $F/2\pi$ is called the first Chern class and is a topological property of the manifold. If the first Chern class is zero, which means that if $dF = 0$, then $F = dA$ globally (for the same A everywhere).

Conversely to the previously discussed result, Calabi and Yau proved that if $c_1(K) = 0$, then there is a unique Kähler metric of $SU(N)$ holonomy.

In that case, we have a covariantly constant spinor, $D_i\eta = 0$, which means that also its integrability condition must hold, $[D_i, D_j]\eta = 0$, which in turn, for the Kähler manifold, means that

$$R_{i\bar{j}} = 0, \tag{29.7}$$

or in other words that the manifold is Ricci flat. That in turn means that on this space (as long as also $\Gamma^{ij}F_{ij}\eta = 0$, the other susy condition), we have not only $\mathcal{N} = 1$ supersymmetry but also the Einstein equations on the compact space satisfied, which leaves only the Einstein equations in four dimensions to be satisfied (the usual Einstein equations of four-dimensional gravity). This means that we have a *consistent* compactification to four dimensions. The resulting space (Kähler manifold of vanishing first Chern class) is called a *Calabi–Yau manifold*.

We already encountered the notion of Kähler space and Kähler potential in a seemingly different context, the theory of $\mathcal{N} = 1$ chiral multiplets in four dimensions, but that is not unrelated. For a general compactification, thus also in particular for a Calabi–Yau compactification, there are scalars parametrizing the deformations of the manifold, which don't require energy, which are called *moduli*. These become massless scalars in four dimensions, and in particular for a compactification preserving $\mathcal{N} = 1$ in four dimensions, they will belong to chiral multiplets, therefore these massless scalars *will also live on a Kähler space, like the Calabi–Yau (CY) manifold itself, whose deformations they represent.*

An important result, due to Strominger, is that even though the compactification preserves only $\mathcal{N} = 1$, the moduli belong to $\mathcal{N} = 2$ multiplets, in particular there are $\mathcal{N} = 2$ vectors and $\mathcal{N} = 2$ hypermultiplets. Thus we have an $\mathcal{N} = 2$ structure on moduli space, and in particular we have special geometry on the space of $\mathcal{N} = 2$ vectors. Before we can describe it further, we need to understand better more notions about topology.

29.2.3 Cohomology, homology, and mass spectra

We have described a bit about cohomology previously, in the context of gauge fields. But we can define also the *de Rham cohomology* for p-forms (antisymmetric tensors with p indices), generalizing it. We can again define closed p-forms, $d\psi = 0$, modulo exact p-forms, $\psi = d\phi$, with respect to the differential operator d (exterior derivative). Again the cohomology classes as equivalence classes of closed p-forms, modulo exact p-forms, defining the *p-th cohomology group* $H^p(K, R)$, whose dimension b_p is called the Betti number. This number is therefore the number of linearly independent p-forms that are closed, but not exact.

It is also the number of linearly independent solutions to the Laplace equation on K, $\Delta_K \phi = 0$, since the Laplacian can be obtained from the d operator as $\Delta = (d + d*)^2 = d * d + dd*$ (using that $d^2 = (d*)^2 = 0$), as we have already described when talking about KK compactification.

There is also a theorem that the same number also equals the number of linearly independent closed p-dimensional surfaces that are topologically nontrivial, that is, the dimension of the *homology group* (the topologically nontrivial surfaces on a manifold form also a group). To understand that better, consider the case of the simplest nontrivial surface, the 2-torus T^2, for which there are two nontrivial one-cycles, corresponding to the two circles defining the torus, thus $b_1(T^2) = 2$.

If we consider p-forms on the 10-dimensional space $M_4 \times K_6$, they will split into n-forms on M_4 and $(p - n)$-forms on K_6: an $A_{M_1...M_p}$ will split into $A_{\mu_1...\mu_n i_{n+1}...i_p}$ ($\mu_1, ..., \mu_n \in M_4, i_{n+1}, ..., i_p \in K_6$). Thus $b_{p-n}(K)$ is the number of linearly independent $(p - n)$-forms on K, as well as the number of linearly independent solutions to $\Delta_K \phi = 0$. Therefore this is at the same time the number of massless n-forms on M_4, since

$$\Box_{10} A = (\Box_4 + \Delta_K)A = \Box_4 A = 0 \tag{29.8}$$

if $\Delta_K A = 0$.

Therefore we can count the number of massless n-forms in four dimensions by counting the number of topologically nontrivial surfaces on K_6.

On a CY manifold CY_N of complex dimension N, there is a theorem that there exists a unique *holomorphic*, everywhere nonzero N-form, Ω (the uniqueness is equivalent with $c_1(K) = 0$). For instance, in the case at hand, of CY_3, we have a unique holomorphic, everywhere nonzero 3-form Ω. Of course, the total number of linearly independent 3-forms on CY_N is $b_3(CY_N)$, and Ω can be expanded in a basis of such b_N forms.

On a CY_3 space, with unique covariantly constant spinor η, the Kähler form is

$$k_{ij} = \bar{\eta}\Gamma_{ij}\eta, \tag{29.9}$$

the complex structure is (the Kähler form with one raised index)

$$J^i{}_j = g^{ik}k_{kj}, \tag{29.10}$$

and the holomorphic 3-form is

$$\Omega_{ijk} = \bar{\eta}\Gamma_{ijk}\eta. \tag{29.11}$$

Therefore all these objects are defined in terms of the unique covariantly constant spinor η.

We can also define for Kähler manifolds (and more generally for complex manifolds) a complex version of cohomology, called *Dolbeault cohomology*, since now we can split the differential operator d into a ∂ (for z^i) and a $\bar{\partial}$ (for $\bar{z}^{\bar{j}}$), and correspondingly we can define (p, q) forms with respect to $(\partial, \bar{\partial})$, giving the cohomology groups $H^{(p,q)}(K)$, of dimensions $h^{p,q}$, called the *Hodge numbers*. Obviously then,

$$b_n = \sum_{p+q=n} h^{p,q}, \tag{29.12}$$

and $h^{0,N} = h^{N,0} = 1$.

29.3 Moduli space of CY_3

There are two types of moduli for CY spaces: complex structure moduli and Kähler moduli.

29.3.1 Complex structure moduli

On $M = CY_3$, there are b_3 topologically nontrivial 3-surfaces, for which we can define a basis (A_I, B^J), where $I, J = 1, ..., b_3/2$, such that $A_I \cap A_J = B^I \cap B^J = 0$ and $A_I \cap B^J = -B^J \cap A_I = \delta_I^J$, where $\cap$ refers to the linking number of the surfaces. Here A_I and B^J are called A-cycles and B-cycles, and in the simplest case of CY space, the 2-torus T^2, they correspond to the two 1-cycles wrapping the two circles forming the torus. In that case, the *linking number* of the two cycles is one, $A \cap B = 1$, since we cannot separate the two 1-cycles without breaking them, and the second cycle passes only once through the first. In general, as seen earlier, the A-cycles have zero linking numbers among themselves, and also the B-cycles have zero linking numbers among themselves.

The basis (A_I, B^J) is unique up to a $Sp(b_3; \mathbb{Z})$ transformation acting on it. This basis is dual to a basis of 3-forms (α_I, β^J) on the 3rd cohomology group $H^3(M, \mathbb{R})$, by

$$\int_{A^I} \beta^J = \delta_I^J; \quad \int_{B^J} \alpha_I = \delta_I^J, \tag{29.13}$$

and the rest of integrals are zero.

We can then define the $b_3 = h^{2,1} + 1$ periods of the holomorphic 3-form Ω by

$$F_I = \int_{A_I} \Omega; \quad Z^J = \int_{B^J} \Omega. \tag{29.14}$$

These scalars belong to b_3 $\mathcal{N} = 2$ vector multiplets, on which we have a special Kähler manifold. We then have

$$F_I = N_{IJ} Z^J, \tag{29.15}$$

or more generally,

$$N_{IJ} = \frac{\partial F_I}{\partial Z^J}, \tag{29.16}$$

which, as we already saw when we defined special geometry, is called the period matrix.

In the case of the simplest CY space, the 2-torus, there is only one complex structure, the parameter τ of the torus equal to the ratios of the two cycles on the torus, R_2/R_1, in the case of a rectangular torus. A general torus, but of fixed overall volume ($V = R_2 R_1$ for the rectangular torus), is defined by defining a parallelogram in $\mathbb{R}^2$, with one side of length along one of the axis, and the other side along a general vector τ in the complex plane, and identifying opposite sides of the parallelogram. Then this τ is the complex structure and is therefore a "shape" modulus.

29.3.2 Kähler structure moduli

We have a Kähler form,

$$J = g_{i\bar{j}} dz^i \wedge d\bar{z}^{\bar{j}}, \tag{29.17}$$

and in string theory we also have the NS–NS B-field. Considering only components in the compact directions $B_{i\bar{j}}$, we can form the *complexified Kähler class K*,

$$K = J + iB. \tag{29.18}$$

This is a 2-form, which can therefore be integrated over a 2-cycle. But there are $b_2 = h^{1,1} + 1$ topologically nontrivial 2-surfaces, with a basis $(A_{I'}, B^{J'})$, where $I', J' = 1, ..., b_2/2$. We can then define the Kähler moduli as

$$X_{I'} = \int_{A_{I'}} K; \quad X^{J'} = \int_{B^{J'}} K, \tag{29.19}$$

which are moduli living in $\mathcal{N} = 2$ hypermultiplets, which is why we denote $X_{I'}, X^{J'}$ by the same letter, as the $Q, \tilde{Q}$ in a hypermultiplet.

In the case of the simplest CY, the 2-torus $T^2 = CY_1$, the Kähler modulus is the overall volume of the 2-torus, plus the B-field we can put on it. Therefore these are "size" moduli.

In conclusion, Kähler structure moduli are "shape" moduli, and complex structure moduli are "size" moduli.

Special geometry on CY_3

As in the general $\mathcal{N} = 2$ case already studied, the vectors are the graviphoton (superpartner of the graviton) and n_v vectors from the vector multiplets, for a total of $n_v + 1$ vectors, acted upon by the $Sp(n_v + 1, \mathbb{Z})$ transformation.

29.4   Type IIB on CY_3

For a compactification of IIB on CY_3, there are the $\mathcal{N} = 2$ multiplets: supergravity, $n_v = b_3 = h_{2,1}$ vectors (complex structure) and $b_2 + 1 = h_{1,1} + 1$ hypers (Kähler structure), as well as the specific complex structure modulus

$$\tau = a + ie^{-\phi}, \tag{29.20}$$

where ϕ is the dilaton (present in all string theories) and a is the axion, present only in type IIB.

Locally, the Kähler potential on the special geometry of the vectors is

$$e^{-k} = i(F_I \bar{Z}^I - Z^I \bar{F}_I), \tag{29.21}$$

as we already saw. However, now we can also define a globally valid form, namely

$$e^{-k} = \langle \Omega | \bar{\Omega} \rangle, \tag{29.22}$$

where the inner product of two 3-forms A and B is defined as

$$\langle A | \bar{B} \rangle \equiv \int_K d^6 x A \wedge B. \tag{29.23}$$

By considering that we can expand Ω in the normalized basis of 3-forms (α_I, β^J) as

$$\Omega = Z^I \alpha_I + F_J \beta^J, \tag{29.24}$$

we find the previously discussed local formula (29.21).

Introducing G-fluxes

As we mentioned, one of the phenomenologically interesting cases developed more recently is the case of type IIB with G-flux, that is nonzero integral of an antisymmetric tensor G. In this case, it means the field

$$G = F^{RR} - \tau H^{NS}, \tag{29.25}$$

where $H^{NS} = dB$ and F^{RR} is the field strength of the other 2-form, A^{RR}, present only in type IIB. This case was defined by Giddings, Kachru, and Polchinski (GKP). Gukov, Vafa, and Witten (GVW) found the superpotential

$$W = \int_{K_6} \Omega \wedge G, \tag{29.26}$$

where Ω is the holomorphic 3-form. Considering the Kähler modulus ρ = complexified volume, and complex structure moduli Z^α and τ, one finds in string theory the tree-level Kähler potentials,

$$K(\rho) = -3\ln[-i(\rho - \bar\rho)],$$

$$K(\tau, Z^\alpha) = -\ln[-i(\tau - \bar\tau)] - \ln\left(-i\int_{K_6} \Omega \wedge \bar\omega\right). \tag{29.27}$$

To find these formulas for the superpotential W and Kähler potential K we need to understand some string theory, but for the rest we can use supergravity. In this case, these contributions are the first ones (tree level) from the point of view of string theory, so in principle there are many more (loop and nonperturbative) that one can have, but they are difficult to calculate.

29.5 Heterotic $E_8 \times E_8$ on CY_3

We now come back to the case we started from, the heterotic $E_8 \times E_8$ supergravity compactified on CY_3, and see what are the main ingredients in a search for a good phenomenology.

First, we considered the case of $H = 0$, but we actually have

$$dH = \text{tr}(F \wedge F) - \text{tr}(R \wedge R), \tag{29.28}$$

where in the first, tr refers to trace over the YM group for F^a_{MN}, and in the second, it refers to trace over the local Lorentz group, and as R stands for R^{XY}_{MN}, for $X, Y = 1, .., 10$. But since we want $H = 0$, we need to satisfy $\text{tr}F \wedge F = \text{tr}R \wedge R$, and the simplest way is to have in some sense "$F = R$." This procedure is called *embedding the spin connection in the gauge group*. Indeed, we saw before that under KK compactification, fields of the same spin are grouped together, for instance, we saw that on S^4 we wrote $g_{\mu m} = B^{AB}_\mu V^{AB}_m$ and $A_{\mu mn} = B^{AB}_\mu (...)^{AB}_{mn}$, instead of having four vectors in $g_{\mu m}$ and other six in $A_{\mu mn}$.

Therefore now, we can identify four-dimensional fields coming from different origins, in particular, fields coming from the 10-dimensional $E_8 \times E_8$ YM connection A^{AB}_M with fields coming from the spin connection ω^{XY}_M, which is an $SO(1,9)$ gauge field. In particular, for $a, b \in SO(6)$ (the local Lorentz group on K_6) and $i \in K_6$ and $A, B \in SO(16) \subset E_8$, we consider the ansatz

$$A^{AB}_i = \begin{pmatrix} 0 & 0 \\ 0 & \omega^{ab}_i \end{pmatrix}, \tag{29.29}$$

and all the other components of the two connections are free.

This ansatz breaks the E_8 group to the subgroup that commutes with $SO(6)$, namely $SO(10)$. A simple way to understand this is to consider, for instance, the group of rotations in our three-dimensional Euclidean space, $SO(3)$, and take a constant vector in three-dimensional space. Then full $SO(3)$ rotations are not a symmetry anymore, but only the subgroup of rotations that leaves the vector invariant (commutes with it), namely

$SO(2) = U(1)$. Similarly in our case, the value of A_i^{AB} discussed previously is a constant vector of sorts on the space of E_8 gauge transformations, hence the gauge group is broken to the subgroup commuting with it, $SO(10)$. To break the gauge group further, we will need Wilson lines, to be defined shortly.

The Ricci tensor R_{ij} is the field strength of the $U(1)$ part of the spin connection. Then embedding the spin connection in the gauge group means we find

$$F_{a\bar{b}} = -2iR_{a\bar{b}}. \qquad (29.30)$$

Finally, we still have to deal with the condition $\Gamma^{ij} F_{ij} \eta = 0$, which is equivalent to the conditions

$$F_{ab} = F_{\bar{a}\bar{b}} = 0, \quad g^{a\bar{b}} F_{a\bar{b}} = 0. \qquad (29.31)$$

The conditions $F_{ab} = F_{\bar{a}\bar{b}} = 0$ mean that we have a *holomorphic vector bundle*, which means that we have holomorphic transition functions, defining gauge transformations between patches (locally, $F = 0$ implies the field is pure gauge, but the gauge transformation need not be the same on different patches, and the transformations between these patches are called transition functions). We might think that, since $F_{ab} = \partial_a A_b - \partial_b A_a$ and $F_{\bar{a}\bar{b}} = \partial_{\bar{a}} A_{\bar{b}} - \partial_{\bar{b}} A_{\bar{a}}$, we can choose $A_a = A_{\bar{a}} = 0$, but we can't, we can at most choose one of them to be zero globally by a gauge transformation. Then $g^{a\bar{b}} F_{a\bar{b}} = 0$, which means that also the $U(1)$ part of A satisfies

$$\int_K F \wedge k \wedge \dots \wedge k = (N-1)!^2 \int_K g^{a\bar{b}} F_{a\bar{b}} = 0, \qquad (29.32)$$

which is a topological invariant. This, together with a condition called the Donaldson–Uhlenbeck–Yau equation, that will not be explained, means that the *holomorphic bundle is stable*, which is a mathematical notion that will also not be explained further. But the bottom line is that there are few enough stable holomorphic vector bundles, which makes our choices simpler.

Finally, we define Wilson lines U_γ by

$$U_\gamma = P \exp \oint_\gamma A \cdot dx \qquad (29.33)$$

where γ is a noncontractible loop and $F_{ij} = 0$, so U_γ depends only on the topological class of γ.

For the same reason explained previously for embedding the spin connection in the gauge group, except now the Wilson line transforms covariantly (like the field strength), the presence of the Wilson line means that the gauge group is broken to the subgroup commuting with it.

This U_γ belongs in general to the gauge group. If we consider only the part that has the $SU(3)$ spin connection in it, and it is a sufficiently general $SU(3)$ element, then the group is broken to the subgroup of E_8 that commutes with $SU(3)$, that is E_6. If it belongs, for instance, to $SU(4)$, then E_8 is broken to $SO(10)$ instead. If we have several Wilson lines, the group is broken to the subgroup that commutes with all the Wilson lines.

In this way we can obtain $E_6, SO(10)$, or $SU(5)$, which are all grand unified (GUT) groups for the Standard Model: In the Standard Model, the couplings of the three components of the gauge group $SU(3) \times SU(2) \times U(1)$ are unified at a large energy scale, suggesting the existence of a larger symmetry, encompassing these groups at that energy, called the grand unified group. Common choices are $SU(5)$ (already excluded by experiment), $SO(10)$ (currently most probable), and E_6. By using more Wilson lines, we can also obtain MSSM, the Minimal Supersymmetric Standard Model, with the Standard Model gauge group, $SU(3) \times SU(2) \times U(1)$. Actually, the MSSM spectrum was found more recently in the case of strongly coupled heterotic string, which is slightly different from the previously discussed construction, and on which we will comment in Chapter 30.

Finally, we should also comment on the fact that until now we have only used one of the E_8 factors in the gauge group, but we have not touched the second. But the second E_8 also has a vital phenomenological function. It is difficult to break the $\mathcal{N} = 1$ susy of MMSM in a manner consistent with experiment. The only known way involves the so-called "hidden sector," a strongly coupled gauge sector that breaks susy nonperturbatively by itself (non-perturbative strongly coupled breaking of susy is easier), and this hidden sector interacts with the "visible sector" (MSSM) only via intermediary fields called "messenger fields." In the case of $E_8 \times E_8$ heterotic theory, the second E_8 corresponds to the hidden sector. Therefore this model is not only large enough to accommodate grand unified groups, has compactifications with natural $\mathcal{N} = 1$ susy, but also has a natural hidden sector.

Important concepts to remember

- Low-energy string theory (at scales much smaller than the string scale) is supergravity.
- There are five perturbative string theories, related nonperturbatively: IIA, IIB, type I $SO(32)$, heterotic $SO(32)$, and $E_8 \times E_8$.
- A phenomenologically useful case is $E_8 \times E_8$ heterotic theory on $M_4 \times K_6$, for which we can obtain the desired $\mathcal{N} = 1$ susy in four dimensions.
- For $H = d\phi = 0$, the susy conditions reduce to $D_i\eta = 0$, $\Gamma^{ij}F_{ij}\eta = 0$ for a unique η.
- The condition of unique $D_i\eta = 0$ reduces to having a Calabi–Yau manifold, a Kähler manifold of $SU(n/2)$ holonomy, or with $c_1(K) = 0$.
- The number of linearly independent massless n-forms on M_4 equals the number of linearly independent solutions of $\Delta_K\phi = 0$, where ϕ is a $(p - n)$-form, which itself equals the Betti number $b_{p-n}(K_6)$, the number of linearly independent topologically nontrivial p-surfaces on K_6.
- On CY_3, there are complex structure moduli, or "shape" moduli, $\int_{A_I} \Omega$ and $\int_{B^J} \Omega$, and Kähler structure moduli, or "size" moduli, $\int_{A_{I'}} K$ and $\int_{B^{J'}} K$.
- On the moduli space of an $\mathcal{N} = 1$ CY compactification, we have $\mathcal{N} = 2$ susy for the vector multiplets, that is, special Kähler geometry.
- For type IIB with G-flux, we can compute in string theory a superpotential and tree-level Kähler potentials.
- For perturbative heterotic CY compactifications, we can embed the spin connection in the gauge group, have a stable holomorphic vector bundle, and use Wilson lines to break the gauge group down to a preferred GUT or the Standard Model gauge group.
- The untouched E_8 group acts as a hidden sector in the MSSM construction.

References and further reading

For more details, see chapters 14-16 in [86] and chapters 9 and 10 in [104].

Exercises

(1) Consider the prepotential $F(Z) = iZ^0 Z^1$. Calculate the scalar potential.
(2) Calculate the kinetic terms for the moduli ρ and τ and find the corresponding canonical scalars as a function of ρ and τ.
(3) If $H \neq 0$ and/or $d\phi \neq 0$, would it be possible in principle to satisfy the $\mathcal{N} = 1$ conditions? What would be the choices?
(4) Consider a space $K = \tilde{K}/Z_n$ where $\tilde{K}$ is homologically trivial. Calculate the first homology group of K, $\pi_1(K)$, the group of inequivalent maps from S^1 to K.

30 Toward realistic embeddings of the Standard Model using supergravity

In this chapter, I will describe the Standard Model (SM) of particle physics, and its extensions, via Grand Unified Theories (GUTs) and the Minimal Supersymmetric Standard Model (MSSM), and Minimal Supergravity (MinSugra), with the hope to show how the compactified low-energy string theory from Chapter 29 could be used to obtain them (MinSugra, MSSM, and SM), and what are the generic features that we obtain. The supersymmetry breaking in the MSSM and MinSugra and the resulting phenomenology will be dealt with in Chapter 31.

30.1 The Standard Model

30.1.1 Particle spectrum

The SM of particle physics is a gauge theory with $SU(3)_C \times SU(2)_L \times U(1)_Y$ local symmetry. It is made up of:

- Gauge bosons: fields in the adjoint representation of the gauge group: eight for $SU(3)$, called G_μ^α, three for $SU(2)$, called W_μ^a, and one for $U(1)$, called B_μ. Electroweak symmetry breaking rearranges the $SU(2)$ and $U(1)$ gauge fields, as we will see, into the observed $W_\mu^\pm$, Z_μ (massive vectors) and A_μ (electromangetic field).
- Quarks, which are charged under $SU(3)_C$.
- Leptons.
- Higgs field, which is responsible for electroweak symmetry breaking.

Any other field is in an extension of the SM. For instance, in the MSSM, there are superpartners for all the SM fields. In the GUTs, there are extra gauge bosons, the "leptoquarks," which combine with the SM gauge fields to form extended gauge groups, such as $SU(5)$ and $SO(10)$. In string theory or supergravity, there are many other fields.

For quarks and leptons, the fields split into three independent generations,

$$
\text{quarks:} \quad \begin{pmatrix} u \\ d \end{pmatrix}, \begin{pmatrix} c \\ s \end{pmatrix}, \begin{pmatrix} t \\ b \end{pmatrix}
$$

$$
\text{leptons:} \quad \begin{pmatrix} e \\ \nu_e \end{pmatrix}, \begin{pmatrix} \mu \\ \nu_\mu \end{pmatrix}, \begin{pmatrix} \tau \\ \nu_\tau \end{pmatrix}. \tag{30.1}
$$

30.1.2 Symmetries

As far as **symmetries**, we have the: - *local* $SU(3)_C \times SU(2)_L \times U(1)_Y$, where Y is hypercharge. The usual electric charge is found as $Q = T_3 + Y$. For the quarks, the electric charge is $+2/3$ for the upper component of the doublet, namely for u, c, t, and $-1/3$ for the lower component of the doublet, namely d, s, b.

We also have -*approximate global* symmetries:

– the lepton numbers L_e, L_μ, L_τ, such that

$$
\begin{aligned}
L_e(e) = L_e(\nu_e) = +1; & L_e(e^+) = L_e(\bar{\nu}_e) = -1 \\
L_\mu(\mu^-) = L_\mu(\nu_\mu) = +1; & L_\mu(\mu^+) = L_\mu(\bar{\nu}_\mu) = -1 \\
L_\tau(\tau^-) = L_\tau(\nu_\tau) = +1; & L_\tau(\tau^+) = L_\tau(\bar{\nu}_\tau) = -1,
\end{aligned} \tag{30.2}
$$

combining into the total lepton number

$$
L = L_e + L_\mu + L_\tau. \tag{30.3}
$$

– the baryon number B:

$$
B(q) = +1/3, \quad B(\bar{q}) = -1/3. \tag{30.4}
$$

There is no experimental evidence of violation of L yet. In any case, even if L is violated, B is violated as well, as $B - L$ is generally assumed to be conserved. In fact, it could even be gauged, as we will see shortly, because it is nonanomalous and unobservable: an anomaly would mean its gauge field could interact with fermions, such as to obtain a one-loop triangle with fermions, and the current in the upper vertex, giving a result proportional to $\epsilon^{\mu\nu\rho\sigma} F_{\mu\nu} F_{\rho\sigma}$, but in the case of $B - L$, no simple interaction to fermions exists.

30.1.3 Spinors and notation

We use the usual notation for Dirac spinors splitting into Weyl spinors as follows. For instance for the electron e splits as

$$
e = \begin{pmatrix} e_L \\ e_R \end{pmatrix}, \tag{30.5}
$$

where $e_L = (1 + \gamma_5)/2e$ and $e_R = (1 - \gamma_5)/2e$.

But we use the Majorana spinor notation, where the degrees of freedom of the two independent fields e_L and e_R are re-assembled into the two Majorna spinors

$$
\mathcal{E} = \begin{pmatrix} e_L \\ \epsilon e_L^* \end{pmatrix}; \quad E = \begin{pmatrix} -\epsilon e_R^* \\ e_R \end{pmatrix}. \tag{30.6}
$$

Therefore $e = \mathcal{E}_L + E_R$, but $E = E_L + E_R$, $\mathcal{E} = \mathcal{E}_L + \mathcal{E}_R$, with $\mathcal{E}_L$ and $\mathcal{E}_R$ related (same degree of freedom), and E_L and E_R related (same degree of freedom).

We will use a notation where we treat all the three families together (for $p = 1, 2, 3$)

$$U_p \equiv (u, c, t)$$
$$D_p \equiv (d, s, b)$$
$$E_p \equiv (e, \mu, \tau)$$
$$\nu_p \equiv (\nu_e, \nu_\mu, \nu_\tau), \tag{30.7}$$

(up, down, electron, and neutrino), as well as for the $SU(2)$ doublets

$$L_p \equiv \begin{pmatrix} \nu_p \\ \mathcal{E}_p \end{pmatrix}$$

$$Q_p \equiv \begin{pmatrix} \mathcal{U}_p \\ \mathcal{D}_p \end{pmatrix} \tag{30.8}$$

(lepton and quark).

30.1.4 Spectrum split into representations

– **The fermionic degrees of freedom** are written in terms of their representations under $(SU(3)_C, SU(2)_L, U(1)_Y)$ as follows.
The right-handed degrees of freedom

$$U_{Rp} : (3, 1, +2/3)$$
$$D_{Rp} : (3, 1, -1/3)$$
$$E_R : (1, 1, -1). \tag{30.9}$$

Note that in the SM, there is no right-handed neutrino (only right-handed electrons). A right-handed neutrino could exist, depending on the neutrino masses. In fact, in the extensions of the SM we will study, there is such an object.
The left-handed degrees of freedom

$$L_{Lp} \equiv \begin{pmatrix} \nu_{Lp} \\ \mathcal{E}_{Lp} \end{pmatrix} : (1, 2, -1/2)$$

$$Q_{Lp} \equiv \begin{pmatrix} \mathcal{U}_{Lp} \\ \mathcal{D}_{Lp} \end{pmatrix} : (3, 2, +1/6) \tag{30.10}$$

We can also write the conjugate parts of the above degrees of freedom (the other component of the Majorana spinor), in the conjugate representation

$$U_{Lp} : (\bar{3}, 1, -2/3)$$
$$D_{Lp} : (\bar{3}, 1, +1/3)$$
$$E_{Lp} : (1, 1, +1)$$
$$L_{Rp} \equiv \begin{pmatrix} \nu_{Rp} \\ \mathcal{E}_{Rp} \end{pmatrix} : (1, 2, +1/2)$$
$$Q_{Rp} \equiv \begin{pmatrix} \mathcal{U}_{Rp} \\ \mathcal{D}_{Rp} \end{pmatrix} : (\bar{3}, 2, -1/6). \tag{30.11}$$

Note that the 2 representation of $SU(2)$ is real, that is, $2 = \bar{2}$, so we will omit the bar in the following.

To these, strictly speaking outside the SM, we can add a right-handed neutrino N_R, which is a singlet under everything: $(1, 1, 0)$.

- **The Higgs field** is a doublet,

$$\phi = \begin{pmatrix} \phi^+ \\ \phi^0 \end{pmatrix} : (1, 2, +1/2), \tag{30.12}$$

with complex conjugate field

$$\tilde{\phi} = \begin{pmatrix} \phi^{0*} \\ -\phi^{+*} \end{pmatrix} : (1, 2, -1/2). \tag{30.13}$$

- **The gauge fields** are the octet of $SU(3)$, G_μ^α:$(8, 1, 0)$, the triplet of $SU(2)$, W_μ^a:$(1, 3, 0)$ and the singlet of $U(1)$, B_μ:$(1, 1, 0)$.

We will denote the $SU(2)$ generators in the adjoint representation as $T^a = \tau^a/2$, where τ^a are the Pauli matrices, and the $SU(3)$ generators in the adjoint representation as $T_\alpha = \lambda_\alpha/2$, where λ_α are the Gell–Mann matrices. Then the electric charge, after the electroweak symmetry breaking is $Q = T_3 + Y$.

Given that $Q = Q_L + Q_R$, where both Q_L and Q_R contain the same degree of freedom (they are conjugate fields), and that the representation for Q_L is $(3, 2, +1/6)$, we can write the covariant derivative of the Majorana spinor Q_p as

$$D_\mu Q_p = \partial_\mu Q_p + \left[-ig_3 G_\mu^\alpha \frac{\lambda_\alpha}{2} - ig_2 W_\mu^a \frac{\tau^a}{2} - i\frac{g_1}{6} B_\mu \right] Q_{Lp}$$
$$+ \left[+ig_3 G_\mu^\alpha \frac{\lambda_\alpha^*}{2} + ig_2 W_\mu^a \frac{\tau_a^*}{2} + i\frac{g_1}{6} B_\mu \right] Q_{Rp}. \tag{30.14}$$

Similarly, we write the covariant derivative on $U = U_L + U_R$, where U_L and U_R represent the same degree of freedom (conjugate fields), with U_L in $(3, 1, +2/3)$, as

$$D_\mu U_p = \partial_\mu U_p + \left[-ig_3 G_\mu^\alpha \frac{\lambda_\alpha}{2} - i\frac{2g_1}{3} B_\mu \right] U_{Rp}$$
$$+ \left[+ig_3 G_\mu^\alpha \frac{\lambda_\alpha^*}{2} + i\frac{2g_1}{3} B_\mu \right] U_{Lp}. \tag{30.15}$$

The other covariant derivatives are left as an exercise.

30.1.5 The Lagrangian

Then the Lagrangian for the SM is split as

$$\mathcal{L}_{\text{SM}} = \mathcal{L}_{\text{kin}} + \mathcal{L}_{\text{Higgs}} + \mathcal{L}_{\text{Yukawa}}, \tag{30.16}$$

plus maybe a ν mass Lagrangian. The kinetic terms are (the field strenths of $G^\alpha_\mu, W^a_\mu, B_\mu$ are $G^\alpha_{\mu\nu}, W^a_{\mu\nu}, B_{\mu\nu}$)

$$
\begin{aligned}
\mathcal{L}_{\text{kin}} = &-\frac{1}{4}G^\alpha_{\mu\nu}G^{\alpha\mu\nu} - \frac{1}{4}W^{a\mu\nu}W^a_{\mu\nu} - \frac{1}{4}B_{\mu\nu}B^{\mu\nu} \\
&- \frac{g_3^2}{64\pi^2}\theta_3\epsilon^{\mu\nu\rho\sigma}G^\alpha_{\mu\nu}G^\alpha_{\rho\sigma} - \frac{g_2^2}{64\pi^2}\epsilon^{\mu\nu\rho\sigma}\theta_2\epsilon^{\mu\nu\rho\sigma}W^a_{\mu\nu}W^a_{\rho\sigma} \\
&- \frac{1}{2}\bar{L}_p\slashed{D}L_p - \frac{1}{2}\bar{E}_p\slashed{D}E_p - \frac{1}{2}\bar{Q}_p\slashed{D}Q_p - \frac{1}{2}\bar{U}_p\slashed{D}U_p - \frac{1}{2}\bar{D}_p\slashed{D}D_p.
\end{aligned}
\tag{30.17}
$$

The Higgs Lagrangian is

$$
\mathcal{L}_{\text{Higgs}} = -(D_\mu\phi)^\dagger D^\mu\phi - \lambda\left[\phi^\dagger\phi - \frac{\mu^2}{2\lambda^2}\right]^2,
\tag{30.18}
$$

where the second term is the symmetry-breaking "mexican hat potential," and the covariant derivative of the Higgs is coupled to the $SU(2) \times U(1)$ gauge fields,

$$
D_\mu\phi = \partial_\mu\phi - ig_2 W^a_\mu\frac{\tau^a}{2}\phi - i\frac{g_1}{2}B_\mu\phi.
\tag{30.19}
$$

The Yukawa terms are

$$
\mathcal{L}_{\text{Yukawa}} = -f_{pq}\bar{L}_p\frac{1-\gamma_5}{2}E_q\phi - h_{pq}\bar{Q}_p\frac{1-\gamma_5}{2}D_q\phi - g_{pq}\bar{Q}_p\frac{1-\gamma_5}{2}U_q\tilde{\phi}.
\tag{30.20}
$$

Note that this is the first term where we did not write diagonal terms, but we used nontrivial matrices f_{pq}, h_{pq}, g_{pq} for the three families. We check that one of these three terms is indeed invariant, as required. We use

$$
\frac{1-\gamma_5}{2} = P_R = P_R^2 = \bar{P}_L P_R,
\tag{30.21}
$$

and we can thus distribute these projectors to the two fermions, obtaining

$$
\left(\bar{Q}_p\frac{1-\gamma_5}{2}D_q\right)\phi = (\bar{Q}_{Lp}D_{Rq})\phi.
\tag{30.22}
$$

But Q_{Lp} is in the representation $(3, 2, +1/6)$, which means that $\bar{Q}_{Lp}$ is in $(\bar{3}, 2, -1/6)$, whereas D_{Rq} is in $(3, 1, -1/3)$ and ϕ is in $(1, 2, +1/2)$, which means this terms in correctly invariant, that is, in $(1, 1, 0)$ $(\bar{3} \times 3 = 1, 2 \times 2 = 1, -1/6 - 1/3 + 1/2 = 0)$. It is left as an exercise to check the other two terms in the Yukawa Lagrangian.

For the potential $V(\phi)$, we can choose a VEV that makes it equal to zero as

$$
\phi = \begin{pmatrix} 0 \\ \frac{v}{\sqrt{2}} \end{pmatrix},
\tag{30.23}
$$

where $v^2 = \mu^2/\lambda$. Including the Higgs field, which is a massive fluctuation (the fluctuation along the component perpendicular to the VEV is a Goldstone boson, which is "eaten" by the vector fields that become massive), we have

$$\phi = \begin{pmatrix} 0 \\ \frac{v+H(x)}{\sqrt{2}} \end{pmatrix}, \tag{30.24}$$

and then we obtain the mass of the Higgs field ($H(x)$ fluctuation) as

$$m_H^2 = 2\lambda v^2 = 2\mu^2. \tag{30.25}$$

The $SU(2)_L \times U(1)_Y$ gauge group is broken by the Higgs field to the $U(1)_Q$, with $Q = T_3 + Y$. Defining the Weinberg angle by

$$\cos\theta_W \equiv \frac{g_2}{\sqrt{g_1^2 + g_2^2}}; \quad \sin\theta_W \equiv \frac{g_1}{\sqrt{g_1^2 + g_2^2}}, \tag{30.26}$$

the physical fields observed at low energies, corresponding to the $W_\mu^\pm$ and Z_μ massive vectors and the massless electromagnetic vector A_μ, are

$$A_\mu = W_\mu^3 \sin\theta_W + B_\mu \cos\theta_W$$
$$Z_\mu = W_\mu^3 \cos\theta_W - B_\mu \sin\theta_W$$
$$W_\mu^\pm = \frac{W_\mu^1 \pm W_\mu^2}{2}. \tag{30.27}$$

The masses of the massive vector bosons are

$$M_W = \frac{g_2 v}{2}, \quad M_Z = \frac{v}{2}\sqrt{g_1^2 + g_2^2}. \tag{30.28}$$

In the presence of the VEV in (30.23), the Yukawa terms become mass terms for the fermions

$$\mathcal{L}_{mF} = -\frac{v}{\sqrt{2}}[f_{pq}\bar{\mathcal{E}}_p E_{Rq} + g_{pq}\bar{\mathcal{U}}_p U_{Rq} + h_{pq}\bar{\mathcal{D}}_p D_{Rq}]. \tag{30.29}$$

We redefine these fermions as follows:

$$\mathcal{E}_L \text{ by } U^{(e)} E_R \text{ by } V^{(e)}$$
$$\mathcal{U}_L \text{ by } U^{(u)} U_R \text{ by } V^{(u)}$$
$$\mathcal{D}_L \text{ by } U^{(d)} D_R \text{ by } V^{(d)}, \tag{30.30}$$

in order to diagonalize f_{pq}, g_{pq}, h_{pq}. Then we write the Dirac spinors

$$e_p = \mathcal{E}_{Lp} + E_{Rp}$$
$$d_p = \mathcal{D}_{Lp} + D_{Rp}$$
$$u_p = \mathcal{U}_{Lp} + U_{Rp}, \tag{30.31}$$

and we obtain diagonal mass terms in terms of these Dirac spinors,

$$\mathcal{L} = -\frac{v}{\sqrt{2}}(f_p \bar{e}_p e_p + g_p \bar{u}_p u_p + h_p \bar{d}_p d_p). \tag{30.32}$$

But then of course we mess up the other terms. Therefore we have two kinds of bases for fermions, the original one where the $SU(2)$ transformation is manifest, the $SU(2)$ eigenstates, and the new one, where the mass terms are diagonal, the mass eigenstates.

30.2 Grand Unified Theories

The idea of a grand unified theory (GUT) is related to the fact that, if we plot the couplings g_1, g_2, g_3 of the three gauge groups, versus the energy scale Λ, extrapolating the renormalization group equations (which we know to be valid at low energy), obtained from loop calculations in the SM (SM fields running in the loops) to high energy, they intersect at a value of about 10^{15}GeV, known as the GUT scale. Therefore it is natural to assume that at this scale, all the gauge fields have a common origin, and that there is a single *grand unified group G* with a common coupling constant, that includes $SU(3) \times SU(2) \times U(1)$. Examples of G include $SU(5), SO(10), F_6$. The essential ingredient here is that there is a "desert" in between accelerator energies and the GUT scale, that is, no new physics (like in particular no new particles) in that region, which could otherwise change the renormalization group equations (RGE), through the new particles running in the RGE loops.

30.2.1 *SU*(5) unification

The first unified group to be proposed was $SU(5)$, by Georgi and Glashow in 1974. Their first observation was that we can organize the spectrum of the SM better inside representations of $SU(5)$, as, for the right-handed fields,

$$D_R + L_R = (3, 1, -1/3) + (1, 2, +1/2) = 5, \tag{30.33}$$

and

$$U_R + E_R + Q_R = (3, 1, +2/3) + (1, 1, -1) + (\bar{3}, 2, -1/6) = \overline{10} = (\bar{5} \times \bar{5})_{a-sym.} \tag{30.34}$$

and correspondingly for the left-handed fields conjugate to it,

$$D_L + L_L = \bar{5}$$
$$U_L + E_L + Q_L = 10. \tag{30.35}$$

So all the fermionic matter belongs to just two representations of $SU(5)$.

All the gauge fields fit into an adjoint representation, 24, of $SU(5)$ ($N^2 - 1 = 24$), which under $SU(3) \times SU(2) \times U(1)$ splits roughly as

$$24 = 8 + 3 + 1(+12 \text{ more}). \tag{30.36}$$

The SM Higgs in the two representation belongs to the fundamental representation of $SU(5)$, which splits as

$$5 = 2 + 3 \tag{30.37}$$

under $SU(3) \times SU(2) \times U(1)$.

Finally, we also need another Higgs that breaks $SU(5)$ to $SU(3) \times SU(2) \times U(1)$. This is an adjoint Higgs, that is, in the 24 representation.

If there is a right-handed neutrino N_R, it will be also a singlet of $SU(5)$.

However, now the $SU(5)$ unification is experimentally excluded, since if true, it would generate a proton decay that is excluded, through the "leptoquark" gauge fields, relating quarks to leptons, and thus leading to proton decay. But experimentally, since we did not observe proton decay, there is a lower bound on the lifetime of the proton, and the value predicted by $SU(5)$ unification is lower than this bound.

30.2.2 $SO(10)$ unification

The simplest model that is still not ruled out by experiment is unification into $SO(10)$. This is even more minimal as the above, since all the fermionic matter is organized in a single representation of $SO(10)$, the spinor representation 16, which moreover includes a right-handed neutrino as well. The $SO(10)$ breaks to $SU(5) \times U(1)$, where the $U(1)$ has quantum numbers of $B - L$, which as we saw could be even a local symmetry (unbroken), since it is non-anomalous (global anomalies are fine, but local ones make the theory inconsistent). Under this breaking, the 16 splits as

$$16 \to 10_{-1} + \bar{5}_3 + 1_{-5}, \tag{30.38}$$

where the lower index signifies the $U(1)$ charge. The fundamental representation splits as

$$10 \to 5 + \bar{5}, \tag{30.39}$$

so the SM Higgs must belong to a 10. The adjoint representation splits as

$$45 \to 24_0 + 10_4 + \overline{10}_{-4} + 1_0, \tag{30.40}$$

and the gauge fields belong to it.

30.2.3   Other groups

Finally, we comment on the cases we saw in low-energy string theory. The gauge group E_8 splits either as $E_8 \supset E_6 \times SU(3)$, or as $E_8 \supset SO(16) \supset SO(10) \times SO(6)$.

Under the split $E_8 \supset SO(10) \times SO(6)$, the adjoint of E_8, the only representation appearing in the 10-dimensional $E_8 \times E_8$ low-energy heterotic string, splits as

$$248 = (45, 1) + (1, 15) + (10, 6) + (16, 4) + (\overline{16}, 4), \tag{30.41}$$

that is, all of the needed representations of $SO(10)$ appear from a single adjoint of E_8, as needed!

30.3   Minimal Supersymmetric Standard Model

In the MSSM, for every field in the SM, there is a superpartner. The fermions and the Higgs become chiral superfields in the MSSM, and the gauge fields become gauge super-fields. The superpartners of the fermions are scalars called sfermions, completing chiral superfields, and the superpartners of the Higgs, called Higgsinos (fermions) also complete chiral superfields. The superpartners of the gauge fields, fermions called gauginos, complete gauge superfields.

One of the best experimental arguments for supersymmetry is related to the above grand unification. As we said, the couplings g_1, g_2, g_3 unify at the GUT scale, but in reality, with current error bars, the three extrapolated lines (using the renormalization group with SM field content) just miss each other. However, as we said, this was derived under the "desert" assumption. If in fact we have the superpartners in the "desert," that is, if we have MSSM instead of the SM, the renormalization group equations are modified, because of the new fields running inside loops, and in fact now unification happens again, within current error bars.

Another argument for supersymmetry is the low mass of the Higgs, or the hierarchy problem. Generically, the scalars like the Higgs couple to fermions and scalar fields, so fermion and scalar field loops for the scalar propagator generate quadratically divergent terms, that would push the scalar mass up to the cut-off mass, say the GUT or Planck mass. One can avoid that using fine-tuning, but that usually means a symmetry: in this case, supersymmetry fixes the problem, since the fermion and scalar loops cancel, if super-symmetry is valid, solving the problem (there still are quantum corrections, since susy is broken, but those are milder).

The chiral superfields for the SM fermions will belong to the same representations as the SM fields. We will denote the basic superfields without a L, R subscript, but we will use a C superscript for the conjugate fields to the SM ones. Then the superfields appearing in the superpotential are

$$
\begin{aligned}
Q &: (3, 2, +1/6) \\
L &: (1, 2, -1/2) \\
E^C &: (1, 1, +1) \\
U^C &: (\bar{3}, 1, -2/3) \\
D^C &: (\bar{3}, 1, +1/3),
\end{aligned}
\tag{30.42}
$$

where the fermions in Q are Q_L, in L are L_L, in U^C are U_L, in D^C are D_L, and in E^C are E_L. If we have a right-handed neutrino, it belongs to a singlet superfield $\nu_R : (1, 1, 0)$.

For the Higgs fields, we have a difference. We cannot get by with a single Higgs doublet ϕ and his complex conjugate, since for superfields we have only chiral fields in the super-potential, not their complex conjugate antichiral fields. So in supersymmetry, we must use independent Higgs fields instead of the ϕ and $\tilde{\phi}$, which are called H_u and H_d, with

$$
\begin{aligned}
H_u &: (1, 2, +1/2) \\
H_d &: (1, 2, -1/2).
\end{aligned}
\tag{30.43}
$$

Then the R-parity invariant, renormalizable, superpotential for the MSSM is

$$
W = \mu H_u H_d + y_u H_u Q U^C + y_d H_d Q D^C + y_l H_d L E^C.
\tag{30.44}
$$

Here R-parity is an extra symmetry, which need not be valid, but is usually assumed in MSSM, defined by

$$
R = (-1)^{3B+L+2S},
\tag{30.45}
$$

where S is spin, and restricts the possible terms in the superpotential. The term with μ gives a mass for the Higgs (remember that the scalar potential contains terms $|\partial W/\partial \phi^i|^2$, so now we get $|\partial W/\partial H_u|^2 + |\partial W/\partial H_d|^2 = \mu^2(|H_u|^2 + |H_d|^2)$, and when the Higgs get a VEV, the terms with y coefficients become fermion masses.

The notation $H_u H_d$ refers to the contraction with the $SU(2)$ invariant tensor for the two representations, the epsilon tensor, so

$$H_2 \cdot H_1 = \epsilon^{ij} H_2^i H_1^j. \tag{30.46}$$

But at accelerator energies, we don't observe supersymmetry, which means there must be susy breaking terms. Such terms are usually called soft susy breaking terms: they do not spoil too much the nice properties of susy for which we introduced it in the first place, like the low Higgs mass and gauge coupling unification, while still breaking susy. They are terms written in usual fields, not superfields, since they don't respect susy: remember that the susy breaking cannot be tree level, or perturbative, in the SM, as the SM doesn't respect the susy sum rules, so we cannot have a spontaneous (through $E_0 \neq 0$) breaking in a supersymmetric action, but rather some nonperturbative breaking, studied in the next chapter, that is described phenomenologically here, and these soft breaking terms are of three types:

- Gaugino masses

$$m_{1/2} \tilde{\lambda}\tilde{\lambda} + h.c. \tag{30.47}$$

- Soft scalar masses

$$m_0 \phi^\dagger \phi \tag{30.48}$$

- A and B terms, which are terms of the same type as the superpotential, just that we replace the superfields with their corresponding scalars (their first components)

$$B_\mu h_u h_d + A h_u q u^C + A h_d q d^C + A h_d l e^C + h.c. \tag{30.49}$$

30.3.1 Minimal supergravity

The Lagrangian for the minimal supergravity model is obtained by coupling the MSSM with $\mathcal{N} = 1$ supergravity, according to the general formula for $\mathcal{N} = 1$ chiral superfields plus gauge superfields. We will not write the Lagrangian here; a few more details will be given in Chapter 31.

30.4 New low-energy string (supergravity) constructions

As we saw in Chapter 29, we can embed the spin connection in the gauge field ("$F = R$") in order to be able to have $H = d\phi = 0$. From the condition $\Gamma^{ij} F_{ij} \eta = 0$, we get stable holomorphic vector bundles. We use CY spaces, with $c_1(K) = 0$, thus with $SU(3)$ holonomy. This breaks E_8 to E_6. We can then break E_6 further using Wilson lines.

But more recently, in 2005, in [106, 107], new constructions were obtained which for the first time obtained just the spectrum of MSSM from low-energy string theory. We give a few of the new characteristics of that construction.

They use the nonperturbative version of the heterotic string previously described.

The string theory at large coupling g_s looks 11 dimensional, with $R_{11} \equiv g_s l_s$ acting as the radius of the extra dimension. We don't know too much about this 11-dimensional theory, called M theory, but we know that its low-energy limit is the unique 11-dimensional supergravity theory.

Thus, since we are interested mainly in compactifying the low-energy of the 11-dimensional theory, in this case we also obtain a role for the 11-dimensional supergravity.

In the case of the the $E_8 \times E_8$ heterotic string, at strong coupling the 11th dimension is not on a circle, but on a circle divided (identified) by a certain $\mathbb{Z}_2$ symmetry, that is, $S^1/\mathbb{Z}_2$, which acts roughly as $x_{11} \to -x_{11}$. This results in an interval $(0, \pi R_{11})$. At each end of this interval, we have a 10-dimensional "wall," called an M9-brane, where gauge fields can live. On each of the two lives a E_8 factor, that is, one E_8 gauge group on the M9-brane at $x_{11} = 0$, and one E_8 gauge group on the M9-brane at $x_{11} = \pi R_{11}$. Therefore now the two group factors are separated spatially, making more obvious the fact that one the factors contains the "visible" sector of the MSSM, while the other is a "hidden" sector, as explained in Chapter 29. This construction is called heterotic M theory

For the visible E_8 factor, the new constructions still use holomorphic vector bundles, coming from the $\mathcal{N} = 1$ susy condition $\Gamma^{ij} F_{ij} \eta = 0$ ($F_{ab} = F_{\bar{a}\bar{b}} = 0 = g^{a\bar{b}} F_{a\bar{b}}$).

But we don't need to embed the spin connection in the gauge group, which in heterotic M theory would correspond to satisfying the $dH = 0$ condition *locally* in x_{11}. In the 11-dimensional M theory, the antisymmetric tensor 3-form H_{IJK} lifts to the antisymmetric tensor 4-form G_{MNPQ}, such that $G_{11IJK} = H_{IJK}$. Instead, the Bianchi identity becomes now

$$(dG)_{11IJKL} = -4\sqrt{2}\pi \left(\frac{k}{4\pi}\right)^{2/3} (J^{(1)}\delta(x_{11}) + J^{(2)}\delta(x_{11} - \pi R_{11}) + J + W)_{IJKL}, \quad (30.50)$$

so the $\mathrm{tr} F \wedge F$ terms is now split, with half of it at one end, and half at the other end. And here J stands for the gravitational contribution, and W for possible other contributions ("M5-branes" in the bulk). The point is that now we need only satisfy a global condition, but not locally, which allows for more choices. One usually splits the gravitational contribution in two halves and writes

$$(dG)_{11IJKL} = -4\sqrt{2}\pi \left(\frac{k}{4\pi}\right)^{2/3} (\tilde{J}^{(1)}\delta(x_{11}) + \tilde{J}^{(2)}\delta(x_{11} - \pi R_{11}) + W)_{IJKL}, \quad (30.51)$$

where

$$\tilde{J}^{(i)} = \frac{1}{16\pi^2} \left(\mathrm{tr} F^{(i)} \wedge F^{(i)} - \frac{1}{2}\mathrm{tr} R \wedge R\right). \quad (30.52)$$

Again, the gauge group is broken (further) by Wilson lines, and one obtains the MSSM gauge group and spectrum.

In these constructions, more information about string theory is used, and a lot of mathematics is required, but again most of it can be understood in terms of 11-dimensional supergravity and the E_8 gauge theories on the two M9-branes (maybe with some information about the M5-branes in the bulk).

Important concepts to remember

- In the SM, we have quarks, leptons, Higgs, and gauge fields.
- In Majorana notation, the fundamental fermionic objects are the left-handed L_{Lp}, Q_{Lp} and the right-handed E_{Rp}, U_{Rp}, D_{Rp}, and their complex conjugates (which contain the same degrees of freedom).
- The gauge fields are in the adjoint, and the Higgs field is a $SU(2)$ doublet (and his complex conjugate).
- A right-handed neutrino singlet is outside the SM, but depending on the form the neutrino masses take, it could be necessary.
- Under electroweak symmetry breaking, the electromagnetic field A_μ and the massive Z_μ are rotations of B_μ ($U(1)_Y$) and W_μ^3, and the massive $W_\mu^\pm$ are $(W_\mu^1 \pm W_\mu^2)/\sqrt{2}$.
- For the fermions, there are $SU(2)$ eigenstates, and mass eigenstates, related by a rotation matrix.
- Under $SU(5)$ unification, the left fermionic degrees of freedom fill up a $\bar{5}$ and the 10, the gauge fields fit into an adjoint 24, which also has 12 "leptoquarks," and the SM Higgs belongs to a 5, the $SU(5)$ Higgs is adjoint (24).
- Under $SO(10)$ unification, all the matter fits into a spinor 16 representation, that is, $16 \rightarrow \bar{5} + 10 + 1$, which includes a right-handed Majorana spinor, the SM Higgs belongs to a fundamental $10 \rightarrow 5 + \bar{5}$, and the gauge fields to an adjoint 45.
- In the E_8 unification needed in the $E_8 \times E_8$ heterotic string, all of the fields fit into a single adjoint 248 representation of E_8.
- In MSSM, the Higgs and the fermions turn into chiral superfields, and the gauge fields to gauge superfields.
- The MSSM superpotential involves μ terms, giving the Higgs mass (there are *two* independent Higgs doublets now), and the Yukawa terms, giving fermion masses.
- The soft breaking terms are gaugino masses, soft scalar masses, and A and B terms.
- Minimal supergravity is MSSM coupled to $\mathcal{N} = 1$ supergravity.
- In the new $E_8 \times E_8$ heterotic constructions, one uses heterotic M theory, with each E_8 factor at an end of the 11th-dimension interval. Instead of the local construction of embedding the spin connection in the gauge group, we only need a global condition now.

References and further reading

For more details about the SM, see for instance [105]. For more details about MSSM, see for instance [12]. For the first string constructions with only the MSSM spectrum, see [106, 107] and references therein.

Exercises

(1) Write explicitly the covariant derivatives $D_\mu E$, $D_\mu L$, and $D_\mu D$.

(2) Check, as in the text, the other two Yukawa couplings ($\bar{Q}_p U_p$ and $\bar{L}_p E_q$)

(3) Calculate the scalar potential coming from the R-parity invariant MSSM superpotential W_{MSSM} (F terms).

(4) In heterotic M theory, for good phenomenology, should we have susy preserved on both M9-branes (stable holomorphic vector bundles)? Why?

Minimal sugra, phenomenology, and models of susy breaking

In this chapter, we will finally consider the phenomenology of supersymmetry breaking (SSB), in particular, as it relates to supergravity, namely to the Minimal Supergravity model and its mechanism of mediation of supersymmetry breaking.

31.1 Masses and MSSM parameters

We saw that, when describing the MSSM, there were four parameters in the (unique, R-parity invariant, renormalizable) superpotential W, namely μ, y_u, y_d, y_l. There were two Higgs doublets, H_u, also called H_2, and H_d, also called H_1.

Since

$$\sum_i \left| \frac{\partial W}{\partial H_1^i} \right|^2 = |\mu|^2 \epsilon_{ij} H_2^{j*} \epsilon_{ik} H_2^k + (1 \leftrightarrow 2) = |\mu|^2 H_2^\dagger H_2 + (1 \leftrightarrow 2), \tag{31.1}$$

we obtain the F-term scalar potential as a mass term,

$$V_F = |\mu|^2 (H_1^\dagger H_1 + H_2^\dagger H_2), \tag{31.2}$$

and of course, we also have a D-term, and the soft susy breaking terms (involving scalar mass m_0, gaugino mass $m_{1/2}$, and A and B_μ terms).

The two Higgs doublets are split according to the electromagnetic charge, that appears as a superscript, as

$$H_1 = H_d = \begin{pmatrix} H_1^0 \\ H_1^- \end{pmatrix} \to \quad c.c. \quad \begin{pmatrix} H_1^{0*} \\ H_1^+ \end{pmatrix},$$

$$H_2 = H_u = \begin{pmatrix} H_2^+ \\ H_2^0 \end{pmatrix} \to \quad c.c. \quad \begin{pmatrix} H_2^- \\ H_2^{0*} \end{pmatrix}. \tag{31.3}$$

Then the scalar (Higgs) potential *for the neutral scalars H_1^0, H_2^0* is

$$V(H_1^0, H_2^0) = m_1^2 |H_1^0|^2 + m_2^2 |H_2^0|^2 + B_\mu (H_1^0 H_2^0 + H_1^{0*} H_2^{0*}) + \frac{g^2 + g'^2}{8} (|H_1^0|^2 - |H_2^0|^2)^2. \tag{31.4}$$

Here the masses m_1 and m_2 come from the soft susy breaking mass terms, plus the μ terms,

$$m_1^2 = m_{H_1}^2 + |\mu|^2, \quad m_2^2 = m_{H_2}^2 + |\mu|^2, \tag{31.5}$$

and the couplings are the ones of the weak and hypercharge groups, as usual, $g = g_2$ (for $SU(2)$) and $g' = g_1$ (for $U(1)_Y$). The quartic Higgs potential comes from the D-terms for $SU(2)$ and $U(1)_Y$.

We note that the potential is not diagonal. It can be diagonalized by a rotation. But, we remember the way the Higgs mechanism (for a single doublet) works in the Standard Model, the real part of the Higgs is the physical one, that corresponds to motion transverse to the symmetry breaking direction, and the imaginary part is the Goldstone boson, eaten by the gauge field, and becoming the longitudinal component of the massive vector.

This means that the real and imaginary parts of the neutral Higgses are rotated by different matrices, while the charged Higgses are rotated together, for a total of

$$\begin{pmatrix} \mathrm{Im}\, H_1^0 \\ \mathrm{Im}\, H_2^0 \end{pmatrix} = \frac{1}{\sqrt{2}} \begin{pmatrix} \cos\beta & \sin\beta \\ -\sin\beta & \cos\beta \end{pmatrix} \begin{pmatrix} Z_L^0 \\ A^0 \end{pmatrix},$$

$$\begin{pmatrix} \mathrm{Re}\, H_1^0 \\ \mathrm{Re}\, H_2^0 \end{pmatrix} = \frac{1}{\sqrt{2}} \begin{pmatrix} \cos\alpha & -\sin\alpha \\ \sin\alpha & \cos\alpha \end{pmatrix} \begin{pmatrix} H^0 \\ h^0 \end{pmatrix},$$

$$\begin{pmatrix} H_1^\pm \\ H_2^\pm \end{pmatrix} = \begin{pmatrix} \cos\beta & \sin\beta \\ -\sin\beta & \cos\beta \end{pmatrix} \begin{pmatrix} W_L^\pm \\ H^\pm \end{pmatrix}, \tag{31.6}$$

where $W_L^\pm, Z_L^0$ are the longitudinal components of the corresponding massive vectors ($W_\mu^\pm$ and Z_μ), $H^\pm$ are charged Higgs, and H^0 and h^0 are physical neutral Higgs scalars, with $m_{h^0} < m_{H^0}$.

Now, since we have two Higgs doublets, we have two Higgs VEVs as well, $v_1 = \langle H_1^0 \rangle$ and $v_2 = \langle H_2^0 \rangle$, and their ratio,

$$\tan\beta = \frac{v_1}{v_2}, \tag{31.7}$$

is a parameter. Together with m_{h^0} (or m_{H^0}), these form the only new *physical* parameters, compared with the Standard Model (of course, the Standard Model has v_1 or v_2 as a parameter, for instance) all the others being expressible in terms of them.

31.2 Minimal Supergravity (MinSugra)

As we explained in Chapter 30, Minimal Supergravity (MinSugra) refers to the MSSM coupled to $\mathcal{N} = 1$ supergravity. For such a case, with

- chiral superfields $\Phi^i = (\phi^i, \psi^i, F^i)$ (scalars ϕ^i, fermions ψ^i, and auxiliary fields F^i), with Kähler potential $K(\Phi^i, \bar{\Phi}^j)$ and superpotential $W(\Phi^i)$,
- vector superfields $V^a = (A_\mu^a, \lambda^a, D^a)$ (gauge fields A_μ^a, gauginos λ^a, and auxiliary fields D^a), perhaps with a nontrivial kinetic function $f_{ab}(\phi^i)$ depending on the scalars in the chiral superfields,
- supergravity superfield $(g_{\mu\nu}, \psi_{\mu\alpha}, M)$ (graviton $g_{\mu\nu}$, gravitino $\psi_{\mu\alpha}$, and auxiliary field M),

we can write some general formulas for the Lagrangian and potential, as we have seen in Chapters 29 and 30, for example.

Note that the Kähler potential and superpotential are subject to Kähler transformations,

$$K(\Phi^i, \bar{\Phi}^{\bar{j}}) \rightarrow K(\Phi^i, \bar{\Phi}^{\bar{j}}) + F(\Phi^i) + \bar{F}(\bar{\Phi}^{\bar{j}}),$$

$$W(\Phi^i) \rightarrow e^{-\kappa_N^2 F(\Phi^i)} W(\Phi^i), \tag{31.8}$$

that are a symmetry of the scalar potential.

To write the scalar potential in a simpler way, we have seen that we can use a generalized Kähler potential (Kähler function)

$$\mathcal{G} = \kappa_N^2 K + \ln |W|^2, \tag{31.9}$$

in terms of which the scalar potential,

$$V = g_{i\bar{j}} F^i \bar{F}^{\bar{j}} - \frac{1}{3} \bar{M} M + \frac{1}{2} (\mathrm{Re}\, f)_{ab} D^a D^b$$

$$= e^{\kappa_N^2 K} \left[D_i W g^{i\bar{j}} D_{\bar{j}} \bar{W} - 3\kappa_N^2 |W|^2 \right] + \frac{1}{2} (\mathrm{Re}\, f)_{ab}^{-1} K_i (t^a \phi)^i (\bar{\phi} t^a)^{\bar{j}} K_{\bar{j}}, \tag{31.10}$$

can be rewritten as

$$V = \kappa_N^2 e^{\mathcal{G}} \left[\mathcal{G}_i \mathcal{G}^{i\bar{j}} \mathcal{G}_{\bar{j}} - 3 \right] + \frac{1}{2} (\mathrm{Re}\, f)_{ab} D^a D^b, \tag{31.11}$$

where, as usual, $K_i = \partial K / \partial \phi^i$, $\mathcal{G}_i = \partial \mathcal{G} / \partial \phi^i$, and so on, and

$$\mathcal{G}^{i\bar{j}} \mathcal{G}_{\bar{j}k} = \delta_k^i. \tag{31.12}$$

Here we have also used the fact that on-shell, the auxiliary fields are

$$M = -3\kappa_N e^{\kappa_N^2 K/2} W,$$

$$F^i = -g^{i\bar{j}} e^{\kappa_N^2 K/2} D_{\bar{j}} \bar{W}$$

$$D^a = (\mathrm{Re}\, f)_{ab}^{-1} K_i (t^b \phi)^i. \tag{31.13}$$

31.3 Susy breaking

We now consider how susy breaking is obtained.

In the global case, as we saw in Chapter 5, the criterion was a positive vacuum energy, $E_0 > 0$, or $\langle \Omega | H | \Omega \rangle \neq 0$.

Moreover, one point that we didn't make before is that susy breaking leads to the existence of a particle called Goldstino, by the (fermionic version of the) Goldstone theorem. The Goldstone theorem was the simple enough statement that for every gauge (local) broken symmetry, there is a Goldstone particle (a boson for a bosonic symmetry, but the Goldstone theorem doesn't rely on the symmetry being bosonic, so if the symmetry is fermionic, there is a fermion) that is massless. The Goldstone particle corresponds to motion in the broken symmetry direction.

For instance, in the bosonic case, we have (roughly for the non abelian case, and exactly in the Abelian case)

$$\delta \phi = \alpha \phi, \tag{31.14}$$

so in the broken symmetry case, there is a VEV for the boson, so

$$\delta\phi = \alpha\langle\phi\rangle + \ldots = \alpha v + \ldots, \tag{31.15}$$

so an alternative definition of the Goldstone boson can be that is the field that changes into a constant (v here) plus more.

Then, in the case of supersymmetry, which is a fermionic symmetry, we have

$$\delta\psi = F\epsilon + \ldots, \tag{31.16}$$

so in the case of broken susy we obtain

$$\delta\psi = \langle F\rangle\epsilon + \ldots, \tag{31.17}$$

so the characteristic of the *Goldstone fermion, or Goldstino*, is that it changes into a constant, defined by $\langle F\rangle$. Indeed, this is consistent with the previous definition, since an auxiliary field VEV $\langle F\rangle$ implies a nonzero vacuum energy, because

$$V = F^2 = \langle F\rangle^2 + \ldots > 0. \tag{31.18}$$

31.4 Susy breaking mechanisms for the MSSM and MinSugra

We saw in Chapter 5 that there is perturbative level susy breaking, that occurs spontaneously, namely the minimum of the potential has a positive energy (e.g., in the case of the Fayet–Iliopoulos model, the potential, coming from the superpotential, does not have a zero energy minimum).

But perturbative susy breaking would imply some mass sum rules that contradict experiments, so it cannot happen in the Standard Model. Therefore susy breaking must be nonperturbative.

In this, phenomenologically good, case there is a separation between the fundamental, or (gauge) unification scale M_X, at which there is a large gauge group, and the susy breaking scale M_S, given by

$$M_S = M_X \exp\left[-\frac{8\pi^2 b}{\mathcal{G}^2(M_X)}\right], \tag{31.19}$$

where $b \sim \mathcal{O}(1)$ is a numerical factor, and $\mathcal{G}(\mu)$ is the running gauge coupling, that is asymptotically free and becomes strong at M_S. The situation is then analogous to the case of the chiral symmetry breaking in QCD, where the fundamental theory that is chiral and UV free leads to an effective low-energy theory, where the coupling is large (nonperturbative), that breaks chiral symmetry.

This means that *some* strong force becomes strong and breaks susy. But we don't see such force in nature, so it must be in a *hidden sector*, that doesn't interact with the Standard Model fields, only with some "messenger fields," that communicate the susy breaking from the hidden sector to the "visible sector" (MSSM).

There are mediation mechanisms that are generally divided into two (or three) groups,

- "**gauge mediation**," which happens via the $SU(3) \times SU(2) \times U(1)_Y$ gauge fields. In this case, the mass splittings between the quarks, leptons, gauge bosons, and their superpartners are of the order of

$$\Delta m \sim \frac{g_i^2}{8\pi^2}, \tag{31.20}$$

the loop counting parameter of the SM fields, with respect to the susy breaking scale. Assuming masses in the range $\sim 100 GeV - 10 TeV$ (so that the superpartners could be observed at accelerators), we obtain $M_S \sim 100 TeV$ (given the gauge couplings g_i).

- "**gravity mediation**," which happens via the auxiliary fields, superpartners of gravity, the M's. In this case, we expect mass splittings between the observed particles and their superpartners of the order of either $\sim \sqrt{G_N} M_S^2 = \kappa_N M_S^2$, leading to $M_S \sim 10^{11} GeV$, or $\sim G_N M_S^3 = \kappa_N^2 M_S^3$, leading to $M_S \sim 10^{13} GeV$. We see that, unlike the gauge mediation case, where the susy breaking scale was low, now it is very high.

- "**anomaly mediation**" is really a special case of gravity mediation, where the gravitino mass $\tilde{m}_g^*$ is $\mathcal{O}(\kappa_N)$ and, due to the renormalization of the couplings, there is a term $\Lambda \frac{\partial}{\partial \Lambda} W(\phi)$ from an anomaly in $T_{\mu\nu}$ (which couples to gravity $g_{\mu\nu}$), which gives the term in the Lagrangian

$$\mathcal{L}_{\text{susy breaking}} = -2\text{Re} \left[\tilde{m}_g^* \Lambda \frac{\partial W(\phi)}{\partial \Lambda} \right]. \tag{31.21}$$

The cases we are most interested in are of gravity mediation (as they are of supergravity type), which generically fall under two categories:

- $\mathcal{O}(\kappa_N)$ mass splittings. There is some hidden sector becoming strongly coupled at a scale Λ, such that $m_W \ll \Lambda \ll m_{\text{Pl}}$, so there is spontaneous supersymmetry breaking. In this case, the mass splittings are $\kappa_N \Lambda^2 \sim TeV$, which gives $\Lambda \sim 10^{11} GeV$, as we said already.

- $\mathcal{O}(\kappa_N^2)$ mass splittings, and the hidden sector is *not spontaneously broken*. But there are *modular superfields* (that include the moduli for compactification), and there is a nonperturbative superpotential W for the moduli superfields. For instance, if there is a "gaugino condensation,"

$$\langle \bar{\lambda}^a \lambda^a \rangle \neq 0 \tag{31.22}$$

in the hidden sector, this will lead to a superpotential for the moduli, and in this case $\kappa_N^2 \Lambda^3 \sim TeV$ implies $\Lambda \sim 10^{13} GeV$, as we previously said.

The susy breaking (in the hidden sector) itself can be

- spontaneous, from some nonperturbative superpotential for the fields, OR
- dynamical, which means that the correct low-energy fields are not the UV ones, but some non-supersymmetric ones.

To understand the last point better, we give some examples.

Example 1 An SQCD model by Seiberg, with quark superfields in the fundamental of $SU(N_c)$ (color group) and fundamental of flavor (global) $SU(N_f)$. The superfield Q is chiral in the fundamental of $SU(N_c)$ and the superfield $\tilde{Q}$ is chiral in the antifundamental of $SU(N_c)$, such that, as usual, the D-term for the gauge group is

$$D^a = Q^\dagger T^a Q - \tilde{Q} T^a \tilde{Q}^\dagger. \tag{31.23}$$

Consider the case of $N_f < N_c$, and take $Q = \tilde{Q}^\dagger$. If there is no tree-level superpotential, Seiberg found that the Wilsonian low-energy effective action is written in terms of the effective gauge group $SU(N_c - N_f)$, and low-energy effective fields that are *gauge invariant combinations of the quarks*, $\tilde{Q}^i Q^j$, with i, j flavor indices and the gauge indices contracted. More precisely, the effective low-energy superpotential was found to be

$$W_{\text{eff}}(Q, \tilde{Q}) = (N_c - N_f) \left(\frac{\Lambda^{3N_c - N_f}}{\det(\tilde{Q}Q)} \right)^{\frac{1}{N_c - N_f}}, \tag{31.24}$$

where the determinant is taken over the flavor indices.

This case is of a *dynamical* nature for the low-energy fields, as the potential is effective and written in terms of low-energy fields that differ from the fundamental ones due to the nonperturbative nature (just like for QCD, the low-energy fields are the nucleons, proton and neutron, and the mesons, pions, and others). However, the potential is written in terms of a *combination* of Q and $\tilde{Q}$ (like the pions and nucleons are gauge invariant combinations of the quarks) but doesn't actually break supersymmetry, as the minimum of the potential is at zero, yet the superpotential is *non-analytic*, which is a feature of its being a dynamical Wilsonian effective action.

Example 2 This is closer to something relevant for the Standard Model. Consider one supersymmetric generation of (supersymmetrized) Standard Model quarks and leptons, but without right-handed leptons and the $U(1)_Y$ group, meaning $Q, \bar{u}, \bar{d}, L$ superfields and gauge group $SU(3) \times SU(2)$, with nonperturbative scales Λ_2 for $SU(2)$ and Λ_3 for $SU(3)$ (so Λ_3 is like Λ_{QCD}, the nonperturbative scale giving, for instance, the masses of the glueballs).

Then one can prove that the low-energy effective theory is written in terms of the gauge invariant quantities

$$X_1 = Q\bar{u}L, \quad X_2 = Q\bar{d}L, \quad Y = Q^2\bar{u}\bar{d}, \tag{31.25}$$

and the effective superpotential is (c is a number)

$$W_{\text{eff}}(X_1, X_2, \Lambda_2, \Lambda_3) = c\frac{\Lambda_3^7}{Y}. \tag{31.26}$$

This gives no supersymmetric vacuum (zero energy), so it is really a case of dynamical susy breaking.

Example 3 One case relevant to the susy breaking in the hidden sector of *gaugino (gluino) condensation*, which means that there is a fermion bilinear due to the nonperturbative gauge group, here taken to be $SU(N)$, with coupling g_o. The condensate is found to be

$$\langle \bar{\lambda}^a \lambda^a \rangle = c \Lambda_{UV}^3 \frac{1}{g_o^2} \exp\left(-\frac{8\pi^2}{Ng_o^2}\right),$$ (31.27)

where c is a constant and Λ_{UV} is the UV cut-off.

31.5 Susy breaking in MinSugra

After understanding the susy breaking mechanisms, we now go back to our phenomeno-logically relevant case, Minimal Supergravity, in order to see how does it communicate the susy breaking to the MSSM (visible sector).

In global supersymmetry, a constant term in the potential,

$$S_{V_0} = \int d^4x \sqrt{-g}[-V_0],$$ (31.28)

breaks susy. But in supergravity (local supersymmetry), we saw that the scalar potential V has a term $-3|W|^2$, so the scalar potential is not positive definite.

In fact, there is a supersymmetric "gravitino mass term" of the form

$$S_{m_{3/2}} = \int d^4x \sqrt{-g} \left[3\frac{m_{3/2}^2}{\kappa_N^2} - m_{3/2}\bar{\psi}_\mu \sigma^{\mu\nu}\sigma_\nu \right].$$ (31.29)

Note that this is indeed a supersymmetric mass term for the gravitino, since in the gauge $\gamma^\mu \psi_\mu = 0$, needed in order to eliminate the unphysical spin 1/2 component of $\psi_{\mu\alpha}$, we have

$$-m_{3/2}\bar{\psi}_\mu \sigma^{\mu\nu}\psi_\nu = -\frac{1}{4}m_{3/2}\bar{\psi}_\mu \left[\gamma^\mu, \gamma^\nu\right]\psi_\nu = +\frac{1}{4}m_{3/2}\bar{\psi}_\mu \{\gamma^\mu, \gamma^\nu\}\psi_\nu = +\frac{1}{2}m_{3/2}\bar{\psi}^\mu \psi_\mu.$$ (31.30)

However, the previously discussed term is actually *not a gravitino mass term as it is*, since the gravitino is massless (it is the superpartner of the massless graviton). That is because there is an AdS background,

$$E_0 = -3\frac{m_{3/2}^2}{\kappa_N^2},$$ (31.31)

and the rules for stability in AdS background are different: unlike in Minkowski space, where the asymptotic stability of a field is that $V''(\phi) \equiv m^2 \geq 0$, in AdS space m^2 must be $\geq$ a negative constant that depends on dimension and spin, bound known as the Breitenlohner–Freedman bound.

But one can add a constant term V_0 to the action, such that

$$-V_0 + 3\frac{m_{3/2}^2}{\kappa_N^2} = 0,$$ (31.32)

that is, we end up in Minkowski background, where the stability *is* for $m^2 \geq 0$, so we obtain a gravitino mass term in Minkowski background, with

$$m_{3/2} = \kappa_N \sqrt{\frac{V_0}{3}}. \tag{31.33}$$

This means that in gravity, the criterion for susy breaking is whether $m_{3/2} \neq 0$ (whether the gravitino is massive or massless).

One important fact that is related to the above statement is that for susy breaking there is a super-Higgs mechanism: the two degrees of freedom of a massless gravitino in four dimensions add to the two degrees of freedom of a Goldstone fermion, or Goldstino, to become the four degrees of freedom of the massive gravitino. So the massless gravitino eats up the Goldstino to become massive, just like, in the usual Higgs mechanism, the massless vector eats the Goldstone boson to become massive.

In the complete Lagrangian of MinSugra there is a term, coming from the coupling of the supergravity multiplet to the chiral multiplets,   '

$$\mathcal{L}_{\text{gravitino}} = \frac{i}{2}\kappa_N^2 e^{\kappa_N^2 K/2} W(\phi)\bar{\psi}_\mu \sigma^{\mu\nu}\psi_\nu + h.c. \tag{31.34}$$

But, remembering that

$$M = -3\kappa_N e^{\kappa_N^2 K/2} W, \tag{31.35}$$

we have that in the vacuum,

$$\langle M \rangle = -2\kappa_N \langle e^{\kappa_N^2 K/2} W \rangle, \tag{31.36}$$

so that we obtain

$$\mathcal{L}_{\text{gravitino}} = -\frac{i}{6}\kappa_N \langle M \rangle \bar{\psi}_\mu \sigma^{\mu\nu}\psi_\nu, \tag{31.37}$$

which means that the gravitino mass is given by the VEV of M,

$$m_{3/2} = \frac{\kappa_N}{3}\langle M \rangle. \tag{31.38}$$

For zero potential, $V = 0$ and no D-term,

$$V = g_{i\bar{j}}F^i \bar{F}^{\bar{j}} - \frac{1}{3}\bar{M}M = 0 \Rightarrow \langle |M| \rangle = \sqrt{3}\sqrt{\langle F_i g^{i\bar{j}}F_j \rangle}. \tag{31.39}$$

Since $m_{\text{Pl}} = 1/\kappa_N$, and defining the SuperSymmetry Breaking (SSB) scale by

$$\sqrt{\langle F_i g^{i\bar{j}}F_j \rangle} \equiv M_{\text{SSB}}^2, \tag{31.40}$$

we obtain that the gravitino mass is related to the SSB scale as

$$m_{3/2} = \frac{\kappa_N}{\sqrt{3}}\sqrt{\langle F_i g^{i\bar{j}}F_j \rangle} = \frac{M_{\text{SSB}}^2}{m_{\text{Pl}}\sqrt{3}}. \tag{31.41}$$

31.6 Polonyi model for susy breaking: hidden sector + gravity mediation

As a simple example (a toy model), relevant for supergravity, of how the susy breaking is generated in a hidden sector and communicated to the visible sector by (super)gravity (or, more precisely, gravitino), where it generates the phenomenological susy breaking terms (soft scalar mass m_0, A and B terms and gaugino mass term), we consider the Polonyi model.

In this model, the *hidden* sector has a chiral superfield with a scalar z and superpotential

$$W_h = \mu m_{\mathrm{Pl}}(z + C), \tag{31.42}$$

where C is a constant $\sim m_{\mathrm{Pl}}$ to be fixed later, and μ is a mass, which must be of the order of observable accelerator energies, $\mu \sim 100 GeV - 10 TeV$.

Then the auxiliary field of z is

$$F_z = \frac{dW_h}{dz} = \mu m_{\mathrm{Pl}} \neq 0, \tag{31.43}$$

so susy is broken (since then $V_0 = |F_z|^2 + ... > 0$).

The complete superpotential involves *other* superfields Φ^i, with an additive term in the superpotential, so

$$W(z, \Phi^i) = W_h(z) + W_o(\Phi^i), \tag{31.44}$$

such that $W_o(0) = 0$.

Then the potential in rigid susy is

$$V = \left| \frac{dW_h(z)}{dz} \right|^2 + \sum_i \left| \frac{\partial W_o}{\partial \phi^i} \right|^2. \tag{31.45}$$

The Kähler potential is canonical,

$$K(z, \bar{z}, \Phi^i, \bar{\Phi}^i) = \bar{z}z + \bar{\Phi}^i \Phi^i. \tag{31.46}$$

Then the *local* susy (supergravity) potential is

$$V(z, \phi^i) = e^{\kappa_N^2(|z|^2 + \bar{\phi}^i \phi^i)} \left[\left| \mu m_{\mathrm{Pl}} + \kappa_N^2 z^* (W_h + W_o) \right|^2 \right.$$
$$\left. + \left| \frac{\partial W_o}{\partial \phi^i} + \kappa_N^2 (W_h + W_o) \right|^2 - 3\kappa_N^2 |W_h + W_o|^2 \right]. \tag{31.47}$$

But we want to impose that we have zero cosmological constant at the minimum of the potential,

$$V(z, 0) = 0 \quad \text{for} \quad z = \langle z \rangle = \min. \tag{31.48}$$

We note that we have

$$W_h(0) = \mu m_{\mathrm{Pl}} C, \quad W_o(0) = 0, \quad \frac{\partial W_o}{\partial \phi^i} = 0, \tag{31.49}$$

where the first two are from what we already imposed, and the last from the minimum condition.

Then we obtain (noting that $\kappa_N = 1/m_{\text{Pl}}$)

$$
\begin{aligned}
V(z,0) &= e^{\kappa_N^2 |z|^2} \left[\left| \mu m_{\text{Pl}}(1 + \kappa_N^2 |z|^2 + \kappa_N^2 z^* C) \right|^2 - 3|\mu|^2 |z + C|^2 \right] \\
&= |\mu|^2 m_{\text{Pl}}^2 e^{\kappa_N^2 |z|^2} \left[1 - \kappa_N^2 |z|^2 - 2\kappa_N^2 C(z + z^*) \right. \\
&\quad \left. + \kappa_N^4 |z|^4 + \kappa_N^4 C^2 |z|^2 + \kappa_N^4 C |z|^2 (z + z^*) - 3\kappa_N^2 C^2 \right],
\end{aligned}
\tag{31.50}
$$

so that

$$
\frac{dV}{dz} = 0 \Rightarrow
$$
$$
2\kappa_N^4 |z|^2 z^* + \kappa_N^4 C^2 z^* + \kappa_N^2 C[z^*(z + z^*) + |z|^2] - \kappa_N^2 z^* - 2\kappa_N C = 0.
\tag{31.51}
$$

Then imposing $V(\langle z \rangle) = 0$ gives

$$
1 + \kappa_N^2 |z|^2 + \kappa_N^2 C|z| = \sqrt{3} \kappa_N |z + C|,
\tag{31.52}
$$

with the solution

$$
\kappa_N \langle z \rangle = \sqrt{3} \mp 1, \quad \kappa_N C = \pm 2 - \sqrt{3}.
\tag{31.53}
$$

In this way, we have fixed the (previously free) constant C and also found $\langle z \rangle$. But then

$$
m_{3/2} = \frac{\kappa_N}{3} \langle M \rangle = \kappa_N^2 \langle e^{\kappa_N^2 |z|^2/2} W_h \rangle,
\tag{31.54}
$$

and since we find

$$
W_h(\langle z \rangle) = \mu m_{\text{Pl}}^2 (\sqrt{3} \mp 1 \pm 2 - \sqrt{3}) = \pm \mu m_{\text{Pl}}^2,
\tag{31.55}
$$

it follows that the gravitino mass, giving the SSB scale in MSSM, is related to the hidden sector scale μ by

$$
m_{3/2} = \pm \mu e^{\frac{\sqrt{3} \mp 1}{2}}.
\tag{31.56}
$$

Finally now, we are interested in the MSSM susy breaking parameters, for which we must consider nonzero ϕ^i's, in the hidden sector susy breaking vacuum, so at $\langle z \rangle$, so we substitute the previously discussed results for $\langle z \rangle$ in the scalar potential, and obtain

$$
\begin{aligned}
V(\langle z \rangle, \phi^i) = e^{\kappa_N^2 |\phi^i|^2/2} &\left\{ \left| \sqrt{3} \frac{m_{3/2}}{\kappa_N} + \kappa_N (\sqrt{3} \mp 1) W \right|^2 \right. \\
&\left. + \left| \sum_i \frac{\partial \tilde{W}}{\partial \phi^i} + \kappa_N^2 \bar{\phi}^i W + m_{3/2} \bar{\phi}^i \right|^2 - 3 \left| \kappa_N^2 \tilde{W} + \frac{m_{3/2}}{\kappa_N} \right|^2 \right\},
\end{aligned}
\tag{31.57}
$$

where we have defined (since the prefactor is a constant)

$$
\tilde{W}(\phi^i) \equiv e^{\kappa_N^2 |\langle z \rangle|^2/2} W_o(\phi^i).
\tag{31.58}
$$

But, for the MSSM construction with phenomenological susy breaking parameters, we are interested in the rigid theory, that is, decoupled from gravity, meaning in the $\kappa_N \to 0$ limit. In that limit, as we can see, $\kappa_N \langle z \rangle$ is finite ($= \sqrt{3} \mp 1$) and $m_{3/2}$ is finite ($= \mu e^{\frac{\sqrt{3} \mp 1}{2}}$, with μ the hidden sector parameter).

Then we can easily find

$$
V(\langle z \rangle, \phi^i) \to \sum_i \left| \frac{\partial W}{\partial \phi^i} \right|^2
$$
$$
+ m_{3/2}^2 \sum_i |\phi^i|^2 + m_{3/2} \left[\sum_i \phi^i \frac{\partial W}{\partial \phi^i} + h.c. + (3 \mp \sqrt{3} - 3)W + h.c. \right],
$$

$$(31.59)$$

where the first term is the global supersymmetric term, the second is a soft supersymmetry breaking scalar mass, and the last (square bracket) is a supersymmetry breaking A term.

Thus, from our list of phenomenological susy breaking terms, we found all but the gaugino masses, which actually can also be found as follows. In the action there is a term (from the interaction of the gauge superfields, containing the gaugino λ to the chiral superfields, containing the scalars ϕ^i)

$$
-\frac{1}{4} \frac{\partial f(z, \phi^i)}{\partial z} F_z \bar{\lambda}_R \lambda_L. \tag{31.60}
$$

But in the vacuum, the auxiliary field F_z takes a VEV,

$$
\langle F_z \rangle = \langle e^{\kappa_N^2 K/2} g^{zi} D_i W \rangle \sim \frac{m_{3/2}}{\kappa_N}, \tag{31.61}
$$

where the order of magnitude relation is due to (31.41).

But this means that, if $\partial f / \partial z \neq 0$, we obtain indeed a gaugino mass in the susy breaking vacuum.

Finally, this means that we have obtained *all* the phenomenological susy breaking terms in the MSSM from the Polonyi model, with susy breaking in the hidden sector communicated to the visible sector by supergravity.

Moreover, since the only parameter of the susy breaking in the messenger field (the gravitino) is the gravitino mass $m_{3/2}$ (in this case itself depending on the only independent parameter in the hidden sector, μ), we have in fact obtained that all the susy breaking terms, which in the phenomenological description were independent parameters, here are proportional to $m_{3/2}$.

Important concepts to remember

- In the MSSM, the four parameters μ, y_u, y_d, y_l and the soft susy breaking terms lead to masses for the two neutral Higgs and off-diagonal terms. When diagonalizing we get, besides the neutral Higgs H^0 (from the SM) and longitudinal components of $W_\mu^\pm$ and Z_μ, also charged Higgs $H^\pm$, an extra neutral Higgs h^0 and an A^0. The extra *physical* parameters are $\tan \beta = v_1/v_2$ and m_{h_0}.
- Minimal Supergravity (MinSugra) is $\mathcal{N} = 1$ supergravity coupled to the MSSM (vector and chiral superfields).
- When breaking susy, we have a Goldstone fermion (or Goldstino), that varies into a constant plus more, $\delta \psi = \langle F \rangle \epsilon + \dots$ and gives nonzero vacuum energy, and through the super-Higgs mechanism gives a mass to the gravitino.

- Susy breaking is a nonperturbative phenomenon in the hidden sector, happening either spontaneously, through a nonperturbative superpotential, or dynamically, where the correct low-energy fields are some nonsupersymmetric ones.
- Susy breaking is mediated by messenger fields, giving: (1) gauge mediation, through the SM gauge fields themselves, (2) gravity mediation, through the supergravity fields of MinSugra, or (3) anomaly mediation, which is a type of gravity mediation, but due to an anomaly in $T_{\mu\nu}$ and a gravitino mass due to the renormalization of the couplings.
- In gauge mediation, mass splittings $\Delta m \sim g_i^2/(8\pi^2)M_S^2 \sim 100GeV - 10TeV$ gives $M_S \sim 100TeV$.
- In gravity mediation, one can have mass splittings of $\sim \kappa_N M_S^2 \sim TeV$, so $M_S \sim 10^{11}GeV$ via spontaneous supersymmetry breaking, or $\sim \kappa_N^2 M_S^3 \sim TeV$, so $M_S \sim 10^{13}GeV$, but no spontaneous breaking: for instance, modular superfields have a W, leading to gaugino condensation $\langle \bar{\lambda}^a \lambda^a \rangle \neq 0$.
- In Seiberg's SQCD model for $SU(N_c)$ with flavor $SU(N_f)$ and $N_c > N_f$, with Q in N_c and $\tilde{Q}$ in $\bar{N}_c$, the low-energy effective fields are gauge invariant combinations of quarks, $\tilde{Q}^i Q^j$, that are dynamical effective fields, and effective gauge group $SU(N_c - N_f)$.
- A gravitino mass term is found by adding a positive constant to the supersymmetric potential that is naturally negative, so the criterion for susy breaking is not $E_0 > 0$ anymore, but rather $m_{3/2} \neq 0$.
- The gravitino mass is related to the susy breaking scale M_{SSB} by $m_{3/2} = M_{SSB}^2/(\sqrt{3}m_{Pl})$.
- The Polonyi model, a simple toy model of susy breaking, has a hidden chiral superfield with mass parameter μ in the superpotential, in which the coupling to supergravity gives $m_{3/2} \sim \mu$ in the vacuum, and one obtains nonzero phenomenological susy breaking parameters, soft scalar masses, A terms, and gaugino masses, all proportional to $m_{3/2}$.

References and further reading

For more on susy breaking and its relation to phenomenology, see Weinberg's book [12], vol. 3, as well as the book by Binetruy [108].

Exercises

(1) Substitute the rotation (31.6) in the Lagrangian to find the Lagrangian for H^0, h^0 and $H^\pm$.

(2) Prove that the on-shell auxiliary fields in MinSugra are given by (31.13).

(3) Prove that the gravitino mass term (31.29) is supersymmetric.

(4) Could the hidden sector superpotential $W_h(z)$ for the Polonyi model be obtained from a nonperturbative susy breaking mechanism in the hidden sector? If yes, of what type?

References

[1] P. J. E. Peebles, "Principles of physical cosmology," Princeton University Press, 1993.

[2] C. W. Misner, K. S. Thorne and J. A. Wheeler, "Gravitation," W. H. Freeman, 1970.

[3] L. D. Landau and E. M. Lifchitz, "Mechanics," Butterworth-Heinemann, 1982.

[4] R. M. Wald, "General relativity," University of Chicago Press, 1984.

[5] P. van Nieuwenhuizen, "Supergravity," Phys. Rep. **68**, 189 (1981).

[6] S. W. Hawking and G. F. R. Ellis, "The large scale structure of space-time," Cambridge University Press, 1973.

[7] S. V. Ketov, "Solitons, monopoles and duality: From sine-Gordon to Seiberg-Witten," Fortsch. Phys. **45**, 237 (1997) [arXiv:hep-th/9611209].

[8] L. Alvarez-Gaumé and S. F. Hassan, "Introduction to S-duality in N=2 supersymmetric gauge theory. (A pedagogical review of the work of Seiberg and Witten)," Fortsch. Phys. **45**, 159 (1997) [arXiv:hep-th/9701069].

[9] P. West, "Introduction to supersymmetry and supergravity," World Scientific, 1990.

[10] J. Wess and J. Bagger, "Supersymmetry and supergravity," Princeton University Press, 1992.

[11] S. J. Gates, M. T. Grisaru, M. Rocek and W. Siegel, "Superspace or one thousand and one lessons in supersymmetry," Front. Phys. **5**, 1 (1983) [arXiv:hep-th/0108200].

[12] S. Weinberg, "The quantum theory of fields," vol. 1: Foundation; vol. 2: Modern Application; vol. 3: Supersymmetry, Cambridge University Press, 1995, 1996 and 2000.

[13] M. Dine, "Supersymmetry and string theory, beyond the Standard Model," Cambridge University Press, 2007.

[14] P. van Nieuwenhuizen, "An introduction to simple supergravity and the Kaluza-Klein program," Les Houches lectures, Course 8, in "Relativity, groups and topology II," pp. 825–932, Elsevier, 1983.

[15] J. Wess and B. Zumino, "A lagrangian model invariant under supergauge transformations," Phys. Lett. B **49**, 52 (1974).

[16] J. Wess and B. Zumino, "Supergauge transformations in four-dimensions," Nucl. Phys. B **70**, 39–50 (1974).

[17] D. Z. Freedman and A. van Proeyen, "Supergravity," Cambridge University Press, 2012.

[18] D. Z. Freedman, P. van Nieuwenhuizen and S. Ferrara, "Progress toward a theory of supergravity," Phys. Rev. D **13**, 3214–3218 (1976).

[19] F. Ruiz Ruiz and P. van Nieuwenhuizen, "Lectures on supersymmetry and supergravity in (2+1)-dimensions and regularization of supersymmetric gauge theories," published in: Tlaxcala 1996, Recent developments in gravitation and mathematical physics (2nd Mexican School on Gravitation and Mathematical Physics, Tlaxcala, Mexico, December 1–7, 1996).

[20] H. Nastase, D. Vaman and P. van Nieuwenhuizen, "Consistency of the $AdS_7 \times S_4$ reduction and the origin of self-duality in odd dimensions," Nucl. Phys. B **581**, 179 (2000) [arXiv:hep-th/9911238].

[21] P. van Nieuwenhuizen, "General theory of coset manifolds and antisymmetric tensors applied to Kaluza-Klein supergravity," pp. 239–323, Supersymmetry and Supergravity '84, Proc. Trieste School, April 1984, Ed. B. de Wit, P. Fayet and P. van Nieuwenhuizen.

[22] M. J. Duff, B. E. W. Nilsson and C. N. Pope, "Kaluza-Klein supergravity," Phys. Rep. **130**, 1 (1986).

[23] E. Bergshoeff, E. Sezgin and P. K. Townsend, "Supermembranes and eleven-dimensional supergravity," Phys. Lett. B **189**, 75 (1987).

[24] P. Fre, "Lectures on special Kahler geometry and electric-magnetic duality rotations," Nucl. Phys. B-Proc. Suppl. **45**, 59 (1996) [arXiv:hep-th/9512043].

[25] B. de Wit and A. van Proeyen, "Special geometry and symplectic transformations," Nucl. Phys. B-Proc. Suppl. **45**, 196 (1996) [arXiv:hep-th/9510186].

[26] E. Cremmer, B. Julia and J. Scherk, "Supergravity theory in eleven-dimensions," Phys. Lett. B **76**, 409 (1978).

[27] B. de Wit and H. Nicolai, "On the relation between $d = 4$ and $d = 11$ supergravity," Nucl. Phys. B **243**, 91–111 (1984).

[28] B. de Wit, H. Nicolai and N. P. Warner, "The embedding of gauged $N = 8$ supergravity Into $d = 11$ supergravity," Nucl. Phys. B **255**, 29–62 (1985).

[29] M. Pernici, K. Pilch and P. van Nieuwenhuizen, "Gauged N=8 D=5 supergravity," Nucl. Phys. B **259**, 460 (1985).

[30] B. de Wit and H. Nicolai, "N=8 supergravity," Nucl. Phys. B **208**, 323 (1982).

[31] B. de Wit and H. Nicolai, "$d = 11$ supergravity with local SU(8) invariance," Nucl. Phys. B **274**, 363–400 (1986).

[32] B. de Wit and H. Nicolai, "The consistency of the S^7 truncation in D=11 supergravity," Nucl. Phys. B **281**, 211–240 (1987).

[33] O. Hohm and H. Samtleben, "Exceptional form of D=11 supergravity," Phys. Rev. Lett. **111**, 231601 (2013) [arXiv:1308.1673 [hep-th]].

[34] O. Hohm and H. Samtleben, "Exceptional field theory I: $E_{6(6)}$ covariant form of M-theory and type IIB," Phys. Rev. D **89**, 066016 (2014) [arXiv:1312.0614 [hep-th]].

[35] O. Hohm and H. Samtleben, "Exceptional field theory. II. $E_{7(7)}$," Phys. Rev. D **89**, 066017 (2014) [arXiv:1312.4542 [hep-th]].

[36] O. Hohm and H. Samtleben, "Exceptional field theory. III. $E_{8(8)}$," Phys. Rev. D **90**, 066002 (2014) [arXiv:1406.3348 [hep-th]].

[37] F. Ciceri, B. de Wit and O. Varela, "IIB supergravity and the $E_{6(6)}$ covariant vector-tensor hierarchy," JHEP **04**, 094 (2015) [arXiv:1412.8297 [hep-th]].

[38] A. Baguet, O. Hohm and H. Samtleben, "Consistent type IIB reductions to maximal 5D supergravity," Phys. Rev. D **92**, 065004 (2015) [arXiv:1506.01385 [hep-th]].

[39] A. Guarino, D. L. Jafferis and O. Varela, "String theory origin of dyonic N=8 supergravity and its Chern-Simons duals," Phys. Rev. Lett. **115**, 091601 (2015) [arXiv:1504.08009 [hep-th]].

[40] A. Guarino and O. Varela, "Dyonic ISO(7) supergravity and the duality hierarchy," JHEP **02**, 079 (2016) [arXiv:1508.04432 [hep-th]].

[41] H. Samtleben, "Lectures on gauged supergravity and flux compactifications," Class. Quant. Grav. **25**, 214002 (2008) [arXiv:0808.4076 [hep-th]].

[42] M. Trigiante, "Gauged supergravities," Phys. Rep. **680**, 1–175 (2017) [arXiv:1609.09745 [hep-th]].

[43] L. J. Romans, "Massive N=2a supergravity in ten-dimensions," Phys. Lett. B **169**, 374 (1986).

[44] R. D'Auria and P. Fre, "Geometric supergravity in d = 11 and its hidden super-group," Nucl. Phys. B **201**, 101–140 (1982). [erratum: Nucl. Phys. B **206**, 496 (1982)].

[45] L. Castellani, R. D'Auria and P. Fre, "Supergravity and superstrings: A geometric perspective, vol. 1: Mathematical foundations," World Scientific, 1991.

[46] L. Castellani, R. D'Auria and P. Fre, "Supergravity and superstrings: A geometric perspective, vol. 2: Supergravity," World Scientific, 1991.

[47] L. Castellani, R. D'Auria and P. Fre, "Supergravity and superstrings: A geometric perspective, vol. 3: Superstrings," World Scientific, 1991.

[48] J. Murugan and H. Nastase, "A non-Abelian particle–vortex duality," Phys. Lett. B **753**, 401–405 (2016) [arXiv:1506.04090 [hep-th]].

[49] T. R. Araujo and H. Nastase, "Non-Abelian T-duality for nonrelativistic holographic duals," JHEP **11**, 203 (2015) [arXiv:1508.06568 [hep-th]].

[50] T. H. Buscher, "A symmetry of the string background field equations," Phys. Lett. B **194**, 59–62 (1987).

[51] T. H. Buscher, "Path integral derivation of quantum duality in nonlinear sigma models," Phys. Lett. B **201**, 466–472 (1988).

[52] E. Bergshoeff, C. M. Hull and T. Ortin, "Duality in the type II superstring effective action," Nucl. Phys. B **451**, 547–578 (1995) [arXiv:hep-th/9504081 [hep-th]].

[53] Clifford Johnson, "D-branes." Cambridge University Press, 2005.

[54] T. Araujo, E. Ó. Colgáin, Y. Sakatani, M. M. Sheikh-Jabbari and H. Yavar-tanoo, "Holographic integration of $T\bar{T}$ \& $J\bar{T}$ via $O(d,d)$," JHEP **03**, 168 (2019) [arXiv:1811.03050 [hep-th]].

[55] E. G. Gimon, A. Hashimoto, V. E. Hubeny, O. Lunin and M. Rangamani, "Black strings in asymptotically plane wave geometries," JHEP **08**, 035 (2003) [arXiv:hep-th/0306131 [hep-th]].

[56] M. J. Duff, R. R. Khuri and J. X. Lu, "String solitons," Phys. Rep. **259**, 213–236 (1995) [arXiv:hep-th/9412184 [hep-th]].

[57] A. A. Tseytlin, "Harmonic superpositions of M-branes," Nucl. Phys. B **475** 149–163 (1996) [arXiv:hep-th/9604035 [hep-th]].

[58] A. A. Tseytlin, "'No force' condition and BPS combinations of p-branes in eleven-dimensions and ten-dimensions," Nucl. Phys. B **487** 141–154 (1997) [arXiv:hep-th/9609212 [hep-th]].

[59] E. Witten, "String theory dynamics in various dimensions," Nucl. Phys. B **443** 85–126 (1995) [arXiv:hep-th/9503124 [hep-th]].

[60] P. K. Townsend, "P-brane democracy," [arXiv:hep-th/9507048 [hep-th]].

[61] A. Achucarro, J. M. Evans, P. K. Townsend and D. L. Wiltshire, "Super p-Branes," Phys. Lett. B **198**, 441–446 (1987).

[62] H. Lu and C. N. Pope, "Interacting intersections," Int. J. Mod. Phys. A **13** 4425–4443 (1998) [arXiv:hep-th/9710155 [hep-th]].

[63] D. Youm, "Partially localized intersecting BPS branes," Nucl. Phys. B **556**, 222–246 (1999) doi:10.1016/S0550-3213(99)00384-3 [arXiv:hep-th/9902208 [hep-th]].

[64] C. M. Hull and P. K. Townsend, "Unity of superstring dualities," Nucl. Phys. B **438**, 109–137 (1995) [arXiv:hep-th/9410167 [hep-th]].

[65] N. A. Obers and B. Pioline, "U duality and M theory," Phys. Rep. **318**, 113–225 (1999) [arXiv:hep-th/9809039 [hep-th]].

[66] H. Nastase, "Introduction to the AdS/CFT correspondence," Cambridge University Press, 2015.

[67] J. M. Maldacena, "The Large N limit of superconformal field theories and supergravity," Adv. Theor. Math. Phys. **2**, 231–252 (1998) [arXiv:hep-th/9711200 [hep-th]].

[68] N. Itzhaki, J. M. Maldacena, J. Sonnenschein and S. Yankielowicz, "Supergravity and the large N limit of theories with sixteen supercharges," Phys. Rev. D **58**, 046004 (1998) [arXiv:hep-th/9802042 [hep-th]].

[69] D. E. Berenstein, J. M. Maldacena and H. S. Nastase, "Strings in flat space and pp waves from N=4 superYang-Mills," JHEP **04**, 013 (2002) [arXiv:hep-th/0202021 [hep-th]].

[70] T. Nishioka and T. Takayanagi, "On Type IIA Penrose Limit and N=6 Chern-Simons Theories," JHEP **08**, 001 (2008) [arXiv:0806.3391 [hep-th]].

[71] G. Itsios, H. Nastase, C. Núñez, K. Sfetsos and S. Zacarías, "Penrose limits of Abelian and non-Abelian T-duals of $AdS_5 \times S^5$ and their field theory duals," JHEP **01**, 071 (2018) [arXiv:1711.09911 [hep-th]].

[72] J. M. Figueroa-O'Farrill and G. Papadopoulos, "Homogeneous fluxes, branes and a maximally supersymmetric solution of M theory," JHEP **08**, 036 (2001) [arXiv:hep-th/0105308 [hep-th]].

[73] M. Blau, J. M. Figueroa-O'Farrill, C. Hull and G. Papadopoulos, "A New maximally supersymmetric background of IIB superstring theory," JHEP **01**, 047 (2002) [arXiv:hep-th/0110242 [hep-th]].

[74] O. Lunin and J. M. Maldacena, "Deforming field theories with U(1) x U(1) global symmetry and their gravity duals," JHEP **05**, 033 (2005) [arXiv:hep-th/0502086 [hep-th]].

[75] S. Frolov, "Lax pair for strings in Lunin-Maldacena background," JHEP **05**, 069 (2005) [arXiv:hep-th/0503201 [hep-th]].

[76] R. Borsato and L. Wulff, "Target space supergeometry of η and λ-deformed strings," JHEP **10**, 045 (2016) [arXiv:1608.03570 [hep-th]].

[77] F. Delduc, M. Magro and B. Vicedo, "An integrable deformation of the $AdS_5 \times S^5$ superstring action," Phys. Rev. Lett. **112**, 051601 (2014) [arXiv:1309.5850 [hep-th]].

[78] I. Kawaguchi, T. Matsumoto and K. Yoshida, "Jordanian deformations of the $AdS_5 x S^5$ superstring," JHEP **04**, 153 (2014) [arXiv:1401.4855 [hep-th]].

[79] T. J. Hollowood, J. L. Miramontes and D. M. Schmidtt, "An Integrable Deformation of the $AdS_5 \times S^5$ Superstring," J. Phys. A **47**, 495402 (2014) [arXiv:1409.1538 [hep-th]].

[80] D. Orlando, S. Reffert, J. i. Sakamoto, Y. Sekiguchi and K. Yoshida, "Yang–Baxter deformations and generalized supergravity—a short summary," J. Phys. A **53**, 443001 (2020) [arXiv:1912.02553 [hep-th]].

[81] S. Ferrara, R. Kallosh and A. Strominger, "N=2 extremal black holes," Phys. Rev. D **52**, R5412–R5416 (1995) [arXiv:hep-th/9508072 [hep-th]].

[82] S. Ferrara and R. Kallosh, "Supersymmetry and attractors," Phys. Rev. D **54**, 1514–1524 (1996) [arXiv:hep-th/9602136 [hep-th]].

[83] S. Ferrara and R. Kallosh, "Universality of supersymmetric attractors," Phys. Rev. D **54**, 1525–1534 (1996) [arXiv:hep-th/9603090 [hep-th]].

[84] A. Sen, "Black hole entropy function and the attractor mechanism in higher derivative gravity," JHEP **09**, 038 (2005) [arXiv:hep-th/0506177 [hep-th]].

[85] D. Astefanesei, H. Nastase, H. Yavartanoo and S. Yun, "Moduli flow and non-supersymmetric AdS attractors," JHEP **04**, 074 (2008) [arXiv:0711.0036 [hep-th]].

[86] M. B. Green, J. H. Schwarz and E. Witten, "Superstring theory: 25th anniversary edition," vol. 2, Cambridge University. Press, 2012.

[87] J. Polchinski, "String theory," vol. 1: "An introduction to the bosonic string," and vol. 2: "Superstring theory and beyond," Cambridge University Press, 2005.

[88] M. Henneaux and L. Mezincescu, "A Sigma Model Interpretation of Green-Schwarz Covariant Superstring Action," Phys. Lett. B **152**, 340–342 (1985)

[89] N. Berkovits, "Super poincare covariant quantization of the superstring," JHEP **04**, 018 (2000) [arXiv:hep-th/0001035 [hep-th]].

[90] N. Berkovits, "ICTP lectures on covariant quantization of the superstring," ICTP Lect. Notes Ser. **13**, 57–107 (2003) [arXiv:hep-th/0209059 [hep-th]].

[91] D. P. Sorokin, "Superbranes and superembeddings," Phys. Rep. **329**, 1–101 (2000) [arXiv:hep-th/9906142 [hep-th]].

[92] I. A. Bandos and D. P. Sorokin, "Superembedding approach to superstrings and super-p-branes," [arXiv:2301.10668 [hep-th]].

[93] H. Nastase, "Cosmology and string theory," Fundam. Theor. Phys. **197**, (2019) Springer, 2019.

[94] H. Bernardo and H. Nastase, "Small field inflation in $\mathcal{N} = 1$ supergravity with a single chiral superfield," JHEP **09**, 071 (2016) [arXiv:1605.01934 [hep-th]].

[95] D. Baumann and L. McAllister, "Advances in inflation in string theory," Ann. Rev. Nucl. Part. Sci. **59**, 67–94 (2009) [arXiv:0901.0265 [hep-th]].

[96] L. McAllister and E. Silverstein, "String cosmology: A review," Gen. Rel. Grav. **40**, 565–605 (2008) [arXiv:0710.2951 [hep-th]].

[97] R. Kallosh, A. Linde, D. Roest and Y. Yamada, "$\overline{D3}$ induced geometric inflation," JHEP **07**, 057 (2017) [arXiv:1705.09247 [hep-th]].

[98] J. M. Maldacena and C. Nunez, "Supergravity description of field theories on curved manifolds and a no go theorem," Int. J. Mod. Phys. A **16**, 822-855 (2001) [arXiv:hep-th/0007018 [hep-th]].

[99] R. Kallosh and A. D. Linde, "Supersymmetry and the brane world," JHEP **02**, 005 (2000) [arXiv:hep-th/0001071 [hep-th]].

[100] G. Obied, H. Ooguri, L. Spodyneiko and C. Vafa, "De Sitter Space and the Swampland," [arXiv:1806.08362 [hep-th]].

[101] H. Ooguri, E. Palti, G. Shiu and C. Vafa, "Distance and de sitter conjectures on the swampland," Phys. Lett. B **788**, 180–184 (2019) [arXiv:1810.05506 [hep-th]].

[102] E. Witten, "A simple proof of the positive energy theorem," Commun. Math. Phys. **80**, 381 (1981)

[103] C. M. Hull, "The positivity of gravitational energy and global supersymmetry," Commun. Math. Phys. **90**, 545 (1983)

[104] K. Becker, M. Becker and J. H. Schwarz, "String theory and M-theory: A modern introduction," Cambridge Univ. Press 2006.

[105] C. Burgess and G. Moore, "The Standard Model: a primer," Cambridge University Press, 2006.

[106] V. Braun, Y.-H. He, B. Ovrut and T. Pantev, "The exact MSSM spectrum from string theory," JHEP **0605**, 043 (2006), hep-th/0512177

[107] V. Bouchard and R. Donagi, "An $SU(5)$ heterotic Standard Model," Phys. Lett. B **633**, (2006) 783, hep-th/0512149.

[108] P. Binetruy, "Supersymmetry: Theory, experiment and cosmology," Oxford Graduate Texts, Oxford University Press, 2006.

Index